AF361703

Holography, 3D Imaging and
3D Display

Editors
Ting-Chung Poon
Virginia Tech
USA

Yaping Zhang
Kunming University of Science and Technology
China

Liangcai Cao
Tsinghua University
China

Hiroshi Yoshikawa
Nihon University
Japan

Editorial Office
MDPI
St. Alban-Anlage 66
4052 Basel, Switzerland

This is a reprint of articles from the Special Issue published online in the open access journal *Applied Sciences* (ISSN 2076-3417) (available at: https://www.mdpi.com/journal/applsci/special_issues/Holography_3D_Imaging_Display).

For citation purposes, cite each article independently as indicated on the article page online and as indicated below:

LastName, A.A.; LastName, B.B.; LastName, C.C. Article Title. *Journal Name* **Year**, *Article Number*, Page Range.

ISBN 978-3-03943-595-1 (Hbk)
ISBN 978-3-03943-596-8 (PDF)

Holography, 3D Imaging and 3D Display

Editors

Ting-Chung Poon
Yaping Zhang
Liangcai Cao
Hiroshi Yoshikawa

MDPI • Basel • Beijing • Wuhan • Barcelona • Belgrade • Manchester • Tokyo • Cluj • Tianjin

Contents

About the Editors

Ting-Chung Poon is a Professor of Electrical and Computer Engineering at Virginia Tech, Virginia, USA. His current research interests include 3-D image processing, and optical scanning holography (OSH). Dr. Poon is the author of the monograph Optical Scanning Holography with MATLAB (Springer, 2007), and is the co-author of, among other textbooks, Introduction to Modern Digital Holography with MATLAB (Cambridge University Press, 2014). Dr. Poon is a Fellow of the Institute of Electrical and Electronics Engineers (IEEE), the Institute of Physics (IOP), the Optical Society (OSA), and the International Society for Optics and Photonics (SPIE). He received the 2016 Dennis Gabor Award of the SPIE for "pioneering contributions to optical scanning holography (OSH), which has contributed significantly to the development of novel digital holography and 3-D imaging."

Yaping Zhang is currently a Professor of Optics and a Director of the Key Lab of Laser Information Processing Technology and Application at Kunming University of Science and Technology, China. Prof. Zhang is an Academic Leader in Optics in Yunan Province, China. She is also a member of the Steering Committee on Opto-Electronic Information Science and Engineering, Ministry of Education, China. She has received over 10 scientific projects, including three projects from the National Natural Science Foundation of China and other projects at the provincial and ministerial-level. She has published over 50 academic papers, with more than 30 published in SCI journals. Prof. Zhang also holds five patents. She was a Committee Member of the OSA FiO+LS conference. Currently, she is a Committee Member of the OSA topical meeting Digital Holography and 3-D Imaging. She is Co-Editor of the Special Issue "Holography, 3-D Imaging and 3-D Display" of Applied Sciences, and she has been a reviewer for 10 SCI journals. Her current research interests include digital holography and its applications to 3-D display, optical scanning holography, and photopolymers. She is a member of the OSA and IEEE.

Liangcai Cao received his BS/MS and Ph.D. degrees from Harbin Institute of Technology and Tsinghua University in 1999/2001 and 2005, respectively. He then became an assistant professor at the Department of Precision Instruments, Tsinghua University. He is now a tenured associate professor and director of the Institute of Opto-Electronic Engineering, Tsinghua University. He was a visiting scholar at UC Santa Cruz and MIT in 2009 and 2014. His current research interests are holographic imaging and display. He is an SPIE Fellow and OSA Fellow.

Hiroshi Yoshikaw received his B.S. degree, M.S. degree, and Ph.D. from Nihon University, all in electrical engineering in 1981, 1983, and 1985, respectively. He joined the faculty at Nihon University in 1985 where he currently holds the position of Professor of Computer Engineering. From Dec. 1988 to Apr. 1990, he was a research affiliate of MIT Media Laboratory. He is a member of SPIE, OSA, ITE (Institute of Television Engineers of Japan), OSJ (Optical Society of Japan), and HODIC. His current research interests are in holographic printing, electro-holography, computer-generated holograms, display holography, and computer graphics.

Editorial

Editorial on Special Issue "Holography, 3-D Imaging and 3-D Display"

Ting-Chung Poon [1], Yaping Zhang [2,*], Liangcai Cao [3] and Hiroshi Yoshikawa [4]

[1]   Bradley Department of Electrical and Computer Engineering, Virginia Tech, Blacksburg, VA 24061, USA; tcpoon@vt.edu

[2]   Faculty of Science, Kunming University of Science and Technology, No. 727, Jingming South Road, Chenggong District, Kunming 650500, China

[3]   Department of Precision Instruments, Tsinghua University, Beijing 100084, China; clc@mail.tsinghua.edu.cn

[4]   Department of Computer Engineering, Nihon University, Chiba 2748501, Japan; yoshikawa.hiroshi@nihon-u.ac.jp

*   Correspondence: yaping.zhang@gmail.com

Received: 23 September 2020; Accepted: 5 October 2020; Published: 11 October 2020

1. Introduction

Modern holographic techniques have been successfully applied in many important areas, such as 3D inspection, 3D microscopy, metrology and profilometry, augmented reality, and industrial informatics. This Special Issue covers selected pieces of cutting-edge research works, ranging from low-level acquisition, to the high-level analysis, processing, and manipulation of holographic information. The Special Issue also serves as a comprehensive review on the existing state-of-the-art techniques in 3D imaging and 3D display, as well as a broad insight into the future development of these disciplines. The Special Issue contains 25 papers in the field of holography, 3-D imaging and 3 D display. All the papers underwent substantial peer review under the guidelines of *Applied Sciences*. The twenty-five papers are divided into three groups: holography, 3-D imaging, and 3-D display.

2. Holography

There are 12 papers under the category of holography and these papers are divided into four topics: digital holography (DH), computer-generated holography (CGH), holographic printing and holographic optical elements (HOEs). Under DH, there are four papers. Tsang et al. report a low complexity method for reconstructing a focused image from an optical scanned hologram that is representing a 3-D object scene. While the proposed method has been applied to holograms obtained from optical scanning holography (OSH), the method also can be applied to any complex holograms such as those obtained from standard phase-shifting holography [1,2]. Rosen et al. review the prime architectures of optical hologram recorders in the family of coded aperture correlation holography (COACH) systems. They also discuss some of the key applications of these recorders in the field of imaging in general [3]. Tahara et al. investigate the quality of reconstructed images in relation to the bit depth of holograms formed by wavelength-selective phase-shifting digital holography. Their results indicate that two-bit resolution per wavelength is required to conduct color 3D imaging [4] Finally, the fourth and the last paper under DH, Xu et al. propose novel generalized three-step phase-shifting interferometry with a slight tilt reference. The proposed slight-tilt reference allows the full and efficient use of the space-bandwidth product of the limited resolution of digital recording devices as compared to the situation in standard off-axis holography [5].

Under CGH, Zhang et al. present a technique for generating 3D computer-generated holograms with scalable samplings, by using layer-based diffraction calculations. They have demonstrated experimentally that the proposed method can reconstruct quality 3D images at different scale factors [6]. Yang et al. investigate a fast computer-generated holographic method for VR and AR near-eye 3-D display. The display system is compact and flexible enough to produce speckle-noise-free high-quality VR and AR 3D images with efficient focus and defocus capabilities [7]. Jiao et al. employ a deep convolutional neural network to reduce the artifacts in a JPEG compressed hologram. Simulation and experimental results reveal that the proposed "JPEG + deep learning" hologram compression scheme can achieve satisfactory reconstruction results for a computer-generated phase-only hologram after compression [8]. In the fourth and the last paper under CGH, Yamaguchi et al. develop a fringe printer for CGHs. The printer can give a plane-type hologram with a 0.35-μm resolution [9].

There are four papers for holographic printing and holographic optical elements (HOEs). Su et al. report the progress in the synthetic holographic stereogram printing technique [10] and Fan et al. propose a stereogram printing based on holographic element-based effective perspective image segmentation and mosaicking [11]. Along the line of HOEs, Zhang et al. employ a multiplexed lens-array holographic optical element (MHOE) for a dual-view integral imaging 3-D display [12], and Ferrara et al. review volume holographic-based solar concentrators recorded on different holographic materials [13].

3. 3-D Imaging

There are six papers under the category of 3-D Imaging. Zhuang et al. report an early study on imaging 3-D objects hidden behind highly scattered media. Experimental results show that the complex amplitude of the object can be recovered from the distorted optical field [14]. Kim et al. suggest an efficient neural network model for shape from focus along with a weight passing method. The proposed network reduces the computational complexity with accuracy comparable with the conventional model [15]. Yang et al. propose absolute phase retrieval using only one coded pattern together with geometric constraints in a fringe projection system. The proposed method is suitable for real-time measurement [16]. Jang et al. investigate sampling based on a Kalman filter for shape from focus in the presence of noise. Experimental results demonstrate more accurate and faster performance than other existing filters [17]. Zhang et al. provide a review on tomographic diffractive microscopy (TDM) and suggest that TDM will play a key role in the exploration of biological cells as well as for the investigation of nano-structure devices [18]. Finally, Wang et al. characterize the phase response of spatial light modulators for coherent imaging [19].

4. 3-D Display

There are seven papers under the category of 3-D display. Pettingill et al. investigate static structures in leaky mode waveguides for transparent, monolithic, holographic, near-eye display [20]. Kim et al. provide their initial results on an electronic tabletop holographic display and hope the work will give guidance for the future development of commercially acceptable holographic display systems [21]. Tsai et al. investigate an optical design for a novel glasses-type 3-D wearable ophthalmoscope and conclude that a wearable ophthalmoscope can be designed optically and mechanically with 3-D technology [22]. Gao et al. investigate a holographic 3-D virtual reality (VR) and augmented reality (AR) display based on 4K-spatial light modulators and show that holographic 3-D VR and AR displays can be good candidates for true 3-D near-eye display as compared to stereoscopic display [23]. Choi et al. propose a new type of the dual-view 3-D display system based on direct-projection integral imaging using a convex-mirror-array and confirm the feasibility of the prosed system in practical applications [24]. Geyer et al. review high-resolution episcopic microscopy (HREM) and provide an overview of scientific publications that have been published in the last 13 years involving HREM imaging [25]. Finally, Falldorf et al. present a novel concept and first experimental results of a new type of 3D display, which is based on the synthesis of spherical waves [26].

Author Contributions: Y.Z. and T.-C.P. drafted the manuscript, and Y.Z. served as the corresponding author. All authors have reviewed and agreed to the published version of the manuscript.

Funding: National Natural Science Foundation of China (11762009, 61565010, 61865007); Natural Science Foundation of Yunnan Province (2018FB101); the Key Program of Science and Technology of Yunnan Province (2019FA025).

Acknowledgments: This Special Issue benefited from the outstanding coordination efforts by Marin Ma, who ensured that the Special Issue editors were engaged in the review process to ensure timely completion of the reviews.

Conflicts of Interest: The authors declare no conflict of interest.

References

1. Poon, T.-C. *Optical Scanning Holography with MATLAB*; Springer US: New York, NY, USA, 2007.
2. Tsang, P.; Poon, T.-C.; Liu, J.-P. Fast Extended Depth-of-Field Reconstruction for Complex Holograms Using Block Partitioned Entropy Minimization. *Appl. Sci.* **2018**, *8*, 830. [CrossRef]
3. Rosen, J.; Anand, V.; Rai, M.; Mukherjee, S.; Bulbul, A. Review of 3D Imaging by Coded Aperture Correlation Holography (COACH). *Appl. Sci.* **2019**, *9*, 605. [CrossRef]
4. Tahara, T.; Otani, R.; Takaki, Y. Wavelength-Selective Phase-Shifting Digital Holography: Color Three-Dimensional Imaging Ability in Relation to Bit Depth of Wavelength-Multiplexed Holograms. *Appl. Sci.* **2018**, *8*, 2410. [CrossRef]
5. Xu, X.; Ma, T.; Jiao, Z.; Xu, L.; Dai, D.; Qiao, F.; Poon, T.-C. Novel Generalized Three-Step Phase-Shifting Interferometry with a Slight-Tilt Reference. *Appl. Sci.* **2019**, *9*, 5015. [CrossRef]
6. Zhang, H.; Cao, L.; Jin, G. Scaling of Three-Dimensional Computer-Generated Holograms with Layer-Based Shifted Fresnel Diffraction. *Appl. Sci.* **2019**, *9*, 2118. [CrossRef]
7. Yang, X.; Zhang, H.; Wang, Q. A Fast Computer-Generated Holographic Method for VR and AR Near-Eye 3D Display. *Appl. Sci.* **2019**, *9*, 4164. [CrossRef]
8. Jiao, S.; Jin, Z.; Chang, C.; Zhou, C.; Zou, W.; Li, X. Compression of Phase-Only Holograms with JPEG Standard and Deep Learning. *Appl. Sci.* **2018**, *8*, 1258. [CrossRef]
9. Yamaguchi, T.; Yoshikawa, H. High Resolution Computer-Generated Rainbow Hologram. *Appl. Sci.* **2018**, *8*, 1955. [CrossRef]
10. Su, J.; Yan, X.; Huang, Y.; Jiang, X.; Chen, Y.; Zhang, T. Progress in the Synthetic Holographic Stereogram Printing Technique. *Appl. Sci.* **2018**, *8*, 851. [CrossRef]
11. Fan, F.; Jiang, X.; Yan, X.; Wen, J.; Chen, S.; Zhang, T.; Han, C. Holographic Element-Based Effective Perspective Image Segmentation and Mosaicking Holographic Stereogram Printing. *Appl. Sci.* **2019**, *9*, 920. [CrossRef]
12. Zhang, H.; Deng, H.; He, M.; Li, D.; Wang, Q. Dual-View Integral Imaging 3D Display Based on Multiplexed Lens-Array Holographic Optical Element. *Appl. Sci.* **2019**, *9*, 3852. [CrossRef]
13. Ferrara, M.; Striano, V.; Coppola, G. Volume Holographic Optical Elements as Solar Concentrators: An Overview. *Appl. Sci.* **2019**, *9*, 193. [CrossRef]
14. Zhuang, B.; Xu, C.; Geng, Y.; Zhao, G.; Chen, H.; He, Z.; Ren, L. An Early Study on Imaging 3D Objects Hidden Behind Highly Scattering Media: A Round-Trip Optical Transmission Matrix Method. *Appl. Sci.* **2018**, *8*, 1036. [CrossRef]
15. Kim, H.; Mahmood, M.; Choi, T. An Efficient Neural Network for Shape from Focus with Weight Passing Method. *Appl. Sci.* **2018**, *8*, 1648. [CrossRef]
16. Yang, X.; Zeng, C.; Luo, J.; Lei, Y.; Tao, B.; Chen, X. Absolute Phase Retrieval Using One Coded Pattern and Geometric Constraints of Fringe Projection System. *Appl. Sci.* **2018**, *8*, 2673. [CrossRef]
17. Jang, H.; Muhammad, M.; Yun, G.; Kim, D. Sampling Based on Kalman Filter for Shape from Focus in the Presence of Noise. *Appl. Sci.* **2019**, *9*, 3276. [CrossRef]
18. Zhang, T.; Li, K.; Godavarthi, C.; Ruan, Y. Tomographic Diffractive Microscopy: A Review of Methods and Recent Developments. *Appl. Sci.* **2019**, *9*, 3834. [CrossRef]
19. Wang, H.; Dong, Z.; Fan, F.; Feng, Y.; Lou, Y.; Jiang, X. Characterization of Spatial Light Modulator Based on the Phase in Fourier Domain of the Hologram and Its Applications in Coherent Imaging. *Appl. Sci.* **2018**, *8*, 1146. [CrossRef]
20. Pettingill, D.; Kurtz, D.; Smalley, D. Static Structures in Leaky Mode Waveguides. *Appl. Sci.* **2019**, *9*, 247. [CrossRef]

21. Kim, J.; Lim, Y.; Hong, K.; Kim, H.; Kim, H.; Nam, J.; Park, J.; Hahn, J.; Kim, Y. Electronic Tabletop Holographic Display: Design, Implementation, and Evaluation. *Appl. Sci.* **2019**, *9*, 705. [CrossRef]
22. Tsai, C.; King, T.; Fang, Y.; Hsueh, N.; Lin, C. Optical Design for Novel Glasses-Type 3D Wearable Ophthalmoscope. *Appl. Sci.* **2019**, *9*, 717. [CrossRef]
23. Gao, H.; Xu, F.; Liu, J.; Dai, Z.; Zhou, W.; Li, S.; Yu, Y.; Zheng, H. Holographic Three-Dimensional Virtual Reality and Augmented Reality Display Based on 4K-Spatial Light Modulators. *Appl. Sci.* **2019**, *9*, 1182. [CrossRef]
24. Choi, H.; Choi, J.; Kim, E. Dual-View Three-Dimensional Display Based on Direct-Projection Integral Imaging with Convex Mirror Arrays. *Appl. Sci.* **2019**, *9*, 1577. [CrossRef]
25. Geyer, S.; Weninger, W. High-Resolution Episcopic Microscopy (HREM): Looking Back on 13 Years of Successful Generation of Digital Volume Data of Organic Material for 3D Visualisation and 3D Display. *Appl. Sci.* **2019**, *9*, 3826. [CrossRef]
26. Falldorf, C.; Chou, P.; Prigge, D.; Bergmann, R. 3D Display System Based on Spherical Wave Field Synthesis. *Appl. Sci.* **2019**, *9*, 3862. [CrossRef]

 *applied sciences*

Article

Fast Extended Depth-of-Field Reconstruction for Complex Holograms Using Block Partitioned Entropy Minimization

Peter Wai Ming Tsang [1,*], Ting-Chung Poon [2,3] and Jung-Ping Liu [3]

[1] Department of Electronic Engineering, City University of Hong Kong, 83 Tat Chee Avenue, Kowloon, Hong Kong, China
[2] Bradley Department of Electrical and Computer Engineering, Virginia Tech, Blacksburg, VA 24061, USA; tcpoon@vt.edu
[3] Department of Photonics, Feng Chia University, No. 100 Wenhwa Rd., Taichung 40724, Taiwan; jpliu@fcu.edu.tw
* Correspondence: eewmtsan@cityu.edu.hk

Received: 3 May 2018; Accepted: 18 May 2018; Published: 21 May 2018

Abstract: Optical scanning holography (OSH) is a powerful and effective method for capturing the complex hologram of a three-dimensional (3-D) scene. Such captured complex hologram is called optical scanned hologram. However, reconstructing a focused image from an optical scanned hologram is a difficult issue, as OSH technique can be applied to acquire holograms of wide-view and complicated object scenes. Solutions developed to date are mostly computationally intensive, and in so far only reconstruction of simple object scenes have been demonstrated. In this paper we report a low complexity method for reconstructing a focused image from an optical scanned hologram that is representing a 3-D object scene. Briefly, a complex hologram is back-propagated onto regular spaced images along the axial direction, and from which a crude, blocky depth map of the object scene is computed according to non-overlapping block partitioned entropy minimization. Subsequently, the depth map is low-pass filtered to decrease the blocky distribution, and employed to reconstruct a single focused image of the object scene for extended depth of field. The method proposed here can be applied to any complex holograms such as those obtained from standard phase-shifting holography.

Keywords: optical scanning holography; extended depth-of-field; automatic focus detection; entropy minimization; block partitioned entropy minimization

1. Background

Optical scanning holography (OSH) [1,2] is one of the most effective techniques for capturing a complex hologram of a physical scene. It is different from existing methods such as phase shifting holography (PSH) [3], parallel phase shifting holography (PPSH) [4], geometric phase shifting digital holography (GPSDH) [5], Fresnel incoherent correlation holography (FINCH) [6,7], Fourier incoherent single channel holography (FISCH) [8], and the consumers scanner approach [9] that requires 2-D digital recording devices (e.g., digital camera) to capture the holographic signal, OSH only utilizes a single-pixel sensor. As such, OSH is a unique holographic recording technique and can even be configured to operate in both coherent and incoherent modes. Operating in the incoherent mode is important as the technique can be used to capture fluorescent specimens holographically. As mentioned in [2], OSH has many applications, such is but not limited to 3-D pattern recognition, 3-D microscopy, 3-D cryptography, and 3-D optical remote sensing. An OSH system can also be implemented to operate at high frame-rate for capturing hologram of a dynamic scene. In general, after a complex hologram is captured, it is often necessary to reconstruct a visible image from the hologram for further inspection and analysis. To reconstruct a hologram, we can back-propagate the hologram onto a sequence of regular spaced reconstruction planes that is parallel to the hologram. If a reconstruction plane is containing object points on the scene, those points will appear as a focused image, otherwise they will take the form of a de-focused haze. The operator or analyst will have to inspect each reconstruction plane, extract the focused image of the object points, if any, and discard the de-focused haze. Although this approach is feasible, the process is time-consuming and the quality of the reconstructed image will be affected by the visual judgment of the operator. For an OSH system that is employed to capture moving objects, it is desirable that the image reconstruction process should not be too lengthy. As such, the algorithms that are used for the reconstruction of optical scanning holograms should be automatic and computationally efficient. Automatic reconstruction of simple binary images from a hologram has been attempted utilizing Weiner filtering [10] and the Wigner distribution [11,12]. Lam and Zhang have proposed a hologram reconstruction method known as "blind sectional reconstruction" (BSR) [13,14]. The method can be divided into 2 stages. First, edge detection is applied to the image in each reconstruction plane. A reconstruction plane that exhibits a local minimum of edge points, as compared with the neighboring planes, will be taken as a focused section containing object points of the scene. This process is sometimes referred to as "automatic focus detection" (AFD). Second, a focused image of the object points that are contained in each of the focused section is obtained through an iterative optimization process, while the de-focused haze will be discarded. Subsequently, the focused images in all the sections can be merged to form an overall, focused image of the 3-D object scene with an extended depth of field. Despite the success of the BSR method, the process is computationally intensive, involving a large amount of memory in the optimization process, and so far, only reconstruction of simple binary images has been demonstrated. However, some enhancement of the method has been made to speed up the calculation and lower the memory requirement [15].

A fast variant of the BSR method was adopted in [16]. Similar to the blind sectional reconstruction, a subset of focused sections is obtained from the sequence of reconstructed planes through AFD. Next, simple edge analysis technique is applied to extract the focused object image (and to reject the de-focused haze) in each focused section. Albeit significant enhancement on the computation speed, the process is relying heavily on the edge analysis method and the manual setting of the parameter(s). The method has successfully demonstrated its capability in reconstructing binary objects from a hologram, but it is unlikely that it can be applied in a more complicated scene.

Recently, a more reliable AFD method, known as entropy minimization, was adopted in [17] for detecting the focused image planes. In this approach, a reconstruction plane is taken as a focused section if it exhibits a local minimum on its entropy value amongst the sequence of reconstructed images. This method is autonomous and does not require manual setting of parameters. Afterwards, a hologram reconstruction method was reported, based on the entropy minimization AFD [18]. For each pixel in each reconstruction plane, a small block with the pixel at the center is defined. The entropy of the pixel is then computed based on the pixels within its block. Next, the depth of each pixel is taken to be the depth of the reconstruction plane that has a minimum entropy value for the pixel of interest. As such, a depth map of the scene is deduced, with which the scene image can be reconstructed. Despite the effectiveness of the method, the amount of computation is overwhelming as the entropy of all the pixels have to be determined for all the reconstruction planes. A faster method has been proposed in [19], whereby the entropy is evaluated for each constituting objects in the scene instead of individual pixels. Briefly, a hologram is back-propagated to a virtual diffraction plane (VDP) that is close to the object space. Segmentation is applied on the VDP to separate the hologram into sub-holograms [20] each representing the fringe patterns of an isolated object in the scene. AFD is then applied to each sub-hologram to locate the focused image plane, and reconstruct the image of the object. However, this method is only applicable if the objects on the x-y plane in the scene are spaced far enough to be separated on the VDP, a requirement which is not guaranteed in practice.

Another fast method for hologram reconstruction has been proposed in [21]. The method is based on the assumption that the imaginary component (or phase angle) of a pixel is close to zero on its focused image plane. On this basis, a focused image of the object scene can be built by selecting, from each reconstruction plane, pixels with small imaginary values.

In this paper we present, partly based on the work in [18], a method for fast hologram reconstruction with extended depth of field (EDF). By EDF, it means that a focused image of all the objects (which may differ in depth and geometrical shape) in the scene can be recovered simultaneously from the hologram. Our proposed method is particular suitable for optical scanned holograms that are unrestricted in size, complexity of object scene, and capturing modes (i.e., coherent and incoherent modes). Briefly the method can be divided into 4 stages. First, the hologram is back-propagated into a sequence of parallel, and uniformly spaced reconstruction planes. Second, each reconstruction plane is evenly partitioned into non-overlapping square image blocks. Third, entropy minimization AFD is employed to locate the focused section of each image block, from which the depth is determined, and further refined with low-pass filtering. Fourth, the collection of depth information from all the image blocks is utilized to reconstruct a focused image of the object scene. In our proposed method, we have assumed that in a typical scene, the depth of pixels within close neighborhood should be similar, an assumption that is generally applicable in practice. As such, the depth map of the scene image can be evaluated on a block-by-block, instead of a pixel-by-pixel basis, resulting in significant increase in the computation efficiency as compared with the method in [18].

2. Optical Scanning Holography

The system of optical scanning holography that we have employed for hologram acquisition is shown in Figure 1. A linearly polarized laser of wavelength λ is passed through a half-wave plate (HWP), which rotates the polarization direction of the laser beam along the bisector of the two principal axes of the electro-optic crystal of the electro-optic modulator (EOM). Since there is birefringence induced along one of the principal axes through external electrical saw-tooth signal, the output beam of the EOM contains two orthogonal polarizations with a frequency difference Ω. The beam of light with orthogonal polarization is split into two beams with polarizing beamsplitter PBS1.

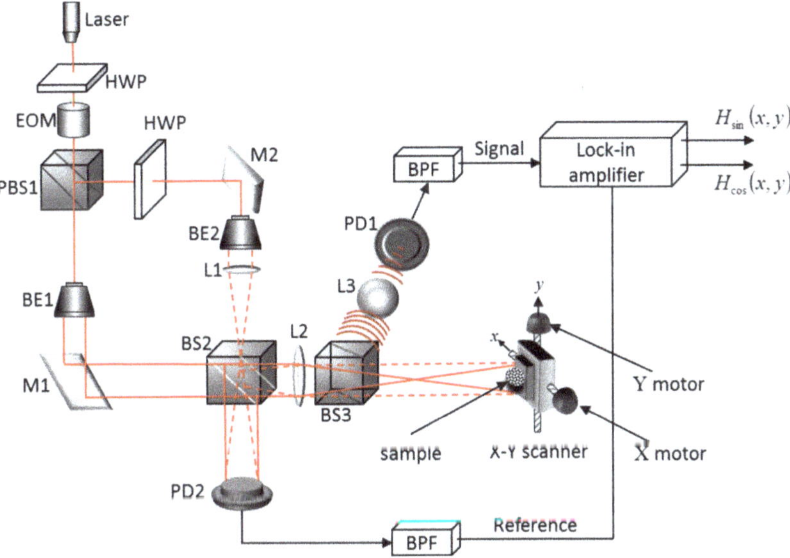

Figure 1. An optical scanning holography (OSH) system.

One beam, expanded by beam expander BE1, is reflected to beamsplitter BS2 by mirror M1. The other beam passes through another HWP, and is reflected by mirror M2 to BS2 through beam expander BE2 and lens L1. Subsequently the pair of beams goes through lens L2, and impinges on the test sample after passing beamsplitter BS3. Note that one beam projecting on the sample is a plane wave (dotted lines) while the other beam is a spherical wave (solid lines). Hence on the sample, we have a time-dependent Fresenl zone plate (TD-FZP) as the interference of a plane wave and a spherical wave gives an FZP and the frequency difference Ω of the two beams gives running fringes within the overall scanning beam on the sample [1,2]. A x-y table is utilized to move the sample along a zigzag scan path. At each position of the scan path, the optical waves scattered by the sample is directed to photodetector PD1 through beamsplitter BS3 and lens L3. The output of PD1 is band-pass filtered at Ω to give the Signal output. At the same time, the pair of beams is also combined in beamsplitter BS2, impinging on photodetector PD2. The output of PD2 is band-pass filtered at Ω and taken to be the Reference for the lock-in amplifier. The lock-in amplifier mixes the Signal and the Reference to form the hologram which is composing of a sine hologram, $H_{\sin}(x,y)$, and a cosine hologram, $H_{\cos}(x,y)$ [1,2]. Mathematically, suppose the 3-D object is divided into N uniformly separated image planes that are parallel to the hologram, and the jth image plane that is located at an axial distance z_j to the hologram is denoted by $I_0(x,y;z_j)$, the hologram acquired with OSH is given by

$$H(x,y) = H_{\cos}(x,y) + iH_{\sin}(x,y) = \sum_{j=0}^{N-1} H_k(x,y;z_j) \tag{1}$$

where $H_k(x,y;z_j) = I_0(x,y;z_j) \otimes F(x,y;z_j)$ with $\otimes$ denoting the operation of convolution, and

$$F(x,y;z_j) = \frac{1}{i\lambda z_j} \exp\left[\frac{i2\pi}{\lambda} \sqrt{(x\delta)^2 + (y\delta)^2 + z_j^2}\right] \tag{2}$$

is the spatial impulse response of propagation of light [1], where δ is the pixel size of the hologram.

3. Proposed Method

The objective of our proposed method is to obtain an extended field of depth (EFD) image (i.e., an image with all the object points in focused, disregard of their geometry and distance from hologram $H(x,y)$). The process can be divided into 4 stages as shown in Figure 2, and outlined as follows. In stage 1, the hologram $H(x,y)$ is back-propagated to a stack of regular spaced reconstruction planes along the axial direction, a process that is referred to as "sectional reconstruction". The image on each reconstruction plane is evenly partitioned into non-overlapping square blocks. In stage 2, the entropies of the image blocks in all the reconstruction planes are evaluated. The depth of each block is taken to be the distance of the reconstruction plane that exhibits the minimum entropy for that particular block position. It can be envisaged that the integration of the depth of all the constituting blocks (each contributed by one of the reconstructed images) will form a blocky depth map. Next, in stage 3, low-pass filtering is applied to smoothen the depth map. Finally, in stage 4, an extended depth of field image of the object scene is obtained by selecting each pixel from the reconstruction image plane that corresponding to the depth of the pixel in the depth map. Details of each stage are described as follows.

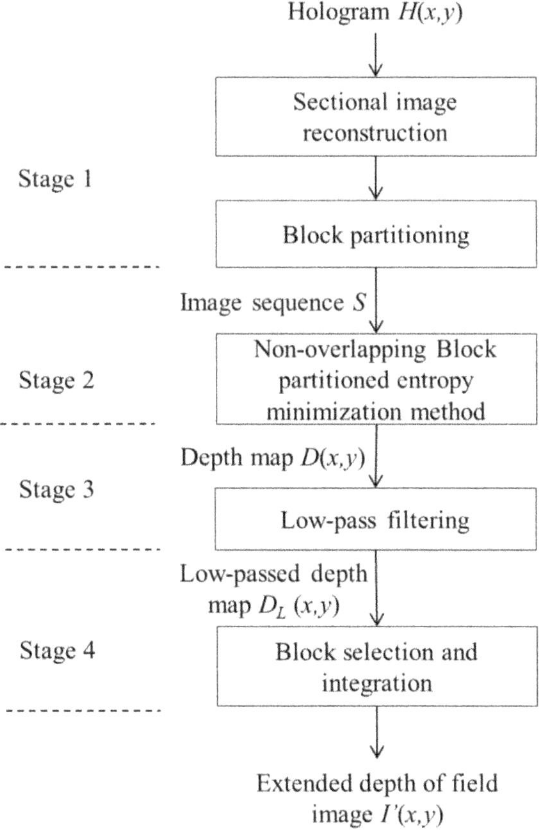

Figure 2. Proposed method.

Stage 1: Sectional reconstruction

In this stage, a stack S of evenly spaced images is reconstructed from hologram $H(x,y)$ through back-propagation. Let $S_j(x,y)\big|_{0 \leq j < N}$ and Z denotes the magnitude of the reconstructed image on the j-th reconstruction plane section, and the maximum range of the depth of the object scene, respectively, we have

$$S_j(x,y) = \left| H(x,y) \otimes F^*(x,y;z_j) \right| \tag{3}$$

where $F^*(x,y;z_j)$ is the conjugate of $F(x,y;z_j)$. The separation between adjacent reconstruction planes is given by

$$\Delta z = z_{j+1} - z_j = \frac{Z}{N} \tag{4}$$

Subsequently, each reconstructed image $S_j(x,y)$ is partitioned into a 2-D array of non-overlapping square image blocks of size $M \times M$, as shown in Figure 3. Without loss of generality, we assume that both the hologram and its reconstructed images are square in size, and having the same size $K \times K$, where K is an integer multiple of M. As such, there are K/M blocks along the horizontal direction, and K/M blocks along the vertical direction. Each image block in $S_j(x,y)$ is represented by the block function $b_j(p,q)\big|_{0 \leq p;q < (K/M)}$, which includes the pixels bounded by the square within the region $pM \leq x < (p+1)M$ and $qM \leq y < (q+1)M$. The EDF image $I'(x,y)$ to be determined is also partitioned in a similar way with each block of pixels denoted by $b_{EFD}(p,q)$.

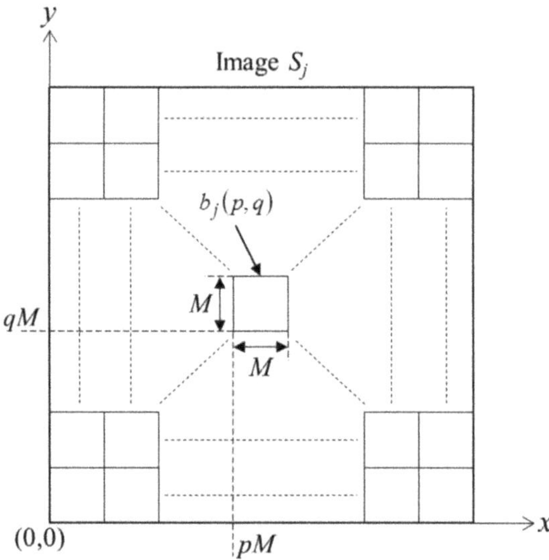

Figure 3. Partitioning an image $S_j(x,y)$ into non-overlapping square blocks of size $M \times M$.

Stage 2: Non-overlapping block partitioned entropy minimization

In this stage, the entropy of all the image blocks that have been partitioned in stage 1 is computed as

$$E_j(p,q) = -P_j(p,q) \log P_j(p,q) \tag{5}$$

where

$$P_j(p,q) = \sum_{x=pM}^{(p+1)M-1} \sum_{y=qM}^{(q+1)M-1} S_j(x,y) \tag{6}$$

An image that is in focus generally exhibits minimum entropy. As such, the depth (or focused plane) of each block $D(p,q)$ can be taken to be the position of the plane that exhibits minimum entropy of the block, i.e.,

$$E_{D(p,q)} = MIN \lfloor E_j(p,q) \rfloor_{0 \leq j < N} \tag{7}$$

where the expression on the right-hand-side of Equation (7) denotes the minimum value of $E_j(p,q)$ for $0 \leq j < N$. Collection of depth of each block results in a crude, blocky depth map of the object scene. Intuitively, we can obtain an EDF image $I(x,y)$ by selecting each block of pixels from the corresponding block with minimum entropy from the stack of reconstructed images. Mathematically, this can be expressed as

$$b_{EDF}(p,q) = b_k(p,q) \tag{8}$$

and

$$I(x,y) = \bigcup_{p,q=0}^{p,q=(K/M-1)} b_{EDF}(p,q) \tag{9}$$

where $k = D(p,q)$, and the block $b_{EDF}(p,q)$ that is selected is referred to as an extended depth of field block. The symbol $\cup$ denotes the union operator that is used to compose the EDF image by patching it with the extended depth of field blocks (i.e., $b_{EDF}(p,q)$). However, as we shall show later, an EDF image reconstructed in this manner will exhibit a blocky, visual unpleasant appearance.

Stage 3: Filtering

To reduce the blocky appearance of the EDF image, we propose to smooth the depth map with a simple low-pass filtering process that can be realized as follows. First, we compute the mean intensity for each EDF block as

$$mean(p,q) = \frac{1}{9} \sum_{s=-1}^{1} \sum_{t=-1}^{1} [b_{EDF}(p+s,q+t)] \tag{10}$$

Next, a corresponding status flag $status(p,q)$ is determined as

$$status(p,q) = \begin{cases} 1 & if\ mean(p,q) \geq T \\ 0 & otherwise \end{cases} \tag{11}$$

where T is a small fixed threshold value corresponding to the minimum intensity that is visible to the human eye. An EDF block with an average intensity higher than T is referred to as a "visible block".

Subsequently, a low-pass filtered depth map is obtained by averaging the depth values of visible blocks within a 3×3 window as

$$D_L(p,q) = \frac{1}{9} \sum_{s=-1}^{1} \sum_{t=-1}^{1} [D(p+s,q+t)] \times status[m(p+s,q+t)] \tag{12}$$

Note that in the filtering process, blocks with too low intensity are discarded as they are likely associated to empty space in the scene, and may not carry reliable depth information.

Stage 4: Reconstruction of the EDF image

Finally, we apply Equation (9) to obtain the EDF image, with the variable $k = D_L(p,q)$, i.e.,

$$b'_{EFD}(p,q) = b_k(p,q)\big|_{k=D_L(p,q)} \tag{13}$$

and

$$I'(x,y) = \bigcup_{p;q=0}^{p;q=(K/M-1)} b'_{EFD}(p,q) \tag{14}$$

In comparison with the method in [18], our proposed scheme has a significantly decrease in the computational loading of the entropy values as they are evaluated on a block-by-block, instead of a pixel-by-pixel basis. With our approach, the computation time for deducing the entropy values has been reduced by M^2 times. The computation time for the reconstruction of the hologram into different planes are similar for both methods. As will be shown later, our proposed method is around 60 times faster than the method in [18].

4. Experimental Results

To evaluate the performance of our proposed method, 2 sets of holograms have been captured with the OSH system that has been configured to operate in the incoherent mode [22], and with the wavelength of the laser beam being 633 nm. The first hologram "A" is a pair of binary Chinese characters "光" and "電", and second hologram "B" is a grey level image of a pair of partially overlapping coins. The pixel size of holograms (i.e., the sampling interval with which each hologram pixel is recorded) "A" and "B" are 5 μm and 15 μm, respectively. In both cases, the range of distance between the object space and the hologram plane is [0.015 m, 0.03 m], but there is no prior knowledge on the depth of individual objects in the scene. The sine and the cosine holograms of the 2 samples are shown in Figure 4a–d.

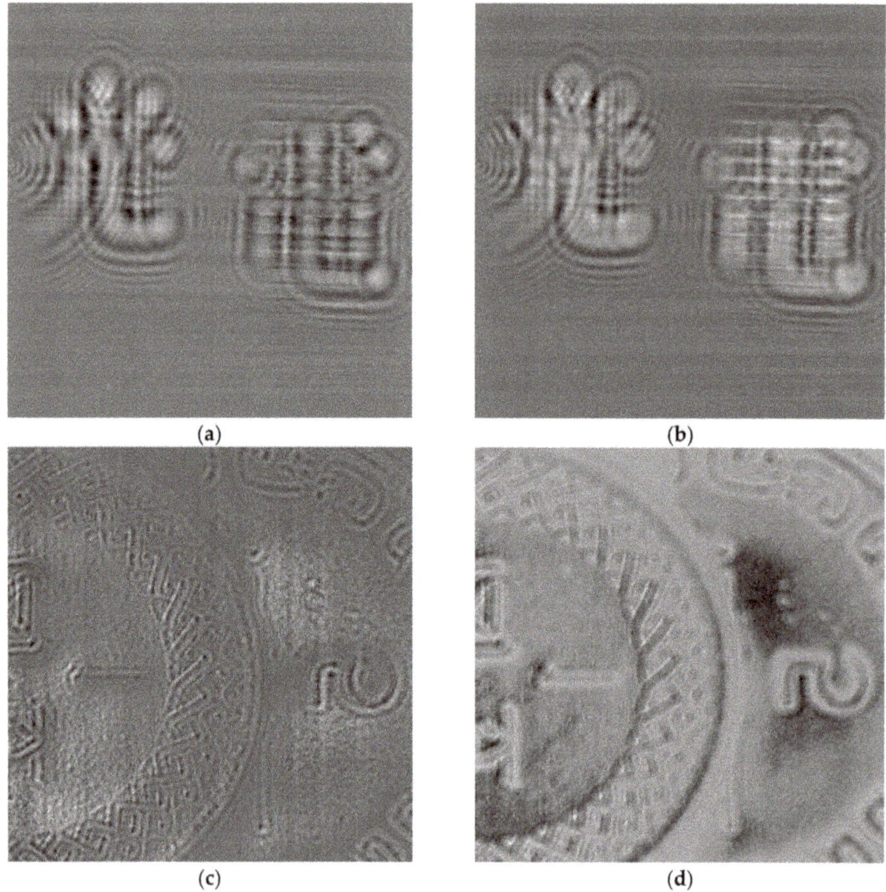

Figure 4. (**a**) Sine hologram of hologram "A"; (**b**) Cosine hologram of hologram "A"; (**c**) Sine hologram of hologram "B"; (**d**) Cosine hologram of hologram "B".

Through a series of trial and error tests conducted with visual judgement, we estimated that there are mainly 2 focused planes that can be identified for sample 'A' at 0.021 m and 0.024 m. For 'B', again there are mainly 2 focused planes at 0.0197 m and 0.027 m. The reconstructed images of sample 'A' at the 2 focused planes are shown in Figure 5a,b. It can be seen that in both figures when certain regions of the image are in focused, the remaining ones are blurred. Similar observation is noted for the reconstructed image of sample 'B' at the 2 focused planes in Figure 5c,d. In both cases, it is not possible to have an EDF image where all the contents are in focus.

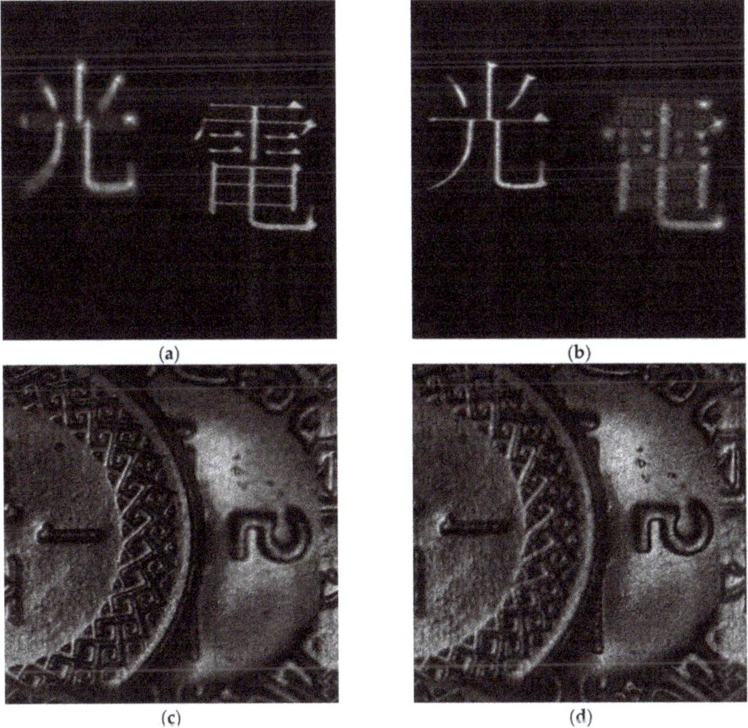

Figure 5. (**a**) Reconstructed image of hologram "A" at 0.021 m; (**b**) Reconstructed image of hologram "A" at 0.024 m; (**c**) Reconstructed image of hologram "B" at 0.0197 m; (**d**) Reconstructed image of hologram "B" at 0.027 m.

Next, we apply stages 1 and 2 of our proposed method to obtain the depth map and the EDF image from the 2 holograms. The number of reconstruction planes and the block size are set to $N = 40$ and $M = 32$, respectively, to provide a sufficiently fine depth resolution of $\Delta z = \frac{Z}{N} = \frac{0.015}{40} = 0.375$ mm, and a reliable measurement of the entropy of each image block. The depth maps of the 2 samples are shown in Figure 6a,b, and the corresponding EDF images obtained with the above settings are shown in Figure 6c,d. We observe that all the objects in the scene represented by each hologram are reconstructed as focused images but the appearance, especially for sample "B", is rather blocky with obvious discontinuities along some of the object boundaries.

Subsequently, we have applied Equation (12) to filter the depth map, and obtained the reconstructed images with Equations (13) and (14). Suppose the dynamic range of the reconstructed image is normalized to the range $[0, 1]$, the threshold T is set to 0.0625, as image intensity lower than this value is hardly visible, and likely to represent empty background or noise signals. The filtered depth maps are shown in Figure 7a,b, and the reconstructed EDF images are shown in Figure 7c,d. In the EDF images of both samples, we observe that the objects in the scenes are reconstructed as sharp focused images, and the blocky appearance has been reduced significantly as compared with those obtained with the unfiltered depth maps. The computation time based on the typical PC is around 5.9 s. The correlation score of the EDF images of holograms 'A' and 'B' (with reference with the ones obtained with manual judgement) are 0.977 and 0.845, respectively. The high correlation scores reflect that the EDF images obtained with our proposed method are close to those derived from manual judgement.

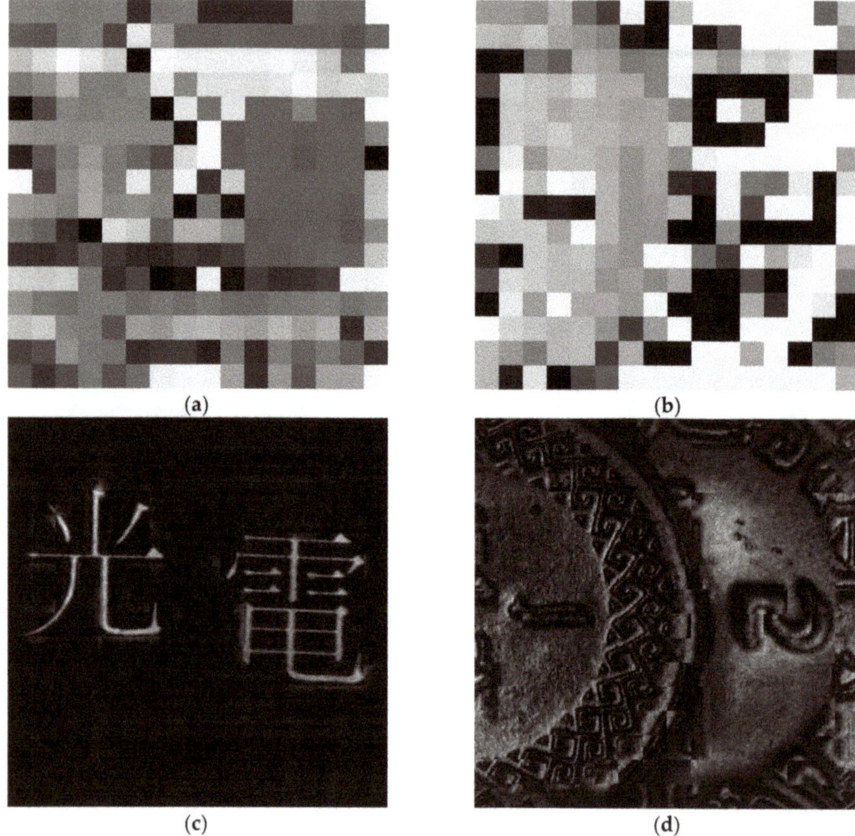

Figure 6. (**a**) Depth map of hologram "A" obtained with our proposed method ($M = 32$) without filtering; (**b**) Depth map of hologram "B" obtained with our proposed method ($M = 32$) without filtering; (**c**) Extended depth of field (EDF) image of hologram "A" reconstructed with our proposed method ($M = 32$) without filtering on the depth map; (**d**) EDF image of hologram "B" reconstructed with our proposed method ($M = 32$) without filtering on the depth map.

We would like to point out that out proposed method can function for quite a wide range of M. In general, the blocky appearance could be decreased with a smaller value of M (i.e., a small block size in computing the entropy values). However, this will jeopardize the accuracy of the depth-map, which may lead to blurriness in certain parts on the EDF image. As an example, we apply our proposed method to the 2 sample holograms (with filtering on the depth maps) with $M = 20$, and the EDF images are shown in Figure 8a,b. It can be seen that while the blocky appearance of EDF image corresponding to sample 'B' (Figure 8b) is reduced, certain parts of the 2 EDF images, especially in Figure 8a, are suffering from mild de-focusing. A proper choice of the range of M is dependent on the setup of the OSH system (such as the size of the hologram and the sampling interval). In our present setup, EDF images of favorable quality can be obtain with values of M in the range $[20, 32]$.

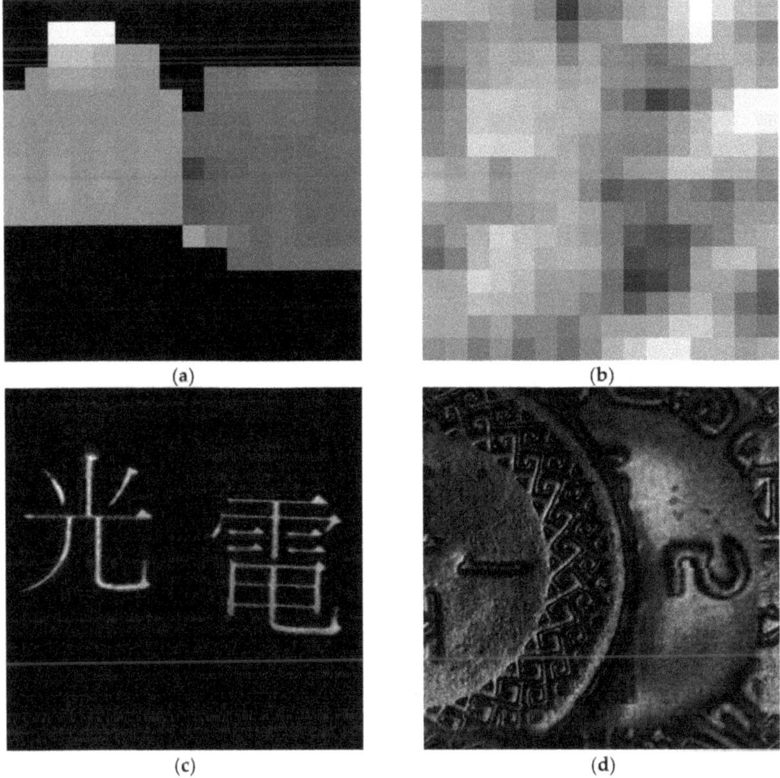

(a)

(b)

(c)

(d)

Figure 7. (**a**) Depth map of hologram "A" obtained with our proposed method (M = 32) with filtering; (**b**) Depth map of hologram "B" obtained with our proposed method (M = 32) with filtering; (**c**) EDF image of hologram "A" reconstructed with our proposed method (M = 32) with filtering on the depth map; (**d**) EDF image of hologram "B" reconstructed with our proposed method (M = 32) with filtering on the depth map.

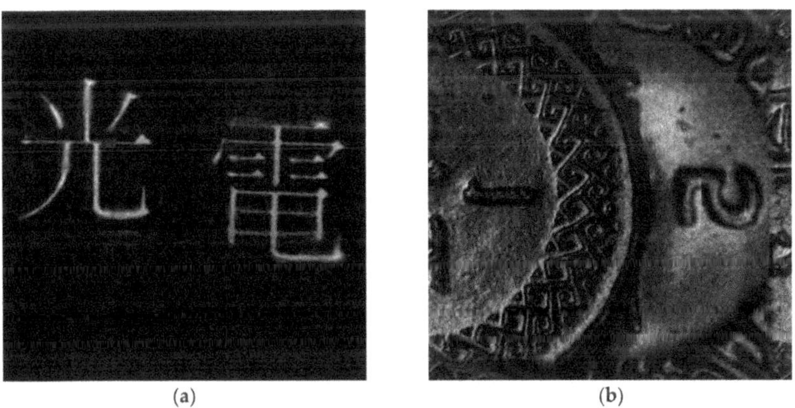

(a)

(b)

Figure 8. (**a**) EDF image of hologram "A" reconstructed with our proposed method (M = 20) with filtering on the depth map; (**b**) EDF image of hologram "B" reconstructed with our proposed method (M = 20) with filtering on the depth map.

Finally, we would like to compare our proposed method with the method in [18]. We have applied the method in [18] to reconstructed the EDF images of the 2 holograms, and the results are shown in Figure 9a,b. We observe that the visual quality of the 2 EDF images are generally favorable, but the EDF image of hologram 'B' is more blocky than the one obtained with the proposed method. The computation time based on the same PC is around 356 s, which is about 60 times longer than the proposed method. The correlation score of the EDF images of holograms 'A' and 'B' (with reference with the ones obtained with manual judgement) are 0.982 and 0.741, respectively, which are more or less similar to the proposed method. As for the visual quality, the EDF image of sample 'B' (e.g., around the characters '1' and '2' on the pair of coins) is better in our proposed method.

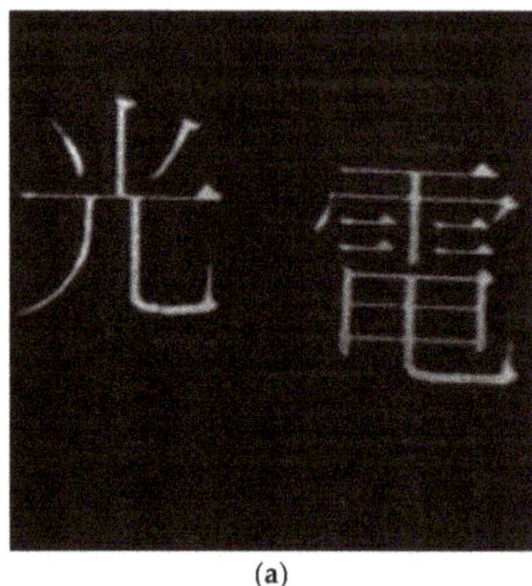

(a)

(b)

Figure 9. (**a**) EDF image of hologram "A" reconstructed with the method in [18]; (**b**) EDF image of hologram "B" reconstructed with the method in [18].

5. Conclusions

In this paper, we have proposed a method for reconstructing the scene image with extended depth of field (EDF) from a hologram that is captured with an OSH system. Our proposed method is based on a scheme that is referred to as "non-overlapping block-based entropy minimization" method, which is applied to determine the depth (focus) map of the scene image in an autonomous, block-by-block manner. From the depth value, each block is reconstructed as a focused image tile. Subsequently, the complete scene image is reconstructed with all the constituting objects appearing as sharp focused images. Further enhancement on the visual quality of the reconstructed image is attained by smoothing the depth map, so that the blocking effect at the object boundaries is reduced. Experimental results reveal that our proposed method is capable of reconstructing EDF images from OSH holograms representing both continuous tone and binary object scenes with favorable quality.

Further research on the proposed method could be conducted through applying different kinds of focus measurements in the auto-focus detection process. An extensive review on the topic has been reported in [23].

Author Contributions: Conceptualization, P.W.M. Tsang, T.-C. Poon and J.-P. Liu; Methodology, P.W.M. Tsang; Software, P.W.M. Tsang; Validation, P.W.M. Tsang, T.-C. Poon and J.-P. Liu; Formal Analysis, P.W.M. Tsang; Investigation, P.W.M. Tsang; Resources, P.W.M. Tsang, T.-C. Poon and J.-P. Liu; Data Curation, P.W.M. Tsang, T.-C. Poon and J.-P. Liu; Writing-Original Draft Preparation, P.W.M. Tsang; Writing-Review & Editing, P.W.M. Tsang, T.-C. Poon and J.-P. Liu; Visualization, P.W.M. Tsang, T.-C. Poon and J.-P. Liu; Supervision, P.W.M. Tsang; Project Administration, P.W.M. Tsang.

Acknowledgments: The authors would like to express their thanks to Wei-Ren Siao for setting up the experiment. This work is partially sponsored by Ministry of Science and Technology of Taiwan under contract number MOST103-2221-E-035-037-MY3.

Conflicts of Interest: The authors declare no conflict of interest.

References

1. Poon, T.-C.; Liu, J.-P. *Introduction to Modern Digital Holography with MATLAB*; Cambridge University Press: Cambridge, UK, 2014.
2. Poon, T.-C. *Optical Scanning Holography with MATLAB*; Springer US: New York, NY, USA, 2007.
3. Zhang, T.; Yamaguchi, I. Three-dimensional microscopy with phase-shifting digital holography. *Opt. Lett.* **1998**, *23*, 1221–1223. [CrossRef] [PubMed]
4. Awatsuji, Y.; Tahara, T.; Kaneko, A.; Koyama, T.; Nishio, K.; Ura, S.; Kubota, T.; Matoba, O. Parallel two-step phase-shifting digital holography. *Appl. Opt.* **2008**, *47*, D183–D189. [CrossRef] [PubMed]
5. Jackin, B.; Narayanamurthy, C.; Yatagai, T. Geometric phase shifting digital holography. *Opt. Lett.* **2016**, *41*, 2648–2651. [CrossRef] [PubMed]
6. Kashter, Y.; Vijayakumar, A.; Miyamoto, Y.; Rosen, J. Enhanced super resolution using Fresnel incoherent correlation holography with structured illumination. *Opt. Lett.* **2016**, *41*, 1558–1561. [CrossRef] [PubMed]
7. Katz, B.; Rosen, J.; Kelner, R.; Brooker, G. Enhanced resolution and throughput of Fresnel incoherent correlation holography (FINCH) using dual diffractive lenses on a spatial light modulator (SLM). *Opt. Express* **2012**, *20*, 9109–9121. [CrossRef] [PubMed]
8. Kelner, R.; Rosen, J.; Brooker, G. Enhanced resolution in Fourier incoherent single channel holography (FISCH) with reduced optical path difference. *Opt. Express* **2013**, *21*, 20131–20144. [CrossRef] [PubMed]
9. Shimobaba, T.; Yamanashi, H.; Kakue, T.; Oikawa, M.; Okada, N.; Endo, Y.; Hirayama, R.; Masuda, N.; Ito, T. In-line digital holographic microscopy using a consumer scanner. *Sci. Rep.* **2013**, *3*, 2664. [CrossRef] [PubMed]
10. Kim, T. Optical sectioning by optical scanning holography and a Wiener filter. *Appl. Opt.* **2006**, *45*, 872–879. [CrossRef] [PubMed]
11. Kim, T.; Poon, T.-C.; Indebetouw, G. Depth detection and image recovery in remote sensing by optical scanning holography. *Opt. Eng.* **2002**, *41*, 1331–1338.
12. Kim, H.; Min, S.; Lee, B.; Poon, T.-C. Optical sectioning for optical scanning holography using phase-space filtering with Wigner distribution functions. *Appl. Opt.* **2008**, *47*, D164–D175. [CrossRef] [PubMed]
13. Zhang, X.; Lam, E.; Kim, T.; Kim, Y.; Poon, T.-C. Blind sectional image reconstruction for optical scanning holography. *Opt. Lett.* **2009**, *34*, 3098–3100. [CrossRef] [PubMed]
14. Zhang, X.; Lam, E.; Poon, T.-C. Reconstruction of sectional images in holography using inverse imaging. *Opt. Express* **2008**, *16*, 17215–17226. [CrossRef] [PubMed]
15. Zhao, F.; Qu, X.; Zhang, X.; Poon, T.-C.; Kim, T.; Kim, Y.; Liang, J. Solving inverse problems for optical scanning holography using an adaptively iterative shrinkage-thresholding algorithm. *Opt. Express* **2012**, *20*, 5942–5954. [CrossRef] [PubMed]
16. Tsang, P.W.M.; Cheung, K.; Kim, T.; Kim, Y.; Poon, T.-C. Fast reconstruction of sectional images in digital holography. *Opt. Lett.* **2011**, *36*, 2650–2652. [CrossRef] [PubMed]
17. Ren, Z.; Chen, N.; Chan, A.; Lam, E. Autofocusing of Optical Scanning Holography Based on Entropy Minimization. In Proceedings of the Digital Holography & 3-D Imaging Meeting, OSA Technical Digest, Shanghai, China, 24–28 May 2015; Paper DT4A.4; Optical Society of America: Washington, DC, USA, 2015.
18. Ren, Z.; Chen, N.; Lam, E.Y. Extended focused imaging and depth map reconstruction in optical scanning holography. *Appl. Opt.* **2016**, *55*, 1040–1047. [CrossRef] [PubMed]
19. Jiao, A.S.M.; Tsang, P.W.M. Enhanced Autofocusing Scheme in Digital Holography Based on Hologram Decomposition. In Proceedings of the 2016 IEEE 14th International Conference on Industrial Informatics (INDIN), Poitiers, France, 19–21 July 2016; Volume 16, pp. 541–545.
20. Jiao, A.S.M.; Tsang, P.W.M.; Poon, T.-C.; Liu, J.-P.; Lee, C.-C.; Lam, Y.K. Automatic decomposition of a complex hologram based on the virtual diffraction plane framework. *J. Opt.* **2014**, *16*, 075401. [CrossRef]
21. Tsang, P.W.M.; Poon, T.-C.; Kim, T.; Kim, Y. Fast reconstruction of digital holograms for extended depths of field. *Chin. Opt. Lett.* **2016**, *14*, 070901. [CrossRef]

22. Liu, J.-P.; Guo, C.; Hsiao, W.; Poon, T.-C.; Tsang, P.W.M. Coherence experiments in single-pixel digital holography. *Opt. Lett.* **2015**, *40*, 2366–2369. [CrossRef] [PubMed]
23. Mir, H.; Xu, P.; Beek, P.V. An extensive empirical evaluation of focus measures for digital photography. In Proceedings of the Digital Photography X, San Francisco, CA, USA, 2–6 February 2014; Volume 9023I, p. 90230I.

Review

Review of 3D Imaging by Coded Aperture Correlation Holography (COACH) †

Joseph Rosen *, Vijayakumar Anand, Mani Ratnam Rai, Saswata Mukherjee and Angika Bulbul

Department of Electrical and Computer Engineering, Ben-Gurion University of the Negev,
Beer-Sheva 8410501, Israel; physics.vijay@gmail.com (V.A.); maniratn@post.bgu.ac.il (M.R.R.);
saswata7@gmail.com (S.M.); angikabulbul@gmail.com (A.B.)
* Correspondence: rosenj@bgu.ac.il
† It is an invited paper for the special issue.

Received: 25 December 2018; Accepted: 30 January 2019; Published: 12 February 2019

Abstract: Coded aperture correlation holography (COACH) is a relatively new technique to record holograms of incoherently illuminated scenes. In this review, we survey the main milestones in the COACH topic from two main points of view. First, we review the prime architectures of optical hologram recorders in the family of COACH systems. Second, we discuss some of the key applications of these recorders in the field of imaging in general, and for 3D super-resolution imaging, partial aperture imaging, and seeing through scattering medium, in particular. We summarize this overview with a general perspective on this research topic and its prospective directions.

Keywords: digital holography; computer holography; spatial light modulators; diffraction gratings; imaging systems

1. Introduction

Holography is a tool for recording a visual scene and reproducing it as close as possible to reality. In holography, even though only intensity sensors are used, it is possible to record the phase pattern of light, and thus the depth information, along with the intensity variation. The phase is detected indirectly by recording an interference pattern of the object wave with a reference wave. Typically, the interference pattern is obtained by combining coherent waves [1]. However, most of the imaging tasks in optics are performed with natural incoherent light. This is true for most microscopes, telescopes, and many other imaging devices. Thus, holography is not widely applied to general incoherent natural light imaging because usually, creating holograms with incoherent light requires uncommon designs.

This review concentrates on the more challenging case of interference with spatially incoherent light. More precisely, the holograms herein are of objects in which there is no statistical correlation between the waves emitted from various points of these objects. The historical roots of incoherent holograms are planted in the mid-nineteen-sixties [2–7], where some of these pioneering systems made use of the self-interference principle, a principle that is extensively used in some of the systems mentioned herein. The self-interference principle indicates that any two beams that originate from the same source point and then split to two waves are mutually coherent and hence they can be mutually interfered. In the case of incoherent illumination, where any two different source points are mutually incoherent, the self interference property becomes the only way to obtain any interference pattern, and thus enables us to record a hologram. The use of the self-interference principle has been continuously developed beyond the sixties by implementing several interesting systems [8–12]. Other methods of recording incoherent holograms like optical scanning holography [13,14] and multiple view projection methods [15,16] do not make use of the self-interference property and are out of the scope of this review. Other reviews which include some of the self-interference methods, together with some of the scanning techniques, can be found in Refs. [17,18].

While the term incoherent holography refers to the method of forming holograms, the term digital holography refers to the recording of the holograms by a digital camera and to the digital reconstruction of holograms. Explicitly, digital holograms recorded by digital cameras are reconstructed in digital computers by digital algorithms. The history of the digital holography also started in the nineteen-sixties [19] and is still an active research field today. Only the topic of incoherent digital holography with coded apertures is discussed herein.

Nowadays, we are in the middle of the era of digital imaging, in which images are recorded by digital cameras and processed by computer software. Digital imaging has accelerated the field of indirect imaging in which a non-image pattern of the observed scene is first recorded in the computer as an intermediate pattern. In the computer, the image of the scene is recovered from the intermediate pattern by digital processing. Digital holography is a typical example of indirect imaging in which the digital camera records one or more holograms, rather than a direct image of the scene. The indirect imaging and digital holography techniques are more complicated than direct imaging, and hence should be justified by as many as possible advantages in comparison to the much simpler direct imaging. The main benefit of digital holography and the initial motivation to begin the research in this field is the ability to image a three-dimensional scene with a single, or very few, camera shots [20]. Other advantages of digital holography have accumulated along the years of research and are discussed in the later sections of this review.

From the end of the nineties, incoherent digital holography techniques began to emerge again with new possibilities [13,21–45]. One notable invention in the field of incoherent digital holography is Fresnel incoherent correlation holography (FINCH) which was also developed based on the self-interference principle [23]. FINCH was able to twist one of the fundamental laws of optics, the Lagrange invariant, in order to enhance its transverse image resolution. Alternative versions of incoherent holography systems called self-interference digital holography (SIDH) techniques, were developed and implemented for applications of adaptive optics [28,32]. More recently, a generalized version of the self-interference incoherent digital holography technique called coded aperture correlation holography (COACH) has been developed [46]. In this survey, we review various designs and applications of the COACH systems and other close techniques.

This review consists of four main sections. The development of COACH architectures, with different modalities and characteristics, is reviewed extensively in the following section. In the third section, various applications based on COACH techniques are discussed. The final section summarizes the review.

2. System Architectures

The general optical configuration of self-interference digital holography is shown in Figure 1. The light emitted from each object point is collected by a beam splitting system, in which the input wave is split into two, or more, and each wave is modulated differently. Since the waves originate from the same object point, they are mutually coherent, and hence they can produce an interference pattern on the image sensor plane. The sensor accumulates the entire interference patterns of all the object points into an incoherent hologram. A single hologram, or several acquired holograms, are fed into a digital computer. In the case of several holograms, they are superposed into a single digital hologram. Finally, the image of the object is reconstructed from the processed hologram by some digital algorithm. Next, we survey several recently proposed systems of the family of COACH methods.

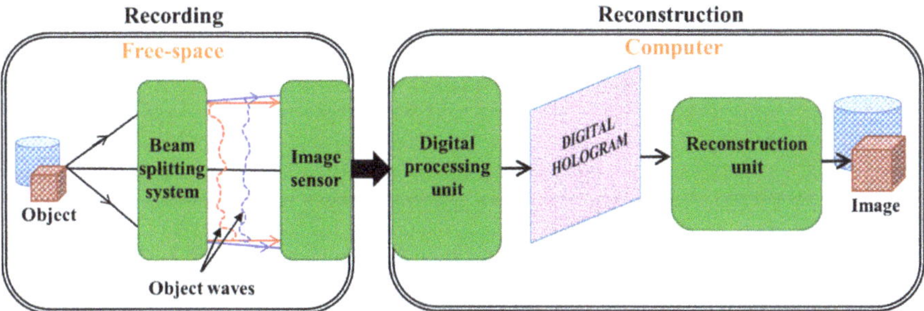

Figure 1. Recording and reconstruction of a hologram in a general self-interference digital holography system.

2.1. Coded Aperture Digital Holography

The use of coded aperture masks is common in X-rays since the end of the sixties. In 1968, Ables and Dicke, separately [47,48], reported the first modern coded aperture for X-ray imaging. The coded aperture, as explained in [49] and illustrated in Figure 2, is a randomly arranged array of pinholes. The goals of such systems are to increase the signal to noise ratio (SNR), the field of view (FOV) and the low power efficiency, which are associated with the regular pinhole imaging [49]. In the coded aperture imaging, instead of imaging the object directly, a non-recognizable pattern is formed on the sensor plane by the superposition of many randomly arranged images, each of which is formed independently by a different pinhole. The coded aperture imaging is not a direct process as in the case of a single pinhole, but an indirect process in which the recorded pattern must be digitally processed to retrieve the image. The coded aperture imaging possesses 3D imaging capabilities as objects located at different distances appear with different magnifications and are shifted by different lateral distances. Therefore, in principle, the different planes of the object can be retrieved [49].

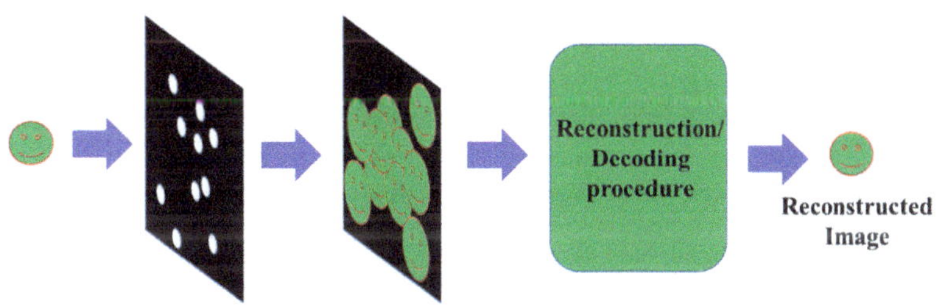

Figure 2. Coded aperture imaging system.

2.1.1. Coded Aperture Correlation Holography Architectures

Recently, a self-interference incoherent digital holography technique using a coded aperture was developed in the optical regime [46] and is shown in Figure 3. Unlike other imaging techniques with coded apertures [49], in COACH, the aperture is a phase-only mask. Thus, the aperture does not absorb the incoming light and consequently, more light power participates in the hologram recording. In addition, the detected pattern is a digital hologram containing the 3D image of the observed scene. COACH can be considered as a generalization of FINCH [23,26] in the sense that instead of the quadratic phase mask of FINCH, a pseudorandom coded phase mask (CPM) modulates the beam. In COACH, the light emitted from each object point splits into two beams in which one of the beams is modulated by the CPM while the other beam is not. The two beams with the common origin are mutually coherent and thus are interfered. Three phase-shifted holograms are recorded for phase values $\theta_{1,2,3} = 0$, $2\pi/3$ and $4\pi/3$ and are superposed into a complex hologram. As in FINCH, this procedure is done in order to remove the twin image and bias terms during the reconstruction. In COACH, unlike FINCH, a one-time training procedure is required. The training is done by recording a library of point spread holograms (PSHs) of a point object positioned at various axial locations. For a 3D object placed within the axial boundaries of the PSH library, three phase-shifted holograms are once again recorded, using the same CPM with the above three phase values, and superposed into a complex object hologram. Any axial plane of the object space is reconstructed by a cross-correlation between the object hologram and the corresponding PSH from the library.

The holograms recorded by COACH cannot be classified as Fourier or Fresnel holograms because neither Fourier transform nor Fresnel back-propagation can reconstruct the image. Instead, this hologram is generated from the interference between the plane and chaotic waves. Therefore, COACH can be considered as a generalized correlation self-interference holography technique in which FINCH is only a special case when the CPM is the special mask of the diffractive spherical lens. Another dissimilarity is that in FINCH, the object image can be reconstructed by a Fresnel back-propagation from the hologram. Therefore, the imaging characteristics of COACH are also different from those of FINCH as analyzed in the following.

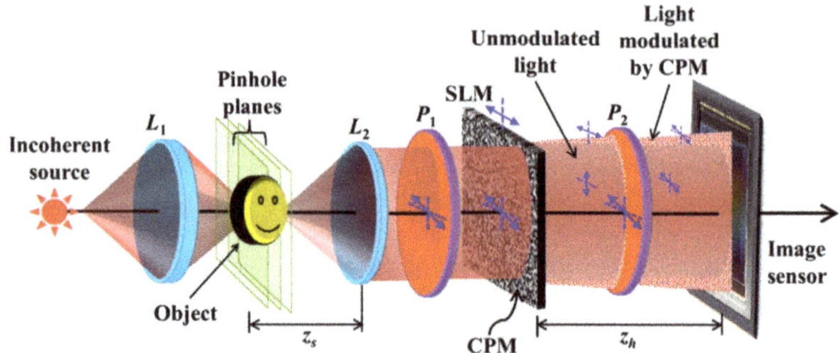

Figure 3. Optical configuration of COACH. CPM—Coded phase mask; L_1, L_2—Refractive lenses; P_1, P_2—Polarizers; SLM—Spatial light modulator; Blue arrows indicate polarization orientations.

While COACH, being a generalized technique, is operational with any random phase mask, it was noticed that for an arbitrary CPM, the reconstruction via a cross-correlation generated a disturbing background noise. In order to reduce the background noise and at the same time to retain the randomness aspect of the CPM, the CPM is synthesized using a modified Gerchberg-Saxton algorithm (GSA) [50] to render a pure phase function for the CPM and a uniform magnitude over some desirable area in the spectrum domain, as shown in Figure 4. The constraint on the CPM plane is required since the CPM is displayed on a pure-phase SLM. On the other hand, the constraint on the spectrum domain (sensor plane) reduces the background noise on the reconstructed image. This is because the noise originated from the bias level of the spectral intensity. This bias level is considerably reduced by the superposition of two, or more, spectral intensities, if this level is approximately the same in all two, or more, spectral intensities. The constraint on the spectrum domain satisfies this last condition of approximately equalizing the bias level.

Based on Figure 3, the mathematical formulation is as follows. For a point object located at the front focal plane of lens L_2, a collimated beam passes through the polarizer P_1 oriented 45° with respect to the active axis of the SLM. As a result, only part of the incident light is modulated by the SLM while the remaining part is not. On the sensor plane, two patterns are detected, namely, a uniform signal from the plane wave and a pseudorandom complex function $G(u,v)$ generated by the CPM modulation.

A second polarizer P_2 oriented 45° along the active axis of the SLM creates interference between the two beams on the sensor plane to yield the following intensity,

$$I_k(u,v) = |A + G(u,v)\exp(i\theta_k)|^2, \qquad k = 1,2,3, \tag{1}$$

where θ_k is the kth phase value of the three-phase shifts. A complex point spread hologram H_{PSH} $= A^*G(u,v)$ is produced by the phase-shifting procedure. An object is placed at the axial location of the point object and the complex object hologram is synthesized from three intensity recordings similar to the process of Equation (1). Any cross-section intensity of the object can be represented as a collection of uncorrelated points. For simplicity, let us assume that the object plane $o(x,y)$ is located at the front focal plane of the lens L_2, a distance f_o from L_2. Each j-th object point, at (x_j,y_j), on the object plane, generates two mutually coherent beams on the sensor plane. One is a tilted plane wave $A_j\exp[i2\pi(ux_j+vy_j)/\lambda f_o]$ and the other is a shifted version of $G(u,v)$ multiplied by the same plane wave as the following: $B_j\exp[i2\pi(ux_j+vy_j)/\lambda f_o]G(u-u_j,v-v_j)$, where $(u_j,v_j) = (x_j,y_j)z_h/f_o$. The intensity pattern on the sensor plane, resulting from the object in the input, is given by

$$\begin{aligned}
I_k(u,v) = \sum_j \Bigg| &A_j \exp\left[\frac{i2\pi(x_ju+y_jv)}{\lambda f_o}\right] \\
&+ \exp(i\theta_k)B_j\exp\left[\frac{i2\pi(x_ju+y_jv)}{\lambda f_o}\right]G(u-u_j,v-v_j)\Bigg|^2
\end{aligned} \tag{2}$$

By the phase-shifting procedure, a complex object hologram H_{OBJ} is synthesized given by,

$$H_{OBJ}(u,v) = \sum_j A_j^* B_j G(u-u_j,v-v_j) \tag{3}$$

A particular plane of the object can be reconstructed by the cross-correlation between the $H_{PSH} = G(u,v)$ and H_{OBJ} as

$$\begin{aligned}
P(u\prime,v\prime) &= \iint \left\{ \sum_j A_j^* B_j G(u-u_j,v-v_j) \right\} G^*(u-u\prime,v-v\prime)dudv \\
&\approx \sum_j A_j^* B_j \Lambda(u\prime-u_j,v\prime-v_j) \propto o(u\prime/M_T,v\prime/M_T),
\end{aligned} \tag{4}$$

where Λ is a δ-like function, approximately equal to 1 around $(0,0)$ and to small negligible values elsewhere. The transverse magnification of COACH is given by $M_T = z_h/f_o$. Since the reconstruction of the object hologram is carried out by a cross-correlation, the lateral and axial resolutions are determined by the lateral and axial correlation distances, respectively. Since the correlation distances are determined by the system aperture size, the lateral and axial resolutions are similar to that of direct imaging with the same numerical aperture (NA). Since the reconstruction is carried out by a cross-correlation with a PSH generated by a pinhole, we have guaranteed that the diameter of the beam from the pinhole has been larger than the diameter of the system's aperture, in order to assure that imaging resolution is limited by the NA.

The experimental demonstration of COACH was carried out with the setup of Figure 3 using a pinhole with a diameter of approximately 100 μm and a National Bureau of Standards (NBS) object of 7.1 *lp/mm* [46]. Three holograms were recorded for the entire objects with the above-mentioned phase shifts and each set of holograms was composed into PSH and complex object holograms. The image of the object was reconstructed by a cross-correlation. The three PSHs for the phase shifts $\theta = 0°$, 120° and 240° are shown in Figure 5a–c, respectively. The object holograms corresponding to the same three phase-shifts are shown in Figure 5d–f. The amplitude and phase of the complex H_{PSH} are shown in Figure 5g,j, respectively. The image of the CPM is shown in Figure 5i. The amplitude and the phase of the complex H_{OBJ} are shown in Figure 5h,k, respectively. The reconstructed image is shown in Figure 5l. The experiment is repeated using two channels and United States Air Force (USAF) object with elements 2 and 3 of group 2 placed at the same location as the NBS object from the lens L_2. A complex hologram is synthesized from three camera shots. The reconstruction result is shown in Figure 5m. The USAF object is shifted by a relative distance of 2 cm from the NBS object. The reconstruction results using H_{PSH}'s recorded at the location of the NBS and the new location of the USAF are shown in Figure 5n,o, respectively.

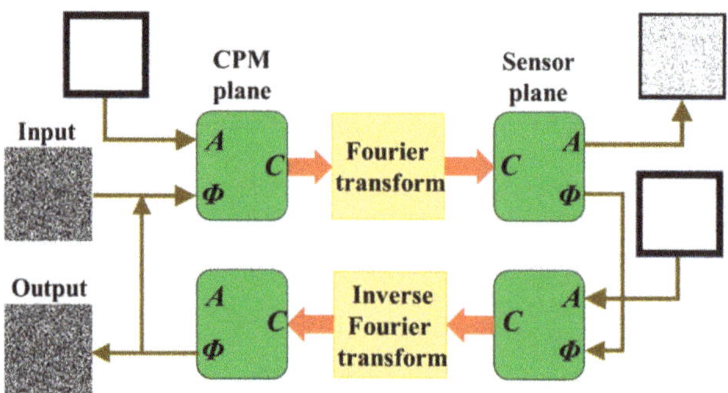

Figure 4. CPM synthesis using the modified Gerchberg-Saxton algorithm.

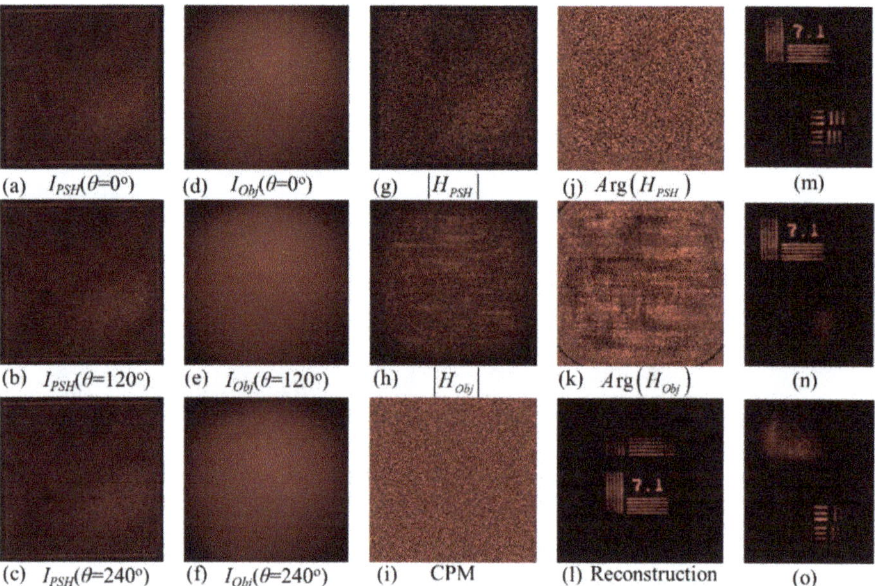

Figure 5. (**a–c**) Intensity patterns for the pinhole for phase shifts $\theta = 0°$, $120°$ and $240°$, (**d–f**) intensity patterns for the object for phase shifts $\theta = 0°$, $120°$ and $240°$, (**g–h**) amplitudes of H_{PSH} and H_{OBJ}, respectively, (**i**) image of the CPM, (**j–k**) phase of H_{PSH} and H_{OBJ}, respectively, (**l**) reconstructed image, (**m**) reconstruction results of two objects when both are on the same plane. Reconstruction results when the two objects are separated by 2 cm and reconstructed using (**n**) H_{PSH} of NBS plane and (**o**) H_{PSH} of USAF plane.

From the reconstruction results, it can be seen that even though the CPMs were synthesized using GSA, there is a substantial amount of background noise. In order to reduce the background noise, additional techniques are necessary. In relation to the topic of optical pattern recognition, it was demonstrated that using a phase-only, instead of a matched filter, the correlation peaks become sharper with reduced side lobes [51]. Therefore, as a first step against the noise, a modified PSH with phase-only Fourier transform, given by, replaced the matched filter based PSH, where $\Im$ and $\Im^{-1}$ are a Fourier and inverse Fourier transform, respectively, and $\bar{\rho}$ is the location vector in the reconstruction plane. The phase-only filter, when used in the correlation with the object hologram, reconstructed sharper images with lower background noise. The reconstruction results [52] of the NBS object 10 *lp/mm* using the matched filter and the phase-only filter are shown in Figure 6a,b, respectively. However, the phase-only filter is less suitable for reconstruction of objects with greyscale transmittance [52], due to the lower immunity of the phase-only filter against noise.

$$\widetilde{H}_{PSH}(\bar{\rho}) = \Im^{-1}\{\exp[i \cdot \arg(\Im\{H_{PSH}(\bar{\rho})\})]\} \tag{5}$$

Averaging is another technique, which involves the recording of multiple holograms followed by averaging over the multiple reconstructions [53]. The SNR of the COACH method was further improved by recording several H_{OBJ} and H_{PSH} under independent CPMs and averaging over the several complex reconstructions. The averaging technique, in the case of COACH, is based on the assumption that any two CPMs synthesized by GSA from different initial random profiles are independent, i.e., their cross-correlation is negligible compared to their auto-correlation. Therefore, for N independent complex PSH holograms and object holograms recorded under statistically independent CPMs, averaging over N reconstructions theoretically increases the SNR by $\sqrt{N}$ [53]. On the other hand, the averaging procedure decreases the time resolution, since instead of 3 intensity patterns, $3N$ intensity patterns are required to reduce the background noise. The results of averaging with 5, 10, 15 and 20 samples are shown in Figure 6c–f, respectively [52].

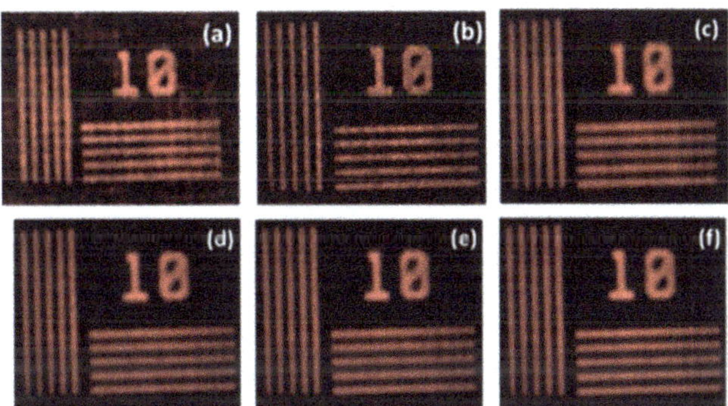

Figure 6. Reconstruction results using (**a**) matched filter, (**b**) phase-only filter. Phase-only filter and averaging with (**c**) 5 samples, (**d**) 10 samples, (**e**) 15 samples and (**f**) 20 samples.

2.1.2. Interferenceless Coded Aperture Correlation Holography

The entire incoherent digital holography techniques discussed so far are based on two-beam interference. In general, two beam interferometers suffer from several limitations, such as the need for vibration isolation, power loss of more than half of the incident optical power and the need to minimize the optical path difference between the optical channels. Two-beam interference is essential in FINCH to read the phase of the wave emitted from each object point. Only the phase function contains the information of an object point location, whereas the wave magnitude is uniform and thus lacks any information. Unlike FINCH, in COACH, the information of a single point location is encoded in both the phase and the magnitude distributions. Both functions have chaotic distributions that can be reconstructed by a cross-correlation with the corresponding PSH. However, to record the phase, self interference is needed, whereas to record the magnitude, a direct detection of the light intensity is enough. Therefore, COACH has the capability to image a 3D scene without two-wave interference. This version of COACH without two-wave interference is termed interferenceless COACH (I-COACH) [54], and it is implemented without splitting the object beam to two beams.

I-COACH is not much different from a direct single-channel imaging system but has the capability to record and reconstruct 3D information in one or a few camera shots. The relaxation of the interference condition improves the SNR during reconstruction and the power efficiency of the imaging process is increased. At this point, it should be mentioned that imaging with a phase-coded aperture has already been proposed by Chi and George [55]. However, Ref. [55] has avoided an imaging 3D scene and because of lack of any noise reduction mechanism, the presented results suffer from substantial background noise. The optical configuration of I-COACH is shown in Figure 7. The polarizer P is oriented along the active axis of the SLM such that all the incident light is modulated. In I-COACH, three CPMs are synthesized from three initial independent random phase profiles and the corresponding intensity patterns are recorded for a point object. Each of the recorded patterns is multiplied by one of the phase constants with the values $\theta_{1,2,3} = 0$, $2\pi/3$ and $4\pi/3$, and the three matrices are combined into a single complex hologram. The same procedure is repeated for the object hologram. The object image is reconstructed by a cross-correlation between the above two holograms.

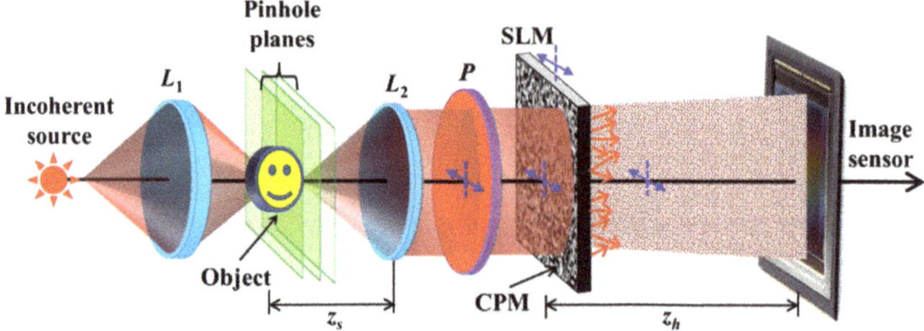

Figure 7. Optical configuration of I-COACH. CPM—Coded phase mask; L_1, L_2—Refractive lenses; P—Polarizer; SLM—Spatial light modulator; Blue arrows indicate polarization orientations.

The experimental results of I-COACH are shown in Figure 8 for a two plane object constructed from element 8 *lp/mm* of the NBS chart and elements 5 and 6 of Group 2 of the USAF chart [54]. The NBS and USAF are separated from each other by 1 cm. The H_{PSH} is recorded at the two planes of the object as with COACH followed by the recording of the H_{OBJ}. Even though the SNR is improved in the case of I-COACH compared to COACH, the technique still needs additional procedures to reduce the background noise. In the present experiment, both phase-only filtering and averaging technique (20 samples) were implemented. The simplicity of the optical configuration without interference but with a unique capability to record and reconstruct 3D information makes I-COACH an attractive candidate for 3D imaging.

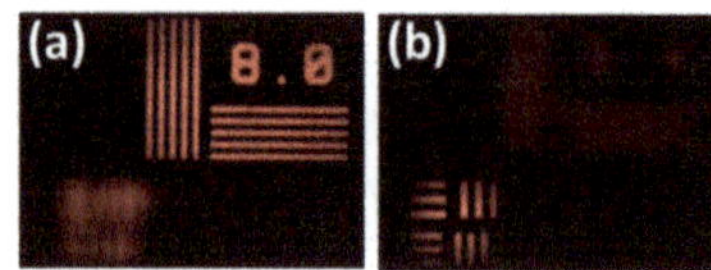

Figure 8. Reconstruction results of I-COACH for (**a**) NBS chart and (**b**) USAF chart.

2.1.3. Single Camera Shot I-COACH

One of the disadvantages of I-COACH is the requirement of several camera shots to reconstruct an object with an acceptable SNR. As mentioned above, the noise reduction techniques in I-COACH are implemented at various levels, starting from the design of the CPMs, reconstructing images by phase-only filtering and recording a few holograms for averaging. The averaging technique reduces the time resolution of the system. In this section, the averaging technique is replaced with a method of recording the holograms with a single camera shot. This method is denoted as the single camera shot I-COACH (SCS-I-COACH) [56].

The Fourier-based GSA (see Figure 4) used in I-COACH for synthesizing the CPM was constrained to yield a uniform magnitude on part of the sensor plane. However, because the image sensor was not located at the Fourier plane of the CPM in the experimental setup, the Fourier relation between the CPM and the sensor plane was not satisfied in the experiment. This means that the Fourier-based GSA does not match the experimental setup, and hence the noise has not been reduced as expected. Therefore, an additional diffractive lens was multiplexed into the CPM and the image sensor was positioned at the focal plane of this lens to fulfill the Fourier relations between the CPM and the sensor plane in the GSA. Although the above modification improved the SNR, the system still required at least two camera shots for creating bipolar holograms. In SCS-I-COACH, bipolar holograms are indeed recorded but the two shots recorded in time are replaced by two shots recorded in space using additional diffractive masks. The integration of two CPMs in the aperture plane using linear and quadratic phase functions is shown in Figure 9. The optical configuration of SCS-I-COACH is shown in Figure 10. An additional constraint is used in the GSA to limit the area of the intensity pattern in the sensor plane. It should be noted that the space sharing of two raw holograms on the sensor area reduces the FOV of the imaging system. Hence, in SCS-I-COACH, the single-shot capability is obtained at the expense of a reduction in FOV and some reduction of SNR.

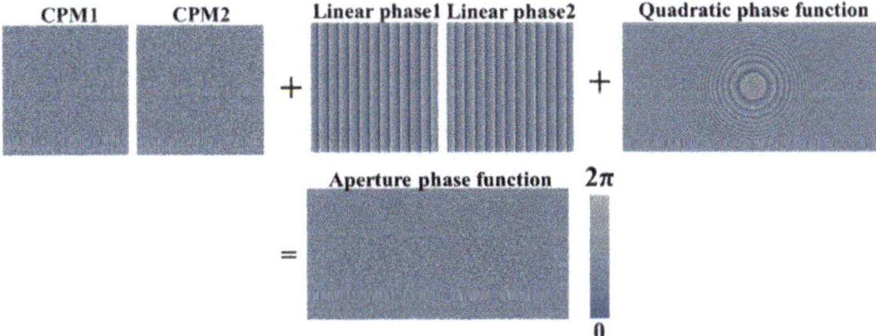

Figure 9. Engineering the aperture of SCS-I-COACH using linear and quadratic phase functions added to the two CPMs.

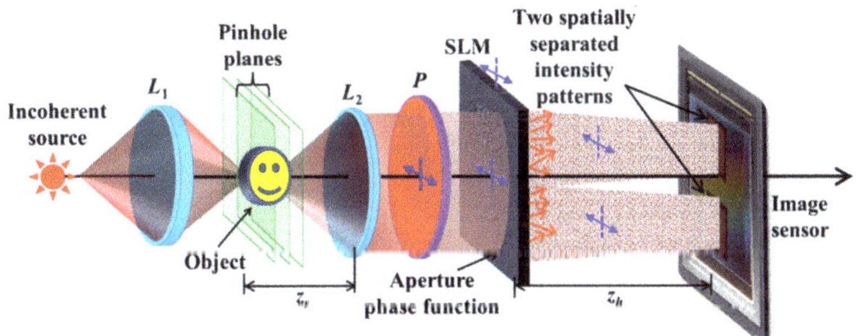

Figure 10. Optical configuration of SCS-I-COACH. L_1, L_2—Refractive lenses; P—Polarizer; SLM—Spatial light modulator; Blue arrows indicate polarization orientations.

In the SCS-I-COACH setup, there are two image responses to the single object point. Therefore, there are two Fourier transforms for CPM1 and CPM2 centered at $(U/4, 0)$ and $(-U/4, 0)$, respectively, where U is the height of the sensor plane. The intensity in the sensor plane is given by,

$$I(u, v) = G_1(u - U/4, v) + G_2(u + U/4, v) \tag{6}$$

where $G_1 = |\Im\{\exp(i\Phi_{CPM1})\}|^2$ and $G_2 = |\Im\{\exp(i\Phi_{CPM2})\}|^2$. The two intensity patterns G_1 and G_2, corresponding to CPM1 and CPM2, respectively, are recorded, extracted and subtracted from each other. The resulting intensity pattern is a bipolar PSH given by $H_{PSH}(z = 0) = G_1(u, v) - G_2(u, v)$. A library of PSHs is created as described earlier by moving the pinhole to various axial locations. Next, the object hologram is recorded for an object placed within the axial boundaries of the PSH library. The object is considered as a collection of independent incoherent source points. If the system is linear and shift invariant, the overall intensity response I_D on the sensor plane is given by a sum of the entire individual shifted impulse responses, as follows

$$I_D = \sum_j G_1\left(u - \frac{z_h u_j}{f_0} - \frac{U}{4}, v - \frac{z_h v_j}{f_0}\right) + \sum_j G_2\left(u - \frac{z_h u_j}{f_0} + \frac{U}{4}, v - \frac{z_h v_j}{f_0}\right) \tag{7}$$

The resulting bipolar object hologram H_{OBJ} obtained by subtracting the individual responses is

$$H_{OBJ} = \sum_j G_1\left(u - \frac{z_h u_j}{f_0}, v - \frac{z_h v_j}{f_0}\right) - \sum_j G_2\left(u - \frac{z_h u_j}{f_0}, v - \frac{z_h v_j}{f_0}\right) \tag{8}$$

The object image is reconstructed by cross-correlating the object hologram with the PSH($z = 0$) as

$$\begin{aligned}
I_{img} = H_{OBJ} \otimes H_{PSH} &= \left[\sum_j G_1\left(u - \tfrac{z_h u_j}{f_0}, v - \tfrac{z_h v_j}{f_0}\right) - \sum_j G_2\left(u - \tfrac{z_h u_j}{f_0}, v - \tfrac{z_h v_j}{f_0}\right)\right] \\
&\quad \otimes [G_1(u, v) - G_2(u, v)] \\
&= \sum_j \Lambda\left(u_0 - \tfrac{z_h u_j}{f_0}, v_0 - \tfrac{z_h v_j}{f_0}\right) \\
&\cong O\left(\tfrac{f_0 u_0}{z_h}, \tfrac{f_0 v_0}{z_h}\right) = O\left(\tfrac{u_0}{M_T}, \tfrac{v_0}{M_T}\right)
\end{aligned} \tag{9}$$

where $\otimes$ represents a two-dimensional correlation and $M_T = z_h/f_0$ is the lateral magnification of the imaging.

The GSA was designed to limit the area of the uniform magnitude on the sensor plane to be only 4.3×4.3 mm out of 14.3×14.3 mm. The linear phase values for the two CPMs were chosen to be $\pm 0.7°$ such that, for $z_h = 25.4$ cm, the intensity patterns are shifted by approximately 3 mm from the optical axis. The system was trained using a pinhole with a diameter of 25 μm. Two objects, element 5 and 6 of group 3 of USAF resolution targets, were used. The relative distance between the two objects was shifted by 4 mm in steps of 1 mm. The intensity responses for the input pinhole and the corresponding bipolar hologram are shown in Figure 11a,b, respectively. The image responses for the input of the two-plane object and the object hologram are shown in Figure 11c,d, respectively. Various object holograms were recorded for the different locations of the two objects, and the resulting bipolar holograms were reconstructed by a cross-correlation with the PSH library. The reconstruction results are shown in Figure 12, demonstrating that although the FOV is limited, the background noise is relatively low for a single camera shot. Because SCS-I-COACH is a single shot recording, it can be used for recording the dynamic scene and for creating holographic videos of moving objects.

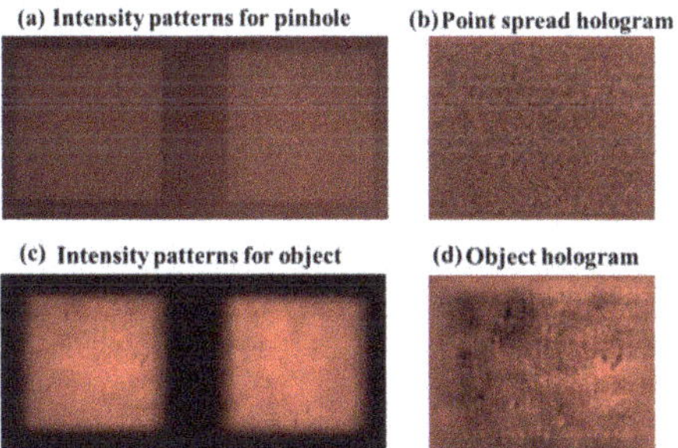

Figure 11. (**a**) System response for the pinhole, (**b**) bipolar PSH, (**c**) system response for the object, and (**d**) bipolar object hologram.

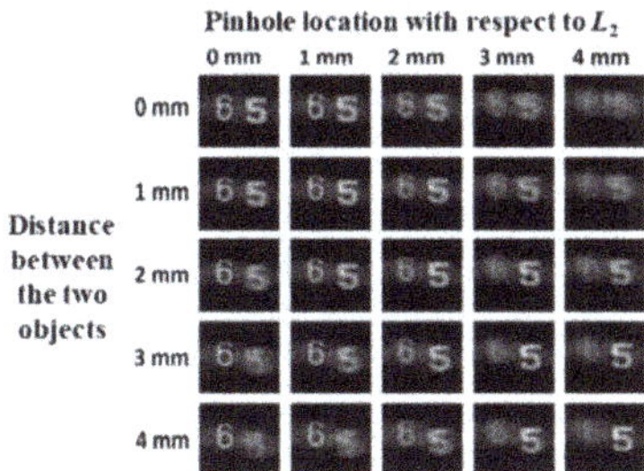

Figure 12. Reconstruction results of SCS-I-COACH.

2.1.4. FOV Extended I-COACH

In this section, we describe a technique to improve the FOV of I-COACH. The FOV of a typical imaging system is limited by the ratio between the finite area of the image sensor and the magnification of the optical system. For a given magnification, the limiting factor of the FOV is the finite size of the image sensor. Different techniques have been developed for enhancing the FOV of an imaging system, such as convolution techniques [57], particle encoding [58] and multiplexing of interferograms [59]. Recently, a robust yet simpler technique was developed for extending the FOV of I-COACH beyond the limit dictated by the image sensor area [60].

The optical configuration of the I-COACH with extended FOV is shown in Figure 13. In this technique, the GSA is directed to generate on the sensor plane an intensity pattern with the size three times larger than the area of the image sensor. The pinhole is shifted to pre-calculated lateral locations in the same axial plane, such that the different sections of the intensity pattern are projected on the image sensor and recorded. The recorded parts of the intensity pattern are stitched into a larger intensity pattern in the computer. The same process of recording and stitching is repeated using a second CPM. Unlike the methods in [57–59], the FOV extension procedure in I-COACH is done only once during the recording of the point holograms, with a relatively longer training, following by a stitching procedure in the computer. Once the synthetic PSH library is created, the recording of the object holograms is as simple as a regular I-COACH system.

I-COACH is a linear space-invariant system. Thus, the response on the camera plane for a 2D object is a collection of independent point objects expressed as $\sum_j a_j I_1\left(\bar{r}_o - M_T\bar{r}_j\right)$, where I_1 is the response for a delta-function in the input, M_T is the transverse magnification given by $M_T = z_h/z_s$, z_s is the distance between the object and lens L_2 and a_j is a constant. The bipolar PSH is $H_{PSH} = I_1\text{-}I_2$, and the bipolar object hologram can be expressed as,

$$H_{OBJ} = \sum_j a_j\left[I_1\left(r_o - M_T r_j\right) - I_2\left(r_o - M_T r_j\right)\right] = \sum_j a_j H_{PSH}\left(r_o - M_T r_j\right) \tag{111}$$

where I_2 is the impulse response with the second independent CPM. The image of the object inside the FOV is reconstructed as

$$
\begin{aligned}
P(\bar{r}_R) &= \iint \sum_j a_j H_{PSH}(\bar{r}_o - M_T\bar{r}_j)\tilde{H}^*_{PSH}(\bar{r}_o - \bar{r}_R)d\bar{r}_o \\
&= \mathfrak{S}^{-1}\left\{\sum_j a_j |\mathfrak{S}\{H_{PSH}\}| \exp(i\varphi - i2\pi M_T\bar{r}_j \cdot \bar{\rho})\exp(-i\varphi)\right\} \\
&= \sum_j a_j \Lambda(\bar{r}_R - M_T\bar{r}_j) \approx o\left(\frac{\bar{r}_R}{M_T}\right)
\end{aligned}
\tag{11}
$$

For a transverse magnification of M_T and sensor area of $D \times D$, the size of FOV is $S \times S$, where $S = D/M_T$. To extend the FOV by a factor of 3, the pinhole is shifted to nine locations resulting of shifts of H_{PSH}, and thus different areas of H_{PSH} are overlapped with the image sensor. The corresponding sections of the intensity patterns are detected separately, each time for the input $\delta(x_s - Sk, y_s - Sl)$, where $k,l = -1,0,1$, and the nine intensities are stitched. Besides, the central object point, the remaining 8 pinholes are outside the FOV of the imaging system. The recorded part of the intensity pattern can be expressed as $I_1(x_o - Dk, y_o - Dl)\text{Rect}[(x_o, y_o)/D]$. The stitched intensity pattern is given by,

$$
\bar{I}_1(x_o, y_o) = \sum_{k=-1}^{1} \sum_{l=-1}^{1} I_1(x_o - Dk, y_o - Dl)\text{Rect}\left[\frac{(x_o, y_o)}{D}\right] * \delta(x_o - Dk, y_o - Dl)
\tag{12}
$$

The process is repeated using the second CPM and the final bipolar PSH is $\bar{H}_{PSH} = \bar{I}_1 - \bar{I}_2$. An object located outside the FOV of the system, around the point (kS, lS), can be expressed as $\sum_j a_j \delta(x_s - x_j + kS, y_s - y_j + lS)$, for $k,l = -1,0,1$ [but $(k,l) \neq (0,0)$)]. The bipolar out-of-FOV object hologram, obtained by subtraction of the two intensity patterns, is given by

$$
H_{OBJ}(x_o, y_o) = \sum_j \bar{H}_{PSH}(x_o - M_T x_j - Dk, y_o - M_T y_j - Dl)\text{Rect}\left[\frac{(x_o, y_o)}{D}\right]
\tag{13}
$$

The object is reconstructed by a cross-correlation between the phase-only filtered synthetic $\bar{H}'_{PSH}$ and the H_{OBJ} as

$$
\begin{aligned}
P(\bar{r}_R) &= \iint \sum_j \bar{H}_{PSH}(x_o - M_T x_j - Dk, y_o - M_T y_j - Dl)\text{Rect}\left[\frac{(x_o, y_o)}{D}\right] \\
&\times \bar{H}'^*_{PSH}(x_o - x_R, y_o - y_R)d\bar{r}_0 = \sum_j a_j \Lambda(\bar{r}_R - M_T\bar{r}_j - D\bar{v}) \approx o\left(\frac{\bar{r}_R}{M_T} - D\bar{v}\right)
\end{aligned}
\tag{14}
$$

where $\bar{v} = (k, l)$. The reconstruction is achieved because there is a high correlation between the synthetic $\bar{H}_{PSH}$ and H_{OBJ} at $(x_R, y_R) = (Dk, Dl)$.

The experiment was carried out using a pinhole with a diameter of 80 μm. The pinhole was shifted to 8 positions outside the FOV, and the intensity patterns were recorded followed by the stitching procedure. The experiment was repeated for an object made up of three transparent digits '6', '0' and '1' from two optical channels. In channel-1, the object '0' was mounted on the optical axis to be within the FOV of the imaging system, whereas in channel-2, the objects '1' and '6' are mounted outside the FOV of the imaging system. The images of the bipolar stitched PSH and zero-padded object holograms are shown in Figure 14a,b, respectively.

Even though the optical configuration used for the FOV extension technique involves a diffractive lens to satisfy the GSA condition between the SLM and sensor plane, there is additional background noise due to the zero padding of the object hologram. To reduce the background noise, the averaging technique was implemented with 20 samples. The reconstruction results using only the central part of the PSH and synthetic PSH for the entire objects are shown in Figure 15a,b, respectively.

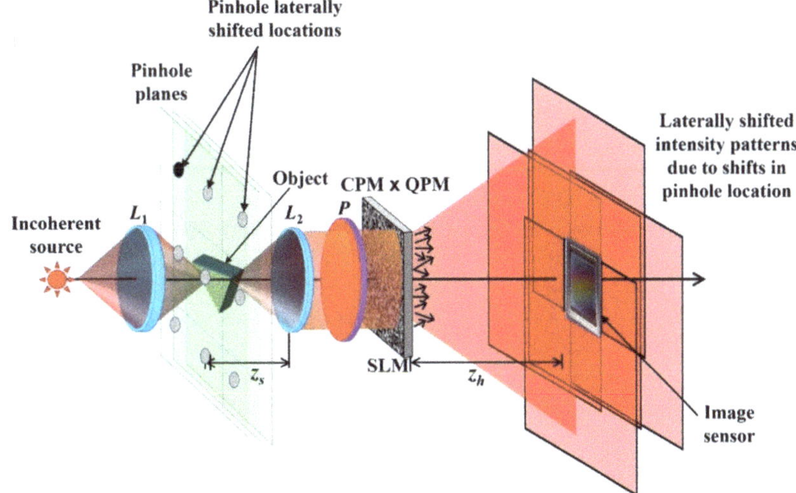

Figure 13. Optical configuration of I-COACH for FOV extension. CPM—Coded phase mask; L_1, L_2—Refractive lenses; P—Polarizer; SLM—Spatial light modulator; QPM—Quadratic phase mask.

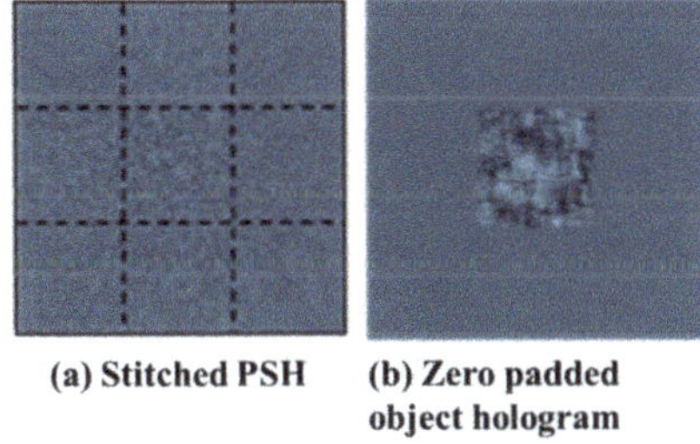

(a) Stitched PSH   **(b) Zero padded object hologram**

Figure 14. Image of the (**a**) stitched PSH and (**b**) zero padded object hologram.

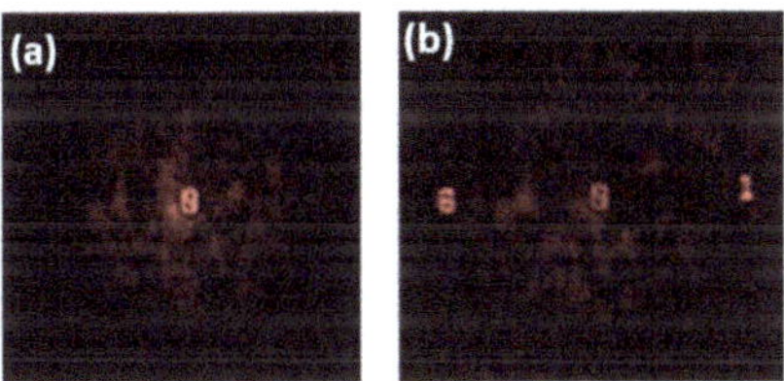

Figure 15. Reconstruction results using (**a**) regular PSH and (**b**) synthetic PSH.

The FOV extension technique is demonstrated in I-COACH system with an extension factor of 3. However, with a longer training process, the FOV can be extended theoretically without limit. While the FOV extension technique is demonstrated on a coded aperture holography system, the technique can be easily adapted to other imaging systems by creating a synthetic point spread function and by stitching the individual responses corresponding to different lateral locations of the pinhole.

2.1.5. Partial Aperture Imaging System

Partial aperture imaging is a technique to image objects through a part of the aperture area with as close as possible resolution capabilities of the full aperture. In this section, the partial aperture imaging capabilities of I-COACH is summarized and compared with equivalent direct imaging systems. The I-COACH systems with partial apertures are denoted as partial aperture imaging systems (PAISs) [61]. In order to retain the resolution of the full aperture, annular apertures are considered. In the design of PAIS, an annular CPM is first synthesized using the GSA. As shown in Figure 16, the constraint on the camera plane is a uniform magnitude over a limited desired area, and the constraint on the CPM plane is a pure annular phase distribution and a zero transparency outside the ring. The obtained CPM is multiplied by an annular diffractive lens (DL) to satisfy the Fourier transform relation between the SLM and the camera planes. As shown in Figure 17, additional phase functions containing quadratic and linear phases are introduced in the aperture plane to deflect and concentrate the light incident outside the annular region away from the image sensor. The scheme of the laboratory imaging system is shown in Figure 18. The hologram recording and reconstruction are similar to those of a regular I-COACH. The light emitted from an object is collected and collimated using lens L_2. The collimated beam is modulated by the annular CPM and Fourier transformed on the sensor plane by the annular DL. The light incident outside the CPM is deflected away from the sensor by the diffractive elements. Like any COACH system, also in PAIS, a training stage is necessary to record the PSH library.

An experiment demonstrating PAIS was carried out [61] by a single optical channel to image an NBS object with 14 *lp/mm*. Three intensity patterns were recorded using three different annular CPMs synthesized from three different initial random phase profiles and composed into a complex hologram using phase shifts $\theta_{1,2,3} = 0$, $2\pi/3$ and $4\pi/3$. PAIS was trained using a pinhole with a diameter of 25 μm with the same three CPMs. Direct imaging was compared against PAIS results using only the annular diffractive lens. The recorded intensity patterns for the point object, object, reconstructions and direct imaging results for different thicknesses (160, 80 and 40 μm of the annular CPMs corresponding to different partial aperture ratios 0.06, 0.03 and 0.014) are shown in Figure 19.

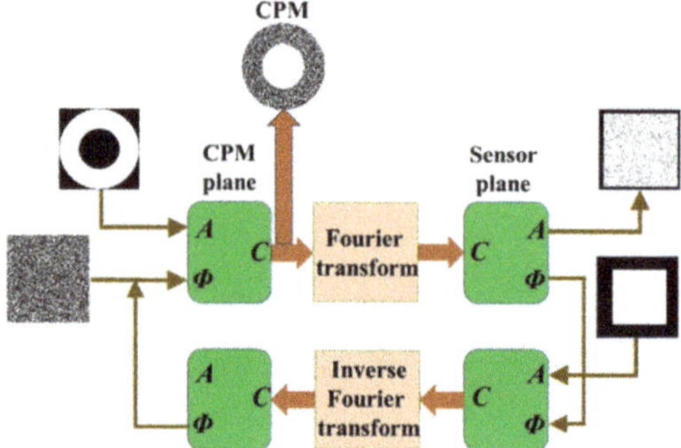

Figure 16. Modified GSA for the synthesis of an annular CPM to render a uniform magnitude over a limited area on the sensor plane.

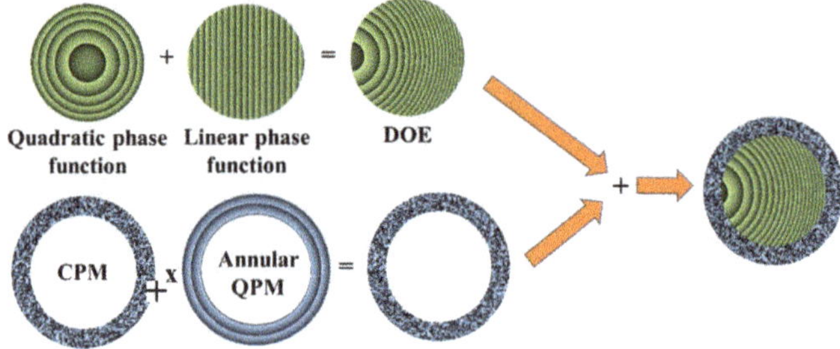

Figure 17. Design of the aperture function for deflecting the light incident outside the annular CPM away from the image sensor; DOE—Diffractive optical element.

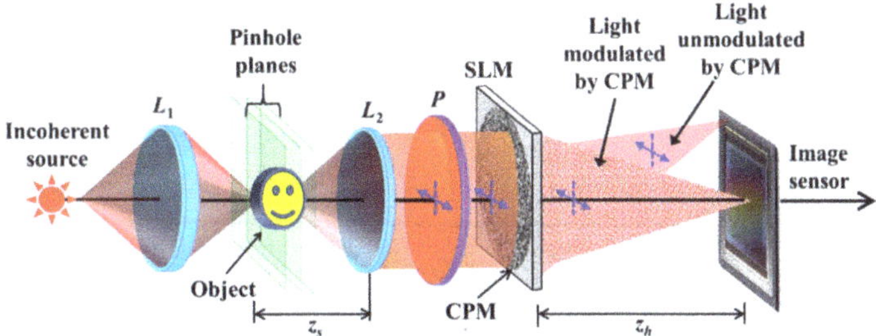

Figure 18. Optical configuration of PAIS. CPM—Coded phase mask; L_1, L_2—Refractive lenses; P—Polarizer; SLM—Spatial light modulator; Blue arrows indicate polarization orientations.

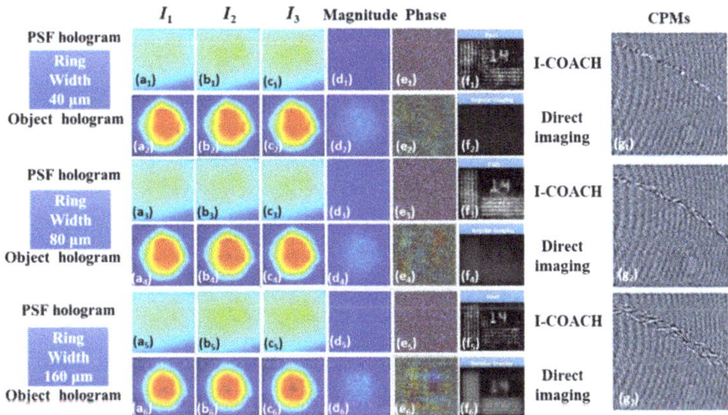

Figure 19. (a_1–a_6), (b_1–b_6) and (c_1–c_6) Intensity patterns recorded for the pinhole and the object using three different CPMs, (d_1–d_6) magnitude and (e_1–e_6) phase of the complex holograms of the respective pinhole and the object, (f_1–f_6) their corresponding reconstructions and direct imaging for ring widths of 40, 80 and 160 µm and (g_1–g_3) show the right top corner of the phase masks with ring widths 40, 80 and 160 µm, respectively.

The experiment was repeated for two plane objects made of two NBS targets 14 *lp/mm* and 16 *lp/mm* separated by 1 cm. The averaged reconstruction results of PAIS with 17 CPM sets were compared with that of direct imaging for annular CPM widths of 40, 80 and 160 µm, respectively, as shown in Figure 20. PAIS has demonstrated better performances compared to direct imaging, and the images could be perceived even with an aperture ratio of only 1.4%, whereas direct imaging fails to provide an image with an aperture ratio of 6%.

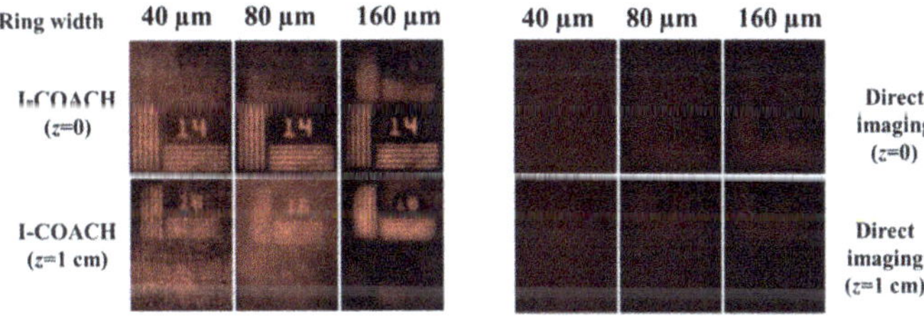

Figure 20. PAIS reconstruction results and direct imaging results for two z planes with ring widths of 40, 80 and 160 µm.

In the current PAIS setup, a minimum aperture ratio of 1.4%, which corresponds to ring width of 5 pixels on the SLM, was studied. The performances of PAIS are better than those of direct imaging. We believe that the proposed PAIS technology might be useful for future optical telescopic systems and for synthetic aperture imaging with short scanning tracks.

3. System Applications

In this section, the various applications utilizing the special characteristics of the COACH techniques are discussed. Due to a limited space of this review, we confine the discussion only to the applications of 3D imaging, super-resolution imaging, and imaging through scatterers.

3.1. D Imaging

One of the foremost characteristics related to the term holography is 3D imaging. The 3D imaging capability of holography has been the main feature, which has made the area of research attractive at the time of its invention. In general, many types of holograms contain 3D information of the observed object such that 3D image can be reconstructed from these holograms. The entire holograms presented in the previous section have 3D imaging capabilities. COACH-based methods have the same axial resolution as direct imaging, but the use of incoherent interferometers makes these methods less attractive. Therefore, the newly introduced incoherent digital holography techniques without two-wave interference, namely, I-COACH, seem the best candidates for 3D imaging. In this section, the imaging by I-COACH of real 3D objects distributed on multiple transverse planes is discussed.

D Imaging with I-COACH

As mentioned in Section 2.1.2, I-COACH is capable of recording 3D objects and reconstructing the 3D image, without two-wave interference. In this subsection, we describe the experiment carried out in [54] as a typical example of 3D holographic imaging by I-COACH. The experimental I-COACH setup for imaging 3D reflective objects is shown in Figure 21. In channel-1, a pinhole with a diameter of 25 μm was used. The light emitted from the object passes through a beam splitter BS_1 and is collimated by a lens L_2 with a focal length of $f_0 = 20$ cm. The collimated light is polarized by P_1 along the active axis of the SLM and modulated by the CPM displayed on the SLM mounted at 20 cm from the lens L_2. The intensity patterns are recorded by a digital camera mounted at $Z_h = 40$ cm from the SLM. In channel-1, the pinhole is shifted to various axial locations and a PSH library is created. In channel-2, an object is critically illuminated by an illumination system using the same beam splitter BS_1. Three different objects, namely, a LED, two one-cent coins, and stapler pins were selected for the study. Three intensity patterns were recorded for each object and averaged with 20 such sets of CPMs. Cross-correlations of the synthesized complex holograms with the PSH library extracted the different planes of the objects. In this case, we did not use the phase-only filtering in order to reconstruct the above objects with greyscale intensity values. The reconstruction results of I-COACH for the three objects after averaging are compared with direct imaging as shown in Figure 22a–c, respectively. The reconstruction results of I-COACH reveal that the performance of averaged I-COACH is close to that of direct imaging.

3.2. Super-resolution Imaging by Coded FINCH Technique

The image resolution of a general optical system is governed by two parameters, namely, wavelength and the NA [62]. Improving the resolution using shorter wavelengths is not always feasible or practical. A more practical way to improve the image resolution is to increase the NA. Increasing the NA requires increasing the diameter of the system aperture or alternatively using special techniques to enhance the resolution without changing aperture diameter. In the following, the lateral resolution enhancement of COACH is discussed.

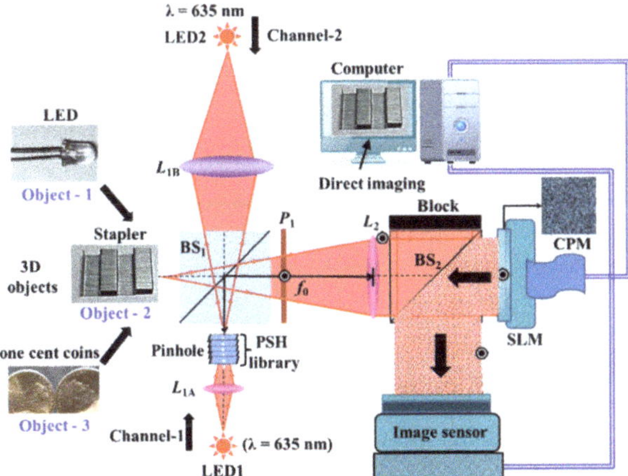

Figure 21. Experimental setup of I-COACH for recording 3D objects. BS_1 and BS_2—Beam splitters; BPF—Band pass filter; SLM—Spatial light modulator; L_{1A} and L_{1B}—identical refractive lenses; CPM—Coded phase mask; LED1 and LED2—identical Light emitting diodes; P_1—Polarizer; ◎—Polarization direction perpendicular to the plane of the page.

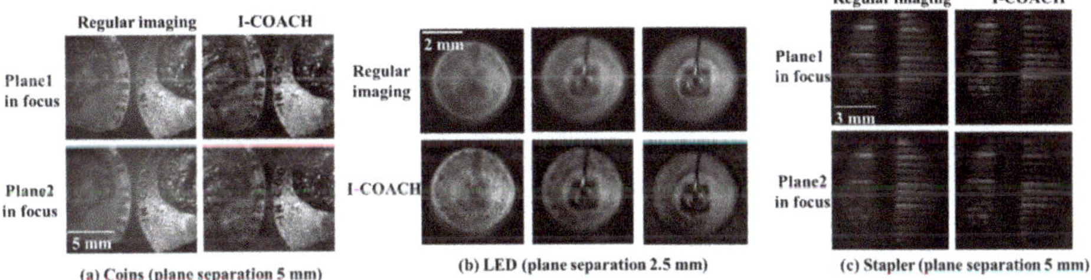

Figure 22. 3D imaging reconstruction results of I-COACH (**a**) Two one-cent coins, (**b**) LED and (**c**) Stapler pins.

The following technique of resolution enhancement is based on the idea that a scattering mask positioned between the observed object and the entrance of an imaging system can increase the effective NA of the system, and thus can improve the resolution. An early implementation of this idea was proposed by Charnotskii et al. [63]. A modern version of this technique, with the ability to control the size of the effective NA, is termed coded FINCH (C-FINCH) and is actually a combination between COACH and FINCH [64]. C-FINCH was designed in order to improve the lateral resolution beyond the inherent limit of FINCH. Moreover, C-FINCH was planned in order to maintain any of the basic advantages of both COACH and FINCH, such as motionless, compactness and the use of a single optical channel.

The optical configuration of the proposed C-FINCH system is shown in Figure 23 [64]. A laser light emitted from a point object is incident on SLM_1 on which a CPM is displayed. The scattering degree of the CPM is calibrated by the GSA. When as much of the area of the CPM spectrum is constrained to be wider, the scattering degree becomes higher [64]. The scattering degree is defined as $\sigma = B/B_{max}$, where B and B_{max} are the constrained area and maximal area in the spectral domain of the GSA, respectively.

The scattering mask displayed on the SLM_1 scatters the incident light and acquires the high spatial frequencies discarded by the system without the CPM, due to the limited NA. The scattered light is collected by lens L_1 and directed into the dual lens FINCH setup. Polarizers P_1 and P_2 are used for the polarization multiplexing scheme in FINCH [27]. A refractive lens L_2 with a focal length of f_o after the SLM$_2$ is used for implementing the dual lens FINCH [31]. DL with a focal length of f_d is displayed on SLM$_2$. As a result of the CPM on SLM$_1$, the two interfering waves are distorted due to the scattering characteristic of the CPM. The two chaotic waves propagate to the image sensor located at a distance of z_h from the SLM$_2$ and create an interference pattern recorded by the sensor. The bias and twin image terms are removed using the phase shifting procedure as in ordinary FINCH [23]. The three recorded raw holograms for $\theta_{1,2,3} = 0°$, 120° and 240° are superposed to obtain a complex hologram. The magnitude of the PSH recorded using a pinhole with a diameter of 5 μm, for different values of the scattering degree σ, is shown in Figure 24.

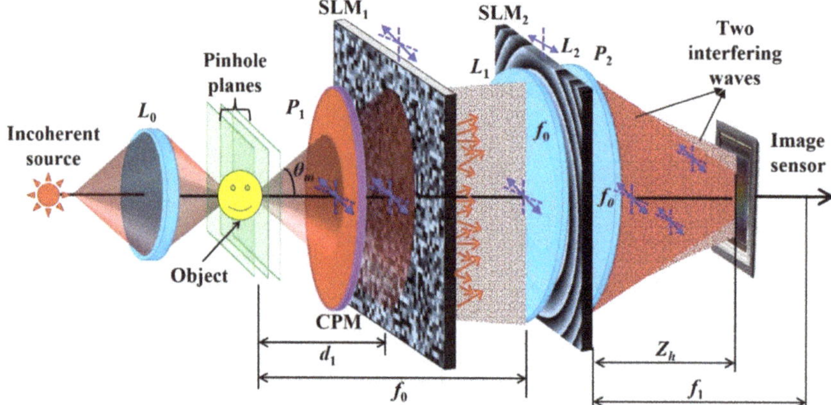

Figure 23. Optical configuration of C-FINCH with a scattering mask. CPM—Coded phase mask; $L_{0,1,2}$—Refractive lenses; $P_{1,2}$—Polarizers; SLM—Spatial light modulator; θ_m—is the maximal angle diffracted from the object acquired by the system; Blue arrows indicate polarization orientations.

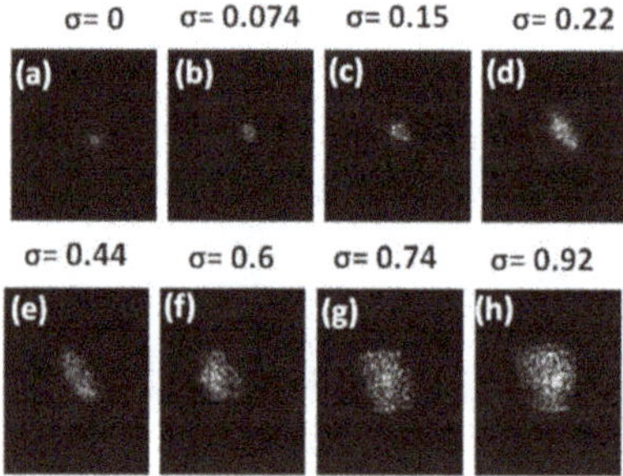

Figure 24. The magnitude of the PSH for different values of the scattering degree σ.

The relation between the sensor plane and SLM$_1$ (CPM) can be approximated to a Fourier relation with a scaling factor of $\lambda d_1 z_h / f_o$, where λ is the central wavelength of the illumination [64]. If SLM$_1$ is illuminated by a beam with a diameter of D, the resolution limit in the sensor plane is given as $\lambda d_1 z_h / (D f_o) \cong M_T \lambda / (2\sin\theta_m)$, where θ_m is the maximal angle between the optical axis and the marginal ray originated from the center of the object, scattered into the system and recorded by the camera. The resolution limit on the object plane is $\lambda / (2\sin\theta_m)$, and the angle θ_m is given by [64],

$$\sin\theta_m = \frac{\lambda\sigma}{2f_o\Delta}(f_0 - d_1) + \frac{w}{2f_0} \tag{15}$$

where Δ is the pixel size of the CPM, w is the diameter of the input aperture of the FINCH system. Since $\sin\theta_m$ is directly dependent on the scattering degree σ, the scattering mask indeed extends the effective NA of the C-FINCH from the initial value $w/2f_0$ of FINCH. Therefore, the scattering mask has expanded the effective NA and improved the resolution of the imaging system.

While the scattering mask improves the lateral resolution, there is a loss of FOV in the object plane. The FOV in the object plane in the absence of the CPM is $V_0 \times V_0$. On the image sensor, the magnification factor projects the FOV to $M_T(V_0 \times V_0)$. In Figure 24, it is seen that with an increase in the scattering degree σ, the area of the interference pattern on the hologram plane increases resulting in a decrease of the FOV. The area of the intensity distribution on the camera is about $D_I \times D_I$, where $D_I = \lambda d_1 \sigma z_h / (f_0\Delta)$. Therefore, the FOV of C-FINCH at the object plane is $V_c \times V_c$, where $V_c = V_0 - \lambda d_1 \sigma z_h / (f_0\Delta \cdot M_T) = V_0 - \lambda d_1 \sigma / \Delta$. Therefore, with an increase in the scattering degree, the lateral resolution is improved while the FOV decreases.

After recording the PSH, the experiment was repeated for a USAF target located at the same axial location as the pinhole and the object holograms were recorded. The image of the target is reconstructed by cross-correlating the object and the point object holograms. The improved resolution of the reconstructed images for different values of the scattering degree σ is shown in Figures 25 and 26. The loss of FOV with an increase in the scattering degree σ is also clearly shown in Figures 25 and 26.

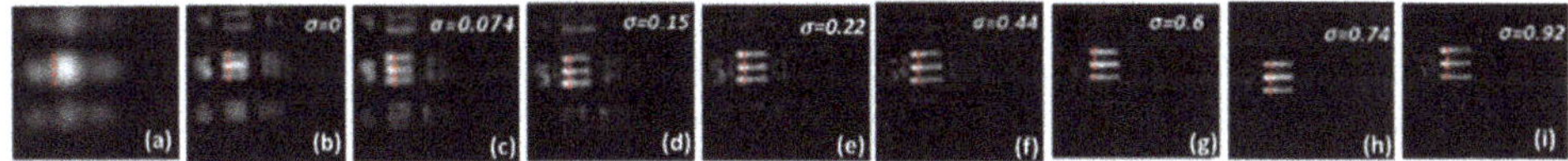

Figure 25. (a) Direct imaging result, (b) reconstruction result of FINCH and (c–i) C-FINCH reconstruction results for different values of the scattering degree σ.

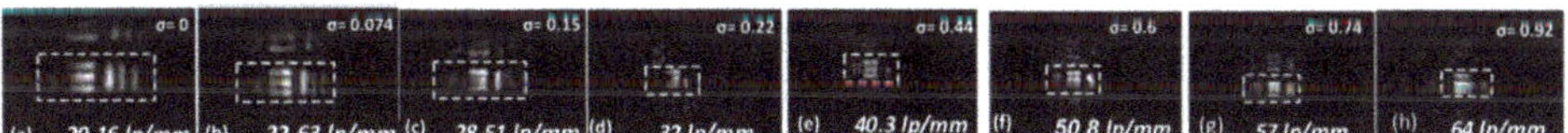

Figure 26. Reconstruction results of C-FINCH for different values of the scattering degree σ.

In conclusion, the techniques of COACH and FINCH can be combined such that the lateral resolution of FINCH is controllably enhanced. A side effect of the resolution enhancement is some loss of the FOV.

3.3. Imaging through Scatterers

Imaging through scatterers is often considered a challenging task [65–68] and it is necessary to find new techniques for imaging through scatterers. In this section, the COACH principles are applied for imaging through scatterers on one hand and on the other hand, we show how scatterers can be used for 3D imaging.

Recently, several techniques of imaging through scatterers using incoherent light have been proposed [65–68]. Based on I-COACH, we have recently proposed an interferenceless incoherent digital holography technique for imaging objects through a thin scattering medium [69]. The optical configuration of the proposed technique is shown in Figure 27.

An incoherent source illuminates a point object and the light emitted from the point object is scattered by the scattering sheet. The scattered light is collected by a lens L_2 with a focal length $f = (1/z_s + 1/z_h)^{-1}$ and is focused on the image sensor. In the absence of the scattering sheet, the optical configuration becomes a single lens imaging system. The recorded intensity pattern for the point object is termed as the point spread hologram I_{PSH} given as,

$$
\begin{aligned}
I_{PSH}(\bar{r}_0) &= C\left|v\left[\tfrac{1}{\lambda z_h}\right]\Im\left[L\left(\tfrac{\bar{r}_s}{z_s}\right)\exp(i\Phi_r)\right]\right|^2 \\
&= C\left|v\left[\tfrac{1}{\lambda z_h}\right]\Im[\exp(i\Phi_r)]\right|^2 * \delta\left(\bar{r}_0 - \tfrac{z_h}{z_s}\bar{r}_s\right)
\end{aligned}
\tag{16}
$$

where C is a constant, $\bar{r}_0 = (x_0, y_0)$ is the transverse location vector on the sensor plane, and Φ_r is the chaotic phase profile of the scatterer. A library of PSHs is recorded at various axial locations. An object is placed at the same axial location and the corresponding recorded intensity pattern is given by

$$
I_{OBJ}(\bar{r}_0) = \sum_j a_j I_{PSH}\left(\bar{r}_0 - \tfrac{z_h}{z_s}\bar{r}_j\right)
\tag{17}
$$

where each a_j is a positive real constant. The Fourier transform of I_{OBJ} is given by

$$
I'_{OBJ} = \Im\{O * I_{PSH}\} = O' \cdot |I'_{PSH}|\exp\left(i\arg\{I'_{PSH}\}\right)
\tag{18}
$$

where, I'_{OBJ}, O' and I'_{PSH} are 2D Fourier transforms of I_{OBJ}, O and I_{PSH}, respectively. The object image I_{IMG} is reconstructed by cross-correlating I_{OBJ} with the filtered version of I_{PSH}, as follows

$$
\begin{aligned}
I_{IMG}(\bar{r}_R) &= \Im^{-1}\{I'_{OBJ}|I'_{PSH}|^\gamma \exp(-i\arg\{I'_{PSH}\})\} \\
&= \sum_j a_j \Lambda\left(\bar{r}_R - \tfrac{z_h}{z_s}\bar{r}_j\right) \approx o\left(\tfrac{\bar{r}_s}{M_T}\right)
\end{aligned}
\tag{19}
$$

where $\bar{r}_R = (x_R, y_R)$ is the transverse location vector on the reconstruction plane and γ $(-1 \leq \gamma \leq 1)$ is chosen to maximize the SNR. The SNR can be improved by using two or three intensity recordings as shown in the case of I-COACH techniques [54,64].

The experiment was carried out using a pinhole with a diameter of $\approx$100 μm and USAF objects (element 6 of group 2 and numeric digit 6 of group 2). The reconstructing filter was given as $|I'_{PSH}|^\gamma \exp(-j\arg\{I'_{PSH}\})$ where the optimal γ for the tested object was $\gamma = -0.3$. The intensity patterns for the pinhole and object, the image of the filter magnitude for $\gamma = -0.3$ and the reconstruction of a single camera shot are shown in Figure 28a–d, respectively. The spacing between the two USAF objects was modified to 3 mm and the experiment was repeated for this case. The I_{OBJ} was recorded again and reconstructed using the I_{PSH} recorded at the two planes of the object. The reconstruction results using the two I_{PSH}s are shown in Figure 29a,b, respectively.

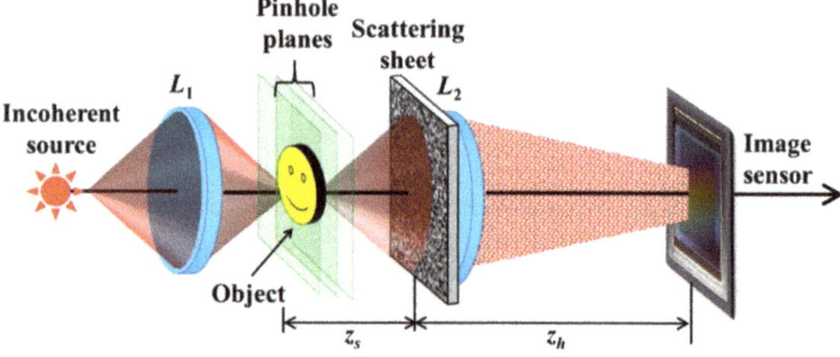

Figure 27. Optical configuration for imaging through a scatterer [69].

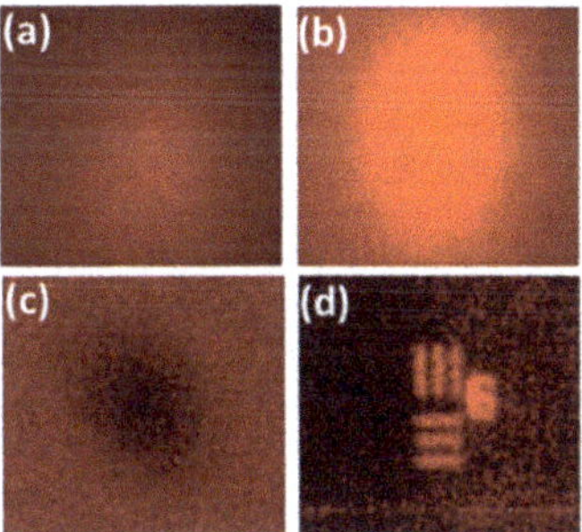

Figure 28. Intensity patterns of (**a**) I_{PSF}, (**b**) I_{OBJ}, (**c**) image of the filter magnitude with $\gamma = -0.3$, and (**d**) single shot reconstruction of an object.

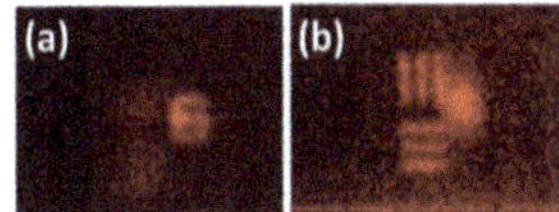

Figure 29. Reconstruction results of the two planes of the object (**a**) at z = 0 and (**b**) separated by 3 mm.

The SNR of the reconstructed images can be improved by recording multiple intensity patterns using different scatterers and averaging over the entire reconstructions as in the case of COACH systems [52]. Alternatively, improvement of the SNR can be achieved using a nonlinear correlator in which both spectral amplitudes of I_{OBJ} and of I_{PSH} are raised to a power of two independent parameters, which are optimized for maximum SNR [70,71].

The I-COACH techniques are simple and useful for imaging through scatterers. However, the need for recording the point spread hologram makes this technique invasive for seeing through biomedical tissues. In another way, this technique can be used to convert any 2D imager into a 3D imager by attaching a scatterer to the lens. Once the system is calibrated by the recording of the point spread holograms for some longitudinal range, a 3D image of the object can be reconstructed by cross-correlation with the PSH library.

4. Discussion and Conclusions

Several different configurations of the COACH recorder have been reviewed herein. Initially, the concept of COACH was suggested as a different 3D incoherent holographic technique with better axial resolution than FINCH. However, along the development of these family of systems, several new advantages have been revealed which might justify using COACH and especially I-COACH even as a 2D imaging system.

One unique advantage of I-COACH concept implemented by PAIS technique is the ability to image through an annular aperture of diameter D with the similar resolution of a full-disc aperture of the same diameter D. To the best of our knowledge, [61] is the first study which shows that an annular aperture can image general targets without substantial reduction of the image resolution. Annular aperture imaging is a new technology to image objects through part of the aperture area with as close as possible resolution capabilities of the full aperture. By using this new imaging method, the area of the optical aperture can be reduced by at least two orders of magnitude without a substantial reduction of the imaging resolution, as long as the reduced aperture is in a shape of a ring along the border of the original aperture. This statement has practical and theoretical importance. In the practical aspect, the method proposed in [61,72] offers much more efficient imaging in the sense of weight and aperture utilization. In the theoretical aspect, it was demonstrated that reducing the aperture by two orders of magnitude still enables us to transfer the same amount of information transmitted by the original clear aperture. PAIS with annular aperture can be adapted for implementation in biomedical optical devices as well as in space-based and ground-based telescopes. The preliminary results shown in [61,72] using a laboratory model are highly promising and might be a significant contribution to the field of imaging.

Another breakthrough in the applications of COACH and I-COACH is the resolution enhancement [64,73]. We have proposed and demonstrated a new technique in which a superresolution can be achieved by inserting a scattering mask between the object and the imaging lens [64]. The technique does not require any change of illumination, optical configuration, and in the case of non-linear I-COACH [70,73] does not need more than a single camera shot. Moreover, the method is a non-scanning and motionless technique. In principle, this technique can be implemented in almost any imaging system by inserting the scattering mask between the object and the entrance pupil. The penalty paid for the resolution enhancement is an additional one-time calibration stage and a longer time reconstruction process to obtain a high SNR. The proposed technique is not limited to only a single lens optical system but can be used in any imaging system such as microscopes, telescopes, and any diffraction limited imaging system. Practically, in microscopes with a working distance of a millimeter or less, it is impossible to introduce SLM between the specimen and the objective, certainly not a reflection SLM as described in this review. However, we believe that a thin constant diffuser might be effective to enhance the resolution of any existing microscope.

Author Contributions: Conceptualization, writing—review and editing, J.R. and A.V.; validation A.V., M.R.R, S.M. and A.B.

Funding: Israel Science Foundation (ISF) (grants no. 1669/16) and from Israel Ministry of Science and Technology (MOST).

Acknowledgments: This study was done during a research stay of J.R. at the Alfried Krupp Wissenschaftskolleg Greifswald.

Conflicts of Interest: The authors declare no conflict of interest.

References

1. Leith, E.N.; Upatnieks, J. Recent advances in holography. *Prog. Opt.* **1967**, *6*, 1–52. [CrossRef]
2. Lohmann, A.W. Wavefront reconstruction for incoherent objects. *J. Opt. Soc. Am.* **1965**, *55*, 1555–1556. [CrossRef]
3. Stroke, G.W.; Restrick, R.C., III. Holography with spatially noncoherent light. *Appl. Phys. Lett.* **1965**, *7*, 229–231. [CrossRef]
4. Cochran, G. New method of making Fresnel transforms with incoherent light. *J. Opt. Soc. Am.* **1966**, *56*, 1513–1517. [CrossRef]
5. Peters, P.J. Incoherent holograms with a mercury light source. *Appl. Phys. Lett.* **1966**, *8*, 209–210. [CrossRef]
6. Worthington, H.R., Jr. Production of holograms with incoherent illumination. *J. Opt. Soc. Am.* **1966**, *56*, 1397–1398. [CrossRef]
7. Bryngdahl, O.; Lohmann, A. Variable magnification in incoherent holography. *Appl. Opt.* **1970**, *9*, 231–232. [CrossRef]
8. Tsuruta, T. Holography Using an Extended Spatially Incoherent Source. *J. Opt. Soc. Am.* **1970**, *60*, 44–48. [CrossRef]
9. Breckinridge, J.B. Two-dimensional white light coherence interferometer. *Appl. Opt.* **1974**, *13*, 2760–2762. [CrossRef]
10. Sirat, G.; Psaltis, D. Conoscopic holography. *Opt. Lett.* **1985**, *10*, 4–6. [CrossRef]
11. Marathay, A.S. Noncoherent-object hologram: Its reconstruction and optical processing. *J. Opt. Soc. Am. A* **1987**, *4*, 1861–1868. [CrossRef]
12. Kim, S.-G.; Lee, B.; Kim, E.-S. Removal of bias and the conjugate image in incoherent on-axis triangular holography and real-time reconstruction of the complex hologram. *Appl. Opt.* **1997**, *36*, 4784–4791. [CrossRef] [PubMed]
13. Schilling, B.W.; Poon, T.-C.; Indebetouw, G.; Storrie, B.; Shinoda, K.; Suzuki, Y.; Wu, M.H. Three-dimensional holographic fluorescence microscopy. *Opt. Lett.* **1997**, *22*, 1506–1508. [CrossRef] [PubMed]

14. Poon, T.-C.; Indebetouw, G. Three-dimensional point spread functions of an optical heterodyne scanning image processor. *Appl. Opt.* **2003**, *42*, 1485–1492. [CrossRef] [PubMed]

15. Shaked, N.T.; Katz, B.; Rosen, J. Review of three-dimensional holographic imaging by multiple-viewpoint-projection based methods. *Appl. Opt.* **2009**, *48*, H120–H136. [CrossRef] [PubMed]

16. Rivenson, Y.; Stern, A.; Rosen, J. Compressive multiple view projection incoherent holography. *Opt. Express* **2011**, *19*, 6109–6118. [CrossRef]

17. Rosen, J.; Brooker, G.; Indebetouw, G.; Shaked, N.T. A Review of Incoherent Digital Fresnel Holography. *J. Hologr. Speckle* **2009**, *5*, 124–140. [CrossRef]

18. Liu, J.-P.; Tahara, T.; Hayasaki, Y.; Poon, T.-C. Incoherent Digital Holography: A Review. *Appl. Sci.* **2018**, *8*, 143. [CrossRef]

19. Goodman, J.W.; Lawrence, R.W. Digital image formation from electronically detected holograms. *Appl. Phys. Lett.* **1967**, *11*, 77–79. [CrossRef]

20. Kim, M.K. *Digital Holography and Microscopy: Principles, Techniques, and Applications*, 1st ed.; Springer: New York, NY, USA, 2011; ISBN 978-1-4419-7793-9.

21. Li, Y.; Abookasis, D.; Rosen, J. Computer-generated holograms of three-dimensional realistic objects recorded without wave interference. *Appl. Opt.* **2001**, *40*, 2864–2870. [CrossRef]

22. Sando, Y.; Itoh, M.; Yatagai, T. Holographic three-dimensional display synthesized from three-dimensional Fourier spectra of real existing objects. *Opt. Lett.* **2003**, *28*, 2518–2520. [CrossRef] [PubMed]

23. Rosen, J.; Brooker, G. Digital spatially incoherent Fresnel holography. *Opt. Lett.* **2007**, *32*, 912–914. [CrossRef] [PubMed]

24. Rosen, J.; Brooker, G. Non-scanning motionless fluorescence three-dimensional holographic microscopy. *Nat. Photonics* **2008**, *2*, 190–195. [CrossRef]

25. Bouchal, P.; Kapitán, J.; Chmelík, R.; Bouchal, Z. Point spread function and two-point resolution in Fresnel incoherent correlation holography. *Opt. Express* **2011**, *19*, 15603–15620. [CrossRef] [PubMed]

26. Rosen, J.; Siegel, N.; Brooker, G. Theoretical and experimental demonstration of resolution beyond the Rayleigh limit by FINCH fluorescence microscopic imaging. *Opt. Express* **2011**, *19*, 26249–26268. [CrossRef] [PubMed]

27. Brooker, G.; Siegel, N.; Wang, V.; Rosen, J. Optimal resolution in Fresnel incoherent correlation holographic fluorescence microscopy. *Opt. Express* **2011**, *19*, 5047–5062. [CrossRef] [PubMed]

28. Kim, M.K. Adaptive optics by incoherent digital holography. *Opt. Lett.* **2012**, *37*, 2694–2696. [CrossRef] [PubMed]

29. Kelner, R.; Rosen, J. Spatially incoherent single channel digital Fourier holography. *Opt. Lett.* **2012**, *37*, 3723–3725. [CrossRef] [PubMed]

30. Katz, B.; Rosen, J.; Kelner, R.; Brooker, G. Enhanced resolution and throughput of Fresnel incoherent correlation holography (FINCH) using dual diffractive lenses on a spatial light modulator (SLM). *Opt. Express* **2012**, *20*, 9109–9121. [CrossRef] [PubMed]

31. Brooker, G.; Siegel, N.; Rosen, J.; Hashimoto, N.; Kurihara, M.; Tanabe, A. In-line FINCH super resolution digital holographic fluorescence microscopy using a high efficiency transmission liquid crystal GRIN lens. *Opt. Lett.* **2013**, *38*, 5264–5267. [CrossRef] [PubMed]

32. Kim, M.K. Incoherent digital holographic adaptive optics. *Appl. Opt.* **2013**, *52*, A117–A130. [CrossRef] [PubMed]

33. Naik, D.N.; Pedrini, G.; Osten, W. Recording of incoherent-object hologram as complex spatial coherence function using Sagnac radial shearing interferometer and a Pockels cell. *Opt. Express* **2013**, *21*, 3990–3995. [CrossRef] [PubMed]

34. Lai, X.; Zeng, S.; Lv, X.; Yuan, J.; Fu, L. Violation of the Lagrange invariant in an optical imaging system. *Opt. Lett.* **2013**, *38*, 1896–1898. [CrossRef] [PubMed]

35. Kelner, R.; Rosen, J.; Brooker, G. Enhanced resolution in Fourier incoherent single channel holography (FISCH) with reduced optical path difference. *Opt. Express* **2013**, *21*, 20131–20144. [CrossRef] [PubMed]

36. Hong, J.; Kim, M.K. Single-shot self-interference incoherent digital holography using off-axis configuration. *Opt. Lett.* **2013**, *38*, 5196–5199. [CrossRef] [PubMed]

37. Wan, Y.; Man, T.; Wang, D. Incoherent off-axis Fourier triangular color holography. *Opt. Express* **2014**, *22*, 8565–8573. [CrossRef] [PubMed]

38. Qin, W.; Yang, X.; Li, Y.; Peng, X.; Yao, H.; Qu, X.; Gao, B.Z. Two-step phase-shifting fluorescence incoherent holographic microscopy. *J. Biomed. Opt.* **2014**, *19*, 060503. [CrossRef] [PubMed]

39. Lai, X.; Xiao, S.; Guo, Y.; Lv, X.; Zeng, S. Experimentally exploiting the violation of the Lagrange invariant for resolution improvement. *Opt. Express* **2015**, *23*, 31408–31418. [CrossRef]

40. Man, T.; Wan, Y.; Wu, F.; Wang, D. Four-dimensional tracking of spatially incoherent illuminated samples using self-interference digital holography. *Opt. Commun.* **2015**, *355*, 109–113. [CrossRef]

41. Muhammad, D.; Nguyen, C.M.; Lee, J.; Kwon, H.-S. Spatially incoherent off-axis Fourier holography without using spatial light modulator (SLM). *Opt. Express* **2016**, *24*, 22097–22103. [CrossRef]

42. Zhu, Z.; Shi, Z. Self-interference polarization holographic imaging of a three-dimensional incoherent scene. *Appl. Phys. Lett.* **2016**, *109*, 091104. [CrossRef]

43. Quan, X.; Matoba, O.; Awatsuji, Y. Single-shot incoherent digital holography using a dual-focusing lens with diffraction gratings. *Opt. Lett.* **2017**, *42*, 383–386. [CrossRef] [PubMed]

44. Quan, X.; Kumar, M.; Matoba, O.; Awatsuji, Y.; Hayasaki, Y.; Hasegawa, S.; Wake, H. Three-dimensional stimulation and imaging-based functional optical microscopy of biological cells. *Opt. Lett.* **2018**, *43*, 5447–5450. [CrossRef] [PubMed]

45. Choi, K.; Yim, J.; Min, S.-W. Achromatic phase shifting self-interference incoherent digital holography using linear polarizer and geometric phase lens. *Opt. Express* **2018**, *26*, 16212–16225. [CrossRef] [PubMed]

46. Vijayakumar, A.; Kashter, Y.; Kelner, R.; Rosen, J. Coded aperture correlation holography—A new type of incoherent digital holograms. *Opt. Express* **2016**, *24*, 12430–12441. [CrossRef]

47. Ables, J.G. Fourier transform photography: A new method for X-ray astronomy. *Proc. Astron. Soc. Aust.* **1968**, *1*, 172–173. [CrossRef]

48. Dicke, R.H. Scatter-Hole Cameras for X-Rays and Gamma Rays. *Astrophys. J.* **1968**, *153*, L101. [CrossRef]

49. Fenimore, E.E.; Cannon, T.M. Coded aperture imaging with uniformly redundant arrays. *Appl. Opt.* **1978**, *17*, 337–347. [CrossRef]

50. Gerchberg, R.W.; Saxton, W.O. A practical algorithm for the determination of phase from image and diffraction plane pictures. *Optik* **1972**, *35*, 227–246.

51. Horner, J.L.; Gianino, P.D. Phase-only matched filtering. *Appl. Opt.* **1984**, *23*, 812–816. [CrossRef]

52. Vijayakumar, A.; Kashter, Y.; Kelner, R.; Rosen, J. Coded aperture correlation holography system with improved performance [Invited]. *Appl. Opt.* **2017**, *56*, F67–F77. [CrossRef]

53. Katz, B.; Wulich, D.; Rosen, J. Optimal noise suppression in Fresnel incoherent correlation holography (FINCH) configured for maximum imaging resolution. *Appl. Opt.* **2010**, *49*, 5757–5763. [CrossRef]

54. Vijayakumar, A.; Rosen, J. Interferenceless coded aperture correlation holography—A new technique for recording incoherent digital holograms without two-wave interference. *Opt. Express* **2017**, *25*, 13883–13896. [CrossRef] [PubMed]

55. Chi, W.; George, N. Optical imaging with phase-coded aperture. *Opt. Express* **2011**, *19*, 4294–4300. [CrossRef] [PubMed]

56. Rai, M.R.; Vijayakumar, A.; Rosen, J. Single camera shot interferenceless coded aperture correlation holography. *Opt. Lett.* **2017**, *42*, 3992–3995. [CrossRef]

57. Li, J.-C.; Tankam, P.; Peng, Z.-J.; Picart, P. Digital holographic reconstruction of large objects using a convolution approach and adjustable magnification. *Opt. Lett.* **2009**, *34*, 572–574. [CrossRef] [PubMed]

58. Zalevsky, Z.; Gur, E.; Garcia, J.; Micó, V.; Javidi, B. Superresolved and field-of-view extended digital holography with particle encoding. *Opt. Lett.* **2012**, *37*, 2766–2768. [CrossRef] [PubMed]

59. Girshovitz, P.; Shaked, N.T. Doubling the field of view in off-axis low-coherence interferometric imaging. *Light Sci. Appl.* **2014**, *3*, e151. [CrossRef]

60. Rai, M.R.; Vijayakumar, A.; Rosen, J. Extending the field of view by a scattering window in I-COACH system. *Opt. Lett.* **2018**, *43*, 1043–1046. [CrossRef]

61. Bulbul, A.; Vijayakumar, A.; Rosen, J. Partial aperture imaging by systems with annular phase coded masks. *Opt. Express* **2017**, *25*, 33315–33329. [CrossRef]

62. Born, M.; Wolf, E. *Principles of Optics*; Pergamon: Oxford, UK, 1980; Chapters 4.4.5, 8.6.2, 8.8; pp. 165, 414, 418–428, 435.

63. Charnotskii, M.I.; Myakinin, V.A.; Zavorotnyy, V.U. Observation of superresolution in nonisoplanatic imaging through turbulence. *J. Opt. Soc. Am. A* **1990**, *7*, 1345–1350. [CrossRef]

64. Kashter, Y.; Vijayakumar, A.; Rosen, J. Resolving images by blurring: Superresolution method with a scattering mask between the observed objects and the hologram recorder. *Optica* **2017**, *4*, 932–939. [CrossRef]
65. Nixon, M.; Katz, O.; Small, E.; Bromberg, Y.; Friesem, A.A.; Silberberg, Y.; Davidson, N. Real-time wavefront shaping through scattering media by all-optical feedback. *Nat. Photonics* **2013**, *7*, 919–924. [CrossRef]
66. Katz, O.; Heidmann, P.; Fink, M.; Gigan, S. Non-invasive single-shot imaging through scattering layers and around corners via speckle correlations. *Nat. Photonics* **2014**, *8*, 784–790. [CrossRef]
67. Edrei, E.; Scarcelli, G. Memory-effect based deconvolution microscopy for super-resolution imaging through scattering media. *Sci. Rep.* **2016**, *6*, 33558. [CrossRef] [PubMed]
68. Antipa, N.; Kuo, G.; Heckel, R.; Mildenhall, B.; Bostan, E.; Ng, R.; Waller, L. DiffuserCam: Lensless single-exposure 3D imaging. *Optica* **2018**, *5*, 1–9. [CrossRef]
69. Mukherjee, S.; Vijayakumar, A.; Kumar, M.; Rosen, J. 3D Imaging through scatterers with interferenceless optical system. *Sci. Rep.* **2018**, *8*, 1134. [CrossRef]
70. Rai, M.R.; Vijayakumar, A.; Rosen, J. Non-linear adaptive three-dimensional Imaging with interferenceless coded aperture correlation holography (I-COACH). *Opt. Express* **2018**, *26*, 18143–18154. [CrossRef]
71. Mukherjee, S.; Rosen, J. Imaging through scattering medium by a single camera shot and adaptive non-linear digital processing. *Sci. Rep.* **2018**, *8*, 10517. [CrossRef]
72. Bulbul, A.; Vijayakumar, A.; Rosen, J. Superresolution far-field imaging by coded phase reflectors distributed only along the boundary of synthetic apertures. *Optica* **2018**, *5*, 1607–1616. [CrossRef]
73. Rai, M.R.; Vijayakumar, A.; Ogura, Y.; Rosen, J. Resolution Enhancement in Nonlinear Interferenceless COACH with a Point Response of Subdiffraction Limit Patterns. *Opt. Express* **2019**, *27*, 391–403. [CrossRef] [PubMed]

Article

Wavelength-Selective Phase-Shifting Digital Holography: Color Three-Dimensional Imaging Ability in Relation to Bit Depth of Wavelength-Multiplexed Holograms [†]

Tatsuki Tahara [1,2,3,*], Reo Otani [4] and Yasuhiro Takaki [5]

[1] Digital Content and Media Sciences Research Division, National Institute of Informatics, 2-1-2 Hitotsubashi, Chiyoda, Tokyo 101-8430, Japan

[2] PRESTO, Japan Science and Technology Agency, 4-1-8 Honcho, Kawaguchi, Saitama 332-0012, Japan

[3] Organization for Research and Development of Innovative Science Technology, Kansai University, Yamate-cho, Suita, Osaka 564-8680, Japan

[4] SIGMAKOKI CO. LTD., 17-2, Shimotakahagi-shinden, Hidaka-shi, Saitama 350-1297, Japan; r.otani@sigma-koki.com

[5] Institute of Engineering, Tokyo University of Agriculture and Technology, 2-24-16 Naka-cho, Koganei, Tokyo 184-8588, Japan; ytakaki@cc.tuat.ac.jp

* Correspondence: tahara@nii.ac.jp; Tel.: +81-3-4212-2178

† 'It is an invited paper for the special issue'.

Received: 22 October 2018; Accepted: 26 November 2018; Published: 28 November 2018

Abstract: The quality of reconstructed images in relation to the bit depth of holograms formed by wavelength-selective phase-shifting digital holography was investigated. Wavelength-selective phase-shifting digital holography is a technique to obtain multiwavelength three-dimensional (3D) images with a full space bandwidth product of an image sensor from wavelength-multiplexed phase-shifted holograms and has been proposed since 2013. The bit resolution required to obtain a multiwavelength holographic image was quantitatively and experimentally evaluated, and the relationship between wavelength resolution and dynamic range of an image sensor was numerically simulated. The results indicate that two-bit resolution per wavelength is required to conduct color 3D imaging.

Keywords: three-dimensional imaging; digital holography; multiwavelength digital holography; color holography; phase-shifting interferometry; phase-shifting digital holography

---

1. Introduction

Holography [1,2] is a technique utilizing interference of light to record a complex amplitude distribution of an object wave. The recorded information is called a "hologram". A three-dimensional (3D) image is reconstructed from the hologram by utilizing the diffraction of light. Holography can be used to record and reconstruct a 3D image of an object or a phase distribution of a wave without having to use multiple cameras or an array of lenses. Furthermore, 3D motion-picture images of any ultrafast physical phenomenon (such as light pulse propagation in 3D space) can be recorded and reconstructed with a single-shot exposure [3,4]. Digital holography (DH) [5–9] is used to record a digital hologram that contains an object wave and reconstructs both the 3D and quantitative phase images of an object by using a computer. DH can potentially be applied to the fields of not only ultrafast optical 3D imaging [10] but also microscopy [6,11,12], particles and flow measurements [13], quantitative phase imaging [14], lensless 3D imaging with incoherent light [15], multidimensional bio-imaging [16], multiwavelength 3D imaging [17], depth-resolved 3D imaging [18], simultaneous recording of multiple 3D images [19], and encryption [20].

Since a full space-bandwidth product of an image sensor is available, phase-shifting DH [21–23] is one way to capture an object wave. Using phase-shifting DH with multiple wavelengths, which is termed color/multiwavelength phase-shifting DH, 3D surface-shape measurements with multiwavelength phase unwrapping [24] and lensless color 3D image sensing [25,26] have been reported. DH using red-, green-, and blue-wavelengths is usually called "RGB digital holography", "three-wavelength DH", and "multiwavelength DH". In this paper, we call DH with multiple wavelengths "two/three-wavelength DH" or "multiwavelength DH" because DH using two-green- and one-blue-wavelengths was simulated. In regard to multiwavelength phase-shifting DH, two types of representative implementations have been reported: temporal division [24] and space-division multiplexing [25,26] of multiple wavelengths. In the case of temporal division, wavelength information is sequentially recorded by changing the wavelengths of light to form a hologram. Mechanical shutters or operations to turn the light sources on and off are required for selecting the recorded wavelength. Three phase-shifted holograms are required at a wavelength [27] and nine holograms are needed for three-wavelength DH. Therefore, temporal division requires much time for multiwavelength 3D imaging. In the case of space-division multiplexing, red-, green-, and blue-wavelengths are simultaneously recorded by using a color image sensor with a Bayer color-filter array. Three exposures are required to obtain a multicolor holographic image. However, both recordable wavelength bandwidth and space-bandwidth product are determined by the array and therefore spatial information and wavelength selectivity is partially sacrificed. In the case of space-division multiplexing, crosstalk between multiwavelength object waves occurs when the wavelength selectivity of the array is insufficient [28]. The field of view (FOV) and spatial resolution of the DH system are decreased by the array due to the sacrifice of the space-bandwidth product of a hologram at each wavelength. The FOV is decreased by 75% compared to phase-shifting DH with a single wavelength.

In the case of color/multiwavelength digital holography, not only temporal-division [24] and space-division multiplexing [25,26], which are generally adopted for multiwavelength imaging in an imaging system, but also spatial division [27–29], temporal frequency-division multiplexing [30–33], and spatial frequency-division multiplexing [34–36] can be merged to record multiple wavelengths. In the case of general imaging systems, wavelength information is temporally or spatially separated. Spatial division is being actively researched because neither temporal nor spatial resolutions are sacrificed. However, in the case of space division, alignment of multiple image sensors is a problem. Numerical correction is reported to solve this problem effectively [29]. On the other hand, holographic multiplexing makes it possible to record multiwavelength/color information by using a monochrome image sensor and to reconstruct it from wavelength-multiplexed image(s). In the 1960s, Lohmann presented the concept of recording a multidimensional image by holographic multiplexing [37], which is based on spatial frequency-division multiplexing [34–36]. This multiplexing enables single-shot multidimensional holographic sensing and imaging; however, it sacrifices the spatial bandwidth available for recording each object wave at each wavelength as the number of wavelengths is increased. As another means of holographic multiplexing, temporal frequency-division multiplexing has been researched, and it provides a wide spatial bandwidth regardless of the number of wavelengths [29–31]. As for temporal frequency-division multiplexing technique, Fourier and inverse Fourier transforms are calculated for each pixel to separate wavelength information. To obtain a color 3D image, however, many wavelength-multiplexed images and an image sensor with a high frame rate are needed.

Since 2013, we have been proposing an interferometric technique which selectively extracts wavelength information by using wavelength-multiplexed phase-shifted interferograms to measure multiwavelength object waves without using a color-filter array [9,38–43]. As for the proposed interferometry, multiwavelength information is multiplexed both on the space and in the spatial-frequency domain, and it is then separated in the polar coordinate plane by using wavelength-dependent phase shifts.

Hereafter, multiwavelength DH based on the proposed interferometry is termed wavelength-selective phase-shifting DH (WSPS-DH). Applying the WSPS-DH provides a full space-bandwidth product of an image sensor at each wavelength regardless of the number of wavelengths measured. Moreover, operations to change the wavelengths of light to form a hologram are not required. When the number of wavelengths is N, only $2N+1$ wavelength-multiplexed images are required for multiwavelength 3D image sensing [9,38–40], while $3N$ holograms are recorded in the temporal division. Recordable wavelength bandwidth is determined by the spectral sensitivity of the monochrome image sensor. Therefore, both wavelength and spatial bandwidth of the WSPS-DH are greater than those of the space-division multiplexing with a color image sensor. After the initially reported WSPS-interferometry was reported, WSPS-DH utilizing $2N$ wavelength-multiplexed images was proposed [41]. In the primitive scheme [38–41], phase ambiguity of 2π was utilized to selectively extract multiwavelength object waves, and then a technique employing arbitrary phase shifts for rigorously retrieving object waves at multiple wavelengths by using $2N+1$ holograms was proposed [9,42,43]. Although Doppler phase-shifting color DH [31,32] has also been proposed as another holographic multiplexing technique with a full space-bandwidth product, it requires the recording of a large number of images. WSPH-DH requires only $2N$ holograms at least [41] by employing two-step phase-shifting interferometry [44–49], while 512 holograms are required for Doppler three-wavelength phase-shifting DH [32]. Therefore, WSPH-DH accelerates measurement speed by more than 80 times when recording three wavelengths [39,41]. The proposed DH has the potential to obtain a multiwavelength holographic 3D image with a small number of recordings without any color absorption. It thus enables multimodal cell imaging with low light intensity when applied to biological microscopy. Moreover, in principle, it alleviates light damage to living cells during multidimensional holographic imaging.

However, it is necessary to consider the influence of bit depth of the recorded holograms on the quality of the reconstructed image. This is because the proposed DH multiplexes holograms at multiple wavelengths on a monochrome image sensor, and available bit depth per wavelength is sacrificed. Furthermore, in the case of the proposed DH, it is worth evaluating whether the dynamic range of holograms is related to wavelength resolution because wavelength information is selectively extracted by using the wavelength dependency of the intensity changes induced by the phase shifts of holograms.

In this paper, we investigate image quality in relation to the dynamic range of holograms formed by wavelength-selective phase-shifting DH. Image quality and wavelength resolution in relation to dynamic range are analyzed with numerically and experimentally obtained holograms.

2. Wavelength-Selective Phase-Shifting Digital Holography (WSPS-DH)

A schematic of WSPS-DH is shown in Figure 1. WSPS-DH is enabled by the wavelength dependency of the intensity change induced by wavelength-dependent phase shifts of interference light. By introducing wavelength-dependent phase shifts to interference light, wavelength information is separated in the polar coordinate plane. A phase shifter such as a mirror with a piezo actuator, a liquid crystal, a birefringent material, a spatial light modulator, an acousto-optic modulator, or an electro-optic modulator is used to generate the phase shifts. When multiwavelength information is recorded, light at wavelengths are not absorbed by a filter, and wavelength-multiplexed phase-shifted holograms are sequentially obtained by changing the phases of interference fringes. Object waves at multiple wavelengths are separately obtained from the recorded wavelength-multiplexed images when phase-shifting interferometry selectively extracts wavelength information [9,38–43]. Diffraction integrals are applied to the extracted object waves, and a multiwavelength holographic 3D image is then reconstructed. Since no light at wavelengths are absorbed by a filter, WSPS-DH is expected to achieve high light-use efficiency. When the WSPS technique is compared to temporal frequency-division multiplexing, the number of recordings can be reduced, and measurement speed can be increased.

Up to now, 2π ambiguity of phase [38–41] or arbitrary symmetric phase shifts [9,42,43] are utilized to extract each object wave rigorously by solving systems of equations. Combining 2π phase ambiguity and arbitrary symmetric phase shifts enables a multiwavelength holographic 3D imaging with only $2N$ wavelength-multiplexed holograms and in total less than 1000 nm of movement of a mirror with a piezo actuator [50,51].

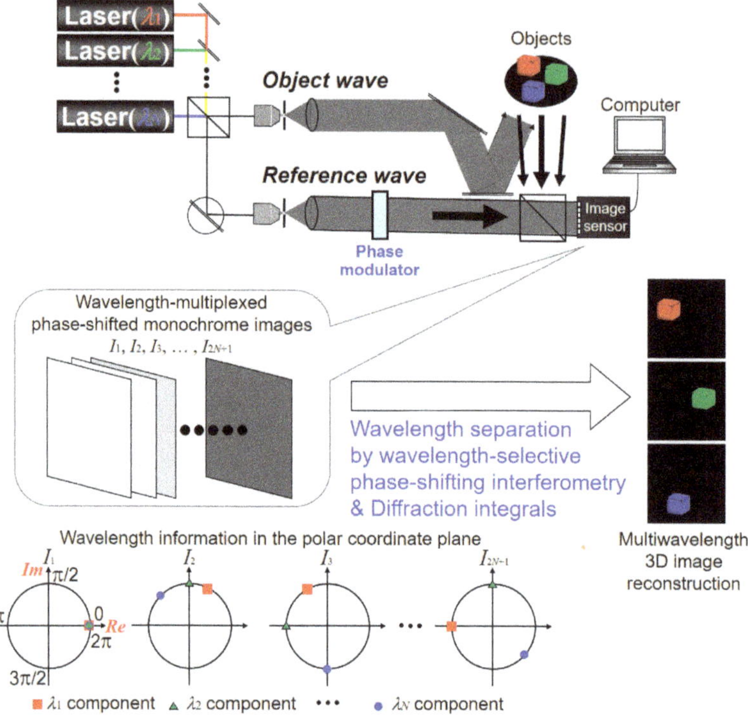

Figure 1. Schematic of wavelength-selective phase-shifting DH (WSPS-DH).

3. Image Quality in Relation to Bit Depth of Wavelength-Multiplexed Holograms

3.1. Experimenal Results

The quality of the reconstructed image in relation to bit depth of wavelength-multiplexed holograms was investigated by using experimentally obtained holograms [40]. The constructed optical system, the method of generating phase shifts, phase shift α, and the specification of the lasers used are described in Reference [40]. Two lasers with oscillation wavelengths of 640 nm and 473 nm, respectively, were set to record five two-wavelength-multiplexed holograms. A monochrome complementary metal-oxide semiconductor (CMOS) image sensor was used to record the holograms. The sensor has 12 bits, 2592 × 1944 pixels, and a pixel pitch of 2.2 μm. Two overhead projector (OHP) transparency sheets were set as a color 3D object. The logo of the International Year of Light and the characters "2015" were drawn on the sheets, and blue- and red-color films were attached to the logo and characters, respectively. A red "2015" sheet and a blue logo one were set at different depths. Five wavelength-multiplexed holograms were obtained by utilizing 2π ambiguity of the phase, and a color 3D image was reconstructed with the algorithm described in Reference [40]. Holograms that have less than 8-bit resolution were generated from the recorded holograms numerically. Object images were reconstructed by using compressed holograms in which bit depth was changed from 1 to 7 bits. Then, the images obtained by holograms without compression were regarded as the true values, and the cross-correlations coefficient (CC) and root-mean-square error (RMSE) of the reconstructed images were calculated.

Color object images obtained from the compressed holograms are shown in Figure 2. The images were reconstructed by using two-wavelength-multiplexed holograms with resolution of more than 2 bits. As bit resolution was decreased, the reconstructed images degraded gradually. However, a clear color object image was reconstructed even when the number of bits was 4. Furthermore, a two-wavelength object image was reconstructed from holograms with 3-bit depth resolution. To investigate the quality of the reconstructed images quantitatively, CC and RMSE of the intensity images at respective wavelengths were calculated. Graphs of CC and RMSE are plotted in Figure 3. Maximum- and minimum-intensity values of the images were set as 255 and 0, respectively. A CC of nearly 0.8 and a RMSE of 1/10 maximum value were obtained when bit depth was 5. From the quantitative evaluations and Figure 2, it can be concluded that quite similar images are reconstructed even when the image sensor had a resolution of less than 8 bits.

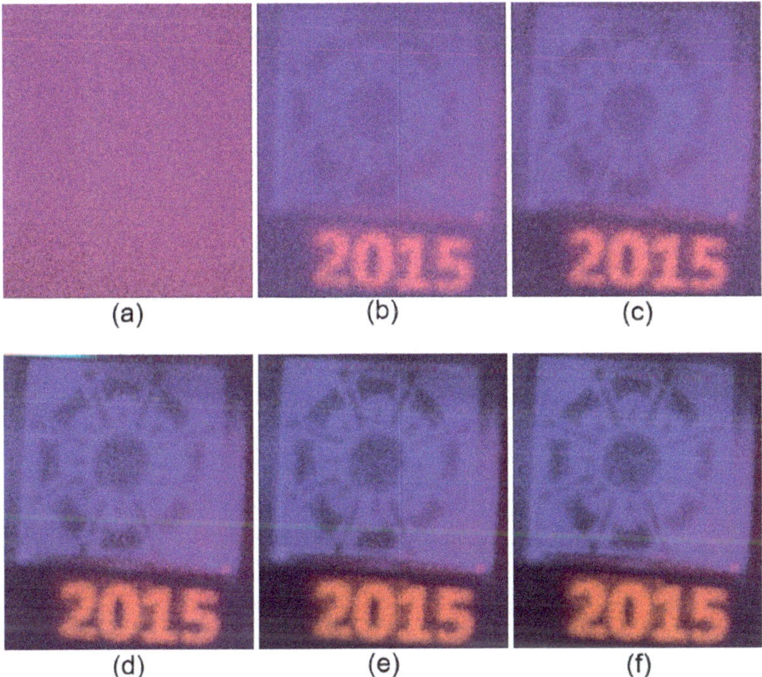

Figure 2. Experimental results: reconstructed images when bit depths of holograms are compressed to (**a**) 2, (**b**) 3, (**c**) 4, (**d**) 5, (**e**) 6, and (**f**) 7 bits.

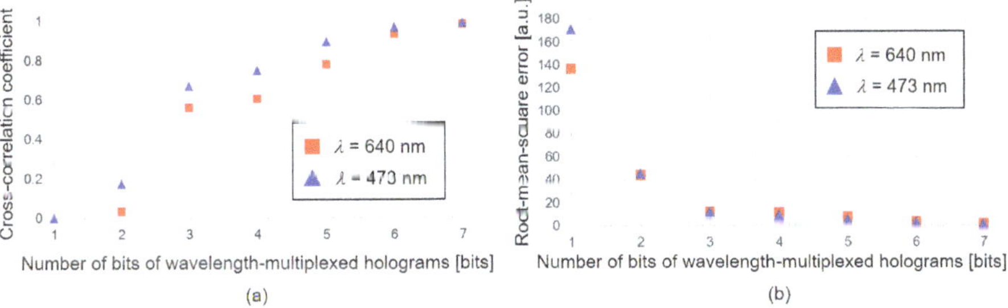

Figure 3. Quantitative evaluations of experimental results: (**a**) cross-correlations coefficient (CC) and (**b**) root-mean-square error (RMSE) of the reconstructed amplitude images.

3.2. Numerical Simulations

To investigate image quality and validate the experimental evaluations, in the case that bit depth of the image sensor was from 1 bit to 16 bits, image quality of reconstructed images was numerically simulated for three-wavelength WSPS-DH. A random pattern was set as the phase distribution of the object wave because scattering object waves were assumed. For color-intensity images, a photographic image of a flower and grass was prepared. For wavelengths of red-, green-, and blue-color light sources, 640 nm, 532 nm, and 488 nm were assumed. In the simulation, the distance between the object and image sensor was set to 150 mm, pixel pitch to 2.2 μm, and the number of pixels of the image sensor to 512 × 512. It was assumed that phase shifts were generated by a mirror with a piezo actuator and the mirror was moved 0 nm, 61 nm, ±244 nm, and ±488 nm sequentially. The intensity ratio between object and reference waves was 1:4 at each wavelength. Resolution of the image sensor was changed from 1 to 16 bits. Six wavelength-multiplexed phase-shifted holograms were obtained numerically and three-wavelength object waves were reconstructed by WSPS-DH [50,51]. Reconstructed images in the numerical simulation are shown in Figure 4. In the same manner as revealed by the experimental results, as bit resolution was decreased, the reconstructed images degraded gradually. However, a clear multicolor object image was reconstructed even when the number of bits was 6. As shown in Figure 5, CCs of the reconstructed amplitude images were more than 0.8 when bit depth of the image sensor was decreased to 6 bits. Furthermore, although the color of the reconstructed image differed from that of the object, a three-wavelength object image was reconstructed even when the image sensor had a 4-bit depth resolution. On the other hand, it was found that RMSE and CC of the reconstructed phase distribution were worse than those of the amplitude images. For phase measurement and 3D shape measurement with multiwavelength phase unwrapping, an image sensor with high dynamic range is required. Using an image sensor with resolution of more than 9 bits will result in performance of less than RMSE of $\lambda/20$ [rad] in phase. Analysis for smooth phase distribution is a future work.

Figure 4. Numerical results: Reconstructed images when bit depths of holograms are (**a**) 2, (**b**) 3, (**c**) 4, (**d**) 5, (**e**) 6, and (**f**) 7 bits.

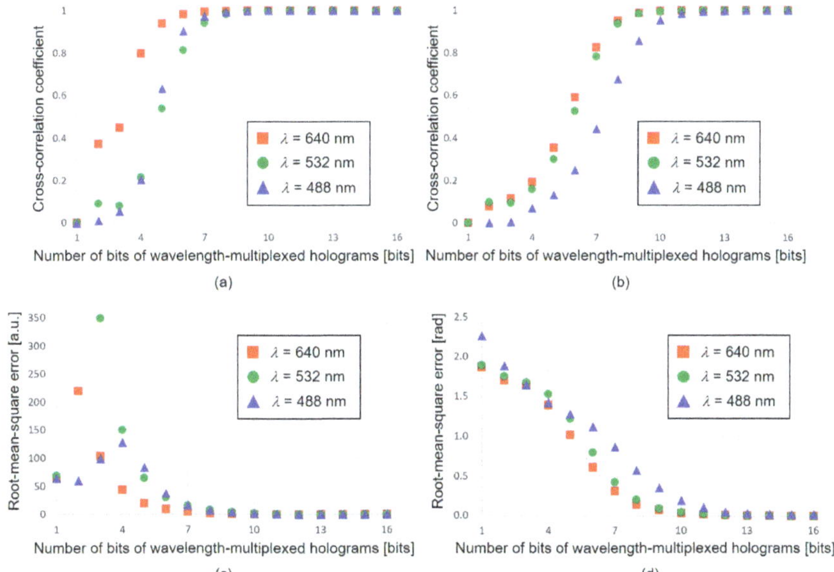

Figure 5. Quantitative evaluations for numerical results: CC of reconstructed (**a**) amplitude and (**b**) phase images and RMSE of reconstructed (**c**) amplitude and (**d**) phase images.

The experimental and numerical results indicate that a resolution of at least 2 bits per wavelength in each hologram is required to obtain a multiwavelength 3D-object intensity image, and a color 3D image with a small color shift can be reconstructed when the sensor has more than 2-bit resolution per wavelength. Measurement error is reduced as bit depth is increased in the same manner as an ordinary imaging system; however, a faithful object intensity image can be reconstructed in a case of resolution of much less than 8 bits. The numerical results also show that using a low-bit image sensor causes a large error in phase measurement; therefore, an image sensor with more than 9 bits is desirable in the case of 3D shape measurement with phase information at multiple wavelengths.

4. Numerical Analysis of the Wavelength Resolution Against Dynamic Range of Holograms

The relation of wavelength resolution to bit resolution of holograms was numerically investigated. It was assumed that the optical setup is based on three-wavelength phase-shifting DH using a monochrome image sensor and a mirror with a piezo actuator under the following conditions. It was assumed that three-wavelength WSPS-DH with six wavelength-multiplexed holograms [50,51] was used and the mirror was moved 0 nm, 61 nm, ±488 nm, and ±732 nm sequentially. A color image and a rough surface were set as amplitude and phase distributions in 3D space, respectively. To investigate image quality quantitatively, CC and RMSE of the reconstructed images were calculated. It was initially assumed that the three wavelengths of light sources were $\lambda_1 = 640$ nm, $\lambda_2 = 532$ nm, and $\lambda_3 = 488$ nm. After that, wavelength λ_1 was set as 607, 589, 561, 556, 552, 546, 540, 534, 533, or 532.5 nm to investigate wavelength resolution of WSPS-DH. The wavelengths were determined from commercially available continuous wave (CW) lasers with long coherence lengths. Pixel pitch was 2.2 µm, and the number of pixels was 512 × 512. The wavelength resolution under the three conditions was investigated: an image sensor having 8-, 12-, and 16-bit resolutions. To investigate wavelength resolution of WSPS-DH under ideal conditions, no random noise such as incoherent stray light and dark-current noise was added to holograms.

Reconstructed images obtained by this numerical simulation are shown in Figure 6, and graphs of calculated RMSE and CC of the amplitude and phase images at λ_2 are plotted in Figure 7. High CC means that faithful images were obtained and low RMSE indicates that multiwavelength 3D image measurement was high precision. The numerical results clarify that high CC was obtained even when the wavelength difference was less than 10 nm when an 8-bit image sensor was used. However, in the case an 8-bit image sensor was used, it was difficult to observe an object image clearly when the wavelength difference was within 2 nm. The difference between phase shifts added at neighboring wavelengths was small and the wavelength dependency of the intensity change induced by the wavelength-dependent phase shifts also became small. It is considered that an 8-bit image sensor could not detect weak wavelength dependency of the intensity change by the quantization. In contrast, in the cases of using an image sensor with 12- and 16-bit resolutions, object waves were successfully reconstructed because the image sensor captured weak wavelength dependency of intensity changes by phase shifts. An image sensor with 16 bits can record smaller intensity changes; therefore, higher CC and lower RMSE were obtained. These results indicate that wavelength resolution can be improved by increasing the bit depth of an image sensor. It is worth noting that this feature is characteristic of WSPS-DH. Thus, a guideline for selecting an appropriate image sensor was confirmed successfully.

Figure 6. Numerical results concerning wavelength resolution in relation to the bit depth of an image sensor. Reconstructed images when (**a–d**) 8-bit, (**e–h**) 12-bit, and (i–l) 16-bit image sensors were used. Wavelength differences are (**a,e,i**) 0.5 nm, (**b,f,j**) 1 nm, (**c,g,k**) 2 nm, and (**d,h,l**) 8 nm.

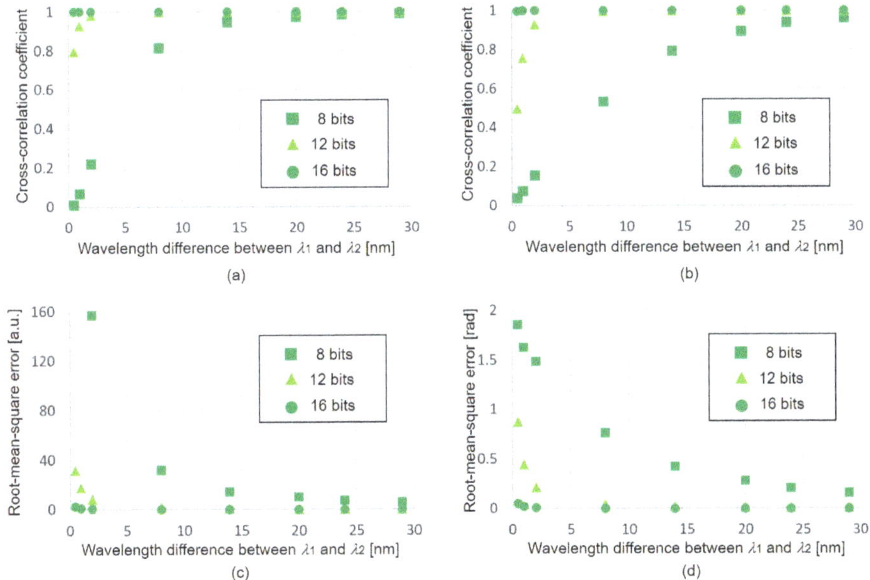

Figure 7. Quantitative evaluations of reconstructed images at λ_2: CC of reconstructed (**a**) amplitude and (**b**) phase images, and RMSE of reconstructed (**c**) amplitude and (**d**) phase images.

5. Discussion

The reason for a color shift in the numerical simulation is discussed. In comparison with the experimental results, the results of the numerical simulation show that the color of the reconstructed images shifts remarkably when bit depth of wavelength-multiplexed holograms is low. Here, the value of the wavelength difference is focused on, and λ_2 is set to 561 nm instead of 532 nm to adjust the wavelength difference for three wavelengths in the numerical simulation described in Section 3.2. The numerical results when setting the wavelengths to 488, 561, and 640 nm are shown in Figure 8. The images indicate that at least 2-bit resolution per hologram at a wavelength is required, and a color 3D image with a small color shift by using an image sensor that has more than 2-bit resolution per wavelength. However, color shift was obviously decreased by increasing the difference of neighboring wavelengths λ_2 and λ_3. This trend can be explained by the fact that, as described in Section 2, WSPS-DH is enabled by the wavelength dependency of the intensity change induced by wavelength-dependent phase shifts of interference light. When the difference between λ_2 and λ_3 was small, the effect for wavelength-dependent intensity change also became small. Selective extraction of object waves at λ_2 and λ_3 were difficult as the bit resolution was decreased because the wavelength-dependent intensity change was small and an image sensor with low bit resolution was not able to detect the change. As a result, the CC of λ_1 was relatively high and the RMSE was relatively low, so the object-intensity image at λ_1 was clearly reconstructed in comparison to those at λ_2 and λ_3. Quantitative evaluations shown in Figure 5 supported this finding because the CC was higher and the RMSE was lower at λ_1. In contrast, in the simulation shown in Figure 8, the wavelength-dependent intensity change became large by increasing wavelength-dependent phase shifts, and therefore each of three object waves was reconstructed from holograms with 4-bit resolution. As a result, the color was improved. From the experimental results presented in Section 3.1 and the numerical results presented in this section, it is clear that the color 3D-image sensing can be achieved when using an image sensor with more than 2-bit resolution per wavelength.

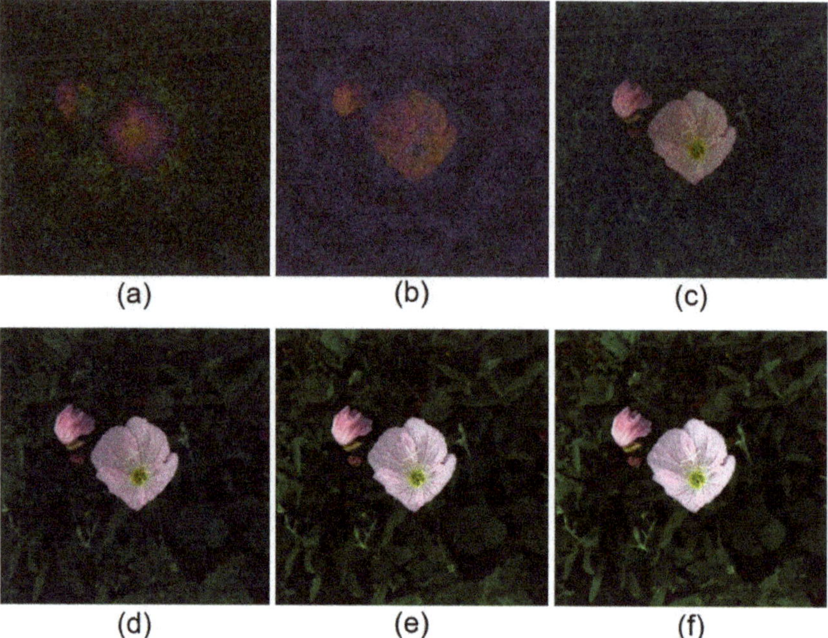

Figure 8. Numerical results presented in Section 3.2 under the assumption that λ_2 is 561 nm instead of 532 nm. Reconstructed images when bit depths of holograms are (**a**) 2, (**b**) 3, (**c**) 4, (**d**) 5, (**e**) 6, and (**f**) 7 bits.

6. Conclusions

The quality of reconstructed images in relation to dynamic range of holograms generated by WSPS-DH was investigated. Quantitative, experimental, and numerical results clarified the required bit resolution to obtain a multiwavelength holographic image and the relationship between the wavelength resolution and dynamic range of an image sensor. Experimental and numerical results indicate that 2-bit resolution per hologram at a wavelength is required to obtain a multiwavelength 3D-object intensity image at least, and a color 3D image with a smaller color shift can be reconstructed when the sensor has more than a 2-bit resolution per wavelength. More than 3 bits per wavelength is sufficient for high-quality multiwavelength 3D imaging. Wavelength resolution can be improved by increasing bit depth of an image sensor, and this finding is characteristic of WSPS-DH. WSPS-DH will perform multiwavelength 3D imaging at high speed for low-light-intensity events. Accordingly, it will contribute to multispectral 3D imaging with high light-use efficiency and high wavelength resolution by using a monochrome image sensor with high dynamic range (such as an electron multiplying charge-coupled device (EM-CCD) camera) and an array of photo multipliers.

Author Contributions: Conceptualization, T.T., R.O. and Y.T.; methodology, T.T.; software, T.T.; validation, T.T.; formal analysis, T.T.; investigation, T.T.; resources, T.T. and Y.T.; data curation, T.T.; writing—original draft preparation, T.T.; writing—review and editing, T.T., R.O. and Y.T.; visualization, T.T.; supervision, T.T.; project administration, T.T.

Funding: This research was partially supported by the Japan Science and Technology Agency (JST), PRESTO, Grant number JPMJPR16P8, Japan, and the Japan Society for the Promotion of Science (JSPS), Grant-in-Aid for Scientific Research (B), grant number 18H01456, Japan.

Conflicts of Interest: The authors declare no conflict of interest.

References

1. Gabor, D. A new microscopic principle. *Nature* **1948**, *161*, 777–778. [CrossRef] [PubMed]
2. Leith, E.N.; Upatnieks, J. Reconstructed wavefronts and communication theory. *J. Opt. Soc. Am.* **1962**, *52*, 1123–1130. [CrossRef]
3. Kubota, T. 48 years with holography. *Opt. Rev.* **2014**, *21*, 883–892. [CrossRef]
4. Kubota, T.; Komai, K.; Yamagiwa, M.; Awatsuji, Y. Moving picture recording and observation of three-dimensional image of femtosecond light pulse propagation. *Opt. Express* **2007**, *15*, 14348–14354. [CrossRef] [PubMed]
5. Goodman, J.W.; Lawrence, R.W. Digital image formation from electronically detected holograms. *Appl. Phys. Lett.* **1967**, *11*, 77–79. [CrossRef]
6. Kim, M.K. *Digital Holographic Microscopy: Principles, Techniques, and Applications*; Springer: New York, NY, USA, 2011.
7. Picart, P.; Li, J.-C. *Digital Holography*; Wiley: Hoboken, NJ, USA, 2013.
8. Poon, T.-C.; Liu, J.-P. *Introduction to Modern Digital Holography with Matlab*; Cambridge University Press: Cambridge, UK, 2014.
9. Tahara, T.; Quan, X.; Otani, R.; Takaki, Y.; Matoba, O. Digital holography and its multidimensional imaging applications: A review. *Microscopy* **2018**, *67*, 55–67. [CrossRef] [PubMed]
10. Takeda, M.; Kitoh, M. Spatiotemporal frequency multiplex heterodyne interferometry. *J. Opt. Soc. Am. A* **1992**, *9*, 1607–1614. [CrossRef]
11. Poon, T.-C.; Doh, K.; Schilling, B.; Wu, M.; Shinoda, K.; Suzuki, Y. Three-dimensional microscopy by optical scanning holography. *Opt. Eng.* **1995**, *34*, 1338–1344. [CrossRef]
12. Takaki, Y.; Kawai, H.; Ohzu, H. Hybrid holographic microscopy free of conjugate and zero-order images. *Appl. Opt.* **1999**, *38*, 4990–4996. [CrossRef] [PubMed]
13. Murata, S.; Yasuda, N. Potential of digital holography in particle measurement. *Opt. Laser Technol.* **2000**, *32*, 567–574. [CrossRef]
14. Watanabe, E.; Hoshiba, T.; Javidi, B. High-precision microscopic phase imaging without phase unwrapping for cancer cell identification. *Opt. Lett.* **2013**, *38*, 1319–1321. [CrossRef] [PubMed]
15. Liu, J.-P.; Tahara, T.; Hayasaki, Y.; Poon, T.-C. Incoherent Digital Holography: A Review. *Appl. Sci.* **2018**, *8*, 143. [CrossRef]
16. Pavillon, N.; Fujita, K.; Smith, N.I. Multimodal label-free microscopy. *J. Innov. Opt. Health Sci.* **2017**, *7*, 13300097.
17. Ferraro, P.; Grilli, S.; Miccio, L.; Alfieri, D.; Nicola, S.; Finizio, A.; Javidi, B. Full color 3-D imaging by digital holography and removal of chromatic aberrations. *J. Disp. Technol.* **2008**, *4*, 97–100. [CrossRef]
18. Williams, L.; Nehmetallah, G.; Banerjee, P. Digital tomographic compressive holographic reconstruction of 3D objects in transmissive and reflective geometries. *Appl. Opt.* **2013**, *52*, 1702–1710. [CrossRef] [PubMed]
19. Zea, A.V.; Barrera, J.F.; Torroba, R. Cross-talk free selective reconstruction of individual objects from multiplexed optical field data. *Opt. Lasers Eng.* **2018**, *100*, 90–97. [CrossRef]
20. Li, W.; Shi, C.; Piao, M.; Kim, N. Multiple-3D-object secure information system based on phase shifting method and single interference. *Appl. Opt.* **2016**, *55*, 4052–4059. [CrossRef] [PubMed]
21. Bruning, J.H.; Herriott, D.R.; Gallagher, J.E.; Rosenfeld, D.P.; White, A.D.; Brangaccio, D.J. Digital wavefront measuring interferometer for testing optical surfaces and lenses. *Appl. Opt.* **1974**, *13*, 2693–2703. [CrossRef] [PubMed]
22. Yamaguchi, I.; Zhang, T. Phase-shifting digital holography. *Opt. Lett.* **1997**, *22*, 1268–1270. [CrossRef] [PubMed]
23. Stern, A.; Javidi, B. Improved-resolution digital holography using the generalized sampling theorem for locally band-limited fields. *J. Opt. Soc. Am. A* **2006**, *23*, 1227–1235. [CrossRef]
24. Cheng, Y.-Y.; Wyant, J.C. Two-wavelength phase shifting interferometry. *Appl. Opt.* **1984**, *23*, 4539–4543. [CrossRef] [PubMed]
25. Yamaguchi, I.; Matsumura, T.; Kato, J. Phase-shifting color digital holography. *Opt. Lett.* **2002**, *27*, 1108–1110. [CrossRef] [PubMed]
26. Kato, J.; Yamaguchi, I.; Matsumura, T. Multicolor digital holography with an achromatic phase shifter. *Opt. Lett.* **2002**, *27*, 1403–1405. [CrossRef] [PubMed]

27. Araiza-Esquivel, M.A.; Martínez-León, L.; Javidi, B.; Andrés, P.; Lancis, J.; Tajahuerce, E. Single-shot color digital holography based on the fractional Talbot effect. *Appl. Opt.* **2011**, *50*, B96–B101. [CrossRef] [PubMed]

28. Desse, J.M.; Picart, P.; Tankam, P. Sensor influence in digital 3λ holographic interferometry. *Meas. Sci. Technol.* **2011**, *22*, 064005. [CrossRef]

29. Leclercq, M.; Picart, P. Method for chromatic error compensation in digital color holographic imaging. *Opt. Express* **2013**, *21*, 26456–26467. [CrossRef] [PubMed]

30. Dändliker, R.; Thalmann, R.; Prongué, D. Two-wavelength laser interferometry using superheterodyne detection. *Opt. Lett.* **1988**, *13*, 339–341. [CrossRef] [PubMed]

31. Barada, D.; Kiire, T.; Sugisaka, J.; Kawata, S.; Yatagai, T. Simultaneous two-wavelength Doppler phase-shifting digital holography. *Appl. Opt.* **2011**, *50*, H237–H244. [CrossRef] [PubMed]

32. Kiire, T.; Barada, D.; Sugisaka, J.; Hayasaki, Y.; Yatagai, T. Color digital holography using a single monochromatic imaging sensor. *Opt. Lett.* **2012**, *37*, 3153–3155. [CrossRef] [PubMed]

33. Naik, D.N.; Pedrini, G.; Takeda, M.; Osten, W. Spectrally resolved incoherent holography: 3D spatial and spectral imaging using a Mach-Zehnder radial-shearing interferometer. *Opt. Lett.* **2014**, *39*, 1857–1860. [CrossRef] [PubMed]

34. Picart, P.; Moisson, E.; Mounier, D. Twin-sensitivity measurement by spatial multiplexing of digitally recorded holograms. *Appl. Opt.* **2003**, *42*, 1947–1957. [CrossRef] [PubMed]

35. Tahara, T.; Akamatsu, T.; Arai, Y.; Shimobaba, T.; Ito, T.; Kakue, T. Algorithm for extracting multiple object waves without Fourier transform from a single image recorded by spatial frequency-division multiplexing and its application to digital holography. *Opt. Commun.* **2017**, *402*, 462–467. [CrossRef]

36. Tayebi, B.; Kim, W.; Sharif, F.; Yoon, B.; Han, J. Single-shot and label-free refractive index dispersion of single nerve fiber by triple-wavelength diffraction phase microscopy. *J. Sel. Top. Quantum Electron.* **2019**, *25*. [CrossRef]

37. Lohmann, A.W. Reconstruction of vectorial wavefronts. *Appl. Opt.* **1965**, *4*, 1667–1668. [CrossRef]

38. Tahara, T.; Kikunaga, S.; Arai, Y.; Takaki, Y. Phase-shifting interferometry capable of selectively extracting multiple wavelength information and color three-dimensional imaging using a monochromatic image sensor. In Proceedings of the Optics and Photonics Japan 2013 (OPJ), Nara, Japan, 13 November 2013.

39. Digital Holography Apparatus and Digital Holography Method. Patent Application Number PCT/JP2014/067556, 1 July 2014.

40. Tahara, T.; Mori, R.; Kikunaga, S.; Arai, Y.; Takaki, Y. Dual-wavelength phase-shifting digital holography selectively extracting wavelength information from wavelength-multiplexed holograms. *Opt. Lett.* **2015**, *40*, 2810–2813. [CrossRef] [PubMed]

41. Tahara, T.; Mori, R.; Arai, Y.; Takaki, Y. Four-step phase-shifting digital holography simultaneously sensing dual-wavelength information using a monochromatic image sensor. *J. Opt.* **2015**, *17*, 125707. [CrossRef]

42. Tahara, T.; Otani, R.; Omae, K.; Gotohda, T.; Arai, Y.; Takaki, Y. Multiwavelength digital holography with wavelength-multiplexed holograms and arbitrary symmetric phase shifts. *Opt. Express* **2017**, *25*, 11157–11172. [CrossRef] [PubMed]

43. Jeon, S.; Lee, J.; Cho, J.; Jang, S.; Kim, Y.; Park, N. Wavelength-multiplexed digital holography for quantitative phase measurement using quantum dot film. *Opt. Express* **2018**, *26*, 27305. [CrossRef] [PubMed]

44. Meng, X.F.; Cai, L.Z.; Xu, X.F.; Yang, X.L.; Shen, X.X.; Dong, G.Y.; Wang, Y.R. Two-step phase-shifting interferometry and its application in image encryption. *Opt. Lett.* **2006**, *31*, 1414–1416. [CrossRef] [PubMed]

45. Liu, J.-P.; Poon, T.-C. Two-step-only quadrature phase-shifting digital holography. *Opt. Lett.* **2009**, *34*, 250–252. [CrossRef] [PubMed]

46. Kiire, T.; Nakadate, S.; Shibuya, M. Digital holography with a quadrature phase-shifting interferometer. *Appl. Opt.* **2009**, *48*, 1308–1312. [CrossRef] [PubMed]

47. Shaked, N.T.; Zhu, Y.; Rinehart, M.T.; Wax, A. Two-step-only phase-shifting interferometry with optimized detector bandwidth for microscopy of live cells. *Opt. Express* **2009**, *17*, 15585–15591. [CrossRef] [PubMed]

48. Liu, J.-P.; Poon, T.-C.; Jhou, G.-S.; Chen, P.-J. Comparison of two-, three-, and four-exposure quadrature phase-shifting digital holography. *Appl. Opt.* **2011**, *50*, 2443–2450. [CrossRef] [PubMed]

49. Vargas, J.; Antonio, J.; Belenguer, T.; Servin, M.; Estrada, J.C. Two-step self-tuning phase-shifting interferometry. *Opt. Express* **2011**, *19*, 638–648. [CrossRef] [PubMed]

50. Tahara, T.; Arai, Y.; Takaki, Y. Three-wavelength phase-shifting digital holography using six wavelength-multiplexed holograms and 2π ambiguity of the phase. In Proceedings of the OSJ-OSA Joint Symposia on Plasmonics and Digital Optics, Tokyo, Japan, 31 October 2016.
51. Tahara, T.; Otani, R.; Arai, Y.; Takaki, Y. Three-wavelength phase-shifting interferometry with six wavelength-multiplexed holograms. In Proceedings of the Digital Holography and Three-Dimensional Imaging, Orlando, FL, USA, 25–28 June 2018.

Article

Novel Generalized Three-Step Phase-Shifting Interferometry with a Slight-Tilt Reference

Xianfeng Xu [1,*], Tianyu Ma [1], Zhiyong Jiao [1], Liang Xu [1], Dejun Dai [2], Fangli Qiao [2] and Ting-Chung Poon [3]

[1]   College of Science, China University of Petroleum (East China), Qingdao 266580, China; matianyu1224@126.com (T.M.); jiaozhy@upc.edu.cn (Z.J.); xliangu@126.com (L.X.)
[2]   Key Laboratory of Marine Science and Numerical Modeling, First Institute of Oceanography, Ministry of Natural Resources of the People's Republic of China, Qingdao 266061, China; djdai@fio.org.cn (D.D.); qiaofl@fio.org.cn (F.Q.)
[3]   Bradley Department of Electrical and Computer Engineering, Virginia Polytechnic Institute and State University, Blacksburg, VA 24061, USA; tcpoon@vt.edu
*   Correspondence: xuxf@upc.edu.cn

Received: 28 October 2019; Accepted: 16 November 2019; Published: 21 November 2019

Featured Application: The method can be used in such areas as optical wave reconstruction, three dimensional microscopic imaging, and interference measurement.

Abstract: A convenient and powerful method is proposed and presented to find the unknown phase shifts in three-step generalized phase-shifting interferometry. A slight-tilt reference of 0.1 degrees is employed. As a result, the developed theory shows that the unknown phase shifts can be simply extracted by subtraction operations. Also, from the theory developed, the tilt angle of the tilt reference can also be calculated, which is important as it allows us to extract the object wave precisely. Numerical simulations and optical experiments were performed to demonstrate the validity and efficiency of the proposed method. The proposed slight-tilt reference allows the full and efficient use of the space-bandwidth product of the limited resolution of digital recording devices as compared to the situation in standard off-axis holography where typically several degrees for off-axis angle is employed.

Keywords: generalized phase-shifting interferometry; slightly tilted reference; numerical simulations

---

1. Introduction

Phase-shifting interferometry (PSI) has been developing for decades and applied to many fields [1–6]. In the development of PSI techniques, there is a transition from the traditional fixed phase shifts to random unknown phase shifts [7,8]. In general, the phase shift of the reference beam introduced into the PSI is typically affected by environmental interference and phase shifter errors [9]. To further simplify the measurement procedures and avoid the negative effects of the environmental disturbances and phase shifter errors, Cai et al. have introduced generalized phase-shifting interferometry (GPSI), where the phase shifts are generally arbitrary, unequal and unknown values, but can be extracted from several interferograms [10,11]. Yoshikawa et al. later have developed GPSI by using statistical distribution of the diffracted wave and normalized holograms [12–14]. To achieve further convenience, non-iterative methods [11,15–18] begin to replace the iterative methods, which always need a long computation time, and sometimes it is difficult to make the extracted values convergent. However, complicated equations and tedious operations in these phase shift extraction processes give rise to some inconvenience in the application of GPSI.

In recent years, interference with a tilt reference [18,19] has been proposed in digital holography for monitoring purposes. In order to improve the convenience and stability of the phase-shift extracting algorithm further in three-step PSI, we propose a simpler and more reliable method to extract the unknown phase shifts with a slightly tilted reference beam during the recording process.

Three-step generalized phase-shifting interferometry (TGPSI) uses three interferograms to extract phase shifts and retrieve the original object wave. In GPSI research, TGPSI can retrieve the object wave stably by only three holograms without any additional measurements. In this paper, we propose the use of a slightly tilted reference so that the extraction of the phase shift can become simpler and more stable without iteration. We derive the theory behind the proposed method and carry out numerical simulations as well as optical experiments to verify our idea.

2. Basic Principle

In the TGPSI method, the complex amplitudes of object wave $O(x, y)$ and reference wave $R(x, y)$ on the recording plane P_H can be written as

$$O(x, y) = A_o(x, y) \exp[-i\varphi_o(x, y)], \tag{1}$$

$$R(x, y) = A_r \exp[-i\varphi_r(x, y)], \tag{2}$$

where for brevity, we have assumed A_r to be constant. Obviously, in Equation (2) the phase of the tilt reference plane wave $\varphi_r(x, y)$ is not a constant but a linear distribution on recording plane P_H given by

$$\varphi_r(x, y) = \frac{2\pi}{\lambda}(x\cos\theta_x + y\cos\theta_y), \tag{3}$$

where λ is the wavelength of the laser and θ_x and θ_y are the off-axis angles along the x and y directions. From Equations (1) and (2), the intensities for the three interferograms are described by

$$I_1 = A_o^2 + A_r^2 + 2A_oA_r\cos(\varphi_o - \varphi_r), \tag{4}$$

$$I_2 = A_o^2 + A_r^2 + 2A_oA_r\cos(\varphi_o - \varphi_r - \Delta\varphi_{r1}), \tag{5}$$

$$I_3 = A_o^2 + A_r^2 + 2A_oA_r\cos(\varphi_o - \varphi_r - \Delta\varphi_{r1} - \Delta\varphi_{r2}), \tag{6}$$

where $\Delta\varphi_{r1}$ and $\Delta\varphi_{r2}$ are unknown phase shifts that need to be determined. Subtracting Equations (5) and (6) from Equation (4) respectively, we have

$$I_2 - I_1 = 4A_oA_r[\sin(\varphi_o - \varphi_r - \Delta\varphi_{r1}/2)\sin(\Delta\varphi_{r1}/2)], \tag{7}$$

$$I_3 - I_1 = 4A_oA_r[\sin(\varphi_o - \varphi_r - \Delta\varphi_{r1}/2 - \Delta\varphi_{r2}/2)\sin(\Delta\varphi_{r1}/2 + \Delta\varphi_{r2}/2)]. \tag{8}$$

From Equations (7) and (8), we derive the following expressions:

$$A_o\sin(\varphi_o - \varphi_r) = \frac{1}{4A_r\sin(\Delta\varphi_{r2}/2)} \times \left\{ \frac{\sin(\Delta\varphi_{r1}/2)}{\sin[(\Delta\varphi_{r1}+\Delta\varphi_{r2})/2]}(I_1 - I_3) - \frac{\sin[(\Delta\varphi_{r1}+\Delta\varphi_{r2})/2]}{\sin(\Delta\varphi_{r1}/2)}(I_1 - I_2) \right\}, \tag{9}$$

$$A_o\cos(\varphi_o - \varphi_r) = \frac{1}{4A_r\sin(\Delta\varphi_{r2}/2)} \times \left\{ \frac{\cos(\Delta\varphi_{r1}/2)}{\sin[(\Delta\varphi_{r1}+\Delta\varphi_{r2})/2]}(I_1 - I_3) - \frac{\cos[(\Delta\varphi_{r1}+\Delta\varphi_{r2})/2]}{\sin(\Delta\varphi_{r1}/2)}(I_1 - I_2) \right\}, \tag{10}$$

which are the imaginary and the real parts of the following complex amplitude:

$$\begin{aligned} O_1(x, y) &= A_o\exp[-i(\varphi_o - \varphi_r)] \\ &= A_o[\cos(\varphi_o - \varphi_r) - i\sin(\varphi_o - \varphi_r)] \end{aligned} \tag{11}$$

Substituting Equations (9) and (10) into (11), we have

$$O_1 = \frac{1}{4A_r \sin(\Delta\varphi_{r2}/2)} \times$$
$$\left\{ \frac{\exp(i\Delta\varphi_{r1}/2)}{\sin[(\Delta\varphi_{r1}+\Delta\varphi_{r2})/2]}(I_1 - I_3) - \frac{\exp[i(\Delta\varphi_{r1}+\Delta\varphi_{r2})/2]}{\sin(\Delta\varphi_{r1}/2)}(I_1 - I_2) \right\} \quad (12)$$

Note that from Equation (11), we can see that this complex field is related to the original object wave O (see Equation (1)) as follows:

$$O = O_1 \exp(-i\varphi_r). \quad (13)$$

In order to extract O, we, therefore, need to know O_1 and φ_r. Let us first concentrate on finding O_1. According to Equation (12), O_1 can be found with known intensities I_1, I_2, I_3 along with unknown, to be determined, phase shifts $\Delta\varphi_{r1}$ and $\Delta\varphi_{r2}$.

It is known that the precise extraction of phase shifts $\Delta\varphi_{r1}$ and $\Delta\varphi_{r2}$ is critical to the reconstruction of the object wave. In TGPSI, many time-consuming iterative algorithms are used to search for the actual values of the phase shift. We propose a powerful and reliable method to extract phase-shift values based on the theory developed so far. To this end, we first analyze the spectrum of the interferograms. Equations (4)–(6) can be rewritten as

$$I_1 = A_o^2 + A_r^2 + A_o A_r \exp(-i\varphi_o)\exp(i\varphi_r) + A_o A_r \exp(i\varphi_o)\exp(-i\varphi_r), \quad (14)$$

$$\begin{aligned} I_2 = A_o^2 + A_r^2 + A_o A_r \exp(-i\varphi_o)\exp(i\varphi_r)\exp(i\Delta\varphi_{r1}) \\ + A_o A_r \exp(i\varphi_o)\exp(-i\varphi_r)\exp(-i\Delta\varphi_{r1}) \end{aligned} \quad (15)$$

$$\begin{aligned} I_3 = A_o^2 + A_r^2 \\ + A_o A_r \exp(-i\varphi_o)\exp(i\varphi_r)\exp(i\Delta\varphi_{r1})\exp(i\Delta\varphi_{r2}) \\ + A_o A_r \exp(i\varphi_o)\exp(-i\varphi_r)\exp(-i\Delta\varphi_{r1})\exp(-i\Delta\varphi_{r2}) \end{aligned} \quad (16)$$

in order to obtain phase shifts $\Delta\varphi_{r1}$ and $\Delta\varphi_{r2}$, Fourier transform (FT) operations on Equations (14)–(16) are needed, and they are

$$F_1(u,v) = F_A(u,v) + A_r^2\delta(u,v) + A_r\left[F_O\left(u + u_p, v + v_p\right) + F_O^*\left(-u + u_p, -v + v_p\right)\right], \quad (17)$$

$$\begin{aligned} F_2(u,v) = F_A(u,v) + A_r^2\delta(u,v) + A_r F_O\left(u + u_p, v + v_p\right)\exp(i\Delta\varphi_{r1}) \\ + A_r F_O^*\left(-u + u_p, -v + v_p\right)\exp(-i\Delta\varphi_{r1}) \end{aligned} \quad (18)$$

$$\begin{aligned} F_3(u,v) = F_A(u,v) + A_r^2\delta(u,v) \\ + A_r F_O\left(u + u_p, v + v_p\right)\exp(i\Delta\varphi_{r1})\exp(i\Delta\varphi_{r2}) \\ + A_r F_O^*\left(-u + u_p, -v + v_p\right)\exp(-i\Delta\varphi_{r1})\exp(-i\Delta\varphi_{r2}) \end{aligned} \quad (19)$$

where symbol '*' presents the conjugate operation. $F_1(u, v)$, $F_2(u, v)$ and $F_3(u, v)$ are the Fourier transforms of I_1, I_2 and I_3, respectively. The first term $F_A(u, v)$ in the right side of Equations (17)–(19) is the spectrum of the object intensity, distributing in the central part of the spectrum plane. Because the reference intensities I_r is a constant, the second term of the three equations give a δ-function, showing a bright point at the origin (0, 0) on the spectrum plane. The last two terms in the right side of Equations (17)–(19) are the shifted spectra of the object wave and its conjugate, where $F_O(u + u_p, v + v_p)$ is the Fourier transform of $A_o exp(-i\varphi_o)exp(i\varphi_r)$ and $F_O''(-u + u_p, -v + v_p)$ is the Fourier transform of $A_o exp(i\varphi_o)exp(-i\varphi_r)$. They are confined in two relatively small areas, but the centers are shifted to points $(-u_p, -v_p)$ and (u_p, v_p) on the spectrum plane due to the tilt reference employed. In some cases of complicated object wave, points $(-u_p, -v_p)$ and (u_p, v_p) may be difficult to locate as they are hidden within the spectra of $F_O(u + u_p, v + v_p)$ or $F_O^*(-u + u_p, -v + v_p)$ due to the slight-tilt reference angle (fraction of a degree). However, we can perform interference of two plane waves as the spectrum of such interference contains δ-functions centered at $(0,0)$, $(-u_p, -v_p)$, and (u_p, v_p), which can be found easily. After $(-u_p, -v_p)$, and (u_p, v_p) are determined, the spectrum at these points can be used to extract unknown phase shifts $\Delta\varphi_{r1}$ and $\Delta\varphi_{r2}$, to be explained below. Indeed, if we isolate the spectra at points $(-u_p, -v_p)$ and (u_p, v_p) in the third term of Equations (17) and (18), we have $A_r F_O\left(u + u_p, v + v_p\right)$ and $A_r F_O\left(u + u_p, v + v_p\right)\exp(i\Delta\varphi_{r1})$, respectively $\Delta\varphi_{r1}$ can be extracted by subtracting the argument of the above two terms and by evaluating at $u = -u_p$ and $v = -v_p$ to give:

$$\Delta\varphi_{r1} = \arg[A_r F_O(0,0)\exp(i\Delta\varphi_{r1})] - \arg[A_r F_O(0,0)]. \tag{20}$$

Alternatively, we can also use their conjugate terms, i.e., $A_r F_O^*\left(-u + u_p, -v + v_p\right)$ and $A_r F_O^*\left(-u + u_p, -v + v_p\right)\exp(-i\Delta\varphi_{r1})$. However, in this case, we need to evaluate the terms at $u = u_p$ and $v = v_p$ to give $\Delta\varphi_{r1} = \arg\left[A_r F_O^*(0,0)\right] - \arg[A_r F_O^*(0,0)\exp(-i\Delta\varphi_{r1})]$.

Similarly, we extract $\Delta\varphi_{r2}$ from the third terms of Equations (18) and (19). Again, we use $u = -u_p$ and $v = -v_p$, and we find

$$\Delta\varphi_{r2} = \arg[A_r F_O(0,0)\exp(i\Delta\varphi_{r1})\exp(i\Delta\varphi_{r2})] - [\arg[A_r F_O(0,0)\exp(i\Delta\varphi_{r1})]] \quad, \tag{21}$$

where arg [.] in the above equations denotes taking the argument of a complex value in the square bracket. Equations (20) and (21) use only a simple subtraction operation to extract unknown phase shifts $\Delta\varphi_{r1}$ and $\Delta\varphi_{r2}$ without any iteration. This is the first contribution of using the proposed small-tilt reference.

Our next goal is to find object wave O. Now with $\Delta\varphi_{r1}$ and $\Delta\varphi_{r2}$ already extracted, we can calculate O_1 from Equation (12). O and O_1 are related by Equation (13) through φ_r as follows:

$$O = O_1 \exp(i\varphi_r) = O_1 \exp\left[i\frac{2\pi}{\lambda}\left(x\sin\theta_x + y\sin\theta_y\right)\right]. \tag{22}$$

We can calculate off-axis angles θ_x and θ_y by the following relationships [1]

$$\sin\theta_x = \lambda\frac{u_p}{Md_x}, \ \sin\theta_y = \lambda\frac{v_p}{Nd_y}, \tag{23}$$

where M and N are the total pixel numbers along the horizontal and vertical directions of the hologram. d_x and d_y are the corresponding pixel size in the two directions. Note that we have been using normalized spatial frequencies, u and v. Hence u_p and v_p are divided by the total length of the Charge-coupled Device (CCD) linear dimension, Md_x and Md_y. Since u_p and v_p have already been located, we can determine the off-axis angles. Once the angles are found, object wave O is completely determined from Equation (22), and finally the complex amplitude of the original object can be calculated by performing backward Fresnel diffraction [1].

The whole process of the proposed new algorithm for finding the unknown phase shifts and object wave reconstruction can be summarized in the following steps:

Step 1: Set and determine the reference tilt angle before the test object is put in the optical path. Experimentally, we fix the tilt angle when dozens of fringes appear on the CCD from the interference of two plane waves. We label I_0 as the intensity pattern of the interference between the two plane waves. We then put in the object and record I_1, I_2 and I_3.

Step 2: Perform FT operations on the fringe pattern from the plane wave interference in Step 1 and search the brightest points on the spectral plane (not the point at the origin of the spectral plane) and locate the coordinates of the found points as $(-u_p, -v_p)$ or (u_p, v_p). Perform FT operations on I_1, I_2 and I_3 and store the complex values of these two points. In most practical situations, $(-u_p, -v_p)$ or (u_p, v_p) can also be found from the spectrum of either one of I_1, I_2 and I_3 without the need of two-plane wave interference.

Step 3: Calculate phase shift $\Delta\varphi_{r1}$ and $\Delta\varphi_{r2}$ by Equations (20) and (21) using the values stored in step 2.

Step 4: Reconstruct O_1 by Equation (12).

Step 5: Find the tilt angle by Equation (23) from the located coordinates of $(-u_p, -v_p)$ or (u_p, v_p).

Step 6: Find O by using Equation (22).

Step 7: Recover the original object image by backward Fresnel diffraction of O.

3. Computer Simulation and Optical Experiments

3.1. Computer Simulation

We have carried out a series of computer simulations to verify the effectiveness of the proposed method before any optical experiments. In these simulations, many elliptical surfaces with different parameters have been used as reflecting target objects. A plane wave of 532 nm illuminates these surfaces and is then reflected back onto the CCD in the recording setup illustrated in Figure 1.

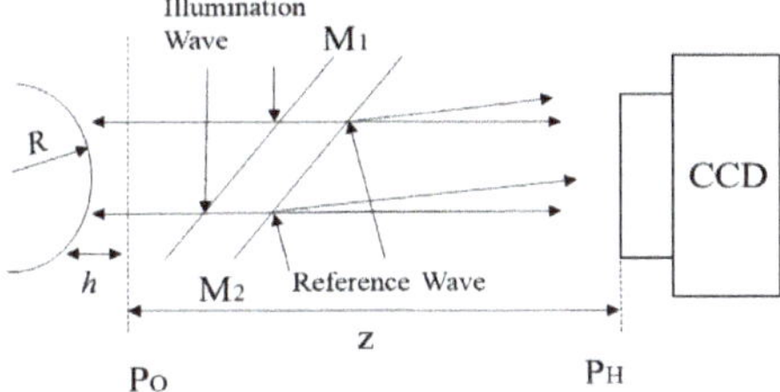

Figure 1. Simulation setup in phase-shifting digital holography with the tilted reference wave.

In the simulation setup in Figure 1 [20], there were 512×512 pixels with $15~\mu m \times 15~\mu m$ size on the CCD and the recording distance was set as $z = 216.5$ mm (all the parameters satisfy the sampling theory to avoid aliasing [1]). With the paraxial approximation of an elliptical wave, we set the phase distribution on the original object plane Po, which is tangential to the surface at the vertex, as $4\pi h(x,y)/\lambda$, with h the surface depth determined by radius of curvature R. We used

$$h(x, y) = \frac{8x^2}{25R} + \frac{y^2}{2R}. \tag{24}$$

Accounting for the possible amplitude variation of an object wave, a Gaussian amplitude intensity distribution of the object wave in plane Po was introduced, decreasing gradually from the center of maximum 1 to the edge of minimum 0.7. Two-dimensional Fresnel diffraction makes the complex amplitude in plane Po give the object wave $O(x, y)$, in the recording plane P_H. Three different interferograms by two reference phase shifts are computer-generated. In this simulation we set two phase-shift values of 1 and 0.6 rad respectively, and we give only a few results with $R = 2000$ mm as example. The reference wave is a plane wave with a slight tilt angle along both the x-axis and the y-axis.

As for optical experimental procedures, in order to accurately determine the small tilt angle of the reference light, we introduce a plane wave as the object wave to obtain I_0. We control the tilt angle of the reference until there are dozens of fringes appearing on the CCD. We then obtain the corresponding spectrum to find $(-u_p, -v_p)$ and (u_p, v_p).

Fast Fourier Transform (FFT) operations are carried out on the interferograms on the computer. Figure 2a shows the pattern of I_0 and its corresponding spectrum is shown in Figure 2d. The two delta functions are located at $(-u_p, -v_p)$ and (u_p, v_p), which have been zoomed in for visual inspection. Figure 2b shows the pattern of I_1 and its corresponding spectrum is shown in Figure 2e. Note that the zero frequency has been suppressed for display purposes in Figure 2d,e. Figure 2c,f show the pattern of I_2 and I_3, respectively.

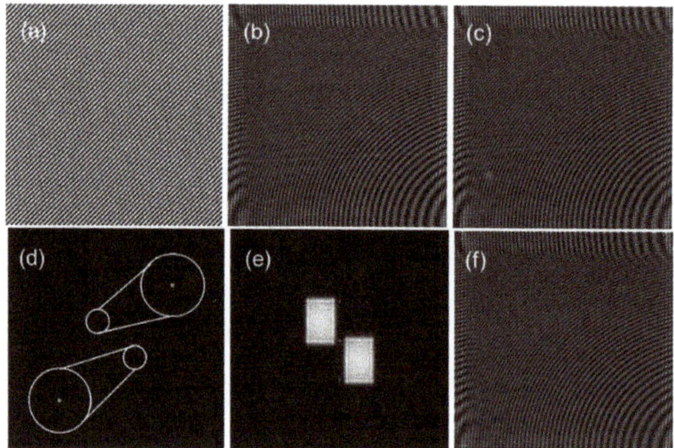

Figure 2. Simulation results: (**a**) I_0; (**b**) I_1; (**c**) I_2; (**d**) spatial spectrum of I_0; (**e**) spatial spectrum of I_1; (**f**) I_3.

By searching the maximum value of the spectrum intensity from the FT of I_0, coordinates $(-u_p, -v_p)$ and (u_p, v_p) are found to be $(-45, 45)$ and $(45, -45)$. Substituting $u_p = 45$, $v_p = 45$, $\lambda = 0.532$ μm, $d_x = 15$ μm, $d_y = 15$ μm and $M = N = 512$ into Equation (23), the reference wave angles in the x and y directions are found to be both approximately 0.18 degrees. After the spectrum coordinates $(-u_p, -v_p)$ and (u_p, v_p) are found, the phase angle of the complex values of the spectra at pixel $(-45, 45)$ or $(45, -45)$ can be calculated and then the two phase shifts are extracted by subtraction operation using Equations (20) and (21).

The phase shifts, $\Delta\varphi_{r1}$ and $\Delta\varphi_{r2}$, calculated by our proposed method are 1.0002 rad and 0.6003 rad, respectively. The two phase shifts with errors of 0.0002 and 0.0003 rad are accurate enough in the non-iterative GPSI method. These phase shift errors come from the digital resolution on the FT spectral plane.

Using the three interferograms and the two extracted phase shifts, wavefront O_1 is retrieved through Equation (12) and then corrected by Equation (22) with the tilt angles of about $\theta_x = 0.18$ degrees and $\theta_y = 0.18$ degrees. Back Fresnel diffraction was carried out to obtain the object on the object plane.

Figure 3a,b shows the amplitude distributions for the original object wave and the reconstructed object wave. In the simulation setup, the intensity distribution was designed as a Gaussian function to simulate an object wave amplitude. That is to say, the object wave recorded has not only phase distribution but also amplitude variation. Although the amplitude of the original object wave is not uniform, the proposed method can recover it ideally, because there is almost no difference between Figure 3a,b. The phase of the original object wave and its reconstructed phase are shown in Figure 3c,d, respectively. It is difficult for us to find any difference between the two figures.

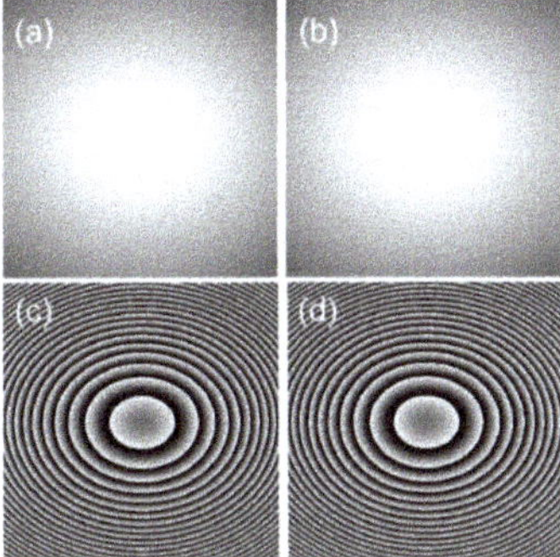

Figure 3. Simulation results: (**a**) intensity of the original object wave; (**b**) intensity of reconstructed object wave; (**c**) original phase distribution; (**d**) reconstructed phase distribution.

To inspect the phase distribution of the object wave quantitatively, the phase across the center of the zones are plotted in Figure 4. For comparison, the original phase and the reconstructed phase are shown in Figure 4a,b, respectively. Obviously Figure 4a,b are almost identical.

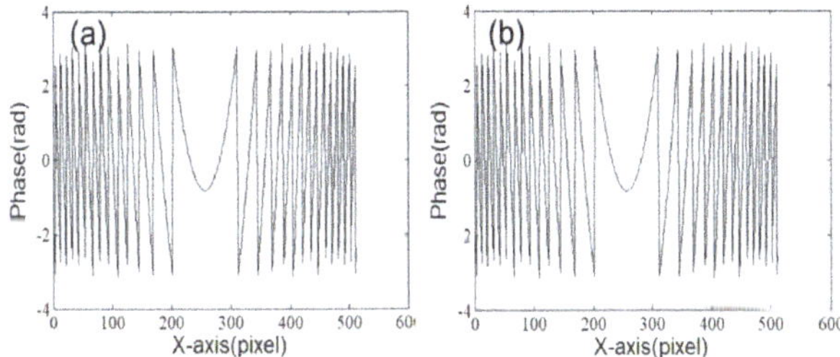

Figure 4. Phase distribution across the center of the zones of the object wave, (**a**) original phase distribution and (**b**) reconstructed phase distribution.

3.2. Optical Experiment

Optical experiments have also been carried out with the optical setup shown in Figure 5 with a 532 nm laser, and the target object is a USAF resolution target. The CCD installed has a recording chip with a resolution of 1392 by 1040 pixels, and the pixel size is 6.45 × 6.45 μm.

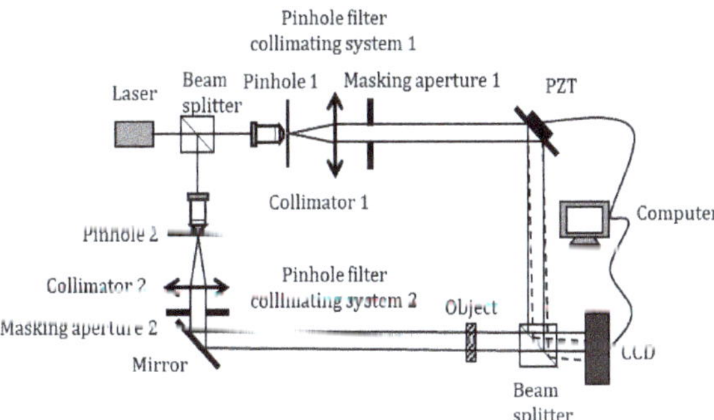

Figure 5. Three step generalized phase-shifting interferometry (TGPSI) recording setup with a slight-tilt reference, where the dotted lines illustrate the tilted reference actually used. A piezoelectric transducer (PZT) is used as a phase shifter.

In our experiment, the phase shift is generated by a phase shifter. The results are illustrated in Figure 6. The three interferograms with unknown phase shifts are shown in Figure 6a–c. Figure 6d is the spectrum of Figure 6a, where the center part is enlarged and shown in the circle. We have used the spectrum of I_1 instead of I_0 to locate the coordinates $(-u_p, -v_p)$ and (u_p, v_p) and they are found to be $(-34, 0)$ and $(34, 0)$. The reason is that there are two bright spots in the spectrum already. The enlarged portion of the spectrum clearly illustrates the two bright spots. Subsequently, the reference slight tilt angle can be calculated by Equation (23) and the result is about $\theta_x = 0.1$ degree in the experiment. Phase shifts are calculated by using the brightest point at these locations on the three Fourier spectra of holograms I_1, I_2 and I_3. The phase shifts calculated using the proposed method are 0.5211 and 1.8018 rad. Figure 6e is the intensity distribution of the reconstructed image, and the reconstructing distance is 10.9 cm. Both the accuracy of the phase shift extraction and the quality of the reconstructed image are excellent.

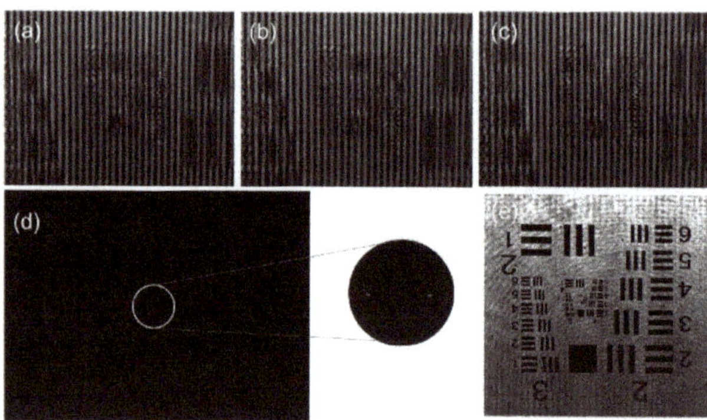

Figure 6. Optical results (**a**) I_1, (**b**) I_2, (**c**) I_3, (**d**) spectrum of I_1 and (**e**) the reconstructed image.

4. Conclusions

We have proposed a method to extract unknown phase shifts in TGPSI with a slight-tilt reference. The major merit of the proposed method is that only two subtractions are used to calculate the unknown phase shifts. The technique is simple and yet powerful as no iteration is needed. It is convenient and accurate. The method needs only three interferograms to complete the phase shift extraction and wavefront reconstruction without iteration. Simulations and optical experiments have verified the feasibility and validity of the proposed method. It should be noted that this method works well when the zero-frequency component in the object spectrum is sufficiently large because the location of this spectrum is easy to find under such condition. For some extreme cases of complicated object waves, a slightly larger tilt angle is needed and the advantage of the efficient use of the space-bandwidth product is not obvious when this method is compared to the off-axis technique. This method is expected to be an attractive alternative for wide-ranging digital holography applications as it allows the full and efficient use of the space-bandwidth product of the limited resolution of digital recording devices because a slight tilt reference of 0.1 degrees is used.

Author Contributions: Conceptualization, X.X.; methodology, X.X., T.M. and Z.J.; software, X.X. and T.M.; validation, T.M. and L.X.; formal analysis, D.D. and F.Q.; investigation, X.X. and T.M.; resources, X.X.; data curation, T.M.; writing—original draft preparation, X.X. and T.M.; writing—review and editing, X.X. and T.-C.P.

Funding: This research was funded by the National Key Research and Development Program of China (2017YFC1404000), Natural Science Foundation of Shandong Province, China (ZR2019MD023) and Fundamental Research Funds for the Central Universities of China (15CX05033A).

Conflicts of Interest: The authors declare no conflict of interest.

References

1. Poon, T.-C.; Liu, J.-P. *Introduction to Modern Digital Holography with MATLAB*; Cambridge University Press: Cambridge, UK, 2014.
2. Bruning, J.H.; Herriott, D.R.; Gallagher, J.E.; Rosenfeld, D.P.; White, A.D.; Brangaccio, D.J. Digital wavefront measuring interferometer for testing optical surfaces and lenses. *Appl. Opt.* **1974**, *13*, 2693–2703. [CrossRef] [PubMed]
3. Yamaguchi, I.; Zhang, T. Phase-shifting digital holography. *Opt. Lett.* **1997**, *22*, 1268–1270. [CrossRef] [PubMed]
4. Xia, P.; Wang, Q.H.; Ri, S.; Tsuda, H. Calibrated phase-shifting digital holography based on a dual-camera system. *Opt. Lett.* **2017**, *42*, 4954–4957. [CrossRef] [PubMed]
5. Zhang, J.Y.Z.; Ren, Y.J.; Zhu, Q.; Lin, Z.Q. Phase-shifting lensless Fourier-transform holography with a Chinese Taiji lens. *Opt. Lett.* **2018**, *43*, 4085–4087. [CrossRef] [PubMed]
6. Wizinowich, P.L. Phase shifting interferometry in the presence of vibration: A new algorithm and system. *Appl. Opt.* **1990**, *29*, 3271–3279. [CrossRef] [PubMed]
7. Xu, X.F.; Cai, L.Z.; Wang, Y.R.; Meng, X.F.; Zhang, H.; Dong, G.Y.; Shen, X.X. Blind phase shift extraction and wavefront retrieval by two-frame phase-shifting interferometry with an unknown phase shift. *Opt. Commun.* **2007**, *273*, 54–59. [CrossRef]
8. Xu, J.C.; Xu, Q.; Chai, L.Q. Iterative algorithm for phase extraction from interferograms with random and spatially nonuniform phase shifts. *Appl. Opt.* **2008**, *47*, 480–485. [CrossRef] [PubMed]
9. Guo, C.S.; Zhang, L.; Wang, H.T.; Liao, J.; Zhu, Y.Y. Phase-shifting error and its elimination in phase-shifting digital holography. *Opt. Lett.* **2002**, *27*, 1687–1689. [CrossRef] [PubMed]
10. Cai, L.Z.; Liu, Q.; Yang, X.L. Generalized phase-shifting interferometry with arbitrary unknown phase steps for diffraction objects. *Opt. Lett.* **2004**, *29*, 183–185. [CrossRef] [PubMed]
11. Xu, X.F.; Cai, L.Z.; Wang, Y.R.; Meng, X.F.; Sun, W.J.; Zhang, H.; Cheng, X.C.; Dong, G.Y.; Shen, X.X. Simple direct extraction of unknown phase shift and wavefront reconstruction in generalized phase-shifting interferometry: Algorithm and experiments. *Opt. Lett.* **2008**, *33*, 776–778. [CrossRef] [PubMed]
12. Yoshikawa, N. Phase determination method in statistical generalized phase-shifting digital holography. *Appl. Opt.* **2013**, *52*, 1947–1953. [CrossRef] [PubMed]
13. Yoshikawa, N.; Kajihara, K. Statistical generalized phase-shifting digital holography with a continuous fringe-scanning scheme. *Opt. Lett.* **2015**, *40*, 3149–3152. [CrossRef] [PubMed]
14. Yoshikawa, N.; Namiki, S.; Uoya, A. Object wave retrieval using normalized holograms in three-step generalized phase-shifting digital holography. *Appl. Opt.* **2019**, *58*, A161–A168. [CrossRef] [PubMed]
15. Zhang, S. A non-iterative method for phase-shift estimation and wave-front reconstruction in phase-shifting digital holography. *Opt. Commun.* **2006**, *268*, 231–234. [CrossRef]
16. Xu, J.C.; Xu, Q.; Chai, L.Q.; Li, Y.; Wang, H. Direct phase extraction from interferograms with random phase shifts. *Opt. Express* **2010**, *18*, 20620–20627. [CrossRef] [PubMed]
17. Guo, C.S.; Sha, B.; Xie, Y.Y.; Zhang, X.J. Zero difference algorithm for phase shift extraction in blind phase-shifting holography. *Opt. Lett.* **2014**, *39*, 813–816. [CrossRef] [PubMed]
18. Xu, X.F.; Zhang, Z.W.; Wang, Z.C.; Wang, J.; Zhan, K.Y.; Jia, Y.L.; Jiao, Z.Y. Robust digital holography design with monitoring setup and reference tilt error elimination. *Appl. Opt.* **2018**, *57*, B205–B211. [CrossRef] [PubMed]
19. Zeng, F.; Tan, Q.F.; Gu, H.R.; Jin, G.F. Phase extraction from interferograms with unknown tilt phase shifts based on a regularized optical flow method. *Opt. Express* **2013**, *21*, 17234–17248. [CrossRef] [PubMed]
20. Cai, L.Z.; Liu, Q.; Yang, X.L. Phase-shift extraction and wave-front reconstruction in phase-shifting interferometry with arbitrary phase steps. *Opt. Lett.* **2003**, *28*, 1808–1810. [CrossRef] [PubMed]

Article

Scaling of Three-Dimensional Computer-Generated Holograms with Layer-Based Shifted Fresnel Diffraction

Hao Zhang *, Liangcai Cao and Guofan Jin

State Key Laboratory of Precision Measurement Technology and Instruments, Department of Precision Instrument, Tsinghua University, Beijing 100084, China; clc@tsinghua.edu.cn (L.C.); jgf-dpi@tsinghua.edu.cn (G.J.)
* Correspondence: haozhang274@tsinghua.edu.cn

Received: 27 March 2019; Accepted: 22 May 2019; Published: 24 May 2019

Abstract: Holographic three-dimensional (3D) displays can reconstruct a whole wavefront of a 3D scene and provide rich depth information for the human eyes. Computer-generated holographic techniques offer an efficient way for reconstructing holograms without complicated interference recording systems. In this work, we present a technique for generating 3D computer-generated holograms (CGHs) with scalable samplings, by using layer-based diffraction calculations. The 3D scene is partitioned into multiple layers according to its depth image. Shifted Fresnel diffraction is used for calculating the wave diffractions from the partitioned layers to the CGH plane with adjustable sampling rates, while maintaining the depth information. The algorithm provides an effective method for scaling 3D CGHs without an optical zoom module in the holographic display system. Experiments have been performed, demonstrating that the proposed method can reconstruct quality 3D images at different scale factors.

Keywords: holography; computer-generated hologram; holographic display; 3D display

---

1. Introduction

The holographic three-dimensional (3D) display is a powerful way of providing depth information for the human eyes, since it can reconstruct the entire whole optical wavefront of the 3D scene [1]. With the development of computing technology and spatial light modulators (SLMs), 3D computer-generated holograms (CGHs) can be reconstructed dynamically without the complicated interference recording systems [2]. The CGH algorithms are used to encode the mathematical representations of the 3D scenes into the holograms, which are closely related to image reconstruction quality and computing efficiency.

Point-based and polygon-based algorithms are commonly used when synthesizing CGHs for various types of 3D scenes [3–6]. These algorithms simulate the wave propagation process from the 3D scene to the CGH plane. The 3D objects are often segmented into many point sources or polygons, and they provide accurate geometric relationships of the 3D scenes. Since the computational load increases with the number of primitives. The computing efficiency is aggravated drastically when calculating the CGH of a complicated 3D scene.

Stereogram-based algorithms can use computer graphics-rendering techniques to process different formats of 3D objects efficiently [7–9]. During the calculations, each holographic element (hogel) of the holographic stereogram is calculated, based on a two-dimensional (2D) parallax image; hence, depth performances are often limited in these stereogram-based CGHs. Recently, several techniques have been developed to improve the depth performances of the stereogram-based algorithms [10–12]. These algorithms have integrated physically-based algorithms and stereogram-based algorithms, in order to reconstruct the accurate depth information of the 3D scenes. However, in stereogram-based algorithms, each hogel corresponds to a viewpoint during the rendering procedure, and this means that we need to render multiple times before implementing the CGH algorithm, which would increase the rendering time. Also, the continuity of the motion parallax and occlusion would be affected by the segmentation setup of the CGH.

Layer-based techniques are efficient in calculating the CGH, by partitioning the 3D scene in multiple layers, according to the depth information [13–16]. Fast Fourier transform (FFT)-based diffraction can be used in simulating the wave propagation processes, from the layers to the CGH plane. However, there are sampling restrictions in the current layer-based 3D CGH algorithms. In FFT-based Fresnel diffraction calculations, the sampling interval on the observation plane is proportional to the propagation distance, while the sampling interval would remain unchanged in convolution-based Fresnel algorithm and angular spectrum method [17]. Hence, the flexibility of the holographic 3D display technique, based on the layer-based algorithm, is limited. Although zero-padding and resampling methods can be used to change the sampling rates, this would take a larger amount of memory and a longer computation time, due to more sampling points that need to be processed.

A double Fresnel transform algorithm was developed, to calculate the wave propagation, with variable magnification, by splitting the propagation distance into two partial steps, which has been used in generating layer-based 3D CGHs with amplitude modulation [18,19]. Since there are minimum propagation distances for the two propagation steps, the propagation distance and the adjustable range of the sampling rates are limited. The wavefront recording plane (WRP) method with non-uniform fast Fourier transform (NUFFT) was used to simplify the 3D CGH calculation [20]. Since the WRP needs to be located close to the object points, the overall depth range is limited during reconstruction.

Chirp Z-transform with Bluestein's algorithm provides a powerful way to perform discrete Fourier transforms with different scaling factors [21,22], which has been used in the Fourier transform-based Fresnel diffraction calculations. Shifted Fresnel diffraction was introduced to simulate the scalable Fresnel diffraction between shifted parallel planes in computational holography [23]. Later, aliasing-reduced shifted and scaled (ARSS) Fresnel diffraction was introduced in the algorithm, to reduce aliasing by a band-limiting function [24], which was also used in lenseless holographic projections [25]. Similar algorithms was also introduced in digital holography to perform magnified reconstructions of the digital holograms [26]. However, the above applications, based on the scaling diffractions, were mainly focused on 2D reconstructions. Also, the sampling technique in 3D CGH still needs further investigation.

In this work, layer-based shifted Fresnel diffraction is implemented to calculate the wave propagation from the partitioned layers to the CGH plane, with adjustable sampling rates. Shifted Fresnel diffraction can simulate the wave propagation between parallel planes with scalable sampling, by using a Bluestein transform. This layer-based processing with scalable diffraction calculation provides an effective way of generating a CGH, with flexible sampling of the 3D scene. By implementing a scalable diffraction calculation into layer-based processing for 3D CGH generation, the reconstructed 3D images are zoomable without an additional optical zoom module. The sampling parameters of the reconstructed 3D images can be adjusted accordingly. Numerical and optical experiments are performed, to demonstrate the effectiveness of our proposed method.

2. Layer-Based Shifted Fresnel Diffraction

The Fresnel diffraction of the 2D optical wave field can be expressed by:

$$U(u,v) = \frac{\exp(jkz)}{j\lambda z}\exp\left[j\frac{k}{2z}\left(u^2+v^2\right)\right]\iint\limits_{\infty} U(x,y)\exp\left[j\frac{k}{2z}\left(x^2+y^2\right)\right]\exp\left[-j2\pi\left(f_x x + f_y y\right)\right]dxdy, \quad (1)$$

where (x, y) and (u, v) denote the coordinates in the source plane and observation plane, respectively. z is the wave propagation distance, $k = 2/\lambda$ is the wave number, and f_x and f_y are calculated as:

$$f_x = \frac{u}{\lambda z}, \ f_y = \frac{v}{\lambda z}. \tag{2}$$

By imposing a 2D Fourier transform, Equation (1) can be represented as:

$$U(u, v) = \frac{\exp(jkz)}{j\lambda z} \exp\left[j\frac{k}{2z}(u^2 + v^2)\right] F\left\{U(x, y) \exp\left[j\frac{k}{2z}(x^2 + y^2)\right]\right\}, \tag{3}$$

where F denotes the Fourier transform. According to the sampling rule of the discrete Fourier transform, the following relation exists:

$$\Delta f_x = \frac{1}{N_x \Delta x}, \ \Delta f_y = \frac{1}{N_y \Delta y}, \tag{4}$$

where N_x and N_y are the sampling numbers in x and y coordinates, Δx and Δy are the sampling intervals in the x and y directions, and Δf_x and Δf_y are the frequency sampling intervals, respectively. Hence, the sampling distances in the observation plane can be deduced as:

$$\Delta u = \frac{\lambda z}{N_x \Delta x}, \ \Delta v = \frac{\lambda z}{N_y \Delta y}. \tag{5}$$

where $N_x \Delta x$ and $N_y \Delta y$ denote the size of the sampling window in the source plane. From Equation (5), we can see that the sampling interval in the observation plane is proportional to the wave propagation distance, and inversely proportional to the size of the sampling window. The sampling ranges in the observation plane can be given by:

$$L_u - N_x \Delta u = \frac{\lambda z}{\Delta x}, L_v = N_y \Delta v = \frac{\lambda z}{\Delta y}, \tag{6}$$

where L_u and L_v are the sampling ranges along the u and v axes in the observation plane. We can see that the sampling parameters in the observation plane cannot be adjusted freely in the Fresnel diffraction calculation, which are determined by the sampling parameters of the source plane and the wave propagation distance.

Shifted Fresnel diffraction can overcome the limitations of traditional Fourier-based Fresnel diffraction, by allowing for arbitrary sampling intervals in the source plane and the observation plane. Figure 1 is the geometric setup of the shifted Fresnel diffraction. (x_0, y_0) is the center of the sampling grid in the source plane (x,y), with a resolution of $P \times Q$, and (u_0, v_0) is the center of the sampling grid in the observation plane (u,v), with a resolution of $M \times N$. Δx and Δy are the sampling intervals in the source plane, and Δu and Δv are the sampling intervals in the observation plane. The coordinates of these two grids are determined by:

$$\begin{aligned} x_p = x_0 + p\Delta x, \ y_q = y_0 + q\Delta y, \\ u_m = u_0 + m\Delta u, \ v_n = v_0 + n\Delta v, \end{aligned} \tag{7}$$

where p, q, m, n are the integer indices:

$$\begin{aligned} -\frac{P}{2} \leq p \leq \frac{P}{2} - 1, -\frac{Q}{2} \leq q < \frac{Q}{2} - 1, \\ -\frac{M}{2} \leq m \leq \frac{M}{2} - 1, -\frac{N}{2} \leq n \leq \frac{N}{2} - 1. \end{aligned} \tag{8}$$

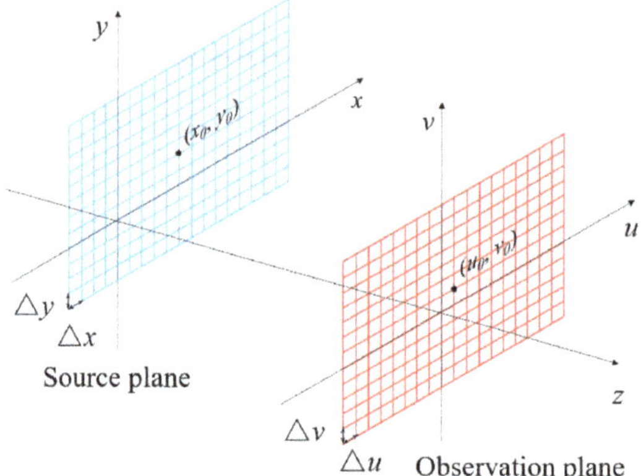

Figure 1. Geometry of shifted Fresnel diffraction.

The sampled 2D Fresnel diffraction can be expressed as:

$$
\begin{aligned}
U(m,n) &= \frac{\exp(jkz)}{j\lambda z}\exp\left[j\frac{k}{2z}\left(u_m{}^2 + v_n{}^2\right)\right]\sum_p\sum_q U(p,q)\exp\left[j\frac{k}{2z}\left(x_p{}^2 + y_q{}^2\right)\right]\exp\left[-j\frac{2\pi}{\lambda z}\left(x_p u_m + y_q v_n\right)\right] \\
&= \frac{\exp(jkz)}{j\lambda z}\exp\left[j\frac{k}{2z}\left(u_m{}^2 + v_n{}^2\right)\right]\exp\left[-j\frac{2\pi}{\lambda z}\left(x_0 m\Delta u + y_0 n\Delta v\right)\right]
\end{aligned}
\tag{9}
$$

$$
\sum_p\sum_q U(p,q)\exp\left[j\frac{k}{2z}\left(x_p{}^2 + y_q{}^2\right)\right]\exp\left[-j\frac{2\pi}{\lambda z}\left(x_p u_0 + y_q v_0\right)\right]\exp\left[-j\frac{2\pi}{\lambda z}\left(pm\Delta x\Delta u + qn\Delta y\Delta v\right)\right].
$$

Let:

$$
\begin{aligned}
2pm &= p^2 + m^2 - (m-p)^2, \\
2qn &= q^2 + n^2 - (n-q)^2.
\end{aligned}
\tag{10}
$$

According to the convolution theorem, the optical wave field $U(m,n)$ can be calculated as:

$$
U(m,n) = C\cdot F^{-1}\{F[a(p,q)]F[b(p,q)]\},
\tag{11}
$$

where:

$$
C = \frac{\exp(jkz)}{j\lambda z}\exp\left[j\frac{k}{2z}\left(u_m{}^2 + v_n{}^2\right)\right]\exp\left[-j\frac{2\pi}{\lambda z}\left(x_0 m\Delta u + y_0 n\Delta v\right)\right]\exp\left[-j\pi\left(\frac{\Delta x\Delta u}{\lambda z}m^2 + \frac{\Delta y\Delta v}{\lambda z}n^2\right)\right],
\tag{12}
$$

$$
a(p,q) = U(p,q)\exp\left[j\frac{k}{2z}\left(x_p{}^2 + y_q{}^2\right)\right]\exp\left[-j\frac{2\pi}{\lambda z}\left(x_p u_0 + y_q v_0\right)\right]\exp\left[-j\pi\left(\frac{\Delta x\Delta u}{\lambda z}p^2 + \frac{\Delta y\Delta v}{\lambda z}q^2\right)\right],
\tag{13}
$$

$$
b(p,q) = \exp\left[j\pi\left(\frac{\Delta x\Delta u}{\lambda z}p^2 + \frac{\Delta y\Delta v}{\lambda z}q^2\right)\right].
\tag{14}
$$

By imposing the required sampling rules in the source plane and the observation plane, diffraction patterns with different scale factors can be reconstructed with the help of shifted Fresnel diffraction. Figure 2 numerically demonstrates the shifted Fresnel diffraction of a square aperture, with three different scale factors. The size of the aperture is 2.4 mm × 2.4 mm. The sampling number in the source plane is 1024 × 1024, and the sampling interval is 8 μm. The wave propagation distance is 300 mm, and the wavelength used in the simulation is 633 nm. The diffraction patterns in the observation plane are computed with different scale factors. Figure 2a–c shows the diffraction patterns, with $\Delta u = 6$ μm, $\Delta u = 8$ μm, and $\Delta u = 10$ μm, respectively.

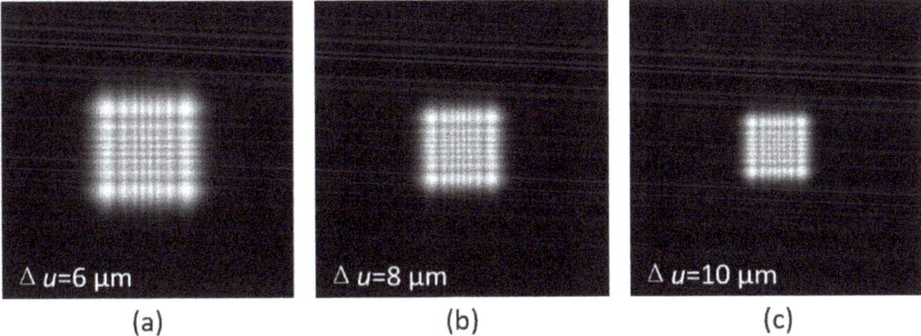

Figure 2. Shifted Fresnel diffraction with different scale factors: (**a**) $\Delta u = 6$ μm, (**b**) $\Delta u = 8$ μm, (**c**) $\Delta u = 10$ μm.

In layer-based 3D CGH synthesis, the 3D scene is firstly partitioned into multiple layers, which are parallel to the CGH plane. Then, the wave fields diffracted from the partitioned layers to the CGH plane are calculated. By using shifted Fresnel diffraction in the diffraction simulations from each layer to the CGH plane, the sampling parameters addressed on the 3D scene can be adjusted accordingly. The diagram of the layer-based shifted Fresnel diffraction is shown in Figure 3. The 3D model is partitioned into different parallel layers, based on its depth information. For each layer, amplitude distribution is extracted according to the rendered image. In order to simulate the diffusive effect of the object surface, random phase distribution is added to each layer. The complex distribution of each layer can be calculated by using shifted Fresnel diffractions with preset sampling rules. The complex distribution of the CGH plane can be acquired after adding the contributions from all the layers [27]. After obtaining the field distribution of the hologram plane, the phase distribution is then extracted for uploading to the phase SLM.

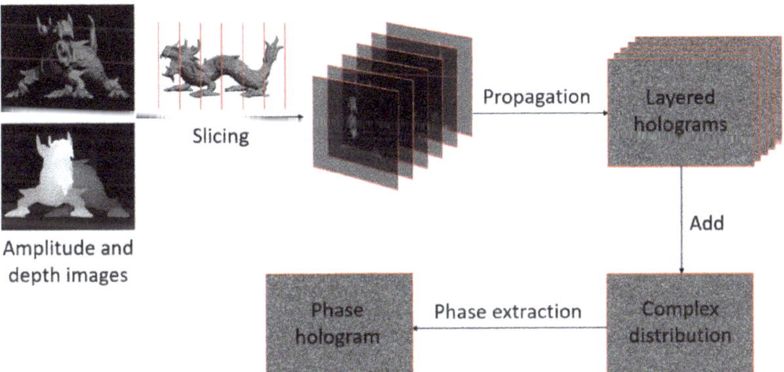

Figure 3. Diagram of the layer-based shifted Fresnel diffraction for 3D computer-generated holograms (CGHs) calculations.

3. Reconstruction Results

To demonstrate the effectiveness of the layer-based shifted Fresnel diffraction for 3D reconstruction, optical experiments are implemented. Figure 4 illustrates the optical diagram of the holographic reconstruction experiment. In the experimental work, the PLUTO phase-only SLM was used for optical reconstructions. The pixel number of the SLM was 1920×1080. The pixel pitch was 8 μm, and the SLM was addressed with 8-bit grayscale levels. The wavelength of the laser used in our experiment was 633 nm. A 4-f system was used in the reconstruction path, for filtering out the zero-order interruption and unwanted orders [28–30].

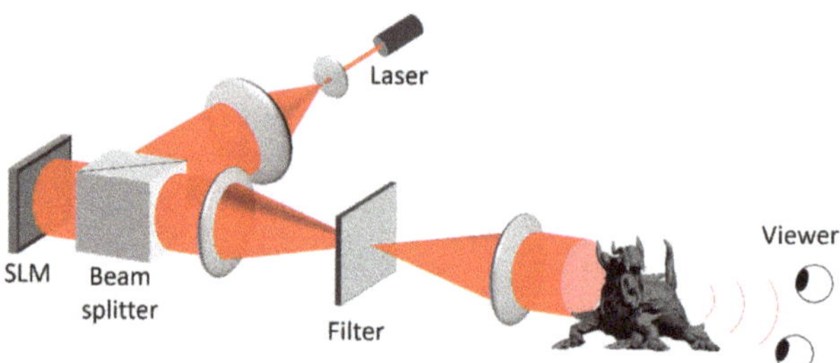

Figure 4. Optical diagram of the holographic reconstruction experiment.

When generating CGHs by using a layer-based algorithm with a shifted Fresnel diffraction, the source planes are the partitioned layers of the 3D scene, and the observation plane is the hologram plane. The original 3D scene used for generating the CGH was a dragon model located 200 mm behind the hologram plane. The dragon model was partitioned into 50 layers, according to its depth information. Since shifted Fresnel diffraction is used in the CGH calculation, the sampling intervals of the object layers can be adjusted. In order to scale the reconstructed 3D scene, the lateral sampling distances of the object layers were set to 6 μm, 8 μm, and 10 μm, respectively. Figure 5a,c,e presents the numerical reconstruction results when focusing on the head of the model. Figure 5b,d,f presents the numerical reconstruction results when focusing on the tail of the model.

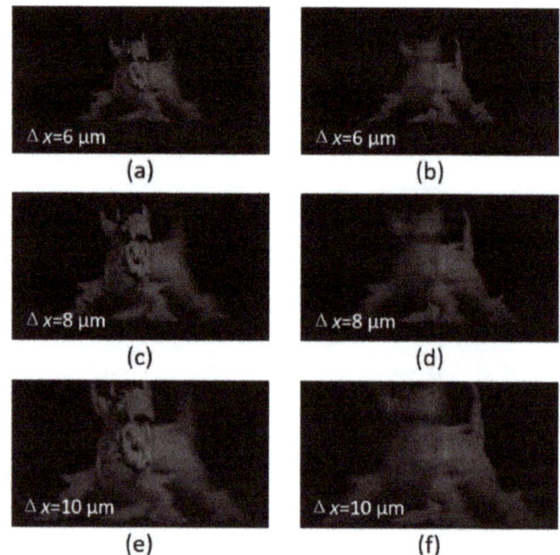

Figure 5. Numerical reconstructions of CGHs with different sampling rates. (**a,c,e**) focus on the head; (**b,d,f**) focus on the tail.

Figure 6 demonstrates the optical reconstruction results of CGHs with different sampling rates, which are consistent with the numerical results shown in Figure 5. According to the numerical and optical demonstrations, we can see that the depth information of the CGHs with different scale factors could be reconstructed with high quality, which demonstrates that the reconstructed images can be scaled with precise control of the sampling rates. Since the scaled sampling was addressed on the object layers of the original 3D object, and because the reconstruction processes of different CGHs are based on the unified sampling rules, smaller sampling interval would lead to a smaller size for the reconstruction image.

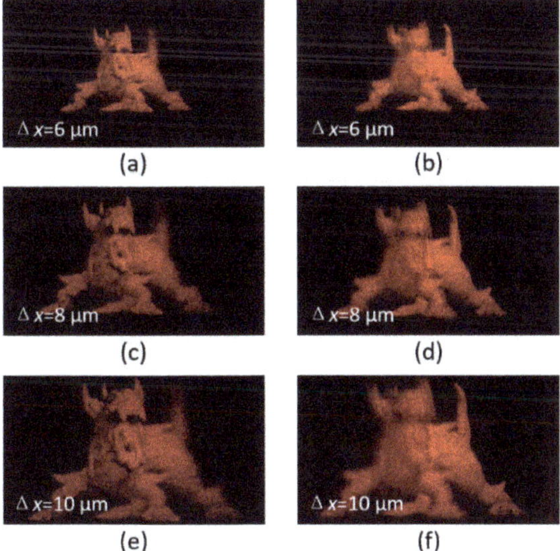

Figure 6. Optical reconstructions of CGHs at different sampling rates. (**a**,**c**,**e**) focus on the head; (**b**,**d**,**f**) focus on the tail.

4. Accuracy Analysis with Different Sampling Parameters

To evaluate the accuracy of the proposed algorithm, a single slit with a size of 2.4 mm was placed at the center of the source plane, to calculate the diffraction fields with different parameters. The sampling interval on the source plane was 8 µm, and the sampling number was 1024. The wavelength was 633 nm. The accuracy was evaluated with a signal-to-noise ratio (SNR) of the wavefield that was compared to the Rayleigh–Sommerfeld diffraction integral.

Figure 7 illustrates the accuracies with different sampling rates in the observation plane as a function of the propagation distance. Due to the paraxial approximation, the accuracies of the shifted-Fresnel diffraction were low for the short propagation distances. The SNR obviously improved with an increase in the propagation distance. On the other hand, the sampling rate in the observation plane also affected the calculation accuracy. Larger sampling intervals would lead to a larger sampling range, according to Equation (6), which induces a calculation that satisfies the paraxial approximation at a longer propagation distance. Additionally, all of the SNRs are larger than 30 dB when the propagation distance is larger than 150 mm.

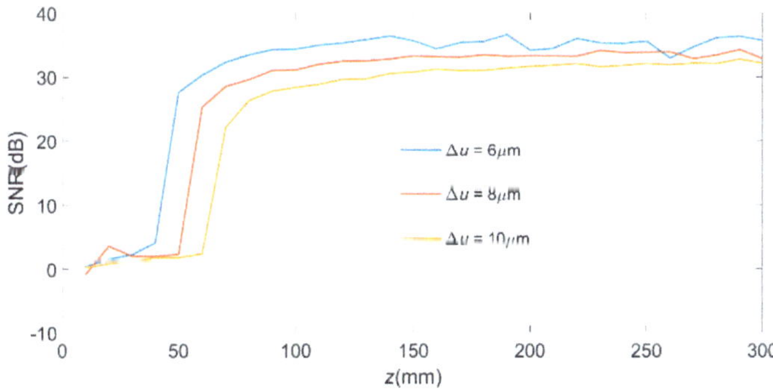

Figure 7. Accuracies of the diffraction calculation with different sampling rates in the observation plane.

Figure 8 illustrates the amplitude profiles of the diffraction fields with different scaling factors, calculated by the proposed method and the Rayleigh–Sommerfeld diffraction integral. In the simulation, the propagation distance was set to 300 mm. From the figure, we can see that the calculation results were in excellent agreement.

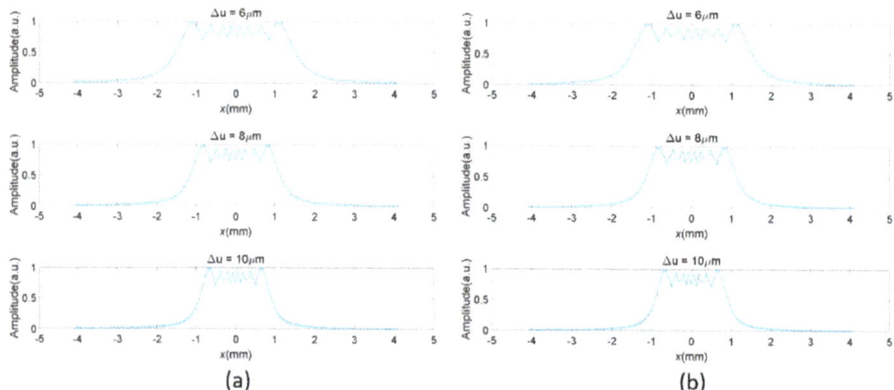

Figure 8. Amplitude profiles of the diffraction fields at different propagation distances: (**a**) Shifted Fresnel diffraction, (**b**) Rayleigh–Sommerfeld diffraction integral.

5. Conclusions

In conclusion, we propose an effective method for scaling 3D CGHs, using layer-based shifted Fresnel diffraction. During the calculation, the wave propagation from the partitioned layers to the CGH plane can be calculated with a flexible sampling rate; hence, the scaling display of the 3D image can be realized by changing the sampling rates of the object layers. A phase-only SLM is used for optical reconstruction, and it demonstrates that the proposed system can scale a 3D scene using different scale factors, without the need for an optical zoom module.

Author Contributions: Conceptualization, H.Z. and L.C.; methodology, H.Z.; software, H.Z.; validation, L.C. and G.J.; writing—original draft preparation, H.Z; writing—review and editing, H.Z., L.C., and G.J.; funding acquisition, H.Z. and L.C.

Funding: This research was funded by the National Natural Science Foundation of China (NSFC), grant number 61875105.

Conflicts of Interest: The authors declare no conflict of interest.

References

1. Benton, S.A.; Bove, V.M. *Holographic Imaging*; Wiley-Interscience: Hoboken, NJ, USA, 2008.
2. Zhang, H.; Zhao, Y.; Cao, L.; Jin, G. Three-dimensional display technologies in wave and ray optics: A review. *Chin. Opt. Lett.* **2014**, *12*, 060002. [CrossRef]
3. Lucente, M.E. Interactive computation of holograms using a look-up table. *ELECTIM* **1993**, *2*, 28–34. [CrossRef]
4. Matsushima, K. Computer-generated holograms for three-dimensional surface objects with shade and texture. *Appl. Opt.* **2005**, *44*, 4607–4614. [CrossRef]
5. Ahrenberg, L.; Benzie, P.; Magnor, M.; Watson, J. Computer generated holograms from three dimensional meshes using an analytic light transport model. *Appl. Opt.* **2008**, *47*, 1567–1574. [CrossRef]
6. Kim, H.; Hahn, J.; Lee, B. Mathematical modeling of triangle-mesh-modeled three-dimensional surface objects for digital holography. *Appl. Opt.* **2008**, *47*, D117–D127. [CrossRef]
7. Fan, F.; Jiang, X.; Yan, X.; Wen, J.; Chen, S.; Zhang, T.; Han, C. Holographic Element-Based Effective Perspective Image Segmentation and Mosaicking Holographic Stereogram Printing. *Appl. Sci.* **2019**, *9*, 920. [CrossRef]

8. Wakunami, K.; Yamaguchi, M. Calculation for computer generated hologram using ray-sampling plane. *Opt. Express* **2011**, *19*, 9086–9101. [CrossRef]

9. Kang, H.; Stoykova, E.; Yoshikawa, H. Fast phase-added stereogram algorithm for generation of photorealistic 3D content. *Appl. Opt.* **2016**, *55*, A135–A143. [CrossRef]

10. Zhang, H.; Zhao, Y.; Cao, L.C.; Jin, G.F. Fully computed holographic stereogram based algorithm for computer-generated holograms with accurate depth cues. *Opt. Express* **2015**, *23*, 3901–3913. [CrossRef]

11. Zhang, H.; Zhao, Y.; Cao, L.; Jin, G. Layered holographic stereogram based on inverse Fresnel diffraction. *Appl. Opt.* **2016**, *55*, A154–A159. [CrossRef]

12. Zhang, H.; Cao, L.; Jin, G. Three-dimensional computer-generated hologram with Fourier domain segmentation. *Opt. Express* **2019**, *27*, 11689–11697. [CrossRef]

13. Chen, J.S.; Chu, D.P. Improved layer-based method for rapid hologram generation and real-time interactive holographic display applications. *Opt. Express* **2015**, *23*, 18143–18155. [CrossRef]

14. Bayraktar, M.; Özcan, M. Method to calculate the far field of three-dimensional objects for computer-generated holography. *Appl. Opt.* **2010**, *49*, 4647–4654. [CrossRef]

15. Zhang, H.; Cao, L.; Jin, G. Computer-generated hologram with occlusion effect using layer-based processing. *Appl. Opt.* **2017**, *56*, F138–F143. [CrossRef]

16. Zhao, Y.; Cao, L.; Zhang, H.; Kong, D.; Jin, G. Accurate calculation of computer-generated holograms using angular-spectrum layer-oriented method. *Opt. Express* **2015**, *23*, 25440–25449. [CrossRef]

17. Voelz, D.G. *Computational Fourier Optics: A MATLAB Tutorial*; SPIE Press: Bellingham, WA, USA, 2011.

18. Zhang, F.; Yamaguchi, I.; Yaroslavsky, L.P. Algorithm for reconstruction of digital holograms with adjustable magnification. *Opt. Lett.* **2004**, *29*, 1668–1670. [CrossRef]

19. Okada, N.; Shimobaba, T.; Ichihashi, Y.; Oi, R.; Yamamoto, K.; Oikawa, M.; Kakue, T.; Masuda, N.; Ito, T. Band-limited double-step Fresnel diffraction and its application to computer-generated holograms. *Opt. Express* **2013**, *21*, 9192–9197. [CrossRef]

20. Chang, C.; Wu, J.; Qi, Y.; Yuan, C.; Nie, S.; Xia, J. Simple calculation of a computer-generated hologram for lensless holographic 3D projection using a nonuniform sampled wavefront recording plane. *Appl. Opt.* **2016**, *55*, 7988–7996. [CrossRef]

21. Bluestein, L. A linear filtering approach to the computation of discrete Fourier transform. *IEEE Trans. Audio Electroacoust.* **1970**, *18*, 451–455. [CrossRef]

22. Rabiner, L.; Schafer, R.; Rader, C. The chirp z-transform algorithm. *IEEE Trans. Audio Electroacoust.* **1969**, *17*, 86–92. [CrossRef]

23. Muffoletto, R.P.; Tyler, J.M.; Tohline, J.E. Shifted Fresnel diffraction for computational holography. *Opt. Express* **2007**, *15*, 5631–5640. [CrossRef]

24. Tomoyoshi, S.; Takashi, K.; Naohisa, O.; Minoru, O.; Yumi, Y.; Tomoyoshi, I. Aliasing-reduced Fresnel diffraction with scale and shift operations. *J. Opt.* **2013**, *15*, 075405.

25. Shimobaba, T.; Makowski, M.; Kakue, T.; Oikawa, M.; Okada, N.; Endo, Y.; Hirayama, R.; Ito, T. Lensless zoomable holographic projection using scaled Fresnel diffraction. *Opt. Express* **2013**, *21*, 25285–25290. [CrossRef]

26. Restrepo, J.F.; Garcia-Sucerquia, J. Magnified reconstruction of digitally recorded holograms by Fresnel–Bluestein transform. *Appl. Opt.* **2010**, *49*, 6430–6435. [CrossRef]

27. Zhang, H.; Cao, L.; Zong, S.; Jin, G. Zoomable three-dimensional computer-generated holographic display based on shifted Fresnel diffraction. In Proceedings of the SPIE 10022, Holography, Diffractive Optics, and Applications VII, Beijing, China, 31 October 2016; Volume 10022, p. 100221D. [CrossRef]

28. Zhang, H.; Xie, J.; Liu, J.; Wang, Y. Elimination of a zero-order beam induced by a pixelated spatial light modulator for holographic projection. *Appl. Opt.* **2009**, *40*, 5031–5841. [CrossRef]

29. Zhang, H.; Tan, Q.; Jin, G. Holographic display system of a three-dimensional image with distortion-free magnification and zero-order elimination. *Opt. Eng.* **2012**, *51*, 075801. [CrossRef]

30. Zhang, H.; Cao, L.; Jin, G. Lighting effects rendering in three-dimensional computer-generated holographic display. *Opt. Commun.* **2016**, *370*, 192–197. [CrossRef]

Article

A Fast Computer-Generated Holographic Method for VR and AR Near-Eye 3D Display

Xin Yang [1,2], HongBo Zhang [3] and Qiong-Hua Wang [1,2,*]

[1] School of Instrumentation and Optoelectronic Engineering, Beihang University, Beijing 100191, China; holooptics@buaa.edu.cn
[2] Beijing Advanced Innovation Center for Big Data-based Precision Medicine, Beihang University, Beijing 100191, China
[3] Department of Computer and Information Sciences, Virginia Military Institute, Lexington, VA 24450, USA; zhangh@vmi.edu
* Correspondence: qionghua@buaa.edu.cn

Received: 4 August 2019; Accepted: 30 September 2019; Published: 4 October 2019

Featured Application: The proposed technique has potential for head mounted near-eye VR or AR applications.

Abstract: A fast computer-generated holographic method with multiple projection images for a near-eye VR (Virtual Reality) and AR (Augmented Reality) 3D display is proposed. A 3D object located near the holographic plane is projected onto a projection plane to obtain a plurality of projected images with different angles. The hologram is calculated by superposition of projected images convolution with corresponding point spread functions (PSF). Holographic 3D display systems with LED as illumination, 4f optical filtering system and lens as eyepiece for near-eye VR display and holographic optical element (HOE) as combiner for near-eye AR display are designed and developed. The results show that the proposed calculation method is about 38 times faster than the conventional point cloud method and the display system is compact and flexible enough to produce speckle noise-free high-quality VR and AR 3D images with efficient focus and defocus capabilities.

Keywords: holographic 3D display; computer generated holography; augmented reality; virtual reality

---

1. Introduction

A holographic display is a promising candidate for 3D display because the phase and amplitude of a 3D scene can be completely reconstructed. It is possible to implement computer-generated holograms (CGH) for a holographic 3D display [1,2]. High resolution holograms such as rainbow holograms and Fresnel holograms have been proposed [3,4]. Many dynamic holographic display solutions have also been implemented [5–10]. Among them, notable frequency domain multiplexing methods are used for color video display through Fresnel hologram with red, green and blue lasers as illumination [7,8]. However, such an optical system is often bulky and the image quality is poor because of the coherent illumination, which affects the practical application of the technology. A lot of efforts have been made, but the progress on the holographic 3D display still has proven difficult due to the following constraints, such as the computational difficulty for the large size of 3D point cloud data, the limited bandwidth of spatial light modulator (SLM), and the speckle noise of the reconstructed 3D images due to the use of coherent illumination sources. The narrow diffraction angle of the current SLM often dictates that the holographic 3D display is restricted to quite small 3D images and narrow viewing angles, with no sign of a significant improvement in the short run

While these difficulties exist, Virtual Reality (VR) and Augmented Reality (AR) displays have become emerging technologies. In the VR display, the human eye views the virtual image through the display device. However, in the AR display, users can view both real scenes and some virtual images simultaneously [11]. In this technique, the virtual image is displayed in front of the human eye through an optical system and AR eyepiece and ambient light from a real object can pass incident on the human eye through the AR eyepiece without distortion. The combination of holography and VR or AR display to achieve a true 3D display is quite intuitive and thus desirable, which can avoid the accommodation convergence conflict problem existing in binocular parallax-based 3D displays [12,13].

The Fresnel hologram is often chosen to produce 3D displays with large depths of field using a coherent source such as laser as illumination. However, it is not quite suitable for holographic VR or AR display because of the large imaging distance between the reconstructed image and the Fresnel hologram. Additionally, for use of a coherent illumination source, speckle noise of holographic 3D display can also bring a negative impact on image quality.

Several hologram computation methods such as point cloud-based, layer-based, and triangular mesh-based algorithms [14–19] have been proposed for speeding up the calculation of the Fresnel hologram. To improve the calculation efficiency, holographic stereogram methods using numerous projection images have been proposed [20,21]. The essence of those methods is to represent the information of one point in the frequency domain by the superimposition of one orthogonal projection image multiplied by the phase of the corresponding direction. Subsequently, the frequency is converted into a spatial domain for an image plane hologram calculation or using Fresnel diffraction for Fresnel hologram calculation. While those methods have merits, a huge number of projected images are required for high-resolution hologram calculation.

In order to achieve an efficient and more compact holographic display system, a hologram calculation method similar to holographic stereogram and an optical display system for holographic near eye VR and AR display are proposed. Within this calculation method, a plurality of images with different angles are projected on a projection plane near the holographic plane. Subsequently, the optical field on the holographic plane is computed by superposition of the convolution of each 2D projected image with specific point spread functions (PSF) corresponding to propagation from the projection plane to the holographic plane in the appropriate directions. Additionally, in order to reduce the speckle noise, LED is chosen as the light source for the optical display system. 4f optical filtering system is utilized for filtering out the unwanted noise. Another lens is used as the eyepiece for VR display and reflective volume holographic optical element (HOE) is used as a combiner for AR display. Overall, the proposed display system is compact and displayed 3D images are free of speckle noise.

2. Methods

The concept of hologram calculation is shown in Figure 1 with two projection planes for ease of explanation since the projection of 3D object is multidirectional. In Figure 1a, A and B are two points in the space, and layer1 and layer2 are two projection planes. On each projection plane, the projected rays from the spatial 3D object points are in the same direction. p_{1A} and p_{1B} are the projected points of the spatial points A and B on layer$_1$, respectively. p_{2A} and p_{2B} are the projection points of space points A and B on layer2, respectively. The projection of 3D object points on one projection plane forms a projected image corresponding to this projecting angle. In the hologram calculation process, the illumination direction of each point on different projection planes is set as the propagation direction, as shown in Figure 1b. The complex amplitude distribution on the holographic plane H is a superposition of light field from all projection planes. When the hologram is reconstructed, the beamlets in these directions are diffracted from the hologram. Similar to integral imaging 3D displays, the overlapping regions of the different beamlets form a 3D object for 3D display [22,23], where the propagation direction of the beamlets determines the outcome of the 3D display.

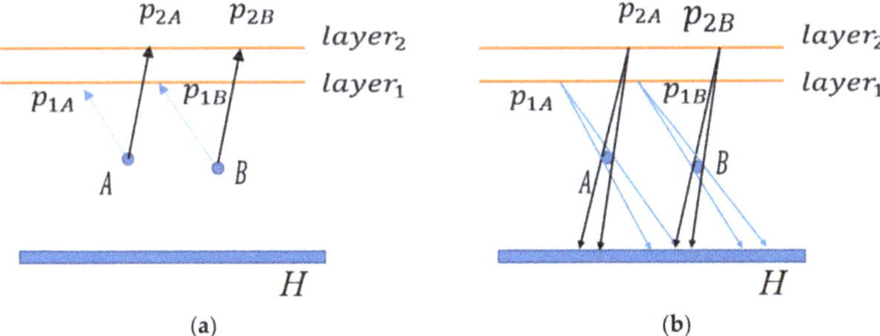

Figure 1. The concept of hologram calculation. (**a**) Projection of 3D object to different layers; (**b**) the calculation of complex amplitude on holographic plane H.

To simplify calculation, the projection layers are all assumed to coincide in the same plane. On this projection plane, projected images from the 3D object are obtained with the designed projection angles. The light in different directions of the 3D object is calculated onto holographic plane H from the projected images. The direction of propagation of light on a projected image is θ_{ix} and angular interval is $\Delta\theta_{ix}$ as shown in Figure 2. The distance from the projection plane to the holographic plane H is z. The boundary of the propagation of the light from point p on the image on the holographic plane falls between x_{ih1} and x_{ih2}, respectively.

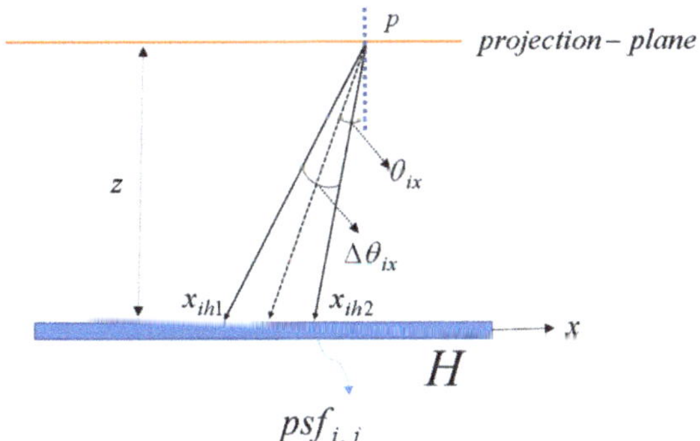

Figure 2. The hologram projection angle and boundary.

Assume the point p is an on-axis point and the PSF of light propagating from the projection plane to the holographic plane can be expressed as:

$$psf_{i,j} = \exp[i\frac{2\pi}{\lambda}(\sqrt{x_h^2 + y_h^2 + z^2})] \tag{1}$$

where $x_{ih1} \leq x_h \leq x_{ih2}$, $y_{jh1} \leq y_h \leq y_{jh2}$. x_{ih1} and x_{ih2} are the boundary of the light field on the holographic plane limited by the projection angle and angle interval in the x direction, respectively. i and j are the index of projected images in x and y directions, respectively. y_{jh1} and y_{jh2} are the boundary of the light field on the holographic plane limited by the projection angle and angle interval in the y direction, respectively.

The boundary of the light field distribution area on the holographic plane limited by the projection angle (θ_{ix}, θ_{jy}) and angle interval ($\Delta\theta_{ix}, \Delta\theta_{iy}$) can be expressed as:

$$x_{ih1} = z\tan(\theta_{ix} + \frac{\Delta\theta_{ix}}{2}); x_{ih2} = z\tan(\theta_{ix} - \frac{\Delta\theta_{ix}}{2})$$
$$y_{jh1} = z\tan(\theta_{jy} + \frac{\Delta\theta_{jy}}{2}); y_{jh2} = z\tan(\theta_{jy} - \frac{\Delta\theta_{jy}}{2}) \tag{2}$$

where θ_{iy} and $\Delta\theta_{jy}$ are the projection angle and angular interval in the y direction, respectively.

The complex amplitude on the holographic plane H is the superposition of the convolution of each projected image with its corresponding PSF, which can be expressed as:

$$U(x,y) = \sum_{i=1:I} \sum_{j=1:J} im_{i,j} \otimes psf_{i,j} \tag{3}$$

where I and J are the number of projected images in x and y directions, respectively. The complex amplitude $U(x,y)$ on the hologram plane can be further encoded as a double phase hologram or as an off-axis amplitude hologram by introducing reference light [24,25]. Thus, the proposed approach has good degree of freedom for potential downstream of visualization tasks.

From above, it is evident that the simplified model consists of one projection plane on which 3D objects yield multiple projection images, as shown in Figure 3. The convolution of each projected image with a PSF yields complex amplitude distribution of the 3D object. Assume that the projection angles are θ_{ix} and θ_{jy} in the x and y directions, respectively. The coordinates of the spatial point $p(x_p, y_p, z_p)$ on the projection plane can be expressed as:

$$\begin{aligned} X_P &= x_p + z_p \tan(\theta_{ix}) \\ Y_P &= y_p + z_p \tan(\theta_{jy}) \end{aligned} \tag{4}$$

where X_p and Y_p are the coordinates of projected point of p on the projection plane. The projected image $im_{i,j}$ can be expressed as:

$$im_{i,j}(X_p, Y_p) = A_P \tag{5}$$

where Ap is the amplitude of object p.

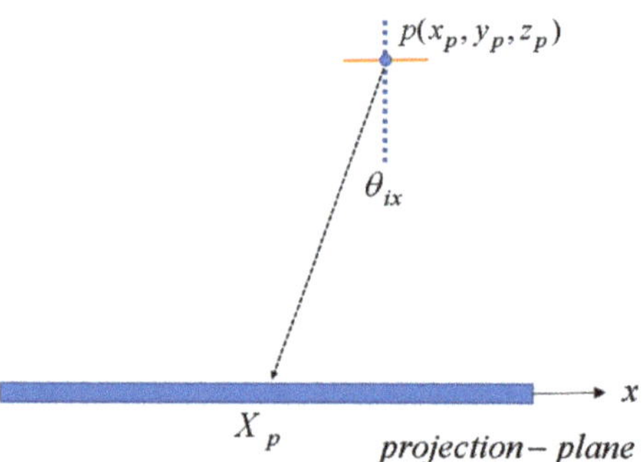

Figure 3. Diagram of the calculation of the projected image.

3. Experiments and Result Analysis

Our experiment utilizes the principle of off-axis amplitude hologram for achieving holographic VR and AR 3D display. Within the experiment, a liquid-crystal-on-silicon (LCoS) SLM is utilized for modulating the holograms. The parameters of the SLM are shown in Table 1. Green light LED from Osram with 1 W power, center wavelength of 528 nm, light-emitting area of 1 mm × 1 mm, bandwidth of 28 nm is used as the illumination source. The center wavelength is used for hologram calculation.

Table 1. Optical parameters in the experiment.

Parameters	Values
SLM horizontal resolution	1920
SLM vertical resolution	1080
SLM pixel size	8 μm
LED size	1 mm × 1 mm
Center wavelength of LED	528 nm
Bandwidth of LED	28 nm

For the center wavelength, the diffraction angle of the SLM is

$$\theta_d = \sin^{-1}\left(\frac{\lambda}{2pix}\right) = 1.89°$$

(6)

where *pix* is the pixel size of the LCoS. Given that the distance between the projection plane and the holographic plane is 10 mm, it is possible to calculate the PSF. It is assumed that the plane wave perpendicular to the holographic plane is used as a reference wave, which means that the phase of the reference wave on the holographic plane is a constant value. Under this situation, the real part of the PSF in Equation (1) on the holographic plane is an amplitude type hologram. This amplitude type hologram can be regarded as new PSF, as shown in Figure 4. The resolution of the PSF is about 84 pixels × 84 pixels, which corresponds to the viewing angle of 3.78° both in *x* and *y* directions from the hologram. For calculation of off-axis amplitude type hologram, only the upper half of the PSF (outlined with red rectangles) is used for the hologram calculation, as shown in Figure 4. The eight PSF sections correspond to eight PSFs of a 3D object projected onto one projection plane with eight different projection angles. Each PSF section has a resolution of 20 pixels × 20 pixels. Each PSF section is named as $psf_{i,j}$ according their positions. Hence in this design, the angular separation between the two adjacent projected images in the *x* and *y* directions is set as 0.9°.

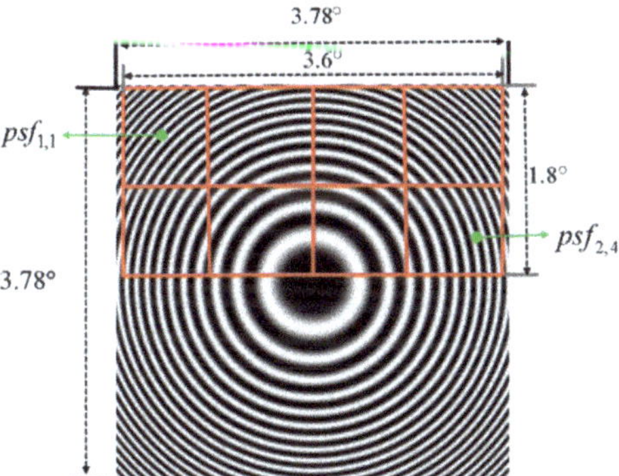

Figure 4. The calculation of point spread functions (PSF).

Figure 5 shows the calculation of the hologram. A 3D object with size of 4.23 mm × 7.05 mm × 3.52 mm and a distance of 5 mm from the center of 3D object to the projection plane is used for hologram calculation. This 3D color model contains 41,022 object points. Two rows and four columns of the projected images of the 3D object are calculated according to Equation (5), and each projected image is firstly converted to grayscale image with a resolution of 600 × 600 and then interpolated and zero padded into a projected image with resolution of 1920 × 1080. The process of convolving the projected images with the corresponding PSF is shown in Figure 5. The projected image $im_{i,j}$ convolves with the corresponding $psf_{i,j}$. Then, the results are superimposed to obtain the hologram.

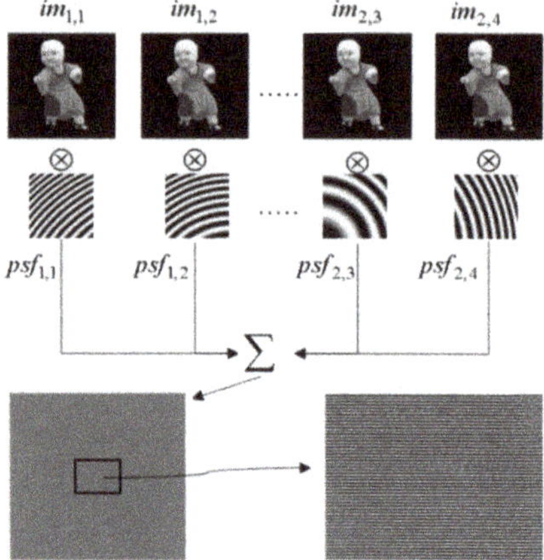

Figure 5. The process of hologram calculation through convolution of corresponding PSF with projected images.

The optical setup of the experiment for holographic 3D VR display is shown in Figure 6. In this setup, the divergent illumination light emitted by the green LED is collimated by the lens$_1$, and the LCoS is illuminated by the reflected light from beam splitter BS. The computer-generated hologram is loaded on the LCoS, and the illumination light is modulated. The modulated light field is filtered by the 4f optical filtering system to form a real 3D image free of zero order and high order image noise finally incident on human eye through the eyepiece lens$_4$. The lenses used for collimating, 4f optical filtering system and eyepiece are all cemented doublet achromatic lenses with focal length of 50 mm and diameter of 30 mm. In this case, the distance from the LCoS to the eye is less than 25 cm, which means this setup is a relatively compact display system.

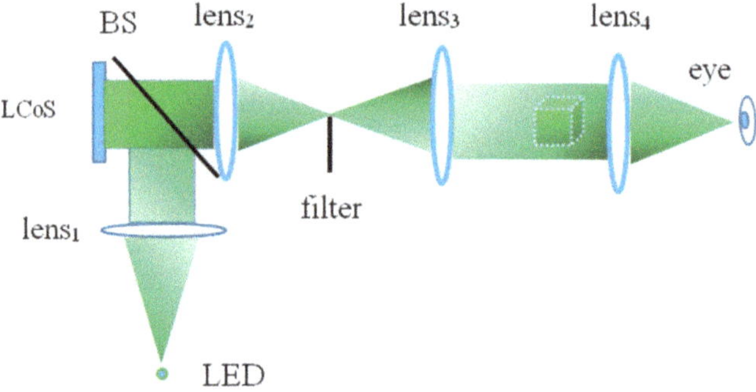

Figure 6. Optical setup for Virtual Reality (VR) display.

The experimental results of the VR setup are shown in Figure 7. Figure 7a shows the reconstructed image of the hologram calculated by the proposed algorithm. For comparison, Figure 7b shows the result of the hologram calculated using the direct point cloud algorithm [14] of the red channel data from the color 3D model. The reproductions are similar, but the computational speed of our proposed method takes only about 2 s on average whereas the 3D cloud point data approach consumes about

76 s on average (38 times slower than our method). The comparison is based on the MATLAB 2015 with a laptop with i7-7700HQ CPU and 16G RAM.

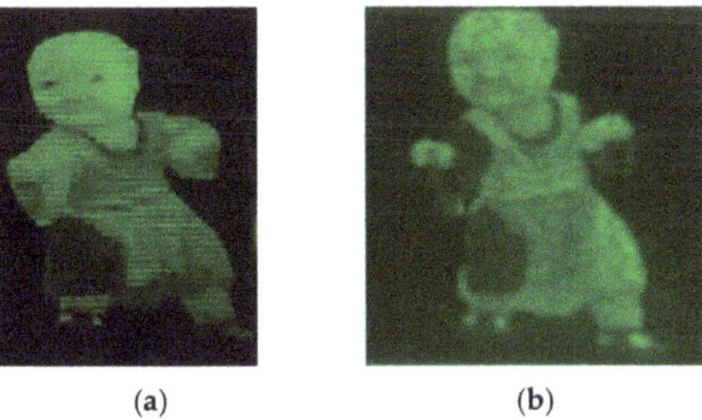

(a)   (b)

Figure 7. Optical reconstructions. (**a**) Proposed method; (**b**) 3D cloud point data method.

Similarly, with the same experimental conditions, we also show another sample to demonstrate the capability of our method that can focus and defocus objects at a scene. For the tested scene, we utilize a stationary coral and a swimming fish 3D animation model (constructed using Unity3D), as shown in Figure 8. For projection of images, eight images at 0.9° angle interval both in the x and y directions are projected by Unity3D to generate projected images for each hologram calculation. In Figure 8a, the camera is aimed at two objects, and both are in focus yielding clear images because distance between the coral and the fish is close. As shown in Figure 8b,c, the distance between fish and coral becomes far in comparison to Figure 8a. Figure 8b,c show different focuses, where b is focused at coral and c is focused at fish.

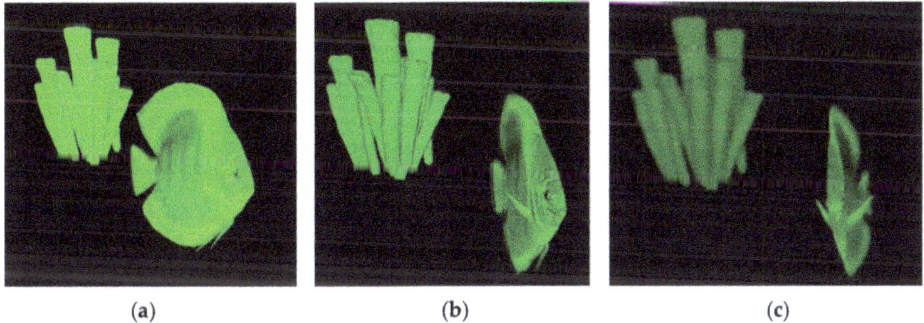

(a)   (b)   (c)

Figure 8. Optical reconstructions of second 3D animation model. (**a**) Both in focus; (**b**) Coral in focus; (**c**) Fish in focus.

The results of Figures 7 and 8 show that the proposed holographic 3D VR display is high in image quality, free of speckle noise, and also able to realize 3D display.

For the AR experiment, volume HOE is used for producing the AR 3D image due to its angle and wavelength selectivity [24–26]. The recording of the HOE is illustrated in Figure 9. The light wave emitted from the laser is filtered and collimated by $lens_1$, and then split into two parts through the beam splitter (BS). The transmitted light passes through $lens_2$ to form a convergent spherical wave, which undergoes interference with the plane wave reflected by $mirror_1$ and $mirror_2$ on the holographic material to form the HOE. The laser used for HOE recording is a single mode semiconductor laser with a wavelength of 532 nm and power of 400 mW. The $lens_2$ is a commercially available aspherical mirror with focal length of 40 mm and a diameter of 50 mm (Thorlabs, AL5040M-A, NA = 0.55). The recording material we used in the experiment is a commercial holographic film (Ultimate Holography, U04), which requires an exposure density of ~600 μJ/cm^2 [27].

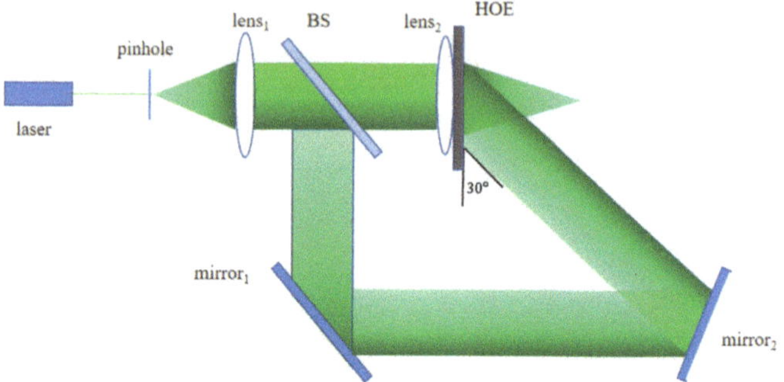

Figure 9. Optical system for holographic optical element (HOE) recording.

Using the optical recording system, the angular selectivity of the reflective volume HOE recorded without using lens$_2$ was tested, and the angular selectivity was within ±5°. Using the SLM described above, the diffraction angle is within ±1.89°, which means that most light diffracted by the SLM can be diffracted by HOE in a high efficiency.

Figure 10a,b are a schematic and a photograph of AR 3D display system, respectively. The angle between the illumination light and the HOE is approximately 30°. The holographic 3D real image reproduced by the optical system is imaged by the HOE and incident at human eye. Ambient light from the real object does not meet the angular selectivity of the fabricated reflective HOE, and thus can directly reach the human eye, forming a holographic AR 3D display.

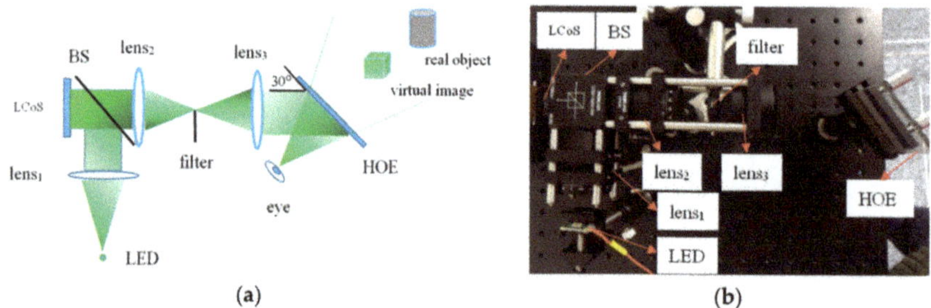

Figure 10. Holographic Augmented Reality (AR) 3D display system. (**a**) Schematic of display system; (**b**) Optical display system.

The diameter of the fabricated HOE for AR 3D display is about 40 mm, which is shown in Figure 11a. The 3D model used for the AR experiment is the same as the second experiment of the VR display. Figure 11b,c shows two reconstructed images taken at different depths. Fish and coral are about 1.2 m and 3 m deep, respectively, which are calibrated with a CCD camera. The calibration process is as follows: Given the camera is focused on a virtual image (e.g., either fish or coral), we move a real object away from the camera. When the captured image of the real object becomes clear, the distance between the real object and the camera is the depth at which the camera can clearly image the real object. The distance is also the distance between the virtual image and camera as well. In our display experiment, when the image of the fish is clear, the image of the coral and external environment are blurred due to the defocus. The results show that the proposed method is able to produce correct depth cues, and hence the accommodation convergence conflict problem can be eliminated.

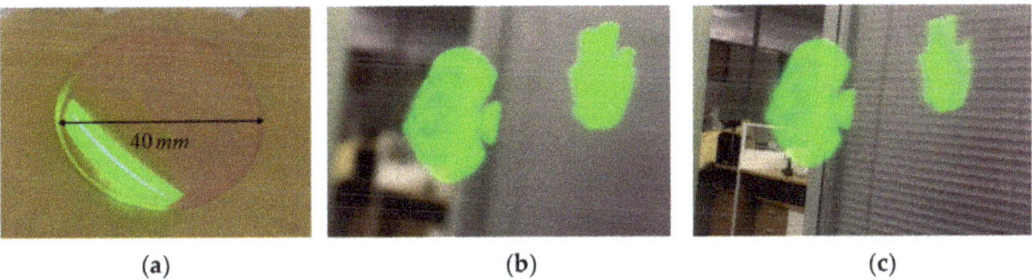

Figure 11. Optical experiment results. (**a**) Fabricated HOE; (**b**) Focused on Fish (1.2 m); (**c**) Focused on coral (3 m).

4. Discussion

The VR results of our method are faster (38 times faster) than the traditional point cloud method. The VR results also show that the proposed method is able to achieve focus and defocus of the 3D images. Similarly, AR results also show that the proposed method is able to achieve a good image quality with 3D display capability as well. More importantly, use of LED as illumination source results in speckle noise-free VR/AR 3D images so that the image quality is better than the use of coherent source as illumination for holographic display. Another advantage of this method is that the depth of the 3D image can be easily controlled through the SLM.

It is worthwhile to point out that the calculation time is determined by the number of views of the projected image and the distance from the projection plane to the holographic plane. The computation time is a trade-off between the number of the projected images and the distance between the projection plane and the holographic plane. Further increase of computational speed is still possible through the use of GPU acceleration for the convolution.

It is also worth noting that the difference between the display modes of AR and VR can be easily changed with different eyepieces, and all the underlying computation should be identical. This offers convenience for easily switching between VR and AR devices for practical applications. The 3D object can be set between the holographic plane and projected plane or the projected plane can be set between the 3D object and holographic plane. In the experiment we used the former method. For an extreme situation, the holographic plane is the projection plane. Under this condition, the convolution in spatial domain is not possible because the PSF cannot be calculated. However, the calculation of the hologram can be performed in the frequency domain [4].

It should be pointed out that use of LED as the illumination source is also associated with potential drawbacks. For example, due to the large light-emitting area of the LED and poor spatial coherence, it is more difficult to produce a large depth of field image. Another disadvantage is that when the increase of the distance between the holographic plane and the 3D object is over a certain threshold (e.g., 20 mm), the blur of the VR/AR image becomes more evident. While this may not be significant for near-eye VR/AR 3D display, a careful control of coherence of the light source as well as more controllable propagation of the light can be potentially useful to solve these problems.

5. Conclusions

We demonstrate a holographic 3D VR and AR technique with a multiple projection image-based method to successfully create high quality near-eye VR and AR 3D display free of speckle noise and without accommodation convergence conflict problems. The proposed calculation method has high computational efficiency and the proposed display system is compact and flexible, which has potentials for head mounted near-eye VR and AR applications.

Author Contributions: X.Y. wrote the program for hologram calculation and completed the relevant experiments; H.Z. helped to discuss the research and modified the document; Q.-H.W. led the project.

Funding: This research was funded by National Key R & D Program of China, under Grant No. 2017YFB1002900 and by National Natural Science Foundation of China under Grant No. 61535007.

Conflicts of Interest: The authors declare no conflict of interest.

References

1. Yaras, F.; Kang, H.; Onural, L. State of the art in holographic display: A survey. *J. Disp Technol.* **2010**, *6*, 443–454. [CrossRef]
2. Matsushima, K.; Arima, Y.; Nakahara, S. Digitized holography: Modern holography for 3D imaging of virtual and real objects. *Appl. Opt.* **2011**, *50*, H278–H284. [CrossRef] [PubMed]
3. Shi, Y.; Wang, H.; Li, Y.; Jin, H.; Ma, L. Practical method for color computer-generated rainbow holograms of real-existing objects. *Appl. Opt.* **2009**, *48*, 4219–4226. [CrossRef] [PubMed]
4. Yang, X.; Wang, H.; Li, Y.; Xu, F.; Zhang, H.; Zhang, J.; Yan, Q. Computer generated full-parallax synthetic hologram based on frequency mosaic. *Opt. Commun.* **2018**, *430*, 24–30. [CrossRef]
5. Cai, X.O.; Wang, H. Study of relationship between recording wavelength and hologram compression. *Opt. Commun.* **2006**, *256*, 111–115. [CrossRef]
6. Zhang, C.; Yang, G.L.; Xie, H.Y. Information compression of computer-generated hologram using BP Neural Network. In *Digital Holography and Three-Dimensional Imaging (DH)*; Optical Society of America: Miami, FL, USA, 2010; p. JMA2.
7. Lin, S.; Kim, E. Single SLM full-color holographic 3-D display based on sampling and selective frequency-filtering methods. *Opt. Express* **2017**, *25*, 11389–11404. [CrossRef] [PubMed]
8. Lin, S.; Cao, H.; Kim, E. Single SLM full-color holographic three-dimensional video display based on image and frequency-shift multiplexing. *Opt. Express* **2019**, *27*, 15926–15942. [CrossRef] [PubMed]
9. Chang, C.; Xia, J.; Yang, L.; Lei, W.; Yang, Z.; Chen, J. Speckle-suppressed phase-only holographic three-dimensional display based on double-constraint Gerchberg-Saxton algorithm. *Appl. Opt.* **2015**, *54*, 6994–7001. [CrossRef]
10. Jeon, W.; Jeong, W.; Son, K.; Yang, H. Speckle noise reduction for digital holographic images using multi-scale convolutional neural networks. *Opt. Lett.* **2018**, *43*, 4240–4243. [CrossRef]
11. He, Z.; Sui, X.; Jin, G.; Cao, L. Progress in virtual reality and augmented reality based on holographic display. *Appl. Opt.* **2019**, *58*, A74–A81. [CrossRef]
12. Huang, H.; Hua, H. Systematic characterization and optimization of 3D light field displays. *Opt. Express* **2017**, *25*, 18508–18525. [CrossRef] [PubMed]
13. Huang, H.; Hua, H. Effects of ray position sampling on the visual responses of 3D light field displays. *Opt. Express* **2019**, *27*, 9343–9360. [CrossRef] [PubMed]
14. Su, P.; Cao, W.; Ma, J.; Cheng, B.; Liang, X.; Cao, L.; Jin, G. Fast computer-generated hologram generation method for three-dimensional point cloud model. *J. Display Technol.* **2016**, *12*, 1688–1694. [CrossRef]
15. Wei, H.; Gong, G.; Li, N. Improved look-up table method of computer-generated holograms. *Appl. Opt.* **2016**, *55*, 9255–9264. [CrossRef]
16. Arai, D.; Shimobaba, T.; Murano, K.; Endo, Y.; Hirayama, R.; Hiyama, D.; Kakue, T.; Ito, T. Acceleration of computer-generated holograms using tilted wavefront recording plane method. *Opt. Express* **2015**, *23*, 1740–1747. [CrossRef] [PubMed]
17. Liu, J.; Liao, H. Fast occlusion processing for a polygon-based computer-generated hologram using the slice-by-slice silhouette method. *Appl. Opt.* **2018**, *57*, A215–A221. [CrossRef]
18. Ji, Y.; Yeom, H.; Park, J. Efficient texture mapping by adaptive mesh division in mesh-based computer generated hologram. *Opt. Express* **2016**, *24*, 28154–28169. [CrossRef]
19. Matsushima, K.; Nakahara, S. Extremely high-definition full-parallax computer-generated hologram created by the polygon-based method. *Appl. Opt.* **2009**, *48*, H54–H63. [CrossRef]
20. Abookasis, D.; Rosen, J. Three types of computer-generated hologram synthesized from multiple angular viewpoints of a three-dimensional scene. *Appl. Opt.* **2006**, *45*, 6533–6538. [CrossRef]
21. Shaked, N.; Katz, B.; Rosen, J. Review of three-dimensional holographic imaging by multiple-viewpoint projection-based methods. *Appl. Opt.* **2009**, *48*, H120–H136. [CrossRef]
22. Zhang, Y.; Fu, Y.; Wang, H.; Li, H.; Pan, S.; Du, Y. High resolution integral imaging display by using a microstructure array. *J. Opt. Technol.* **2019**, *86*, 100–104. [CrossRef]
23. Zhang, H.; Deng, H.; Li, J.; He, M.; Li, D.; Wang, Q. Integral imaging-based 2D/3D convertible display system by using holographic optical element and polymer dispersed liquid crystal. *Opt. Lett.* **2019**, *44*, 387–390. [CrossRef]

24. Maimone, A.; Georgiou, A.; Kollin, J. Holographic near-eye displays for virtual and augmented reality. *ACM Trans. Graph.* **2017**, *36*, 8501–8516. [CrossRef]
25. Ting, C.H.; Wokunami, K.; Yamamoto, K.; Huang, Y.P. Reconstruct holographic 3D objects by double phase hologram. *Proc. SPIE* **2015**, *9495*, 1–5.
26. Zhou, P.; Li, Y.; Liu, S.; Su, Y. Compact design for optical-see-through holographic displays employing holographic optical elements. *Opt. Express* **2018**, *26*, 22866–22876. [CrossRef] [PubMed]
27. Gentet, P.; Gentet, Y.; Lee, S. Ultimate 04 the new reference for ultra -Realistic color holography. In Proceedings of the 2017 International Conference on Emerging Trends& Innovation in ICT (ICEI), Pune, India, 3–5 February 2017; pp. 162–166.

Article

Compression of Phase-Only Holograms with JPEG Standard and Deep Learning

Shuming Jiao [1,2,3,†], Zhi Jin [1,2,†], Chenliang Chang [4], Changyuan Zhou [1,2], Wenbin Zou [1,2,*] and Xia Li [1]

[1] Shenzhen Key Lab of Advanced Telecommunication and Information Processing,
   College of Information Engineering, Shenzhen University, Shenzhen 518060, China;
   albertjiaoee@126.com (S.J.); jinzhi_126@163.com (Z.J.); zhouchangyuan@email.szu.edu.cn (C.Z.);
   lixia@cuhk.edu.cn (X.L.)

[2] Shenzhen Key Laboratory of Advanced Machine Learning and Applications, Shenzhen 518060, China

[3] Tsinghua Berkeley Shenzhen Institute (TBSI), Shenzhen 518000, China

[4] Jiangsu Key Laboratory for Opto-Electronic Technology, School of Physics and Technology,
   Nanjing Normal University, WenYuan Road 1, Nanjing 210023, China; 06245@njnu.edu.cn

* Correspondence: wzouszu@sina.com

† These authors contributed equally to this work.

Received: 22 June 2018; Accepted: 24 July 2018; Published: 30 July 2018

Abstract: It is a critical issue to reduce the enormous amount of data in the processing, storage and transmission of a hologram in digital format. In photograph compression, the JPEG standard is commonly supported by almost every system and device. It will be favorable if JPEG standard is applicable to hologram compression, with advantages of universal compatibility. However, the reconstructed image from a JPEG compressed hologram suffers from severe quality degradation since some high frequency features in the hologram will be lost during the compression process. In this work, we employ a deep convolutional neural network to reduce the artifacts in a JPEG compressed hologram. Simulation and experimental results reveal that our proposed "JPEG + deep learning" hologram compression scheme can achieve satisfactory reconstruction results for a computer-generated phase-only hologram after compression.

Keywords: hologram; holography; phase-only; compression; deep learning; JPEG, convolutional neural network

---

1. Introduction

Computer Generated Holography (CGH) allows the recording and reconstruction of a desired complex light wavefront including both amplitude and phase information. With the development of computer and optical technologies, CGH has been extensively applied in many fields such as three-dimensional dynamic holographic display [1–3], holographic projection [4–7], virtual and augmented reality [8,9], optical tweezers [10] and optical information security [11,12]. The research works on CGH generation, conversion and compression algorithms receive much attention in the past decade. For example, the enormous amount of calculation in the generation of a complex hologram from a 3D object model consisting of many points is a huge challenge and the fast CGH calculation problem has been investigated from different perspectives [13–17]. Another major concern is that holographic display devices such as spatial light modulator (SLM) usually cannot display both the amplitude and phase part of a complex hologram simultaneously. Fast and high-quality phase-only hologram calculation from an object image is favorable for commonly used phase-only type SLMs. A phase-only hologram (Figure 1) can be calculated from an object image with various methods such as Gerchberg–Saxton iterative algorithm [18–20], error diffusion algorithm [21–24], random phase mask method [25–27] and down-sampling mask method [28,29]. Compared with other methods, error diffusion method [21–24] has several advantages including no forward and backward field propagation, reduction of high frequency noise in the result and direct converting a complex hologram into a phase only one [8]. In this paper, the error diffusion method is employed for the computer generation of phase-only holograms from object images.

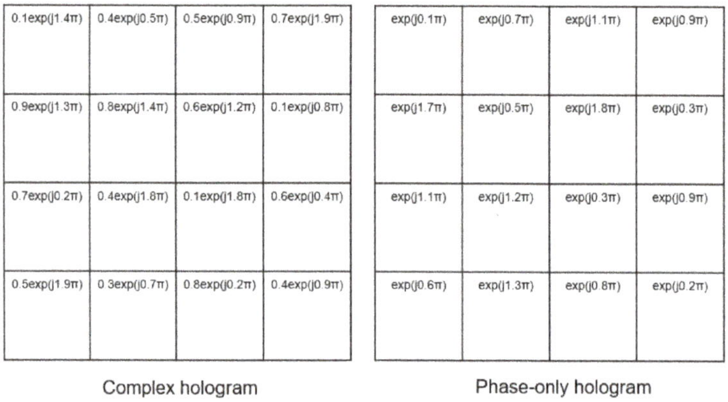

0.1exp(j1.4π)	0.4exp(j0.5π)	0.5exp(j0.9π)	0.7exp(j1.9π)		exp(j0.1π)	exp(j0.7π)	exp(j1.1π)	exp(j0.9π)
0.9exp(j1.3π)	0.8exp(j1.4π)	0.6exp(j1.2π)	0.1exp(j0.8π)		exp(j1.7π)	exp(j0.5π)	exp(j1.8π)	exp(j0.3π)
0.7exp(j0.2π)	0.4exp(j1.8π)	0.1exp(j1.8π)	0.6exp(j0.4π)		exp(j1.1π)	exp(j1.2π)	exp(j0.3π)	exp(j0.9π)
0.5exp(j1.9π)	0.3exp(j0.7π)	0.8exp(j0.2π)	0.4exp(j0.9π)		exp(j0.6π)	exp(j1.3π)	exp(j0.8π)	exp(j0.2π)

Complex hologram   Phase-only hologram

Figure 1. An example of complex hologram (**left**); and phase-only hologram (**right**) with a size of 4×4 pixels.

In addition to the numerical generation of complex or phase-only holograms, the compression of hologram data is a critical issue. Since the unit pixel size of a hologram is typically rather small (a few micrometers), the entire hologram may contain a huge number of pixels and demands considerable storage cost and transmission bandwidth. It is necessary to develop efficient hologram compression algorithms to reduce the size of data representing a hologram. In the past decade, much research effort has been made on hologram compression [30–34] and a literature review can be found in [30]. In essence, a hologram can be regarded as a special kind of two-dimensional (2D) image. For natural 2D images (e.g., photographs), standardized image compression algorithms such as JPEG [35] have been widely employed. However, some image properties of holograms significantly differ from photographs. For example, a photograph is usually locally smooth while a hologram consists of many high-frequency fringe patterns. Consequently, JPEG compression often cannot give optimal compression performance for holograms [30]. Instead of directly adopting JPEG standard, many customized compression algorithms are proposed for holograms such as Fresnelets [31], Wavelet-Bandelets Transform [32], vector quantization [33], enhanced wavelet transform [34] and rational covariance extension approach [36,37]. These customized hologram image compression methods can yield very satisfactory performance for certain types of holograms. However, on the downside, these customized algorithms are not compatible with commonly used JPEG standard, which is supported very extensively by almost every computer system and digital device while these customized compression methods are not very generally adopted. A hologram compression method is very favorable if it has compatibility with JPEG standard and can achieve good compression efficiency at the same time. In this work, we propose a phase-only hologram compression and decompression scheme incorporating both JPEG standard and deep learning post-processing. In this scheme, the compression steps are identical to JPEG compression standard and the decompression steps are implemented with JPEG decompression plus enhancement by a deep convolutional network.

This article is organized as follows. In Section 2, the error diffusion method for phase-only hologram generation is briefly described. In Section 3, the working principles of JPEG image compression standard are discussed and our proposed artifact reduction scheme for JPEG compressed holograms using deep convolutional network is presented. In Section 4, simulation and experimental results are demonstrated to verify the effectiveness of our proposed scheme. In Section 5, a brief conclusion is provided.

2. Computer Generated Phase-Only Hologram with Error Diffusion Method

Error diffusion algorithm [21–24] is a non-iterative and high-quality method for calculating a phase-only hologram from an object image on computer. The working principles of phase-only hologram calculation with error diffusion method are described below. To start with, $O(x,y)$ denotes the object image and a complex Fresnel hologram $H(x,y)$ can be calculated based on the Fresnel diffraction formulas illustrated by Equations (1) and (2).

$$f(x,y;z) = \frac{exp(i\frac{2\pi z}{\lambda})}{i\lambda z} exp\left[i\frac{\pi\left(x^2+y^2\right)}{\lambda z}\right] \tag{1}$$

$$H(x,y) = O(x,y) * f(x,y;z) \tag{2}$$

where $f(x,y;z)$ denotes the impulse function for Fresnel transform, λ denotes the wavelength of illumination light and z denotes the distance between the object image plane and hologram plane. As shown in Figure 2, a Fresnel zone plate can be generated at the hologram plane for each object point with a Fresnel impulse function. The final calculated hologram is a superposition of the Fresnel zone plates from all the points (or pixels) in the object image. More detailed explanation about the Fresnel hologram calculation can be found in [13–17].

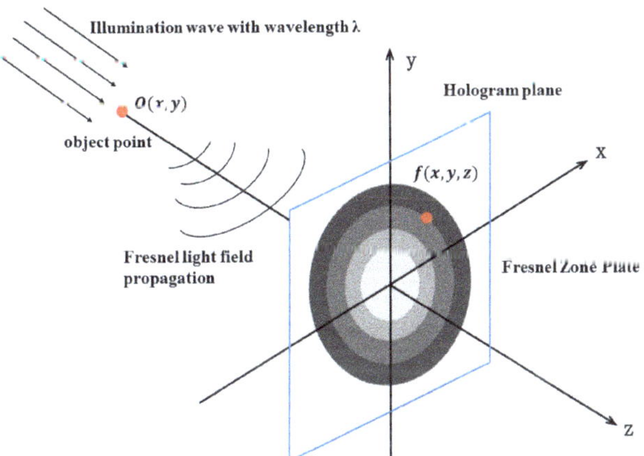

Figure 2. Fresnel light filed propagation from one object point.

The complex hologram $H(x,y)$ can be converted to a phase-only hologram $P(x,y)$ with error diffusion algorithm [21–24] to achieve high-quality holographic display on a phase-only spatial light modulator. In the error diffusion method, the complex value of each hologram pixel is forced to be unity amplitude and the resulting complex error is diffused to neighboring pixel values. The following operations (Equations (3)–(8)) are performed on each individual holographic pixel sequentially in row-by-row and column-by-column scanning manner, as illustrated in Figure 3.

$$E\left(x_j,y_j\right) = H\left(x_j,y_j\right) - P\left(x_j,y_j\right) \tag{3}$$

$$H\left(x_j,y_j+1\right) \leftarrow H\left(x_j,y_j+1\right) + w_1 E\left(x_j,y_j\right) \tag{4}$$

$$H\left(x_j+1, y_j-1\right) \leftarrow H\left(x_j+1, y_j-1\right) + w_2 E\left(x_j, y_j\right) \tag{5}$$

$$H\left(x_j+1, y_j\right) \leftarrow H\left(x_j+1, y_j\right) + w_3 E\left(x_j, y_j\right) \tag{6}$$

$$H\left(x_j+1, y_j+1\right) \leftarrow H\left(x_j+1, y_j+1\right) + w_4 E\left(x_j, y_j\right) \tag{7}$$

$$P\left(x_j, y_j\right) = Phase\left[H\left(x_j, y_j\right)\right] \tag{8}$$

where $P(x_j, y_j)$ denotes the phase-only pixel value at position (x_j, y_j) after the complex pixel value $H(x_j, y_j)$ is phase truncated (the magnitude is forced to be unity), $E(x_j, y_j)$ denotes the complex error and four different weighting coefficients, $w_1 = 7/16$, $w_2 = 3/16$, $w_3 = 5/16$ and $w_4 = 1/16$, are imposed on four different directions, as shown in Figure 3.

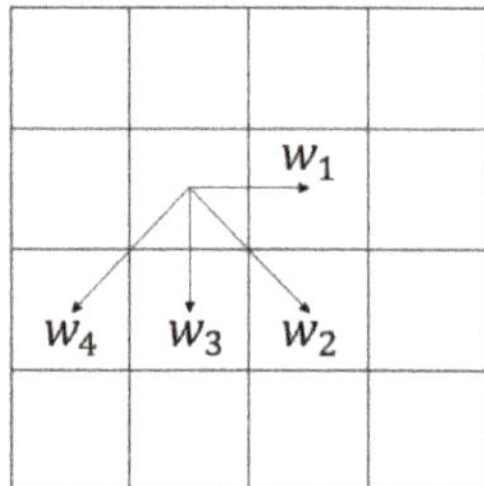

Figure 3. Propagation of errors to four different neighboring pixels with corresponding weighting coefficients in error diffusion algorithm.

After each pixel in a complex hologram is processed, the entire hologram will become a phase-only type hologram. More details about the error diffusion methods for phase-only hologram generation can be found in [21–24]. In practice, bi-directional error diffusion, a slightly modified version of the above unidirectional error diffusion, can yield slightly better performance [21].

3. JPEG Image Compression and Proposed Artifact Reduction Scheme by Deep Convolutional Network

Each pixel in a phase-only hologram has an intensity value ranging from 0 to 2π and a phase-only hologram can be regarded as a gray-scale intensity image. Various image compression algorithms can be attempted for the compression of phase-only holograms.

JPEG is a very commonly used lossy compression scheme for digital images, which was firstly proposed in 1992 by Joint Photographic Experts Group [35]. In JPEG scheme, the original image (e.g., a gray-level photograph or a phase-only hologram) is first divided into 8×8 pixel blocks and each individual block undergoes discrete cosine transform (DCT). Then, the coefficients in the transformed domain of each block are quantized with more quantization levels for low frequency components and less quantization levels for high frequency components. Subsequently, zigzag coding, entropy coding and Huffman coding are performed on the quantized coefficients. Finally, the original image becomes a compressed bit stream with much smaller data size than the original uncompressed image. A reconstructed image with certain image degradation can be obtained when the JPEG binary bit stream is decompressed with the inverse procedures as compression.

JPEG compression can significantly reduce the data size of an original natural image and only introduce minor quality degradation in the decompressed image. The reason is that a natural image (e.g., a photograph) is usually locally smooth and most of its information is concentrated in the low frequency coefficients of DCT spectrum. The discarded high frequency information in the quantization step has negligible effect on the image quality. However, phase-only holograms have substantially different image characteristics compared with photographs [38]. One feature of a hologram is that it contains many high frequency fringe patterns. If JPEG compression is directly applied to a phase-only hologram, the decompressed result after compression will suffer from heavy damage and the holographically reconstructed image from the JPEG compressed hologram will be severely degraded as well.

In this work, we employed deep convolutional neural networks to reduce the artifacts of JPEG compressed phase-only holograms. In the past few years, deep learning methods receive much attention and exert considerable impact in many fields such as image quality enhancement [39–42]. A previous work [39] proposed a 20-layer network DnCNN to reduce the Gaussian noise on the noisy images. Although DnCNN is designed for denoising task, it shows promising performance on image compression artifacts reduction and super-resolution as well. Assisted by the prior knowledge, a previous work [40,42] proposed denoiser-based and total variation (TV)-based solutions, respectively for image restoration task. Very recently, deep learning has been introduced to the holographic research area and gained success in different applications [43–49].

The network structure employed in our task is illustrated in Figure 4a, based on the previous works of restoring JPEG compressed photographs [50]. The overall flowchart of our proposed "JPEG + deep learning" hologram compression scheme is illustrated in Figure 4b. The proposed CNN consists of four convolutional layers, and each layer has a specific function. The first layer is used for feature extraction, and with 9×9 filter size it generates 64 feature maps. By adding a bias term to the outputs and employing ReLU as the activation function, the convolution operation in this layer can be formulated as:

$$F_1(\mathbf{I}) = ReLU(W_1 * \mathbf{I} + B_1) \tag{9}$$

where W and B represent the weights and biases, respectively; "$*$" denotes the convolution operation; and F represents the nonlinear mapping process. Different from the previous approach [51] where zero-padding is not adopted, in the proposed network, all the filters whose size is larger than 1 adopt the zero-padding strategy to maintain the size of the reconstructed images unchanged.

Since the features extracted by the first layer are in low quality, the second convolutional layer is used for feature enhancement, and, with 7×7 filter size, it generates 32 feature maps. Because more nonlinear mappings are involved, more potential features can be extracted, and the third layer is used for nonlinear mapping. With 1×1 filter size, it generates 16 feature maps. The last layer is used for image reconstruction and the corresponding filter size is 5×5. After the first three convolutional layers, features in the inputs are successfully extracted and enhanced by optimizing the output of the network to generate the final recovered image. The operation of these layers can be formulated as:

$$F_i(\mathbf{I}) = ReLU(W_i * F_{i-1}(\mathbf{I}) + B_i) \tag{10}$$

where i indicates the ith convolutional layer; $F_i(\mathbf{I})$ and $F_{i-1}(\mathbf{I})$ are the outputs of the ith and $(i-1)$th convolutional layer, respectively; and W_i and B_i represent the weights and biases for the ith layer, respectively. Since there is no activation function after the last conventional layer, the final restored images can be expressed as

$$F_4(\mathbf{I}) = W_4 * F_3(\mathbf{I}) + B_4 \tag{11}$$

In addition, aiming to learn the optimal values for the weight and bias of each layer, the L2 distance between the reconstructed hologram images $F(\mathbf{I}; \Theta)$ and corresponding uncompressed ground truth hologram images $\mathbf{I}_{GT}$ needs to be minimized. Given a set of m uncompressed images $\mathbf{I}_{GT}^i$, the corresponding compressed images $\mathbf{I}^i$, the Mean Squared Error (MSE) loss function is:

$$Loss(\Theta) = \frac{1}{m} \sum_{i=1}^{m} \|F(\mathbf{I}^i; \Theta) - \mathbf{I}_{GT}^i\|_2^2 \tag{12}$$

where Θ contains the parameters of the network, including both weights and biases. The network weighting parameters are trained separately for phase-only holograms generated at different distances.

Hence, the artifacts caused by JPEG compression in phase-only holograms such as high frequency noise and blocking effect can be suppressed by the proposed convolutional network.

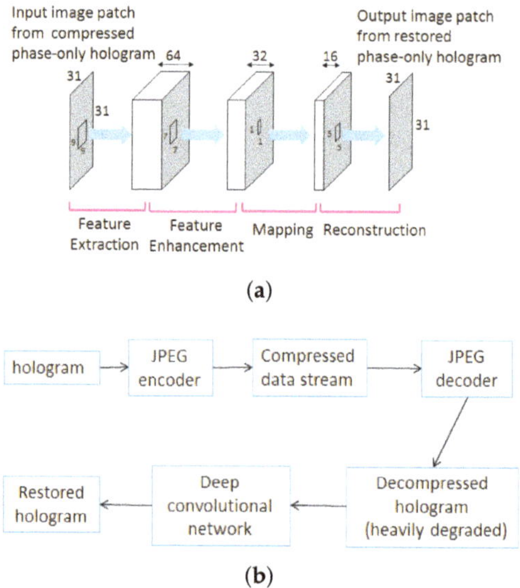

Figure 4. (**a**) Structure of deep convolutional network for the quality enhancement of JPEG compressed holograms in our work; and (**b**) overall flowchart of our proposed "JPEG+deep learning" hologram compression scheme.

4. Results and Discussion

In our work, ten pairs of compressed and uncompressed phase-only holograms (1024 × 1024 pixel size) calculated from ten different object images (512 × 512 pixel size) with error diffusion method were employed as the training holograms (Figure 5). Two compressed phase-only holograms generated from "Pepper" and "Cameraman" images, respectively, were employed as the test holograms (Figure 6). Data augmentation was performed: firstly, we rotated the training images by 90°, 180° and 270°, respectively; and, secondly, we flipped them horizontally. Taking into account the training complexity, the small patch training strategy was adopted, so that training images were split into 31 × 31 patches with the stride of 10 and there were 960, 384 training sub-images in total. Since different holograms usually have many similar image blocks and fringe patterns, the replication properties in hologram images allowed a small number of examples to contain most universal hologram features, which has been revealed in previous works [52]. Hence, the total training sub-images was sufficient for our proposed four-layer network. In this work, stochastic gradient descent backpropagation solver was used with the batch size 128. The initial learning rate was 0.0001 and decayed every five epochs by a factor of 10. There were 25 epochs in total. We used the deep learning platform caffe [53] on an NVIDIA GTX TITAN X GPU with 3072 CUDA cores and 12 GB of RAM to implement all the operations in our network. The training time of a four-layer network was 5 h and the test time for an image with 512 × 768 was around 0.32 s.

In the computer calculation of phase-only holograms, the optical wavelength of illumination light was set to be 532 nm, the pixel size was 8 μm and the distance between object image and hologram plane was 0.3 m or 0.5 m. The ten training holograms and two testing holograms were then compressed by JPEG when the quality factor is set to 1. The JPEG compressed hologram was restored by our proposed deep convolutional network. One example of the original hologram, JPEG compressed hologram and restored hologram with our proposed scheme is shown in Figure 7. It can be observed that some high frequency patterns are deteriorated in the hologram after JPEG compression and restored to certain extent after the quality enhancement using our proposed network.

Figure 5. Ten object images employed for generating training holograms.

Figure 6. Two object images ("Peppe" and "Cameraman") employed for generating testing holograms.

The reconstructed images from original uncompressed holograms, JPEG compressed holograms and restored holograms were compared in numerical simulations. The reconstructed images are shown in Figure 8. The PSNR, SSIM [54], multi-scale SSIM (MS-SSIM) [55], visual information fidelity (VIF) [56] and information fidelity criterion (IFC) [57] values of the reconstructed results from compressed holograms and restored holograms, in comparison with the reconstructed results from uncompressed holograms, are presented in Table 1. The compression ratio is approximately 7 and data size of original hologram, 1024 KB (1 MB), is reduced to around 142 KB to 144 KB after JPEG compression. All these assessment values reveal that our proposed scheme can significantly enhance the reconstructed image quality from JPEG compressed holograms.

Table 1. PSNR (dB), SSIM, MS-SSIM, VIF and IFC values of reconstructed results from JPEG compressed holograms and restored holograms with our proposed scheme.

		Cameraman		Pepper	
		0.5 m	0.3 m	0.5 m	0.3 m
Compression ratio		7.2113	7.1111	7.2113	7.1608
Reconstructed image from compressed hologram	PSNR	19.10	17.83	18.92	17.64
	SSIM	0.1651	0.0967	0.2007	0.1036
	MS-SSIM	0.6091	0.4628	0.6907	0.5252
	VIF	0.3946	0.2183	0.6396	0.3430
	IFC	0.4522	0.2519	0.5543	0.2835
Reconstructed image from restored hologram	PSNR	28.86	26.86	29.88	27.16
	SSIM	0.6036	0.4465	0.6767	0.4852
	MS-SSIM	0.8798	0.8022	0.9138	0.8343
	VIF	0.5378	0.4316	0.6841	0.6306
	IFC	0.9027	0.5098	1.2064	0.6379

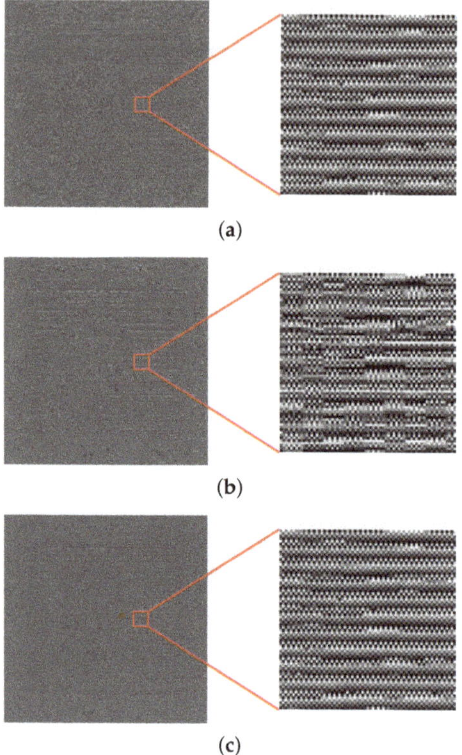

Figure 7. (**a**) Original hologram; (**b**) JPEG compressed hologram; and (**c**) restored hologram with our proposed scheme.

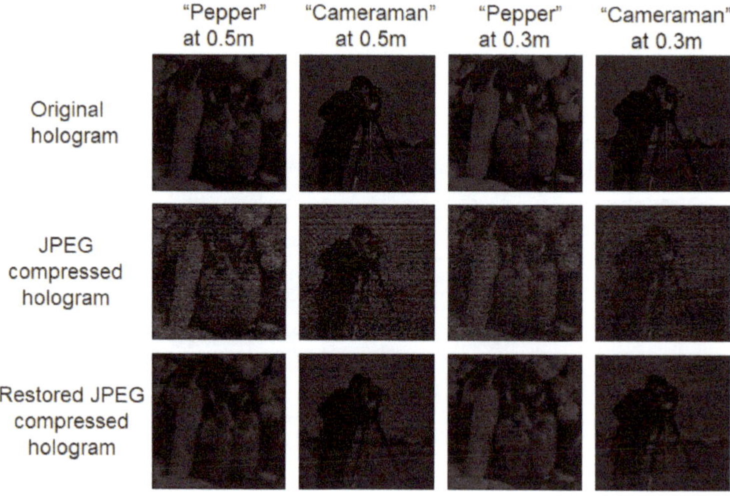

Figure 8. Reconstructed images from original holograms, JPEG compressed holograms and restored holograms in numerical simulation.

Our proposed scheme was further verified in optical experiments and the calculated holograms were optically reconstructed using the optical setup [6] shown in Figure 9. The hologram was reconstructed with a phase-only Holoeye PLUTO spatial light modulator (SLM). The system parameters were the same as the numerical simulation (wavelength: 532 nm; pixel size: 8 μm; hologram size: 1024 × 1024 pixels; object-hologram distance: 0.3 m/0.5 m). The phase-only holograms loaded into the SLM were assigned with an additional carrier phase to transversally shift the desired reconstructions away from the un-diffracted (zero-order) beam component. The projected beam from the SLM is transmitted through a 4-f optical filtering system containing a low-pass iris with 3.3 mm diameter to block the zero-order noise. The optically reconstructed results are shown in Figure 10. It can be observed that the quality of reconstructed images after hologram enhancement with deep learning are improved, compared with the ones from compressed holograms without restoration.

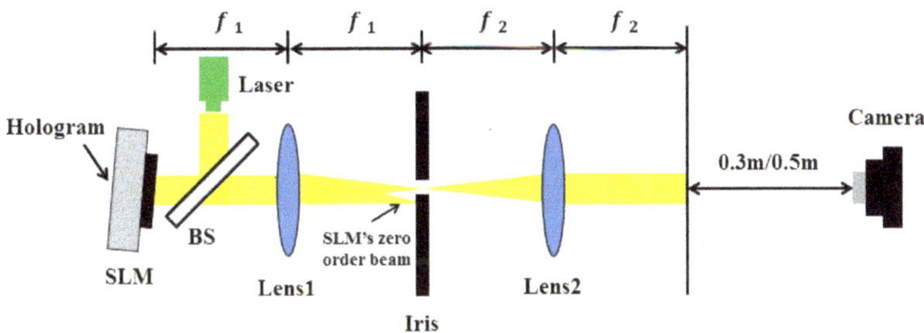

Figure 9. Optical setup for hologram reconstruction experiment.

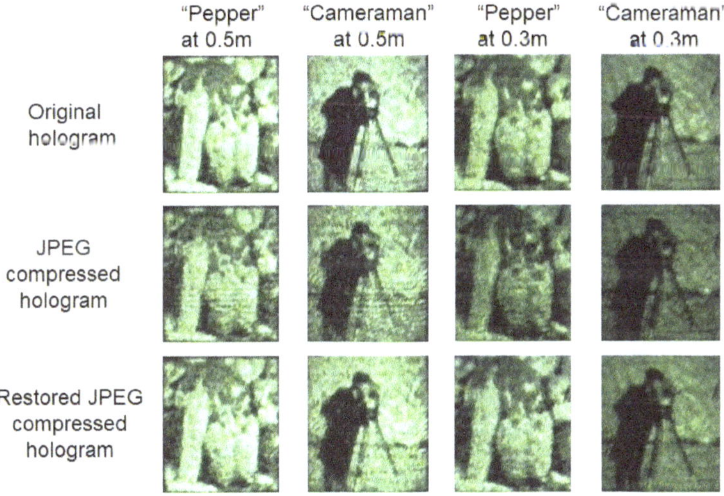

Figure 10. Reconstructed images from original holograms, JPEG compressed holograms and restored holograms in optical experiments.

5. Conclusions

It is essential to reduce the enormous amount of data representing a computer-generated phase-only hologram in the processing, transmission and storage process. JPEG photograph compression standard can be attempted for hologram compression with an advantage of universal compatibility, compared with customized hologram compression algorithms. Deep convolutional networks can be employed to reduce the artifacts for a JPEG compressed hologram. Simulation and experimental results reveal that our proposed "JPEG + deep learning" hologram compression scheme can maintain the reconstructed image quality when the data size of original hologram is significantly reduced.

Author Contributions: Conceptualization, S.J.; Funding acquisition, W.Z. and X.L.; Investigation, S.J., Z.J. and C.C.; Methodology, S.J. and Z.J.; Resources, C.C.; Software, S.J. and Z.J.; Writing—original draft, S.J.; Writing—review and editing, S.J., Z.J., C.C. and C.Z.

Funding: This work was supported by the NSFC Project under Grants 61771321, 61701313 and 61472257; Chinese Postdoctoral Science Foundation Grants 2017M622763 and 2017M622778; and Pearl River Talent Plan of Guangdong Province (Postdoctoral Scheme, 2016); in part by the Natural Science Foundation of Shenzhen under Grants KQJSCX20170327151357330, JCYJ20170818091621856, and JSGG20170822153717702; and in part by the Interdisciplinary Innovation Team of Shenzhen University.

Acknowledgments: We appreciate the help from Tomoyoshi Shimobaba and Yuki Nagahama in Chiba University, Chiba, Japan.

Conflicts of Interest: The authors declare no conflict of interest.

References

1. Yaraş, F.; Kang, H.; Onural, L. Real-time phase-only color holographic video display system using LED illumination. *Appl. Opt.* **2009**, *48*, 48–53. [CrossRef] [PubMed]
2. Gao, H.; Li, X.; He, Z.; Su, Y.; Poon, T.C. Real-time dynamic holographic 3D display. *Inf. Disp.* **2012**, *28*, 17–20.
3. Li, X.; Wang, Y.; Liu, J.; Jia, J.; Pan, Y.; Xie, J. Color holographic display using a phase-only spatial light modulator. In *Digital Holography and Three-Dimensional Imaging*; Optical Society of America: San Jose, CA, USA, 2013.
4. Buckley, E. Holographic laser projection. *J. Disp. Technol.* **2011**, *7*, 135–140. [CrossRef]
5. Qu, W.; Gu, H.; Tan, Q. Holographic projection with higher image quality. *Opt. Express* **2016**, *24*, 19179–19184. [CrossRef] [PubMed]
6. Qi, Y.; Chang, C.; Xia, J. Speckleless holographic display by complex modulation based on double-phase method. *Opt. Express* **2016**, *24*, 30368–30378. [CrossRef] [PubMed]
7. Shimobaba, T.; Makowski, M.; Kakue, T.; Oikawa, M.; Okada, N.; Endo, Y.; Hirayama, R.; Ito, T. Lensless zoomable holographic projection using scaled Fresnel diffraction. *Opt. Express* **2013**, *21*, 25285–25290. [CrossRef] [PubMed]
8. Maimone, A.; Georgiou, A.; Kollin, J.S. Holographic near-eye displays for virtual and augmented reality. *ACM Trans. Graph.* **2017**, *36*, 85. [CrossRef]
9. Gao, Q.; Liu, J.; Duan, X.; Zhao, T.; Li, X.; Liu, P. Compact see-through 3D head-mounted display based on wavefront modulation with holographic grating filter. *Opt. Express* **2017**, *25*, 8412–8424. [CrossRef] [PubMed]
10. Liesener, J.; Reicherter, M.; Haist, T.; Tiziani, H.J. Multi-functional optical tweezers using computer-generated holograms. *Opt. Commun.* **2000**, *185*, 77–82. [CrossRef]
11. Wang, Y.Y.; Wang, Y.R.; Wang, Y.; Li, H.J.; Sun, W.J. Optical image encryption based on binary Fourier transform computer-generated hologram and pixel scrambling technology. *Opt. Lasers Eng.* **2007**, *45*, 761–765. [CrossRef]
12. Zhuang, Z.; Jiao, S.; Zou, W.; Li, X. Embedding intensity image into a binary hologram with strong noise resistant capability. *Opt. Commun.* **2017**, *403*, 245–251. [CrossRef]
13. Nishitsuji, T.; Shimobaba, T.; Kakue, T.; Ito, T. Review of Fast Calculation Techniques for Computer-generated Holograms with the Point Light Source-based Model. *IEEE Trans. Ind. Inf.* **2017**, *13*, 2447–2454. [CrossRef]
14. Shimobaba, T.; Masuda, N.; Ito, T. Simple and fast calculation algorithm for computer-generated hologram with wavefront recording plane. *Opt. Lett.* **2009**, *34*, 3133–3135. [CrossRef] [PubMed]
15. Shimobaba, T.; Ito, T.; Masuda, N.; Ichihashi, Y.; Takada, N. Fast calculation of computer-generated-hologram on AMD HD5000 series GPU and OpenCL. *Opt. Express* **2010**, *18*, 9955–9960. [CrossRef] [PubMed]
16. Jia, J.; Wang, Y.; Liu, J.; Li, X.; Pan, Y.; Sun, Z.; Zhang, B.; Zhao, Q.; Jiang, W. Reducing the memory usage for effectivecomputer-generated hologram calculation using compressed look-up table in full-color holographic display. *Appl. Opt.* **2013**, *52*, 1404–1412. [CrossRef] [PubMed]

17. Jiao, S.; Zhuang, Z.; Zou, W. Fast computer generated hologram calculation with a mini look-up table incorporated with radial symmetric interpolation. *Opt. Express* **2017**, *25*, 112–123. [CrossRef] [PubMed]

18. Gerchberg, R.W.; Saxton, W.O. A practical algorithm for the determination of the phase from image and diffraction plane pictures. *Optik* **1972**, *35*, 237–246.

19. Chang, C.; Xia, J.; Yang, L.; Lei, W.; Yang, Z.; Chen, J. Speckle-suppressed phase-only holographic three-dimensional display based on double-constraint Gerchberg—Saxton algorithm. *Appl. Opt.* **2015**, *54*, 6994–7001. [CrossRef] [PubMed]

20. Chen, C.Y.; Li, W.C.; Chang, H.T.; Chuang, C.H.; Chang, T.J. 3-D modified Gerchberg—Saxton algorithm developed for panoramic computer-generated phase-only holographic display. *JOSA B* **2017**, *34*, 42–48. [CrossRef]

21. Tsang, P.W.M.; Poon, T.C. Novel method for converting digital Fresnel hologram to phase-only hologram based on bidirectional error diffusion. *Opt. Express* **2013**, *21*, 23680–23686. [CrossRef] [PubMed]

22. Tsang, P.W.M.; Jiao, A.S.M.; Poon, T.C. Fast conversion of digital Fresnel hologram to phase-only hologram based on localized error diffusion and redistribution. *Opt. Express* **2014**, *22*, 5060–5066. [CrossRef] [PubMed]

23. Liu, J.P.; Wang, S.Y.; Tsang, P.W.M.; Poon, T.C. Nonlinearity compensation and complex-to-phase conversion of complex incoherent digital holograms for optical reconstruction. *Opt. Express* **2016**, *24*, 14582–14588. [CrossRef] [PubMed]

24. Pang, H.; Wang, J.; Zhang, M.; Cao, A.; Shi, L.; Deng, Q. Non-iterative phase-only Fourier hologram generation with high image quality. *Opt. Express* **2017**, *25*, 14323–14333. [CrossRef] [PubMed]

25. Naydenova, I. *Advanced Holography: Metrology and Imaging*; InTech: London, UK, 2011.

26. Tsang, P.W.M.; Chow, Y.T.; Poon, T.C. Generation of edge-preserved noise-added phase-only hologram. *Chin. Opt. Lett.* **2016**, *14*, 100901. [CrossRef]

27. Tsang, P.W.M.; Chow, Y.T.; Poon, T.C. Generation of patterned-phase-only holograms (PPOHs). *Opt. Express* **2017**, *25*, 9088–9093. [CrossRef] [PubMed]

28. Tsang, P.W.M.; Chow, Y.T.; Poon, T.C. Generation of phase-only Fresnel hologram based on down-sampling. *Opt. Express* **2014**, *22*, 25208–25214. [CrossRef] [PubMed]

29. Cheung, W.K.; Tsang, P.W.M.; Poon, T.C.; Zhou, C. Enhanced method for the generation of binary Fresnel holograms based on grid-cross downsampling. *Chin. Opt. Lett.* **2011**, *9*, 120005. [CrossRef]

30. Dufaux, F.; Xing, Y.; Pesquet-Popescu, B.; Schelkens, P. Compression of digital holographic data: An overview. In *Applications of Digital Image Processing*; International Society for Optics and Photonics: San Diego, CA, USA, 2015.

31. Darakis, E.; Soraghan, J.J. Use of Fresnelets for phase shifting digital hologram compression. *IEEE Trans. Image Process.* **2006**, *15*, 3804–3811. [CrossRef] [PubMed]

32. Bang, L.T.; Ali, Z.; Quang, P.D.; Park, J.H.; Kim, N. Compression of digital hologram for three-dimensional object using Wavelet-Bandelets transform. *Opt. Express* **2011**, *19*, 8019–8031. [CrossRef] [PubMed]

33. Tsang, P.; Cheung, K.W.; Poon, T.C. Low-bit-rate computer-generated color Fresnel holography with compression ratio of over 1600 times using vector quantization. *Appl. Opt.* **2011**, *50*, 42–49. [CrossRef] [PubMed]

34. Cheremkhin, P.A.; Kurbatova, E.A. Compression of digital holograms using 1-level wavelet transforms, thresholding and quantization of wavelet coefficients. In *Digital Holography and Three-Dimensional Imaging*; Optical Society of America: JeJu Island, Korea, 2017.

35. Wallace, G.K. The JPEG still picture compression standard. *Commun. ACM* **1992**, *34*, 30–44. [CrossRef]

36. Ringh, A.; Karlsson, J.; Lindquist, A. Multidimensional Rational Covariance Extension with Applications to Spectral Estimation and Image Compression. *SIAM J. Control Opt.* **2018**, *54*, 1950–1982. [CrossRef]

37. Zorzi, M. An interpretation of the dual problem of the THREE-like approaches. *Automatica* **2015**, *62*, 87–92. [CrossRef]

38. Jiao, S.; Zou, W. Processing of digital holograms: segmentation and inpainting. In *Holography, Diffractive Optics, and Applications VII*; International Society for Optics and Photonics: Beijing, China, 2016; Volume 10022, p. 1002206.

39. Zhang, K.; Zuo, W.; Chen, Y.; Meng, D.; Zhang, L. Beyond a gaussian denoiser: Residual learning of deep cnn for image denoising. *IEEE Trans. Image Process.* **2017**, *26*, 3142–3155. [CrossRef] [PubMed]

40. Zhang, K.; Zuo, W.; Gu, S.; Zhang, L. Learning deep cnn denoiser prior for image restoration. In Proceedings of the IEEE Conference on Computer Vision and Pattern Recognition (CVPR), Honolulu, HI, USA, 21–26 July 2017.

41. Yin, J.-L.; Chen, B.-H.; Li, Y. Highly Accurate Image Reconstruction for Multimodal Noise Suppression Using Semisupervised Learning on Big Data. In *IEEE Transactions on Multimedia*; IEEE: Piscataway, NJ, USA, 2018.

42. Yuan, G.; Ghanem, B. L0TV: A new method for image restoration in the presence of impulse noise. In Proceedings of the IEEE Conference on Computer Vision and Pattern Recognition (CVPR), Boston, MA, USA, 7–12 June 2015.

43. Nguyen, T.; Bui, V.; Lam, V.; Raub, C.B.; Chang, L.C.; Nehmetallah, G. Automatic phase aberration compensation for digital holographic microscopy based on deep learning background detection. *Opt. Express* **2017**, *25*, 15043–15057. [CrossRef] [PubMed]

44. Pitkaaho, T.; Manninen, A.; Naughton, T.J. Performance of Autofocus Capability of Deep Convolutional Neural Networks in Digital Holographic Microscopy. In *Digital Holography and Three-Dimensional Imaging*; Optical Society of America: JeJu Island, Korea, 2017.

45. Shimobaba, T.; Kuwata, N.; Homma, M.; Takahashi, T.; Nagahama, Y.; Sano, M.; Hasegawa, S.; Hirayama, R.; Kakue, T.; Shiraki, A.; et al. Convolutional neural network-based data page classification for holographic memory. *Appl. Opt.* **2017**, *56*, 7327–7330. [CrossRef] [PubMed]

46. Sinha, A.; Lee, J.; Li, S.; Barbastathis, G. Lensless computational imaging through deep learning. *Optica* **2017**, *4*, 1117–11253. [CrossRef]

47. Rivenson, Y.; Zhang, Y.; Günaydın, H.; Teng, D.; Ozcan, A. Phase recovery and holographic image reconstruction using deep learning in neural networks. *Light Sci. Appl.* **2018**, *7*, 17141. [CrossRef]

48. Ren, Z.; Xu, Z.; Lam, E.Y. Learning-based nonparametric autofocusing for digital holography. *Optica* **2018**, *5*, 337–344. [CrossRef]

49. Horisaki, R.; Takagi, R.; Tanida, J. Deep-learning-generated holography. *Appl. Opt.* **2018**, *57*, 3859–3863. [CrossRef] [PubMed]

50. Dong, C.; Deng, Y.; Loy, C.C.; Tang, X. Compression artifacts reduction by a deep convolutional network. In Proceedings of the IEEE International Conference on Computer Vision (ICCV 2015), Santiago, Chile, 11–18 December 2015; pp. 576–584.

51. Yu, K.; Dong, C.; Loy, C.C.; Tang, X. Deep convolution networks for compression artifacts reduction. *arXiv* **2016**, arXiv:1608.02778.

52. Jiao, A.S.M.; Tsang, P.W.M.; Poon, T.C. Restoration of digital off-axis Fresnel hologram by exemplar and search based image inpainting with enhanced computing speed. *Comput. Phys. Commun.* **2015**, *193*, 30–37. [CrossRef]

53. Jia, Y.; Shelhamer, E.; Donahue, J.; Karayev, S.; Long, J.; Girshick, R.; Guadarrama, S.; Darrell, T. Caffe: Convolutional architecture for fast feature embedding. In Proceedings of the 22nd ACM international conference on Multimedia, Orlando, FL, USA, 3–7 November 2014; pp. 675–678.

54. Wang, Z.; Bovik, A.C.; Sheikh, H.R.; Simoncelli, E.P. Image quality assessment: From error visibility to structural similarity. *IEEE Trans. Image Process.* **2004**, *13*, 600–612. [CrossRef] [PubMed]

55. Wang, Z.; Bovik, A.C. Mean squared error: Love it or leave it? A new look at signal fidelity measures. *IEEE Signal Process. Mag.* **2009**, *26*, 98–117. [CrossRef]

56. Sheikh, H.R.; Bovik, A.C. Image information and visual quality. *IEEE Trans. Image Process.* **2006**, *15*, 430–444. [CrossRef] [PubMed]

57. Sheikh, H.R.; Bovik, A.C.; de Veciana, G. An information fidelity criterion for image quality assessment using natural scene statistics. *IEEE Trans. Image Process.* **2005**, *14*, 2117–2128. [CrossRef] [PubMed]

applied
sciences

Article

High Resolution Computer-Generated Rainbow Hologram

Takeshi Yamaguchi *,† **and Hiroshi Yoshikawa** †

Department of Computer Engineering, College of Science and Technology, Nihon University,
7-24-1 Narashinodai, Funabashi-shi, Chiba 274-8501, Japan; yoshikawa.hiroshi@nihon-u.ac.jp
* Correspondence: yamaguchi.takeshi89@nihon-u.ac.jp; Tel.: +81-47-469-5391
† These authors contributed equally to this work.

Received: 10 September 2018; Accepted: 13 October 2018; Published: 17 October 2018

Abstract: We have developed an output device for a computer-generated hologram (CGH) named a fringe printer, which can output a 0.35-µm plane-type hologram. We also proposed several CGH with a fringe printer. A computer-generated rainbow hologram (CGRH), which can reconstruct a full color 3D image, is one of our proposed CGH. The resolution of CGRH becomes huge (over 50 Gpixels) due to improvement of the fringe printer. In the calculation, it is difficult to calculate the whole fringe pattern of CGRH at the same time by a general PC. Furthermore, since the fine pixel pitch provides a wide viewing angle in CGRH, object data, which are used in fringe calculation, should be created from many viewpoints to provide a proper hidden surface removal process. The fringe pattern of CGRH is calculated in each horizontal block. Therefore, the object data from several view points should be organized for efficient computation. This paper describes the calculation algorithm for huge resolution CGRH and its output results.

Keywords: holography; computer-generated hologram; rainbow hologram; high resolution; fringe printer; wide viewing angle

---

1. Introduction

A hologram has 3D information such as the binocular parallax, convergence, accommodation, and so on. Therefore, a reconstructed image of a hologram provides natural spatial effects. A computer-generated hologram (CGH) whose interference fringes are calculated on a computer can reconstruct a virtual object. At the moment, since the hologram requires a fine pixel pitch that is almost the same as the wavelength of the light, the output device of CGH is not common. The electron beam writer provides an excellent quality CGH [1]. Furthermore, there were some papers whose CGHs were output by a laser lithography system [2]. However, both the equipment and running costs of the electron beam writer are very expensive. On the other hand, Sakamoto et al. had proposed a CGH printer with a CD-R writer [3]. Since the recording material is CD-R, the running cost of this CD-R printer is very inexpensive. However, the size of this CGH is restricted by the size of the CD-R.

We have developed the output device of the plane-type CGH, which is called the fringe printer [4]. This printer consists of a laser, an SLM and an X-Y translation stage with stepper motors. The pixel pitch of this printer is 0.35 µm, and the size of CGH depends on the X-Y translation stage. Our group has investigated several types of CGHs [5,6]. The Fresnel-type CGH is very popular in CGH [7–9]. However, all light is diffracted by the Fresnel fringe pattern, and it is difficult to reconstruct the full-color 3D image. Therefore, we proposed a computer-generated rainbow hologram (CGRH) [10]. The rainbow hologram, which was proposed by Benton [11], can reconstruct the full-color 3D image by sacrificing the vertical parallax. Due to the development of the fringe printer, the resolution of CGRH increases to over 50 Gpixels. The calculation system requires large memory when the entire fringe pattern is calculated. It is difficult to calculate on a common PC.

In this paper, we have proposed the large CGRH calculation. Since the computation of CGRH is separated on the vertical, it can be calculated independently. In addition, we have described the deterioration of image quality caused by this separation and proposed the solution. Moreover, the CGRH output by the fringe printer has a wide viewing angle, and the reconstructed image should have the correct occlusion. Therefore, object data created from different viewpoints are organized to fit the separate fringe pattern calculation. Furthermore, we have revealed the problem that is caused on the hologram plane and proposed the solution. As a result, we obtain a full-color reconstructed image from about a 56-Gpixel CGRH by a common PC.

2. Rainbow Hologram

The rainbow hologram [11] is well known as a full-color reconstructed image by white light. By sacrificing the vertical parallax, the rainbow hologram, which is a transmission hologram, can reconstruct the full-color image.

Figure 1 shows the schematic image of the rainbow hologram recording. The viewing area of the reconstructed image from the master hologram is limited by the horizontal slit and recorded as a transfer hologram. Figure 2 shows the reconstruction of the rainbow hologram with white light. The observer can see the monochromatic image from the same position of the horizontal slit. If the observer changes the viewing position on the vertical direction, the color of the reconstructed image also changes. The full-color image can be obtained with three slits corresponding to red, green and blue.

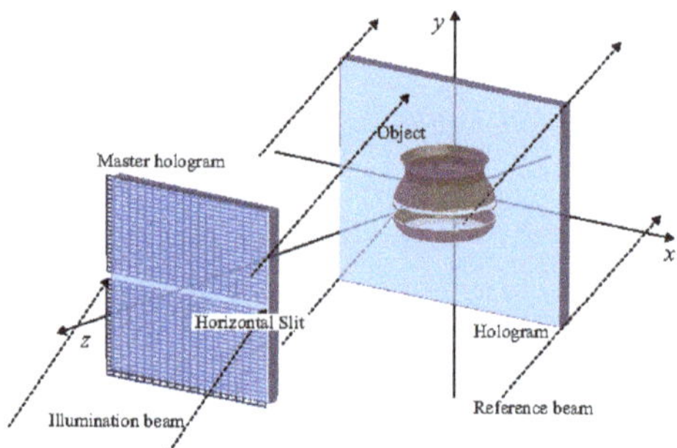

Figure 1. Recording of the rainbow hologram.

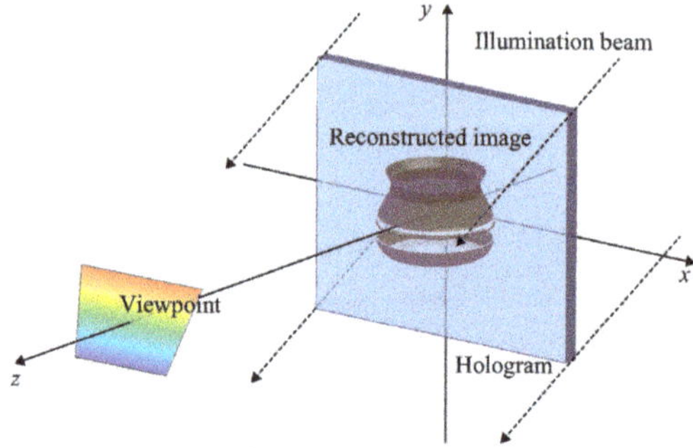

Figure 2. Reconstruction of the rainbow hologram by white light.

3. Computer-Generated Rainbow Hologram

The calculation algorithm of CGRH had been proposed in [10]. Therefore, this section elucidates the gist of the proposed method. The calculation flow can be divided into two steps. In the first step, the 1-D complex amplitude named the sub-line is calculated from sliced object data. Figure 3 shows the calculation geometry of the sub-line. Since the reconstructed image of CGRH only has horizontal parallax, one sub-line should be formed by a single horizontal slice of the object data. The object data are supposed as a collection to the self-illuminated points and can be divided by the position of each point.

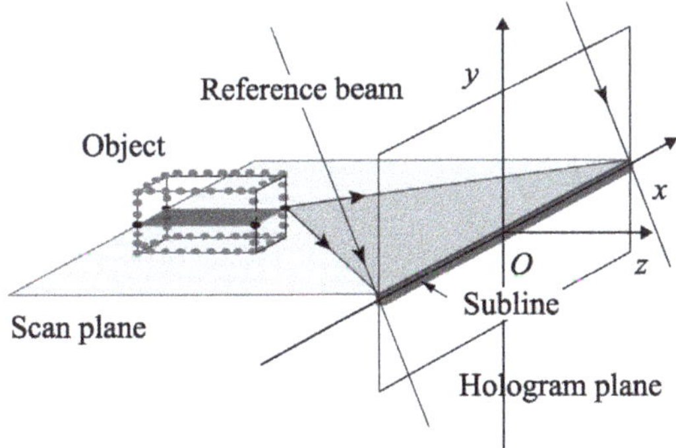

Figure 3. Calculation geometry for the sub-line.

In the second step, we add an off-axis reference, which has a vertical incident angle, to the sub-line. Figure 4 shows the second step of CGRH computation. The 1-D hologram (sub-line) does not have vertical diffracted light. Therefore, the proposed calculation used a reference beam, which only changes the phase on the vertical direction, and part of fringe pattern named the holo-line was calculated. Since the proposed calculation uses the same sub-line in each holo-line, only 1-D object beam calculation is required.

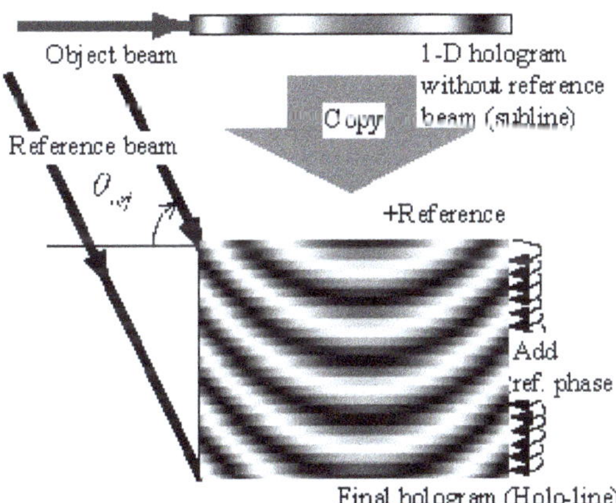

Figure 4. Two steps to calculate the holo-line of a computer generated rainbow hologram.

When the calculated CGRH is reconstructed, a cylindrical lens is required to converge the diffracted light to the viewing position. However, this is an undesirable reconstruction when using the lens. Therefore, we added the lens effect to the reference light. This change provides the wavelength selection of the rainbow hologram, and diffracted light by the entire calculated fringe pattern is observed at the same position.

3.1. Calculation of the Fringe Pattern

Computation of the holo-line follows the interference computation between the wavefront of the object beam and the wavefront of the reference beam. The intensity pattern on the hologram plane $I(x,y)$ is described as:

$$I(x,y) = |O(x) + R(y)|^2,\qquad(1)$$

where $O(x)$ is the complex amplitude of the object beam on each horizontal slice plane and $R(y)$ is the complex amplitude of the reference beam. Since the reference beam of the rainbow hologram is usually a collimated beam, R of Equation (1) only changes in the y direction.

The location of the i-th object point is specified as $(x_i,\ y_i,\ z_i)$. Each point has a real-valued amplitude a_i and a relative phase ϕ_i. The object light of a single sub-line is defined as:

$$O(x) = \sum_{i=1}^{N} \frac{a_i}{r_i}(\cos\phi_i + j\cdot\sin\phi_i),\qquad(2)$$

$$r_i = \sqrt{(x_i - x)^2 + z_i^2},\qquad(3)$$

where N is the number of object points for the single sub-line and r_i is the distance between the i-th object point and the point (x) on the hologram. The wavenumber is defined as $k = 2\pi/\lambda$, where λ is the free space wavelength of the light. The rainbow hologram displays a 3D image on or close to the hologram plane. The object point that is close to the hologram plane makes a high intensity fringe pattern according to Equation (2). Therefore, Equation (2) is changed into the following equation, which can avoid divergence.

$$O(x) = \sum_{i=1}^{N} \frac{a_i}{r_i + \alpha}(\cos\phi_i + j\cdot\sin\phi_i),\qquad(4)$$

where α is a positive constant value.

The complex amplitude $R(y)$ is determined by:

$$R(y) = a_{ref}(\cos\phi_r + j\cdot\sin\phi_r),\qquad(5)$$

$$\phi_r = ky\cdot\sin\theta_{ref},\qquad(6)$$

where θ_{ref} is the vertical incident angle of the illumination beam and ϕ_r is the phase component of the reconstructed beam.

The intensity pattern of Equation (1) becomes:

$$I(x,y) = |O(x)|^2 + |R(y)|^2 + 2\cdot\mathrm{Re}\{O(x)\}\cdot\mathrm{Re}\{R(y)\} + 2\cdot\mathrm{Im}\{O(x)\}\cdot\mathrm{Im}\{R(y)\},\qquad(7)$$

where $\mathrm{Re}\{C\}$ and $\mathrm{Im}\{C\}$ take the real and imaginary part of the complex number C, respectively. The first and second term are the DC term, which does not contribute to the reconstructed image. For the calculation of CGRH, the only the third and fourth term are calculated.

To converge the diffracted light to the horizontal slit, the phase component of the reference beam in Equation (6) is changed into the following equation.

$$\phi_r = kr\cdot(\sin\theta_{ref} - \sin\theta_s),\qquad(8)$$

$$\theta_s = \tan^{-1}\frac{y}{V_d},\qquad(9)$$

where θ_s is the convergence angle to the horizontal slit and V_d is the distance between the hologram plane and the viewpoint.

3.2. Calculation Area

The maximum diffraction angle of the CGH depends on the pixel pitch of output device. In this paper, we used a fringe printer, which can output a 0.35-μm pixel pitch CGH for the output device. Since the object point that is close to the hologram requires a fine pixel pitch, the calculation area on the hologram plane should be limited by the coordinate of the object point according to the following equation.

$$d \le \frac{\lambda}{2(\sin\theta_{obj} - \sin\theta_{ref})} \tag{10}$$

where d is the pixel pitch of the CGH and θ_{obj} and θ_{ref} are the incident angles of the object beam and the reference beam, respectively. In the actual calculation, we employed the virtual window (as shown in Figure 5), which only passes through the object light, satisfying Equation (10).

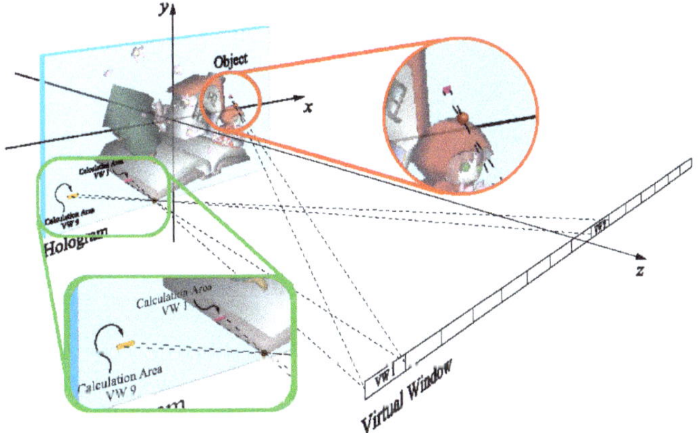

Figure 5. Addition of the calculation area by the virtual window.

4. Point Cloud Creation

The object to be recorded is approximated as a collection of self-illuminated points. Each point has x, y, z coordinates, a phase component and a real-valued amplitude of R, G, B. Since CG data are usually used as the object, we have to convert CG data to the point cloud data [7].

For the creation of the point cloud from CG data, the OpenGL library is employed as the 3D-API [12]. By using a color buffer and a z-buffer, which are calculated by OpenGL, the point cloud is calculated. However, the color buffer of the calculated point cloud is made from a single viewpoint. Therefore, when the viewer changes position, the absence of the hidden surface's data causes a problem: hole and overlap. In order to solve this problem, we employed multi-view rendering, which makes multiple point cloud data from several viewpoints. Accordingly, the reconstructed image shows the correct occlusion in every viewpoint.

4.1. Shift Point Data

Since the object point that is close to the hologram plane is calculated by Equation (4) to avoid divergence, the intensity of the object beam becomes weak, and the reconstructed image has dark bands on the hologram plane, as shown in Figure 6c. Figure 6a is a CG object used in this research, and the CG object is placed as shown in Figure 6b; the horizontal white arrow shows the hologram plane. Therefore, the object points that are close by less than a certain distance from the hologram plane need to be shifted a bit far from the hologram plane, as shown in Figure 7. Figure 8 shows the original point cloud data and the shifted point cloud data. As is evident from Figure 8, the shifted point cloud data seem almost the same as the original data when the observer sees from front.

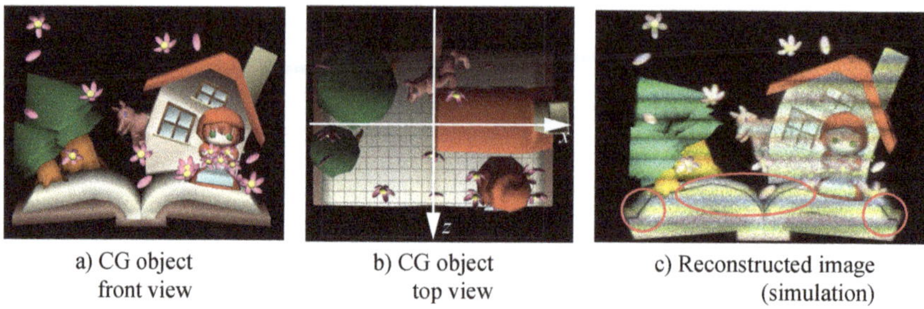

a) CG object
front view

b) CG object
top view

c) Reconstructed image
(simulation)

Figure 6. CG object and simulated image including the dark bands problem.

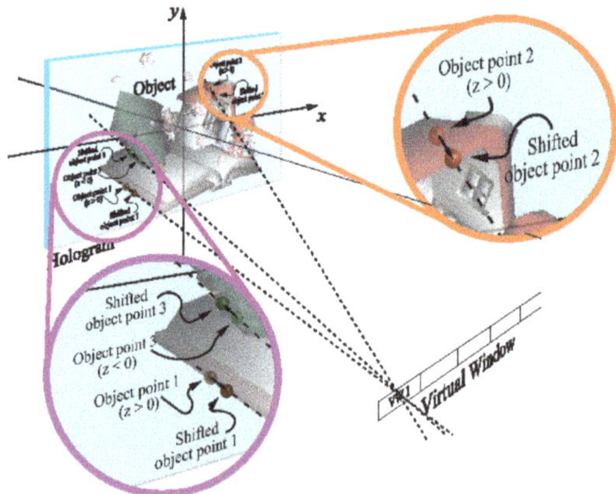

Figure 7. Shift object points in the direction of the virtual window.

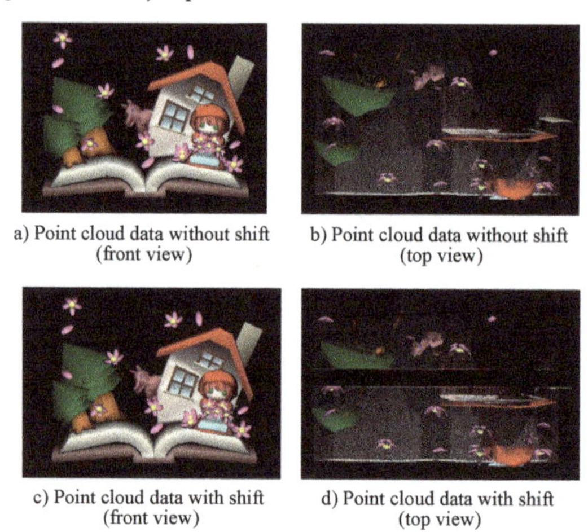

a) Point cloud data without shift
(front view)

b) Point cloud data without shift
(top view)

c) Point cloud data with shift
(front view)

d) Point cloud data with shift
(top view)

Figure 8. Point cloud data with and without shift.

4.2. Reallocation of the Point Cloud

According to the performance improvement of the output device, the resolution of CGH becomes very high (over 50 Gpixels). It is difficult to calculate the whole fringe pattern of CGRH at the same time by a general PC. In the calculation of CGRH, each holo-line is independent, as described in Reference [10]. If the object points are only divided with the vertical coordinate, the reconstructed image has distortion. When the hologram is reconstructed, the observer sees diffraction light from the point of the hologram plane, which has to be linearly aligned with the viewpoint and object point. Therefore, the object points are divided with the straight line connecting the viewpoint with the each edge of the holo-line, as shown in Figure 9.

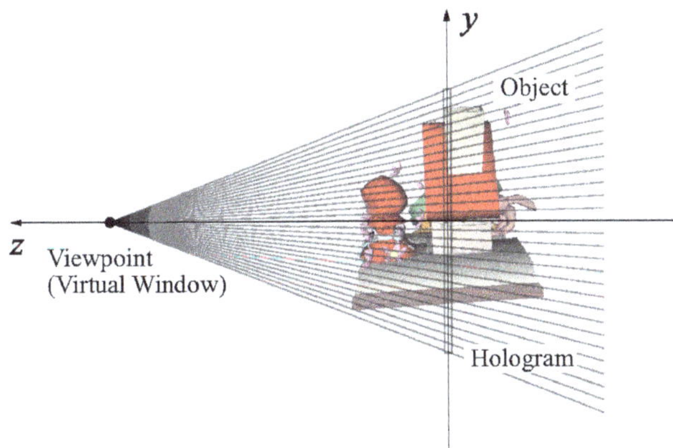

Figure 9. Split object data for each horizontal calculation block.

5. Normalization of the Fringe Pattern

The calculated fringe pattern should be normalized by the entire fringe pattern. Otherwise, each segment reconstructs the difference in the contrast of the image. Especially since the horizontal block of CGRH is calculated for different numbers of object points, the difference in the contrast of the image is very large. In this research, the calculated fringe pattern is stored as the floating-point number and normalized by a range of 3σ after the calculation of the whole fringe pattern [13]. This σ is the standard deviation calculated from the intensity of the fringe pattern.

6. Results

6.1. Fringe Printer

Figure 10 shows the schematic image of the fringe printer. The fringe printer consists of a laser, an X-Y translation stage, liquid crystal on silicon (LCoS) as the spatial light modulator and optical parts. A fractional part of the entire holographic fringe is displayed on the LCoS, and its demagnified image is recorded on a holographic plate. Then, the plate is translated by the X-Y stage to expose the next segment of the fringe. Table 1 shows the fringe printer's parameter. Holographic material VRP-M, which is manufactured by Slavich, is used as the holographic recording material. VRP-M is a silver halide material. After development, VRP-M is bleached to improve the diffraction efficiency.

Table 1. Parameters of the improved fringe printer.

Fringe Printer	Value
Moving Area (mm^2)	200 × 200
Focal Length (L2) (mm)	200
Focal Length (L3) (mm)	10
Demagnification Rate	1/20
Wavelength (laser) (nm)	473
LCoS	**Value**
Resolution (pixel)	1920 × 1080
Pitch (μm)	7.0
Gray-scale Level (bit)	8

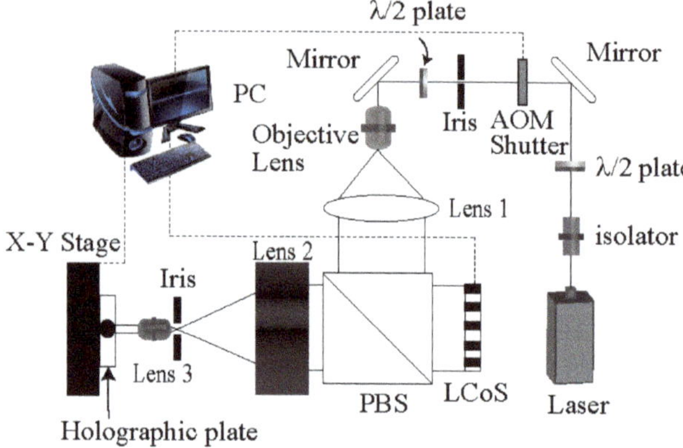

Figure 10. Schematic of the fringe printing system.

6.2. Reconstructed Images

Table 2 shows the parameter of CGRH. For the full-color reconstruction, we calculate three fringe patterns, which use different wavelengths, as shown in Table 2, from point cloud data. The calculated fringe patterns are synthesized to an 8-bit gray-scale bitmap pattern. We used a Windows 10 64-bit PC, which has Intel Core i7 7700 K CPU and 32 GB memory. By using the proposed method, an over 50-Gpixel CGRH was calculated with uniform contrast. Since the normalization had to be calculated after the entire fringe pattern calculation, the previous method only created about a quarter size of the memory size, because the fringe pattern is calculated as a float value. After the point cloud data creation, it took 14.5 min to process and to prepare the fringe calculation such as the shift of object points, the addition of the calculation area and sorting the point cloud data for each holo-line. The computation time of the fringe pattern was approximately 50 min. Figure 11 shows the reconstructed images from several viewpoints. A white LED light, called HOLOLIGHT [14], was used as the illumination light. HOLOLIGHT can easily make light that is close to the collimated light. Reconstructed images show good color reproduction compared with the original CG in the horizontal direction. In the change of the viewpoint in vertical direction, the color spectrum changes from red to blue, as shown in Figure 11a,e.

Figure 12 shows the numerical reconstruction for with and without shift object points, which are placed on or close to the hologram plane. To apply shift object data, the dark bands of Figure 12b are solved in Figure 12a. Figure 13 shows the reconstruction from the different normalizations. To change the normalization of each holo-line to the entire hologram, the contrast of the reconstructed image became uniform. These numerical simulations employed the algorithm of [15].

Table 2. Parameter of the hologram.

Parameter of the Hologram	Value
Resolution (pixels)	288,080 × 194,560
Size (mm^2)	101 × 68.1
Pitch (µm)	0.35
Wavelength (R, G, B) (nm)	633, 532, 473
Sub-line resolution (pixels)	512
Number of holo-lines	380
Average number of object points	approximately 335,000

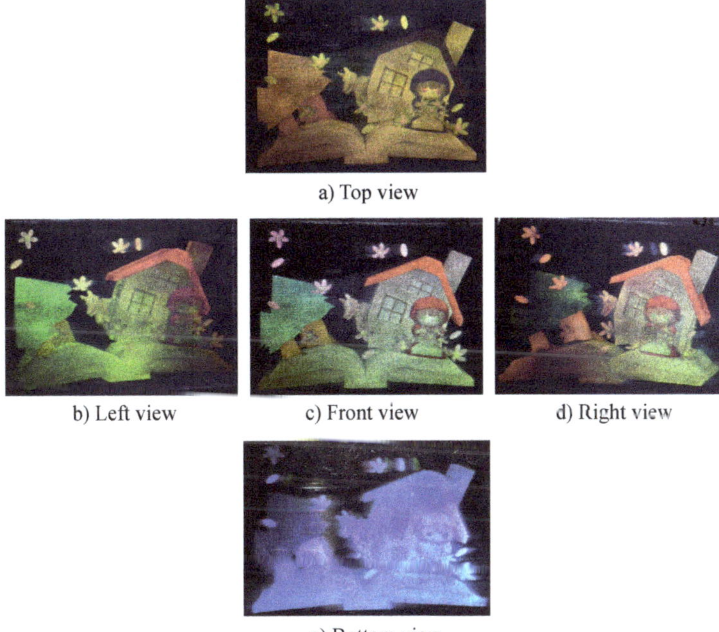

a) Top view

b) Left view   c) Front view   d) Right view

e) Bottom view

Figure 11. Reconstructed images from several viewpoints.

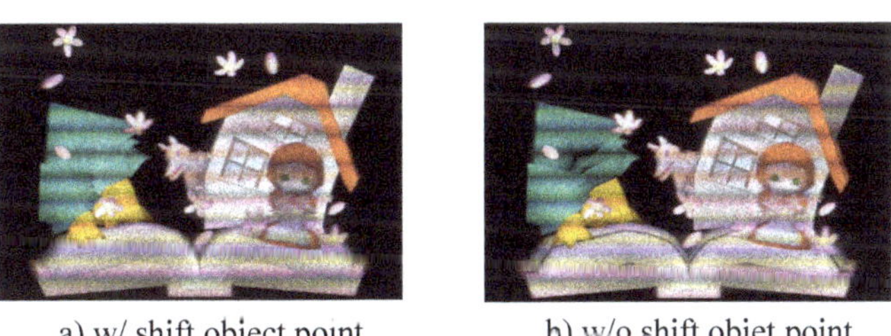

a) w/ shift object point   b) w/o shift objet point

Figure 12. Comparison of with and without shift object data (numerical reconstruction).

a) Normalization as a whole

b) Normalization as a each Holo-line

Figure 13. Comparison of the normalizations (numerical reconstruction).

7. Conclusions

In this research, we have calculated and output an over 50-Gpixel CGRH that reconstructs a full-color 3D image. The reconstructed images shows good color reproduction. For the calculation of the large CGRH, we proposed split calculation for each holo-line. To apply this calculation model, several object data, which were created from different viewpoints, were managed. In addition, to employ the shift object data and the normalization as a whole, the output CGRH provides good reconstructed images.

Author Contributions: Conceptualization, T.Y. and H.Y.; methodology, T.Y. and H.Y.; software, T.Y.; validation, T.Y. and H.Y.; formal analysis, T.Y. and H.Y.; investigation, T.Y.; resources, T.Y. and H.Y.; data curation, T.Y.; writing—original draft preparation, T.Y.; writing—review and editing, T.Y. and H.Y.; visualization, T.Y.; supervision, T.Y.; project administration, T.Y.

Funding: This research received no external funding.

Acknowledgments: The authors would like to thank Yayoi Nakaguchi for CG data creation.

Conflicts of Interest: The authors declare no conflict of interest.

Abbreviations

The following abbreviations are used in this manuscript:

CGH   Computer-generated hologram
CGRH   Computer-generated rainbow hologram
SLM   Spatial light modulator
PBS   Polarized beam splitter
LCoS   Liquid crystal on silicon

References

1. Hamano, T.; Yoshikawa, H. Image-type CGH by means of e-beam printing. *Proc. SPIE* **1998**, *3293*, 2–14.
2. Nishi, H.; Matsushima, K. Rendering of specular curved objects in polygon-based computer holography. *Appl. Opt.* **2017**, *56*, F37–F44. [CrossRef] [PubMed]
3. Sakamoto, Y.; Morishima, M.; Usui, A. Computer-generated holograms on a CD-R disk. *Proc. SPIE* **2004**, *5290*, 42–49.
4. Yoshikawa, H.; Yamaguchi, T.; Iwamoto, T. Improvement on printed image fidelity of fringe printer for computer-generated holograms. In Proceedings of the 11th International Symposium on Display Holography, Aveiro, Portugal, 25–29 June 2018.
5. Yamaguchi, T.; Matuoka, M.; Fujii, T.; Yoshikawa, H. Development of fringe printer and its practical applications. In Proceedings of the 8th International Symposium on Display Holography, Shenzhen, China, 13–17 July 2009; pp. 572–582.
6. Yamaguchia, T.; Ozawa, H.; Yoshikawa, H. Computer-generated Alcove hologram to display foating image with wide viewing angle. *Proc. SPIE* **2011**, *7959*, 795719.
7. Waters, J.P. Holographic image synthesis utilizing theoretical methods. *Appl. Phys. Lett.* **1966**, *9*, 405–407. [CrossRef]

8.   Lucente, M. Interactive Computation of Holograms Using a Look-up Table. *J. Electr. Imaging* **1993**, *2*, 28–34. [CrossRef]

9.   Matsushima, K.; Nakahara, S. Extremely high-definition full-parallax computer-generated hologram created by the polygon-based method. *Appl. Opt.* **2009**, *48*, H54–H63. [CrossRef] [PubMed]

10. Yoshikawa, H.; Taniguchi, H. Computer Generated Rainbow Holograms. *Opt. Rev.* **1999**, *6*, 118–123. [CrossRef]

11. Benton, S.A. Hologram Reconstructions with Extended Incoherent Sources. *J. Opt. Soc. Am.* 1969, *59*, 1545A.

12. Fujii, T.; Yoshikawa, H. OpenGL-based Rendering of Computer-Generated Rainbow Hologram. In Proceedings of the Nicograph International, Pattaya, Thailand, 30–31 May 2008.

13. Nakaguchi, Y.; Yamaguchi, T.; Yoshikawa, H. Image quality improvement of large size image type computer-generated hologram. In Proceedings of the 2012 International Workshop on Advanced Image Technology, Ho Chi Minh City, Vietnam, 9–10 January 2012; pp. 572–575.

14. Nemoto, A.; Ikeda, T. A high-directional LED system to form holographic light patterns. *SPIE Newsroom* **2016**. [CrossRef]

15. Yoshikawa, H.; Yamaguchi, T.; Fujita, H. Computer Simulation of Reconstructed Image from Rainbow Hologram. In Proceedings of the OSA Technical Digest (CD) (Optical Society of America, Vancouver, BC, Canada, 18–20 June 2007.

 *applied sciences*

Review

Progress in the Synthetic Holographic Stereogram Printing Technique

Jian Su [1,†], **Xingpeng Yan** [1,*,†], **Yingqing Huang** [2], **Xiaoyu Jiang** [1], **Yibei Chen** [1] **and Teng Zhang** [1]

[1]   Department of Information Communication, Academy of Army Armored Forces, Beijing 100072, China; zgysujian@163.com (J.S.); jiangxiaoyu2007@gmail.com (X.J.); zgychenyibei@163.com (Y.C.); pofeite007@gmail.com (T.Z.)

[2]   Academy of Army Armored Forces, Beijing 100072, China; huangyingqing1105@163.com

*   Correspondence: yanxp02@gmail.com

†   These authors contributed equally to this work.

Received: 23 April 2018; Accepted: 11 May 2018; Published: 23 May 2018

Abstract: The synthetic holographic stereogram printing technique can achieve a three-dimensional (3D) display of a scene. The development and research status of the synthetic holographic stereogram printing technique is introduced in this paper. We propose a two-step method, an infinite viewpoint camera method, a single-step Lippmann method, a direct-write digital holography (DWDH) method and an effective perspective images' segmentation and mosaicking (EPISM) method. The synthetic holographic stereogram printing technique is described, including the holographic display with large format, the large field of view with no distortion, the printing efficiency, the color reproduction characteristics, the imaging quality, the diffraction efficiency, the development of a holographic recording medium, the noise reduction, and the frequency response analysis of holographic stereograms.

Keywords: holographic printing; holographic stereogram; holographic element (hogel)

---

1. Introduction

Holography is a three-dimensional (3D) display technology based on the interference and diffraction of light waves, and it was first proposed and realized by Gabor [1]. For almost two decades, computer technology and holography have been integrated more closely, and holographic printing techniques have developed rapidly. Compared with the conventional optical holography, the hologram of virtual scenes can be stored into the holographic recording medium with the help of holographic printing techniques. Holographic printing techniques have been widely used in various fields [2–6], such as military, architecture, commerce, automotive industry, and entertainment.

According to the different sources of interference patterns and different approaches of recording, holographic printing techniques can be categorized into three types: synthetic holographic stereogram printing, holographic fringe printing, and wavefront printing. In synthetic holographic stereogram printing, the 3D scene is reconstructed by sequential perspective images that are captured by a tracking camera or modeled by a computer with rendering techniques. After being interfered with by the reference beam, the information of perspective images is stored in the holographic recording medium. During the reconstruction of holographic stereograms, stereoscopic vision occurs when different perspective images with parallax information is viewed, and the parallax is changing when eyes are moving [7,8]. In holographic fringe printing, interference patterns are calculated by the computer, displayed by the spatial light modulator (SLM) and printed on the holographic recording medium directly [9–14]. Thin transmission holograms, such as the Fresnel hologram, the rainbow hologram, the image hologram, and the holographic stereogram, can be achieved [15–17]. In wavefront printing, the holograms are calculated by a computer, but the algorithm is different from that of holographic fringe printing. Holographic fringe patterns are diffracted and propagated to the recording medium, interfered with by the reference beam, and reflection volume-type holograms can be achieved [18–26].

Compared with holographic fringe printing and wavefront printing, since there is no complex diffraction calculation, synthetic holographic stereogram printing is the only holographic printing technique in application currently, and it is developing rapidly. In addition to the laboratory research, there are several companies worldwide engaging in synthetic holographic stereogram printing and providing related business services, such as Zebra Imaging Inc. in the United States as well as Geola Inc. and XYZ Imaging Inc. in Europe [3]. It is beneficial to readers to have a systematic review of the synthetic holographic stereogram printing technique. This paper introduces research on the synthetic holographic stereogram printing technique and other related topics.

2. Development of Synthetic Holographic Stereogram Printing Technique

Synthetic holographic stereogram printing was first proposed by DeBitetto [27] and promoted by King et al. [28]. The earliest synthetic holographic stereograms are horizontal-parallax-only (HPO) holographic stereograms. When the viewer is not located at the slit plane of the stereogram, there will be image distortions. The problem will not occur in full parallax holographic stereograms. With the two-step method, we can achieve a real–virtual combined holographic stereogram, i.e., the reconstructed 3D scene is viewed inside or outside of the holographic recording medium. The production and reconstruction of the two-step method is shown in Figure 1 [29]. Full parallax perspective images of the scene are acquired under incoherent illumination, and Fresnel holograms of them are recorded in different hogels (holographic elements) successively (see Figure 1a). When viewing the master hologram, eyes should be close to the positions of aperture pupils (see Figure 1b). To separate the aperture pupil plane and the viewing plane, the master hologram (H_1 plate) should be recopied to the transfer hologram (the H_2 plate) with the image hologram photographing method, i.e., there are double exposures in the two-step method (see Figure 1c). When viewing the transfer hologram from different virtual pupil positions, different perspective images can be captured by human eyes, and the stereoscopic vision is formed (see Figure 1d).

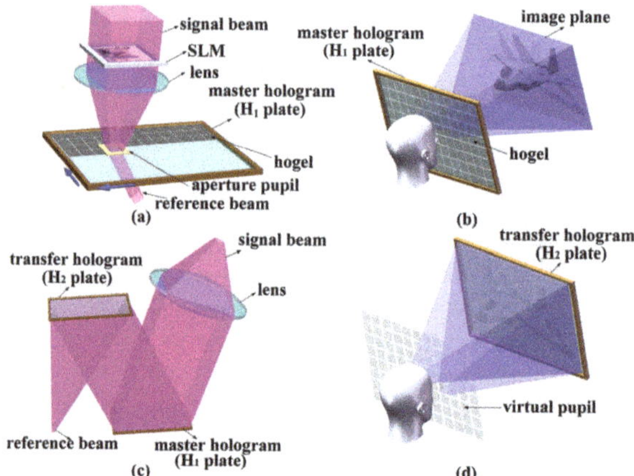

Figure 1. Production and reconstruction of the two-step method. (**a**) The production of the master hologram. (**b**) The reconstruction of the master hologram. (**c**) The reproduction of the master hologram to the transfer hologram. (**d**) The reconstruction of the transfer hologram. (Reprinted with permission from ref [29], [OSA]).

Since the 1990s, the media lab of Massachusetts Institute of Technology (MIT), the imaging science and engineering laboratory of Tokyo Institute of Technology, and the General Optics

Laboratory (Geola) have been researching the synthetic holographic stereogram printing technique [30]. To simplify the process of the two-step method, they proposed the infinite viewpoint camera method [31,32], the single-step Lippmann method [33,34], and the direct-write digital holography (DWDH) method [3,35,36].

MIT studied the problem of image distortions in HPO holographic stereograms and proposed two predistortion techniques, i.e., the infinite camera predistortion method and the perspective slicing predistortion method. The infinite viewpoint camera method for synthetic holographic stereogram printing was proposed, whose core idea was to transform the perspective images into parallax related images. The perspective images are first captured by an infinite camera, and the light arriving at the hogels can be considered as bundles of parallel light approximately since the distance between the camera and the holographic recording medium is far enough. The transformation is just an operation on a series of arrays as shown in Figure 2 [29]. Suppose there are $s \times t$ $(s = 1, 2, \cdots, M, t = 1, 2, \cdots, N)$ perspective images, and each image contains $i \times j$ $(i = 1, 2, \cdots, m, j = 1, 2, \cdots, n)$ pixels. The perspective image matrix is expressed as $P_{st}(i,j)$. All pixels at the same location of each $P_{st}(i,j)$ are extracted to form a new matrix $H_{ij}(s,t)$, which denotes a parallax-related image. During the holographic printing, a convex lens is used to control the refraction direction of light rays. The resolution of reconstructed images equates to the number of hogels. A double-frustum algorithm for rapidly generating image data for full parallax holographic stereograms was proposed by Halle et al. [37]. There is a special renderer with an orthoscopic camera and a pseudoscopic camera together, and each exposing image is synthesized by using rendering images of these two cameras.

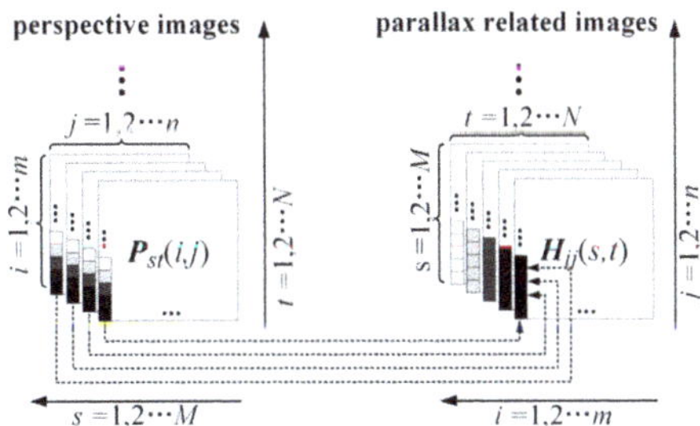

Figure 2. Transformation principle of perspective images in the infinite viewpoint camera method. (Reprinted with permission from ref [29], [OSA]).

In the single-step Lippmann method, the exposing images for hogels are acquired by the perspective projection, not by camera sampling, and the principle of this method is shown in Figure 3 [29]. Based on the center of each hogel, scene points are projected to the position of liquid crystal display (LCD), respectively (see Figure 3a). According to the viewer's position, the occlusion relation of scene points in space should be considered and the hidden surfaces should be removed. The calculated images are then displayed on the LCD, converged through the lens, and recorded on the hogels (see Figure 3b). During the reconstruction of the scene, light rays are diffracted from each hogel in the same way as the image calculation; thus, the viewer can perceive the scene (see Figure 3c). The double-frustum algorithm can be also applied to the single-step Lippmann method.

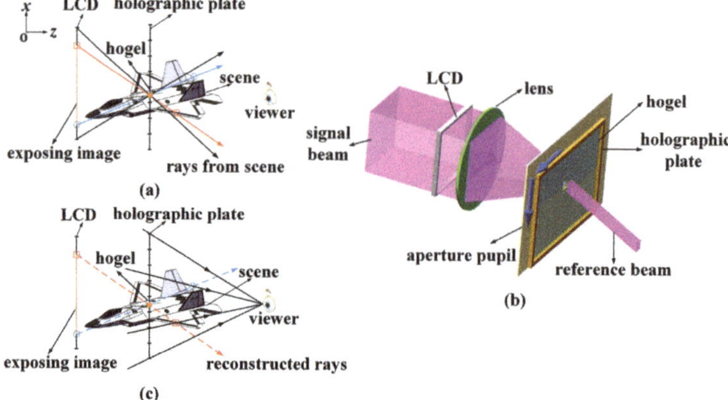

Figure 3. Principle of the single-step Lippmann method. (**a**) Calculation of an exposing image. (**b**) Optical setup of the method. (**c**) Reconstruction geometry for the holographic stereogram. (Reprinted with permission from ref [29], [OSA]).

Geola Inc. has realized full parallax holographic stereogram printers with pulse lasers [35], while the commercial products provided by Geola Inc. are only HPO holographic stereograms. The DWDH method was proposed by Bjelkhagen and Brotherton-Ratcliffe. The core idea of the DWDH method is the $H_1 - H_2$ conversion, i.e., the image transformation from the camera film plane to the SLM plane, which is usually referred as "I-to-S" transforming [38]. The principle of the DWDH method is shown in Figure 4. There are six principal planes, i.e., the hologram plane, the SLM plane, the projected SLM image plane, the camera plane, the film plane, and the projected film plane. There is the assumption of a small-aperture camera and a "point" hogel, and each image of the SLM plane is acquired by the pixel swapping technique from different images of film planes according to the ray-tracing principle. The spatial location relations of these planes are easily established, and there is an exact pixel matching relationship between the SLM plane and film plane.

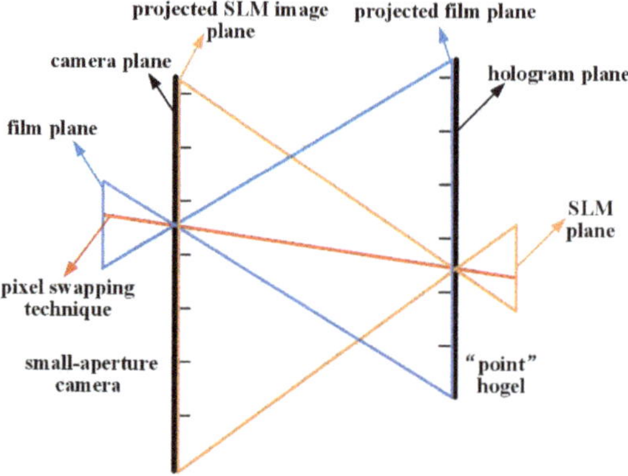

Figure 4. Principle of the direct-write digital holography (DWDH) method.

In the two-step method, the process is relatively complex with double exposures, and it is difficult to make a large-size hologram since we can hardly achieve large-format collimating light during the second exposure. There is only single exposure in the infinite viewpoint camera method, in the single-step Lippmann method, or in the DWDH method. Recently, a novel method based on effective perspective image segmentation and mosaicking (EPISM) was proposed by our group [29,38]. The principle of the EPISM method is shown in Figure 5. On the basis of ray-tracing principle and the reversibility of light propagation, the viewing frustum effect of human eyes is analyzed and simulated (see Figure 5a). With the segmentation and mosaicking of effective perspective images, synthetic effective perspective images for single-step exposure can be achieved (see Figure 5b). By using the EPISM holographic stereogram printing method, the image processing is simple and the reconstructed images are of high resolution.

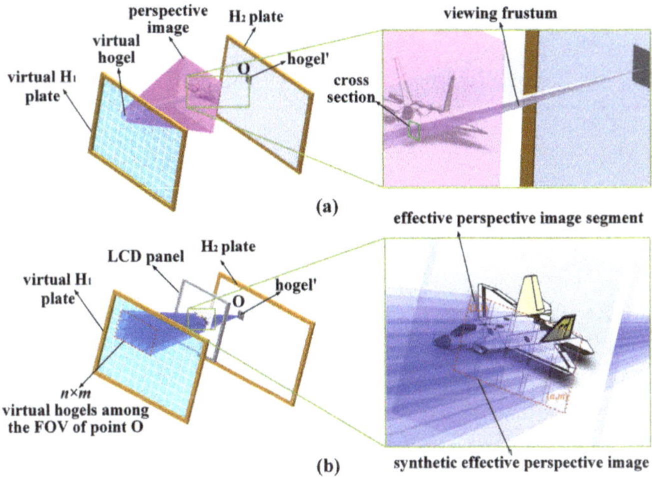

Figure 5. Principle of the effective perspective image segmentation and mosaicking (EPISM) method. (a) The extraction of effective perspective image segment corresponding to a single virtual hogel. (b) The synthetic effective perspective image mosaicked by effective images segments of multiple virtual hogels. (Reprinted with permission from ref [29], [OSA]).

Principles of different synthetic holographic stereogram printing methods have been described. Except for the two-step method, there is only single exposure in the other printing methods. According to different models to generate the hogel images, the methods mentioned above can be classified as two types [39]. In the first type, hogel images are generated by a camera placed on the viewing zone of the holographic stereogram. As shown in Figure 6, the hologram plane coincides with the camera image plane, and the camera is placed on every point of view on the viewing zone. Perspective images of the 3D scene are first captured by the camera, and they are then rearranged to generate hogel images. According to different transformation algorithms, there are examples of the infinite viewpoint camera method and the DWDH method. Although the hologram plane is not matched with the camera image plane in the EPISM method, the EPISM method can be also considered as the first type since the camera is placed on the viewing zone of the holographic stereogram. In the second type, hogel images are generated by a camera placed on the hologram plane. As shown in Figure 7, the camera is placed on every hogel of the holographic stereogram and the SLM plane coincides with the camera image plane. Hogel images are generated directly, and there are examples of the single-step Lippmann method and double-frustum algorithm. The camera's field of view (FOV) is coincided with the hogel's FOV, and the hogel image's resolution coincides with the SLM's resolution.

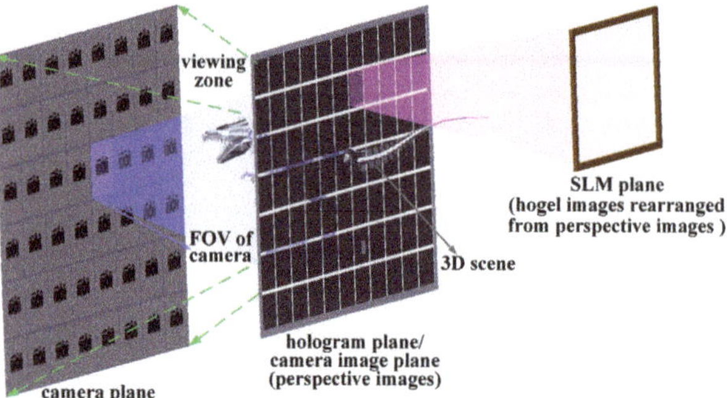

Figure 6. Hogel images are generated by a camera placed on the viewing zone of the holographic stereogram.

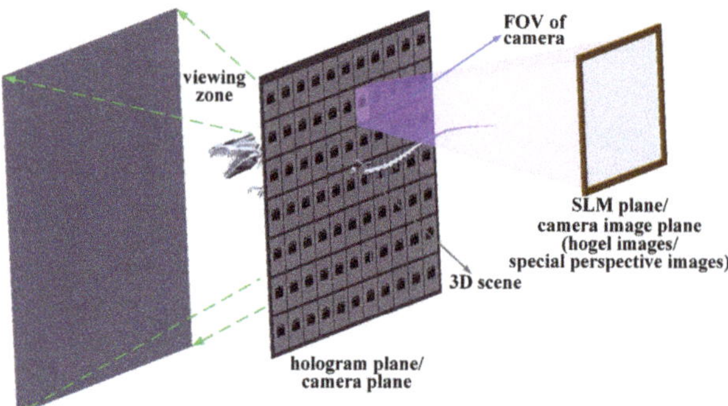

Figure 7. Hogel images are generated by a camera placed on the hologram plane.

For either the DWDH method or the EPISM method, the core idea is the $H_1 - H_2$ conversion. However, they are different, as shown in Figure 8. In the DWDH method, there are six principal planes, i.e., the hologram plane, the SLM plane, the projected SLM image plane, the camera plane, the film plane, and the projected film plane. There is the assumption of a small-aperture camera and a "point" hogel, and each image of the SLM plane is acquired by the pixel swapping technique from different images of film planes according to the ray-tracing principle. Consequently, there is an exact pixel matching relationship between the SLM plane and the film plane. Once the parameters of the printing system (such as the distance between the camera plane and the hologram plane as well as the pixel intervals of the film plane and SLM plane) are determined, the relationship between the hogel sizes of the camera plane and the hologram plane are fixed. In our proposed EPISM method, there are only three principal planes, i.e., the virtual master hologram plane, the SLM plane, and the transfer hologram plane, and only the hogels in the transfer hologram are considered as "point" hogels. The virtual master hologram plane and the transfer hologram plane in the EPISM method can be considered equally to the camera plane and the hologram plane in the DWDH method correspondingly. The images for hogels in the transfer hologram are acquired by the pixel mosaicking technique, not the pixel swapping technique. The DWDH method is based on single pixel mapping, whereas the EPISM method is based on the image block operation. The calculation burden in the EPISM method is much less than that of the DWDH method, especially for a full parallax holographic stereogram. Moreover, since it is the pixel mosaicking technique, the hogel sizes in the virtual master hologram and the transfer hologram can be arbitrary, without any limitations.

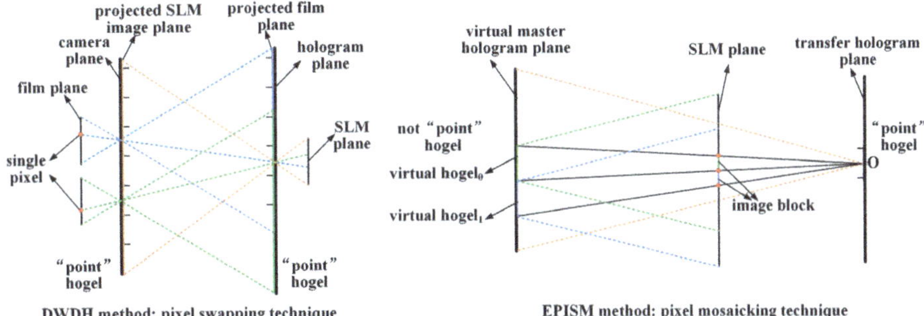

Figure 8. Differences between the DWDH method and the EPISM method.

3. The Research Status of Synthetic Holographic Stereogram Printing

3.1. Holographic Display with a Large Format, a Large Field of View, and No Distortion

The MIT group studied the principle of synthetic holographic stereogram, which has a large format, a large FOV, and no distortions during image reconstruction. Ultragram was proposed to improve the distortion properties in HPO holographic stereograms [30,31]. A holographic display in the m^2 level was achieved with hogels spliced by Benton et al. [40]. Zebra Imaging Inc. was founded by scientists from the media lab of MIT, and has successfully achieved holographic stereogram printing with high quality [41]. Zebra Imaging Inc. has provided tens of thousands of holographic maps for the US army since 2006. Holographic advertisement and holographic map produced by Zebra Imaging Inc. are shown in Figure 9.

Figure 9. Holographic advertisement (**a**) and holographic map (**b**) produced by Zebra Imaging Inc.

From the perspective of diffraction efficiency and the FOV, holographic stereogram printing systems based on a diffuser or a diffraction optical element (DOE) were compared by Zherdev et al. Two embodiments of diffractive lens, i.e., the composite holographic lens (CHL) and the amplitude diffractive lens (ADL), were used to increase the FOV, and the FOV could be up to 120° [42].

Since the objective lens, the telecentric lens, and other optical elements are used in the holographic stereogram printing systems, there will be radial distortions when the images are projected by the SLM [43]. The radial distortions caused by the optical elements during the reconstruction of a full parallax and full color white light viewable holographic stereogram were studied by Park et al., while the peak signal-to-noise ratio (PSNR) and the structural similarity (SSIM) were used as imaging quality metrics [43].

3.2. Printing Efficiency

Printing efficiency is critical in holographic stereogram printing as it is the hogel-by-hogel printing method. The relationship between the total time of holographic printing and the factors such as hogel size, the light sensitivity of recording material, laser output power, exposure time, moving time, and waiting time was studied by Morozov et al. [44]. Spatial splitting technology and time sequential technology were proposed to improve the printing efficiency, and for a hologram with size 10×10 cm, the overall printing time 250 min for conventional method reduced to 67 and 32 min, respectively.

A multichannel holographic printing method was proposed by Rong et al. [45]. Each perspective image of the 3D scene was segmented into nonoverlapping subimages, and subimages were rearranged to the encoding images firstly. The SLM was then partitioned into multi-areas that were independent spatially, and each encoding image was loaded to the corresponding channel of the SLM for exposing. With the proposed method, the printing time could be reduced significantly for large-scale holograms.

An array of small lenses and reduction optics were used to realize the parallel exposure by Yamaguchi et al. [46]. Since 12 hogels were simultaneously exposed by each exposure, the printing efficiency could be improved greatly, as the printing speed was about 10 times faster than the conventional method.

Compared with the continuous wave laser, the pulsed laser is not sensitive to vibrations or tiny temperature fluctuations, so the waiting time could be negligible during the exposure. Many researches and companies have successfully realized the synthetic holographic stereogram printing with pulsed lasers [47–49]. Geola Inc. and XYZ Imaging Inc. have cooperated to produce holographic printers, which could print differently sized, high-quality, and colorful holograms automatically. The first RGB pulsed laser holographic printer made by Geola Inc. in 2001 and the printed hologram are shown in Figure 10 [3,47].

Figure 10. The first RGB pulsed laser holographic printer made by Geola in 2001 (**a**) and the printed hologram (**b**). (Reprinted with permission from ref [47], [SPIE]).

3.3. Color Reproduction Characteristic

With the development of the holographic recording medium, color reflection holography has been the focus of research in recent years. The diffraction wavelength selectivity of reflection holography and the colorimetric principle were analyzed, and the expression of reflection hologram colorimetric system was proposed by Wang et al., meanwhile the color reproduction quality of reconstructed images was evaluated [50]. Based on the spectral measurement of reproduced light, a new method of color management for a full-color holographic stereogram printing was proposed by Yamaguchi et al. [51]. The color shifts with the variations of illumination beam angle were also discussed, and it was shown that the specified colors could still be reproduced within a certain range of incident angle variation.

Based on the sensitometric characteristic of the holographic recording medium, multi-exposure [52] or space-division exposure [51] could be used in full-color holographic printing. In the multi-exposure method, the hogels were exposed by a single laser beam which combined the lasers of RGB colors simultaneously, and the recording effect depended on the dynamic range of recording material since there were three different gratings recorded in the region of the hologram [53]. In the space-division exposure method, the hogels were divided into different parts spatially, and each part was exposed with a single-color laser, and the diffracted light intensity would then decrease to one-third of that of monochromatic exposure as the recording density decreased.

A high-density light-ray recording method for a full-color, full-parallax holographic stereogram was proposed by Yamaguchi et al. [54]. RGB lasers were used for the production and reconstruction of colorful holographic stereogram, and the relationship between the hogel size and the reconstruction effects, such as the angular resolution and the visibility of the hogel's array structure, was discussed. Experiment results showed that the angular resolution of reproduced light rays was 1.08° when the hogel size was 50 × 50 μm, which could satisfy the demand of the human visual system.

3.4. Imaging Quality

The relationship between the hogel size and the lateral resolution of the stereogram was analyzed by Lucente [55]. A hogel overlapping method for enhancing the lateral resolution of holographic stereogram was proposed by Hong et al. [56]. The shifting distance of hogel was smaller than the hogel size, and the maximum number of recordable overlapped hogels depended on the dynamic range of the holographic recording medium. Perspective images captured from different viewing points for the conventional and overlapping methods are shown in Figure 11. However, with such hard pupils overlapped method, the reconstructed images were blurred because of spectral aliasing.

A band-limited diffuser was localized before the holographic recording medium to produce a high-resolution holographic stereogram by Yamaguchi et al., but some additional noises were introduced [57].

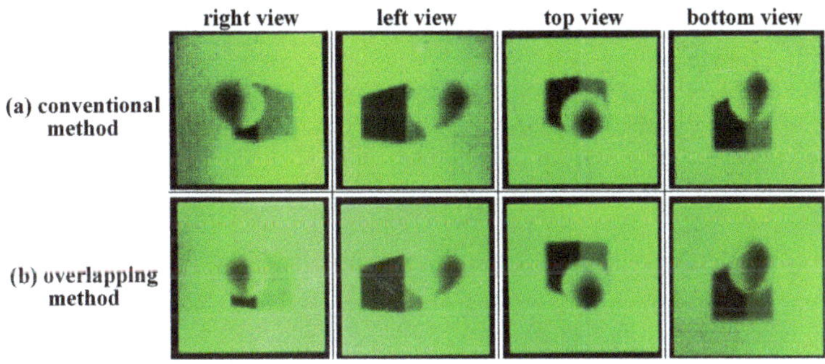

Figure 11. Perspective images captured from different viewing points by Hong's group: (a) conventional method and (b) overlapping method. (Reprinted with permission from ref [56], [OSA]).

3.5. Diffraction Efficiency

Diffraction efficiency of the hologram is related to various factors, such as the properties of the recording medium, the light intensity ratio between the signal beam and the reference beam, the polarization properties of the beams, and the follow-up processing. Diffraction efficiency is highest when the light intensity ratio between the signal beam and the reference beam is about 1. A pseudorandom band-limited diffuser was used to modulate the signal beam and broaden the frequency spectrum of the Fourier-plane of the lens by Klug et al., and this resulted in an improvement of diffraction efficiency [58].

Additionally, in the optical hologram, from the perspective of the recording medium's properties, some scholars have added nanoparticles to the recording materials to improve the diffraction efficiency, and the research results could also be applied to the holographic printing. For instance, 13 nm silica nanoparticles were added to the methacrylate photopolymers by Suzuki et al., and a noticeable reduction of scattering loss and a high-contrast refractive-index change were achieved [59]. A gold-nanoparticle-doped photopolymer was used to suppress side lobes of angle selectivity of volume holographic gratings by Cao et al., achieving a higher diffraction efficiency [60,61].

3.6. The Development of a Holographic Recording Medium

Three main types of holographic recording medium are silver-halide materials, dichromate gelatin materials, and photopolymer materials [3]. With such conventional materials, the holograms cannot be modified once the information is recorded into the materials, which means the lack of image-updating capability, leading to restricted use and high cost. Consequently, many scholars are committed to the study of updatable holographic recording medium. A holographic stereogram was recorded into a photorefractive polymer material achieved by Peyghambarian et al., resulting in a dynamically updatable holographic display, which could refresh images every 2 s [62,63]. Images from the updatable holographic 3D display is shown in Figure 12. Poly-doped organic compound was used in a holographic stereographic technique by Tsutsumi et al., resulting in an updatable holographic 3D display with fast response time and high diffraction efficiency [64,65]. A permanent holographic recording photosensitive material with high diffraction efficiency of about 90% was investigated by Gao et al. [66,67].

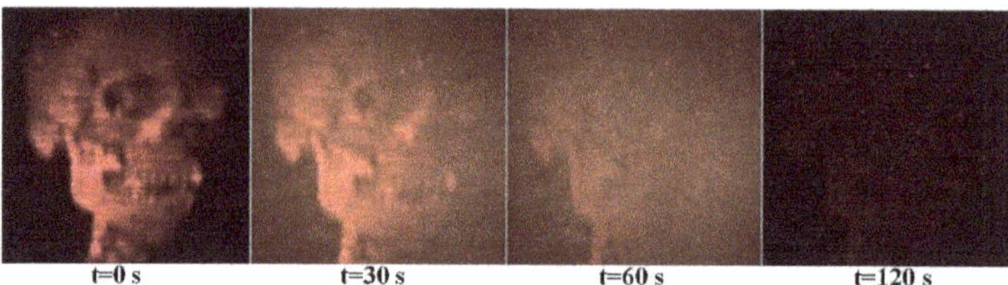

Figure 12. Images from the updatable holographic 3D display achieved by Peyghambarian's group. (Reprinted with permission from ref [62], [Springer Nature]).

3.7. Noises Reduction

In synthetic holographic stereogram printing, since it is usually set up on an anti-vibration system with an optical table, the main source of the noises comes from the movement of the motorized X-Y stage with the holographic recording medium hogel by hogel. In order to enhance the robustness of the printing system against the ambient noises, an anti-vibration algorithm was designed and applied by Lee et al., reducing the vibration significantly [68]. The schematic and prototype of the holographic stereogram printer is shown in Figure 13.

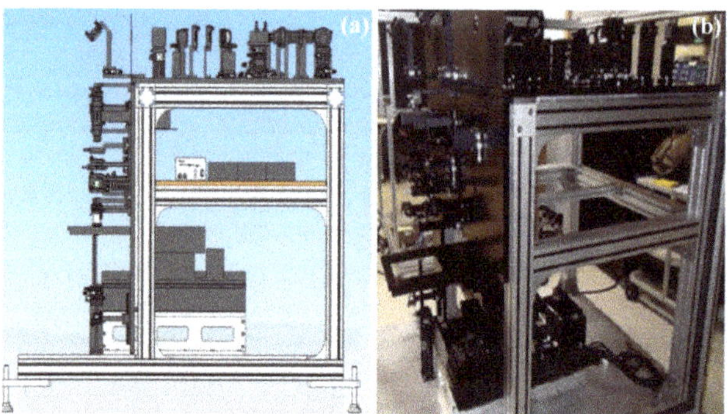

Figure 13. Schematic and prototype of the holographic stereogram printer produced by Lee's group. (Reprinted with permission from ref [68], [SPIE]).

The problem of speckle noises in synthetic holographic stereogram printing was also studied by Yamaguchi et al. [57]. A diffuser was used in the printing system to equalize the intensity distribution of the signal beam inside the hogel, but it would generate speckles. A moving diffuser was used to suppress the granularity noises and a multiple exposure method was used to suppress the high-frequency noises.

3.8. Frequency Response Analysis of the Holographic Stereogram

Holographic stereogram is an optical system, especially a diffraction limited system. It is difficult to understand the working mechanism of the system comprehensively if just from the perspective of geometrical optics. Considering from the angular spectrum, some scholars establish the angular regulation model, analyze the regulation mechanism, and discuss the frequency response of the holographic stereogram [69].

The modulation transfer function (MTF) of HPO image-plane holographic stereograms was constructed by St Hilaire et al., and the optimum sampling of the slit plane with fixed depth object points was also discussed [70]. The optical transfer function (OTF) of 3D display systems was investigated by Helseth et al., and the influence of the Stiles–Crawford effect of human eyes was also considered [71]. In our previous study, the wavefront reconstructed errors in the holographic stereogram were expressed as defocusing aberrations, and the frequency responses of the full-parallax holographic stereogram were studied when the Rectangle, the shaped Gaussian, and the shaped Blackman window functions were used as the exit pupil functions, respectively. The design criterion of the exit pupil function was then discussed [72]. Moreover, exit modeling and OTF analysis of the EPSIM-based holographic stereogram were carried out [38]. Characteristics of the OTF with respect to the exit pupil size and the aberration were investigated in detail. Hogel sizes, i.e., the sampling interval of original perspective images and the printing interval of synthetic effective perspective images, were optimized for the reconstruction. Optical setup of the holographic stereogram printing system and images of optical reconstruction from different perspectives are shown in Figure 14.

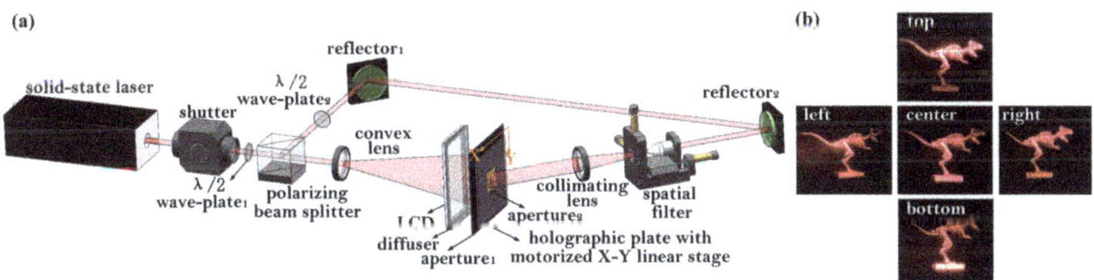

Figure 14. Optical setup of the holographic stereogram printing system (**a**) and images of optical reconstruction from different perspectives produced by our group (**b**). (Reprinted with permission from ref [38], [Springer Nature]).

4. Conclusions

Holographic printing techniques can be used for a realistic 3D display, and synthetic holographic stereogram printing is the only holographic printing technique in application currently. In this paper, the development and research status of the synthetic holographic stereogram printing technique is introduced in detail. We predict that the synthetic holographic stereogram printing technique will rapidly develop and that holographic stereogram printers will enter our daily lives for personal use in the future.

Author Contributions: J.S. structured and wrote the paper; X.Y. selected topic and supervised the progress of the whole work; Y.H. guided other authors to complete the paper; X.J. and Y.C. supplied the materials to the paper; T.Z. helped to proofread the manuscript.

Acknowledgments: The authors are grateful to the National Key Research and Development Program of China (Grant No.2017YFB1104500), the National Natural Science Foundation of China (Grant No.61775240), and the

Foundation for the Author of National Excellent Doctoral Dissertation of the People's Republic of China (FANEDD) (Grant No.201432).

Conflicts of Interest: The authors declare no conflict of interest.

References

1. Gabor, D. A new microscopic principle. *Nature* **1948**, *161*, 777–778. [CrossRef] [PubMed]
2. Lucente, M. The first 20 years of holographic video—And the next 20. In Proceedings of the SMPTE 2nd Annual International Conference on Stereoscopic 3D for Media and Entertainment, New York, NY, USA, 21–23 June 2011.
3. Bjelkhagen, H.I.; Brotherton-Ratcliffe, D. Ultrarealistic imaging: The future of display holography. *Opt. Eng.* **2014**, *53*, 112310. [CrossRef]
4. Yamaguchi, M. Full-parallax holographic light-field 3-D displays and interactive 3-D touch. *Proc. IEEE* **2017**, *105*, 947–959. [CrossRef]
5. Park, J.S.; Stoykova, E.; Kang, H.J. White light viewable silver-halide holograms in design applications. *Bulg. Chem. Commun.* **2016**, *48*, 37–40.
6. Zheng, H.; Sun, G.; Yu, Y. A review of holographic printing technologies. *Laser Optoelectron. Prog.* **2012**, *49*, 110002. (In Chinese) [CrossRef]
7. Yamaguchi, M. Light-field and holographic three-dimensional displays [Invited]. *J. Opt. Soc. Am. A* **2016**, *33*, 2348–2364. [CrossRef] [PubMed]
8. Kang, H.; Stoykova, E.; Berberova, N.; Park, J.; Nazarova, D.; Park, J.S.; Kim, Y.; Hong, S.; Ivanov, B.; Malinowski, N. Three-dimensional imaging of cultural heritage artifacts with holographic printers. In Proceedings of the SPIE 19th International Conference and School on Quantum Electronics: Laser Physics and Applications, Sozopol, Bulgaria, 26–30 September 2016; Volume 10226, p. 102261I.
9. Yoshikawa, H.; Yamaguchi, T. Review of holographic printers for computer-generated holograms. *IEEE Trans. Ind. Inform.* **2016**, *12*, 1584–1589. [CrossRef]
10. Jolly, S.; Smalley, D.E.; Barabas, J.; Bove, V.M. Direct fringe writing architecture for photorefractive polymer-based holographic displays: Analysis and implementation. *Opt. Eng.* **2013**, *52*, 055801. [CrossRef]
11. Yoshikawa, H. Research activities on digital holographic displays in Japan. In Proceedings of the SPIE Three-Dimensional Imaging, Visualization, and Display 2011, Orlando, FL, USA, 25–29 April 2011; Volume 8043, p. 804305.
12. Li, Y.; Wang, H.; Ma, L.; Shi, Y. Three-dimensional imaging and display of real-existing scene using fringe. In Proceedings of the SPIE International Conference on Optics in Precision Engineering and Nanotechnology (icOPEN2013), Singapore, 9–11 April 2013; Volume 8769, p. 87691I.
13. Yoshikawa, H.; Yamaguchi, T. Computer-generated holograms for 3D display (Invited Paper). *Chin. Opt. Lett.* **2009**, *7*, 1079–1082.
14. Su, J.; Yan, X.; Huang, Y.; Chen, Y.; Jiang, X. Resolution matching in laser direct printing of a computer-generated hologram. *J. Opt. Soc. Am. B* **2017**, *34*, B1–B8. [CrossRef]
15. Yoshikawa, H.; Takei, K. Development of a compact direct fringe printer for computer-generated holograms. In Proceedings of the SPIE Practical Holography XVIII: Materials and Applications, San Jose, CA, USA, 18–22 January 2004; Volume 5290, pp. 114–121.
16. Yamaguchi, T.; Yoshikawa, H. Computer-generated image hologram. *Chin. Opt. Lett.* **2011**, *9*, 120006. [CrossRef]
17. Sohn, I.B.; Choi, H.K.; Yoo, D.; Noh, Y.C.; Noh, J.; Ahsan, M.S. Three-dimensional hologram printing by single beam femtosecond laser direct writing. *Appl. Surf. Sci.* **2018**, *427*, 396–400. [CrossRef]
18. Kang, H.; Stoykova, E.; Yoshikawa, H.; Hong, S.; Kim, Y. *Comparison of System Properties for Wave-Front Holographic Printers*; Osten, W., Ed.; Springer: Berlin/Heidelberg, Germany, 2014; pp. 851–854, ISBN 978-3-642-36359-7.
19. Oi, R.; Chou, P.-Y.; Jackin, B.J.; Wakunami, K.; Ichihashi, Y.; Okui, M.; Huang, Y.-P.; Yamamoto, K. Three-dimensional reflection screens fabricated by holographic wavefront printer. *Opt. Eng.* **2018**, *57*, 061605.
20. Yamaguchi, T.; Miyamoto, O.; Yoshikawa, H. Volume hologram printer to record the wavefront of three-dimensional objects. *Opt. Eng.* **2012**, *51*, 075802. [CrossRef]

21. Kim, Y.; Stoykova, E.; Kang, H.; Hong, S.; Park, J.; Park, J.; Hong, J. Seamless full color holographic printing method based on spatial partitioning of SLM. *Opt. Express* **2015**, *23*, 172–182. [CrossRef] [PubMed]

22. Kang, H.; Stoykova, E.; Kim, Y.; Hong, S.; Park, J.; Hong, J. Color holographic wavefront printing technique for realistic representation. *IEEE Trans. Ind. Inform.* **2017**, *12*, 1590–1598. [CrossRef]

23. Kang, H.; Stoykova, E.; Kim, Y.; Hong, S.; Park, J.; Hong, J. Color wavefront printer with mosaic delivery of primary colors. *Opt. Commun.* **2015**, *350*, 47–55. [CrossRef]

24. Cao, L.; Wang, Z.; Zhang, H.; Jin, G.; Gu, C. Volume holographic printing using unconventional angular multiplexing for three-dimensional display. *Appl. Opt.* **2016**, *55*, 6046–6051. [CrossRef] [PubMed]

25. Wakunami, K.; Oi, R.; Senoh, T.; Ichihashi, Y.; Yamamoto, K. Wavefront printing technique with overlapping approach toward high definition holographic image reconstruction. In Proceedings of the SPIE Three-Dimensional Imaging, Visualization, and Display 2016, Baltimore, MD, USA, 17–21 April 2016; Volume 9867, p. 98670J.

26. Ichihashi, Y.; Yamamoto, K.; Wakunami, K.; Oi, R.; Okui, M.; Senoh, T. An analysis of printing conditions for wavefront overlapping printing. In Proceedings of the SPIE Practical Holography XXXI: Materials and Applications, San Francisco, CA, USA, 28 January–2 February 2017; Volume 10127, p. 101270L.

27. DeBitetto, D.J. Holographic panoramic stereograms synthesized from white light recordings. *Appl. Opt.* **1969**, *8*, 1740–1741. [CrossRef] [PubMed]

28. King, M.C.; Noll, A.M.; Berry, D.H. A new approach to computer-generated holography. *Appl. Opt.* **1970**, *9*, 471–476. [CrossRef] [PubMed]

29. Su, J.; Yuan, Q.; Huang, Y.; Jiang, X.; Yan, X. Method of single-step full parallax synthetic holographic stereogram printing based on effective perspective images' segmentation and mosaicking. *Opt. Express* **2017**, *25*, 23523–23544. [CrossRef] [PubMed]

30. Kang, H.; Stoykova, E.; Park, J.; Hong, S.; Kim, Y. *Holographic Printing of White-Light Viewable Holograms and Stereograms*; Mihaylova, E., Ed.; InTech: London, UK, 2013; pp. 171–201, ISBN 978-953-51-1117-7.

31. Halle, M.W.; Benton, S.A.; Klug, M.A.; Underkoffler, J.S. Ultrgram: A generalized holographic stereogram. In Proceedings of the SPIE Practical Holography V, San Jose, CA, USA, 1–7 February 1991; Volume 1461, pp. 142–155.

32. Halle, M.W. The Generalized Holographic Stereogram. Master's Thesis, Massachusetts Institute of Technology, Cambridge, MA, USA, 1991.

33. Yamaguchi, M.; Ohyama, N.; Honda, T. Holographic three-dimensional printer: New method. *Appl. Opt.* **1992**, *31*, 217–222. [CrossRef] [PubMed]

34. Yamaguchi, M.; Endoh, H.; Honda, T.; Ohyama, N. High-quality recording of a full-parallax holographic stereogram with a digital diffuser. *Opt. Lett.* **1994**, *19*, 135–137. [CrossRef] [PubMed]

35. Bjelkhagen, H.I.; Brotherton-Ratcliffe, D. *Ultra-Realistic Imaging: Advanced Techniques in Analogue and Digital Colour Holography*; CRC Press: Boca Raton, FL, USA, 2013; ISBN 978-1-4398-2800-7.

36. Brotherton-Ratcliffe, D.; Rodin, A. Holographic Printer. U.S. Patent 7,161,722, 9 January 2007.

37. Halle, M.W.; Kropp, A.B. Fast computer graphics rendering for full parallax spatial displays. In Proceedings of the SPIE Practical Holography XI and Holographic Materials III, San Jose, CA, USA, 8–14 February 1997; Volume 3011, pp. 105–112.

38. Su, J.; Yan, X.; Jiang, X.; Huang, Y.; Chen, Y.; Zhang, T. Characteristic and optimization of the effective perspective images' segmentation and mosaicking (EPISM) based holographic stereogram: An optical transfer function approachg. *Sci. Rep.* **2018**, *8*, 4488. [CrossRef] [PubMed]

39. Sánchez, A.M.; Prieto, D.V. Design, development and implementation of a low-cost full-parallax holoprinter. In Proceedings of the SPIE Practical Holography XXXII: Displays, Materials, and Applications, San Francisco, CA, USA, 27 January–1 February 2018; Volume 10558, p. 105580H.

40. Benton, S.A.; Bove, V.M. *Holographic Imaging*; John Wiley & Sons: Hoboken, NJ, USA, 2008; ISBN 978-0-470-06806-9.

41. Newswanger, C.; Klug, M. Holograms for the masses. In Proceedings of the 9th International Symposium on Display Holography (ISDH), Cambridge, MA, USA, 25–29 June 2012; Volume 415, p. 012082.

42. Zherdev, A.Y.; Odinokov, S.B.; Lushnikov, D.S.; Shishova, M.V.; Gurylev, O.A.; Kaytukov, C.B. High-aperture diffractive lens for holographic printer. In Proceedings of the SPIE, Holography, Diffractive Optics, and Applications VII, Beijing, China, 12–14 October 2016; Volume 10022, p. 100220I.

43. Park, J.; Kang, H.; Stoykova, E.; Kim, Y.; Hong, S.; Choi, Y.; Kwon, S.; Lee, S. Numerical reconstruction of a full parallax holographic stereogram with radial distortion. *Opt. Express* **2014**, *22*, 20776–20788. [CrossRef] [PubMed]

44. Morozov, A.V.; Putilin, A.N.; Kopenkin, S.S.; Borodin, Y.P.; Druzhin, V.V.; Dubynin, S.E.; Dubinin, G.B. 3D holographic printer: Fast printing approach. *Opt. Express* **2014**, *22*, 2193–2206. [CrossRef] [PubMed]

45. Rong, X.; Yu, X.; Guan, C. Multichannel holographic recording method for three-dimensional displays. *Appl. Opt.* **2011**, *50*, B77–B80. [CrossRef] [PubMed]

46. Yamaguchi, M.; Endoh, H.; Koyama, T.; Ohyama, N. High-speed recording of full-parallax holographic stereograms by a parallel exposure system. *Opt. Eng.* **1996**, *35*, 1556–1559. [CrossRef]

47. Brotherton-Ratcliffe, D.; Zacharovas, S.J.; Bakanas, R.J.; Pileckas, J.; Nikolskij, A.; Kuchin, J. Digital holographic printing using pulsed RGB lasers. *Opt. Eng.* **2011**, *50*, 091307.

48. Bakanas, R.; Jankauskaitė, V.; Bulanovs, A.; Zacharovas, S.; Vilkauskas, A. Comparison of diffraction patterns exposed by pulsed and CW lasers on positive-tone photoresist. *Appl. Opt.* **2017**, *56*, 2241–2249. [CrossRef] [PubMed]

49. Brotherton-Ratcliffe, D.; Vergnes, F.M.R.; Rodin, A.; Grichine, M. Holographic Printer. U.S. Patent 7,800,803, 21 September 2010.

50. Wu, Q.; Wang, H.; Shi, Y.; Li, Y. Color reproduction quantitative analysis of color reflection holography. *Chin. J. Lasers* **2016**, *43*, 213–221. (In Chinese)

51. Yang, F.; Murakami, Y.; Yamaguchi, M. Digital color management in full-color holographic three-dimensional printer. *Appl. Opt.* **2012**, *51*, 4343–4352. [CrossRef] [PubMed]

52. Takano, M.; Shigeta, H.; Nishihara, T.; Yamaguchi, M.; Takahashi, S.; Ohyama, N.; Kobayashi, A.; Iwata, F. Full-color holographic 3D printer. In Proceedings of the SPIE Practical Holography XVII and Holographic Materials IX, Santa Clara, CA, USA, 20–24 January 2003; Volume 5005, p. 126–136.

53. Bjelkhagen, H.I.; Mirlis, E. Color holography to produce highly realistic three-dimensional images. *Appl. Opt.* **2008**, *47*, 123–133. [CrossRef]

54. Maruyama, S.; Ono, Y.; Yamaguchi, M. High-density recording of full-color full-parallax holographic stereogram. In Proceedings of the SPIE, Practical Holography XXII: Materials and Applications, San Jose, CA, USA, 19–24 January 2008; Volume 6912, p. 69120N.

55. Lucente, M. Diffraction-Specific Fringe Computation for Electro-Holography. Ph.D. Thesis, Massachusetts Institute of Technology, Cambridge, MA, USA, 1994.

56. Hong, K.; Park, S.-G.; Yeom, J.; Kim, J.; Chen, N.; Pyun, K.; Choi, C.; Kim, S.; An, J.; Lee, H.-S.; et al. Resolution enhancement of holographic printer using a hogel overlapping method. *Opt. Express* **2013**, *21*, 14047–14055. [CrossRef] [PubMed]

57. Utsugi, T.; Yamaguchi, M. Reduction of the recorded speckle noise in holographic 3D printer. *Opt. Express* **2013**, *21*, 662–674. [CrossRef] [PubMed]

58. Klug, M.A.; Halle, M.W.; Lucente, M.; Plesniak, W.J. A compact prototype one-step Ultragram printer. In Proceedings of the SPIE Practical Holography VII: Imaging and Materials, San Jose, CA, USA, 31 January–5 February 1993; Volume 1914, pp. 15–24.

59. Suzuki, N.; Tomita, Y. Silica-nanoparticle-dispersed methacrylate photopolymers with net diffraction efficiency near 100%. *Appl. Opt.* **2004**, *43*, 2125–2129. [CrossRef] [PubMed]

60. Li, C.; Cao, L.; Wang, Z.; Jin, G. Hybrid polarization-angle multiplexing for volume holography in gold nanoparticle-doped photopolymer. *Opt. Lett.* **2014**, *39*, 6891–6894. [CrossRef] [PubMed]

61. Li, C.; Cao, L.; Li, J.; He, Q.; Jin, G.; Zhang, S.; Zhang, F. Improvement of volume holographic performance by plasmon-induced holographic absorption grating. *Appl. Phys. Lett.* **2013**, *102*, 061108. [CrossRef]

62. Tay, S.; Blanche, P.-A.; Voorakaranam, R.; Tunç, A.V.; Lin, W.; Rokutanda, S.; Gu, T.; Flores, D.; Wang, P.; Li, G.; et al. An updatable holographic three-dimensional display. *Nature* **2008**, *451*, 694–698. [CrossRef] [PubMed]

63. Blanche, P.-A.; Bablumian, A.; Voorakaranam, R.; Christenson, C.; Lin, W.; Gu, T.; Flores, D.; Wang, P.; Hsieh, W.-Y.; Kathaperumal, M.; et al. Holographic three-dimensional telepresence using large-area photorefractive polymer. *Nature* **2010**, *468*, 80–83. [CrossRef] [PubMed]

64. Tsutsumi, N.; Kinashi, K.; Tada, K.; Fukuzawa, K.; Kawabe, Y. Fully updatable three-dimensional holographic stereogram display device based on organic monolithic compound. *Opt. Express* **2013**, *21*, 19880–19884. [CrossRef] [PubMed]

65. Tsutsumi, N.; Kinashi, K.; Sakai, W.; Nishide, J.; Kawabe, Y.; Sasabe, H. Real-time three-dimensional holographic display using a monolithic organic compound dispersed film. *Opt. Mater. Express* **2012**, *2*, 1003–1010. [CrossRef]
66. Gao, H.; Liu, P.; Liu, J.; Zheng, Z.; Yao, Q.; Zhou, W.; Xu, F.; Yu, Y.; Zheng, H. Study on permanent holographic recording in trimethylol propane triacrylate-based photopolymer films with high diffraction efficiency. *J. Opt. Soc. Am. B* **2017**, *34*, B22–B27. [CrossRef]
67. Gao, H.; Liu, P.; Zeng, C.; Yao, Q.; Zheng, Z.; Liu, J.; Zheng, H.; Yu, Y.; Zeng, Z.; Sun, T. Holographic storage of three-dimensional image and data using photopolymer and polymer dispersed liquid crystal films. *Chin. Phys. B* **2016**, *25*, 094205. [CrossRef]
68. Lee, B.; Kim, J.-H.; Moon, K.; Kim, I.-J.; Kim, J. Holographic stereogram printing under the non-vibration environment. In Proceedings of the SPIE Three-Dimensional Imaging, Visualization, and Display 2014, Baltimore, MD, USA, 5–9 May 2014; Volume 9117, p. 911704.
69. Plesniak, W.; Halle, M.; Bove, V.M.; Barabas, J.; Pappu, R. Reconfigurable image projection holograms. *Opt. Eng.* **2006**, *45*, 2365–2372. [CrossRef]
70. St Hilaire, P. Modulation transfer function and optimum sampling of holographic stereograms. *Appl. Opt.* **1994**, *33*, 768–774. [CrossRef] [PubMed]
71. Helseth, L.E. Optical transfer function of three-dimensional display systems. *J. Opt. Soc. Am. A* **2006**, *23*, 816–820. [CrossRef]
72. Jiang, X.; Pei, C.; Liu, J.; Zhao, K.; Yan, X. Optimization of exit pupil function: Improvement on the OTF of full parallax holographic stereograms. *J. Opt.* **2013**, *15*, 125402. [CrossRef]

Article

Holographic Element-Based Effective Perspective Image Segmentation and Mosaicking Holographic Stereogram Printing

Fan Fan [1,2], Xiaoyu Jiang [1,*], Xingpeng Yan [1], Jun Wen [1], Song Chen [1], Teng Zhang [1] and Chao Han [1]

[1] Department of Information Communication, Academy of Army Armored Forces, Beijing 100072, China; fanfan1912@163.com (F.F.); yanxp02@gmail.com (X.Y.); zqywenjun@gmail.com (J.W.); chensong_w@hotmail.com (S.C.); pofeite007@gmail.com (T.Z.); hansjohannes@163.com (C.H.)
[2] Institute of Construction and Development, Academy of Army Research, Beijing 100012, China
* Correspondence: jiangxiaoyu2007@gmail.com

Received: 5 February 2019; Accepted: 25 February 2019; Published: 4 March 2019

Abstract: Effective perspective image segmentation and mosaicking (EPISM) method is an effective holographic stereogram printing method, but a mosaic misplacement of reconstruction image occurred when focusing away from the reconstruction image plane. In this paper, a method known as holographic element-based effective perspective image segmentation and mosaicking is proposed. Holographic element (hogel) correspondence is used in EPISM method as pixel correspondence is used in direct-writing digital holography (DWDH) method to generate effective perspective images segments. The synthetic perspective image for holographic stereogram printing is obtained by mosaicking all the effective perspective images segments. Optical experiments verified that the holographic stereogram printed by the proposed method can provide high-quality reconstruction imagery and solve the mosaic misplacement inherent in the EPISM method.

Keywords: holographic printing; holographic stereogram; holographic element

1. Introduction

Holographic stereogram printing technology is used in many fields. Since Dennis Gabor invented holographic technology in 1948, holographic printing techniques can be categorized into three types: synthetic holographic stereogram printing, holographic fringe printing, and wavefront printing. The most widely used technology is holographic stereogram printing.

Holographic stereogram printing was first proposed by DeBitetto [1] in 1969. The perspective images sampled by a camera are exposed to a holographic recording medium by using a slit, generating a horizontal parallax holographic stereogram. In 1970, King [2] proposed a two-step horizontal parallax holographic stereogram printing technique. This two-step printing technique first records the master holographic plate, Then the second-step of the process is to record the reproduced image of the master holographic plate onto the transfer holographic plate. The production of a holographic stereogram that reproduces with white light and generates an orthoscopic real image is realized.

In 1990, Yamaguchi [3–5] proposed to print a full parallax holographic stereogram in a single step; based on this, a series of studies were carried out. In 1991, Halle [6–12] proposed a single-step holographic stereogram printing technique called Ultragram; this method processes the perspective image obtained by the sampling camera to generate a holographic stereogram that can be used for single-step printing. In this manner, arbitrary depth, full parallax, and undistortion holographic stereogram printing can be realized.

In 2001, Brotherton-Ratcliffe [13] proposed a technique for holographic stereogram printing by using pulsed lasers. In 2008, a single-step direct-writing digital holography printing technology [14,15] was proposed and used in Geola's holographic printing system. This technique employs a pixel corresponding method, replaces the image pixels loaded onto the spatial light modulator (SLM) with the pixels of the perspective image, and accurately reproduces the recorded scene. In 2017, Su and Yuan [16] proposed a method called effective perspective image segmentation and mosaicking (EPISM) method to simulate two-step method by a single-step process. Further research on the EPISM method was carried out by Su and Yan [17–19] to improve image quality and printing efficiency. In addition, many researchers have made valuable research in recent years [20,21].

The EPISM method uses the correspondence between the observation point and hogel to obtain an effective perspective images segments, and the resolution of the perspective image can be fully used by the EPISM method, but the use of an observation point will cause a mosaic misplacement of the reconstruction image when focusing away from the reconstruction image plane. To solve this problem, by using the idea of pixel correspondence of the DWDH method, the hogel correspondence is used as pixel correspondence to generate effective perspective images segments, and holographic element-based EPISM method is proposed.

This paper is divided into the following sections: In Section 2, the basic principles of the DWDH method, the EPISM method, and the holographic element-based EPISM method are introduced. In Section 3, the basic algorithms of holographic element-based EPISM method are given. In Section 4, the proposed method is verified by optical experiments, and the optical experiment results are compared with the experiment results of the EPISM method. In Section 5, conclusions are presented.

2. The Basic Principles

2.1. The Basic Principle of the DWDH Method and the EPISM Method

2.1.1. The Basic Principle of the DWDH Method

The DWDH method converts sampled perspective images into a rearranged image for exposing. This algorithm is actually a pixel transformation from the film plane of the camera to the SLM plane of hogel. It is usually called 'I to S' transformation. The principle of the DWDH method is shown in Figure 1, There are six main planes: hologram plane, SLM plane, projected SLM plane, camera plane, film plane, and projected film plane.

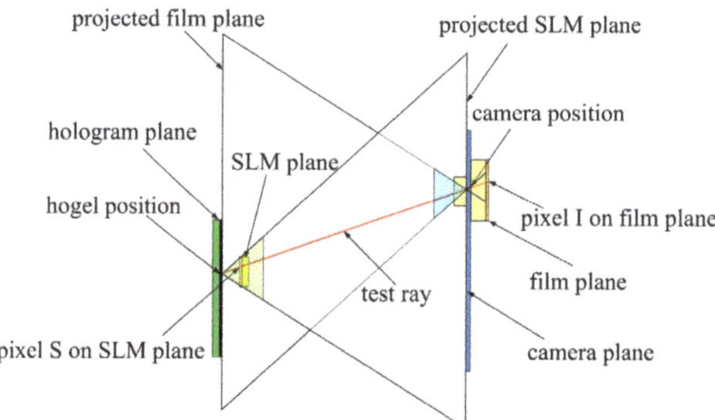

Figure 1. Ray-tracing principle of "I to S" transformation by DWDH method.

In this conversion, the camera lens and the print head of the holographic optical printer are regarded as a point, and a test ray passes through the camera lens and hogel. The pixel where the ray intersects the SLM plane is replaced by the pixel where the ray intersects the film plane, the 'I to S' transformation of a pixel is completed. All pixels on the SLM are replaced by pixels on the film plane, a rearranged image for exposing is obtained. All the hogels on the hologram plane were exposed, and the DWDH holographic stereogram was obtained.

The pixel correspondence of the DWDH method can accurately reproduce the 3D scene, but the resolution of the sampled image is determined by the number of hogels; high-quality holographic stereograms often require hundreds of thousands of sampled images.

2.1.2. The Basic Principle of the EPISM Method

The EPISM method is proposed to achieve a two-step holographic stereogram printing effect by a single-step process. In the EPISM method, a liquid crystal display (LCD) panel is used as SLM. To achieve this purpose, by simulating the propagation process of information from different perspective images, the exposing synthetic perspective images for hogels on H_2 plate can be computer-generated directly. As shown in Figure 2, the EPISM method takes the center of the hogel on H_2 plate as an observation point. According to ray-tracing principle, when observing the hogel on virtual H_1 plate at point O, a viewing frustum with the observation point as the vertex and the hogel as the bottom is obtained. Since the reproduction image of the hogel on virtual H_1 plate is the corresponding perspective image, and the reproduction position is the position of the LCD panel when recording, the intersection of frustum and the reproduced image is the effective perspective image segments of the hogel for the observation point O. Mosaicking all effective perspective images segments together obtains a synthetic perspective image for recording onto the hogel of H_2. Record all the synthetic perspective images corresponding to the hogel on H_2 plate, and a full parallax holographic stereogram based on the EPISM method is received, as shown in Figure 2.

The resolution of the sampled image is determined by the resolution of the LCD panel in the EPISM method. Since the observation point is used on the hogel on the H_2 plate to generate effective perspective images segments, the influence of the hogel size on the effective perspective images segments is ignored, and this ignorance leads to slight mosaic misplacement in the position away from the reconstructed image plane, and the larger the dimension of the hogel on H_2 plate, the more obvious the mosaic misplacement.

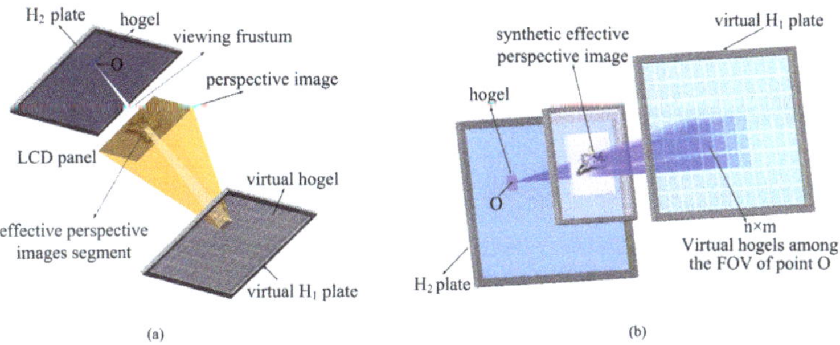

Figure 2. The primitive principle of the proposed method. (**a**) The extraction of effective perspective image segment corresponding to a single virtual hogel. (**b**) The synthetic effective perspective image mosaicked by effective images segments of multiple virtual hogels.

2.2. The Basic Principle of the Holographic Element-Based Effective Perspective Image Segmentation and Mosaicking (EPISM) Methods

Using holographic elements to replace pixels, the effective perspective images segments of adjacent hogels on virtual H_1 plate are mosaicked as shown in Figure 3, then the aliasing perspective images segments of adjacent hogel cannot be simulated by the LCD panel.

To avoid the aliasing of effective perspective images segments of adjacent hogels, the pixel correspondence in the DWDH method is used to establish the hogel correspondence. The DWDH method uses the hogel (as a point) position and the pixel coordinates on the SLM plane to determine the camera position and the pixel coordinates on the film plane. As shown in Figure 4, if the dimension of hogel is considered, the pixel on the SLM plane in the DWDH method should be replaced by a segment of the LCD panel in holographic element-based EPISM method, and the dimension of the segment of the LCD panel is equal to the dimension of the hogel on virtual H_1 plate. The position of the hogel on virtual H_1 plate and the effective perspective images segments of its reconstruction images are determined by the position of the hogel on the H_2 plate and the segment of the LCD panel. A synthetic perspective image based on the holographic element can be obtained by mosaicking all the perspective images segments, and the aliasing of perspective images segments is avoided. Based on this idea, holographic element-based EPISM method is proposed.

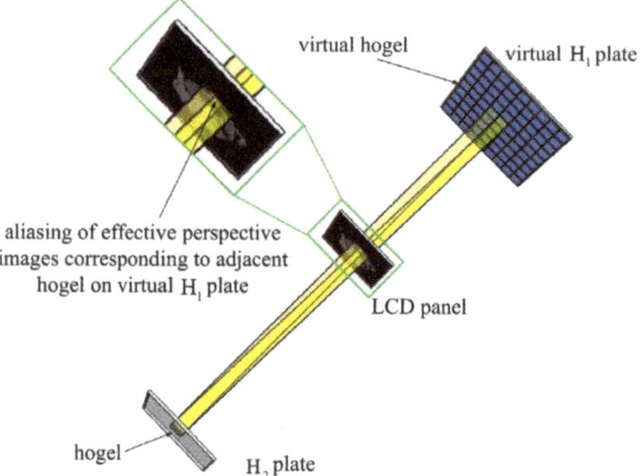

Figure 3. The effective perspective images segments of adjacent hogels are aliased When holographic element is used for the corresponding.

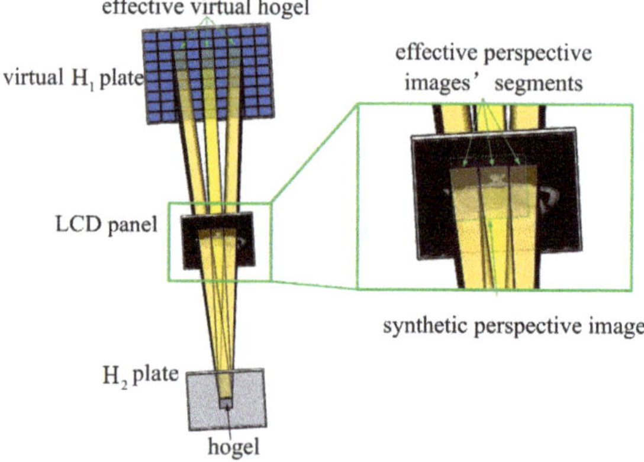

Figure 4. The holographic elements are used as pixels to solve the aliasing problem of adjacent hogels.

The principle of this method is described as follows: as shown in Figure 5, to treat a hogel as a pixel, this requires the same hogel dimensions on the virtual H_1 and H_2 plates, and the dimension of the effective perspective images segments are also the same as the hogel dimension, denoted as l. The dimension of the LCD panel is fixed, and the LCD panel is partitioned according to the dimension of the effective perspective images segments. It is necessary to choose a suitable distance to make the rectangle area formed by the hogel on H_2 plate and the segment of the LCD panel coincide with the hogel on the virtual H_1 plate. The reproducing image of the hogel on the virtual H_1 plate intersecting the segment of LCD generates the effective perspective image segments. For this purpose, as shown in

Figure 6, when the distance between the virtual H_1 plate and the LCD panel is denoted as L_1, we need to make L_1 be an integer multiple of L_2 which is the distance between the H_2 plate and the LCD panel. This ensures that the rectangular area formed by the hogel on H_2 plate and the segment of the LCD panel just falls on the hogel of virtual H_1 plate.

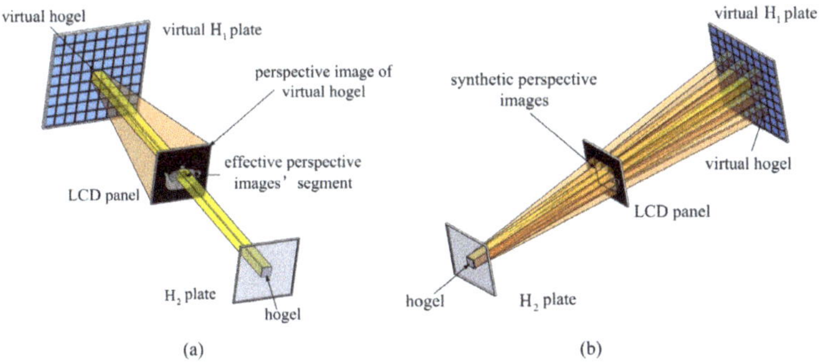

Figure 5. Holographic element-based effective perspective image segmentation and mosaicking methods (**a**) The extraction of effective perspective image segment corresponding to a single virtual hogel (**b**) The synthetic effective perspective image mosaicked by effective images segments of multiple virtual hogels.

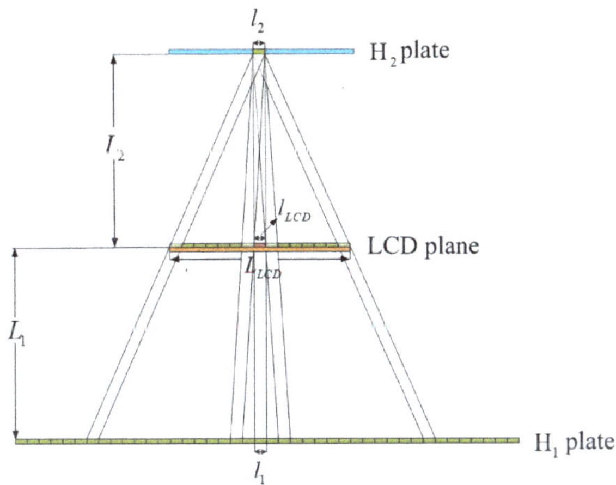

Figure 6. Parameter setting of holographic element-based EPISM methods.

3. The Basic Algorithm of the Holographic Element-Based EPISM Method

3.1. The Selection of the Hogel on Virtual H_1 Plate

First, hogel number of the H_1 plate should be fixed. As shown in Figure 6, to make full use of the LCD resolution, choosing the appropriate field of view (FOV) of the hogel on the H_2 plate and L_2 makes the display area of the LCD completely contained by the FOV. The display area dimensions of the LCD panel is denoted as L_{LCD}. Using l_1 and l_2 to represent the dimension of hogel on H_1 plate, we denote the dimension of effective perspective images segments as l_{LCD}. According to the geometric relationship, there is $l_1 = l_2 = l_{LCD} = l$. As shown in Figure 7, to determine the number of the hogel on the virtual H_1 plate, we first determine the number of segments on the LCD panel, which is presented by n_{LCD}. Choosing the hogel on H_1 plate corresponds to the hogel on the H_2 plate, then

$$n_{\mathrm{LCD}} = \begin{cases} \frac{L_{\mathrm{LCD}}}{l}, & \text{if} \quad \frac{L_{\mathrm{LCD}}}{l} \text{ is odd} \\ \frac{L_{\mathrm{LCD}}}{l} - 1, & \text{if} \quad \frac{L_{\mathrm{LCD}}}{l} \text{ is even} \end{cases} \tag{1}$$

where n_{LCD} should be odd here, n_{hogel} represents the segment number on both sides of the center segment on the LCD panel, $n_{\mathrm{hogel}} = \frac{n_{\mathrm{LCD}}-1}{2}$.

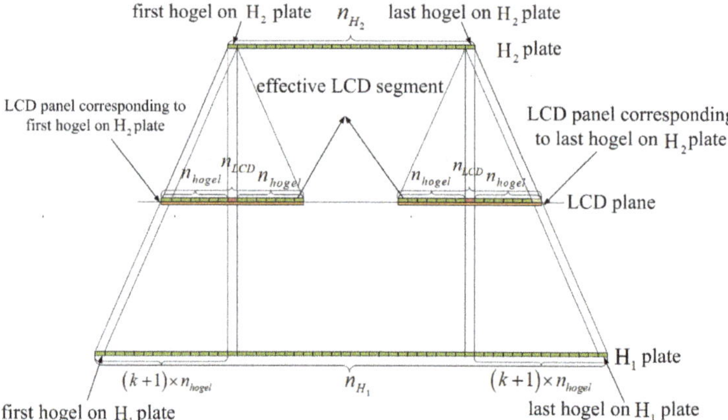

Figure 7. Determination of the hogel number for holographic element-based EPISM methods at $k = 1$.

Introducing an index k, k represents the ratio of the distance from the LCD panel to virtual H_1 plate and H_2 plate, and $k = \frac{L_1}{L_2}$. According to the geometric relationship, when $k = 1$, the hogel corresponding to the adjacent segment of the LCD panel is separated by one hogel. When L_1 and L_2 are not equal, let k be a positive integer; this means that the adjacent segment of the LCD panel has corresponding hogels on H_1 plate. The interval hogel number between adjacent hogels is exactly equal to k.

The number of hogels on the H_1 plate can be represented as $n_{H_1} = 2 \times n_{\mathrm{hogel}} \times (k+1) + n_{H_2}$.

3.2. Effective Perspective Image Segmentation and Mosaicking

As shown in Figure 8, for the ith hogel on the H_2 plate, the order of the hogel on the virtual H_1 plate that corresponds to the ith hogel on the H_2 plate should be $i + n_{\mathrm{hogel}} \times (k+1)$. For the ith hogel on the H_2 plate, the segment on the center of the LCD panel is denoted as the $0th$ segment, and let $m = \left(-n_{\mathrm{hogel}}, n_{\mathrm{hogel}}\right)$, the order of hogel on the virtual H_1 plate that corresponds to the mth segment of the LCD panel and ith hogel on the H_2 plate is $i + n_{\mathrm{hogel}} \times (k+1) + m \times (k+1)$, and this is the hogel we are looking for. Left endpoint of the hogel on the virtual H_1 plate that corresponds to the $0th$ segment of the LCD panel is denoted as x_0; according to the previous results, we have $x_0 = \left[i + n_{\mathrm{hogel}} \times (k+1) - 1\right] \times l_1$, left end the hogel on the H_1 plate that corresponds to the mth segment of LCD denoted as s, $s = \left[i + m \times (k+1) + n_{\mathrm{hogel}} \times (k+1) - 1\right] \times l_1$, $h = (x_0 - s) \times \left(1 - \frac{1}{k+1}\right) \times l_1$ represents the distance between the corresponding effective perspective images segments and the location of hogel on H_1 plate. The left end coordinates of the segment are denoted as e,

$$e = \left[\left(h - \frac{1}{2}\right) \times l_1 + \frac{L_{\mathrm{LCD}}}{2}\right] \times 100 + 1 \tag{2}$$

and the right end coordinates of the segment denote as f,

$$f = \left[\left(h - \frac{1}{2}\right) \times l_1 + \frac{L_{LCD}}{2}\right] \times 100 + e_p \tag{3}$$

where $e_p = l_1 \times 100$, according to the left and right ends of the segment, determine the mth effective perspective images segments. By mosaicking all the selected effective perspective images segments in sequence, we obtain the synthetic perspective image corresponding to the ith hogel on H_2 plate. By exposing the hogel on H_2 plate in sequence, the holographic stereogram is obtained.

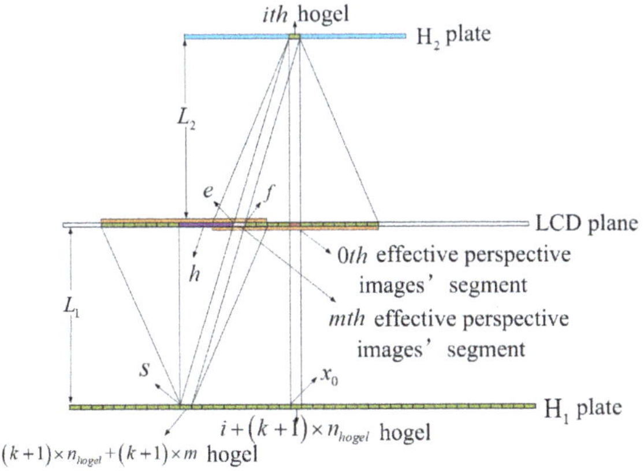

Figure 8. Diagram of effective perspective images segments.

4. Experimental Verification

An LCD panel was used for exposing holographic stereogram, for convenience in calculation, the corresponding resolution on 1 cm × 1 cm display area of LCD panel was 100 × 100 pixels. To this end, the Panasonic LCD panel (VVX09F035M20) had been selected, its size was 8.9 inches, and the resolution was 1920 × 1200. Select 10 cm × 10 cm area in the center of LCD panel as the effective display area. Thus, the resolution of the synthetic perspective image for exposure is 1000 × 1000 pixels.

A mapped teapot model is used as the 3D scene. The depth was 4.8 cm, the height was 3 cm, the width was 4.2 cm, and it was tipped 40°. Let $k = 1$, $l_1 = l_2 = 0.2$ cm, $L_1 = L_2 = 18.6$ cm. A camera was set in front of the teapot, and the FOV was 30°. The size of H_2 plate was 6 cm × 6 cm, $n_{H_2} = 30$, and the hogel number of H_2 plate was 30 × 30 = 900. According to the previous formula, $n_{H_1} = 126$, the number of the camera position should be 126 × 126 = 15,876.

As shown in Figure 9, the synthetic perspective images of the holographic element-based EPISM method are given, image(6, 16) is the synthetic perspective image corresponding to the order number of the sixth row and the sixteenth column's hogel on H_2 plate. As EPISM methods, the synthetic perspective images are pseudoscopic images, and the reconstructed scene can reproduce the recording scene truthfully in correct depth.

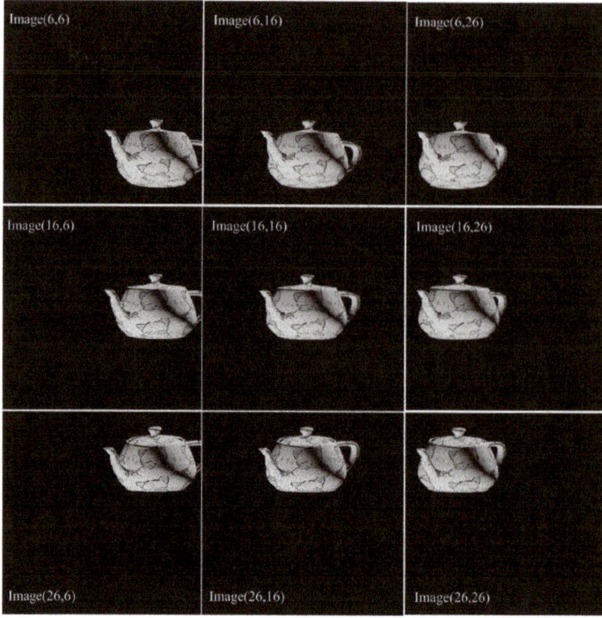

Figure 9. Synthetic perspective image of the holographic element-based EPISM methods.

As shown in Figure 10, the optical setup using for holographic-based EPISM method holographic stereogram is presented. A 639 nm custom-made red laser was used as the laser source for holographic stereogram printing. The max power of laser source was 1.2w. An electric shutter was used to control the exposure time. Its model was Sigma Koki SSH-C2B. The direction of laser beam was changed by a reflector. A non-polarizing beam splitter(NPBS) was set in the new direction of the laser beam, and the laser beam was divided into two laser beams perpendicular to each other. The laser beam vertical to the original direction was denoted as signal beam, and the laser beam in the same direction as the original direction was denoted as reference beam. An attenuator was set in the same direction of signal beam to adjust the intensities, and a spatial filter was used to expand the signal beam. The LCD panel introduced earlier was used to load the synthetic perspective image, a diffuser was used to scatter the signal beam to the hogel aperture, and the diffuser was placed close to the LCD panel. The reference beam direction was changed by a reflector, and an attenuator was used to adjust the intensities of the reference beam. Another reflector was used to change the reference beam to the backside of holographic plate, and a spatial filter was used to expand the reference beam. Then, the expanded beam was changed into a uniform plane wave by a collimating lens; the focal length of the lens is 75 mm. A manual ultrafine silver-halide plate for He-Ne laser was used as holographic plate; its grain size was about 10–12 nm. The holographic plate placed 18.6 cm away from LCD panel, two diaphragms with apertures are placed on both sides of the LCD panel, the signal beam and reference beam passing through the aperture interfere on the holographic plate to generate hogel. A X-Y motorized stage(KSA300, MC600) was used to carry the holographic plate; it can move both on the horizontal and vertical rail. The step of the X-Y motorized stage was l_2. A computer was used to time-synchronously control the shutter, the LCD panel, and the motorized stage.

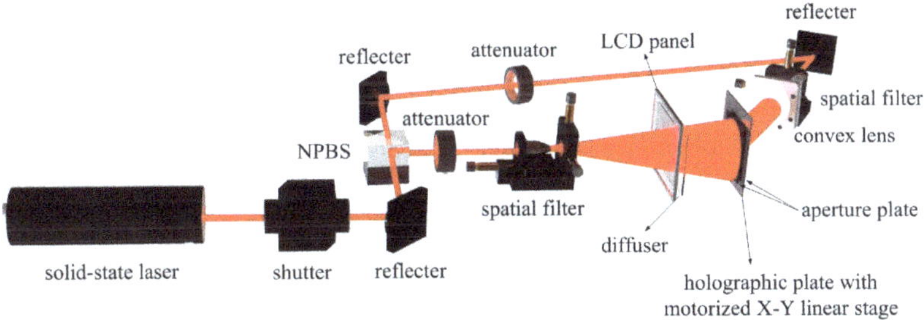

Figure 10. The optical setup of the synthetic holographic stereogram printing system.

The synthetic perspective image used to expose the hogel on the H_2 plate needs to be flipped horizontally, because the LCD panel corresponding to the virtual H_1 plate and the LCD panel corresponding to the H_2 plate were in the opposite direction.

Equation $T_e = E/(P_s + P_r)$ was used to express the exposure time, where P_s is the intensity of signal beam energy, P_r is the intensity of reference beam energy, and E denotes as the light sensitivity of holographic plate. In this experiment, $E = 1250 \ \mu J/cm^2$, $P_s = 10 \ \mu J/cm^2$, $P_r = 300 \ \mu J/cm^2$. The energy ratio between P_s and P_r was selected as 1:30; this ratio can greatly reduce the printing time on the premise of guaranteeing the image quality, and the exposure time was $T_e = 4$ s. The waiting time was 4 times as much as exposure time to reduce the vibration result, and the waiting time $T_w = 16$ s. The printing time was the sum of exposure time and waiting time, the hogel's printing time $T_h = T_e + T_w = 20$ s, and total printing time $T_t = T_h \times N_{H_2} \times N_{H_2} = 18,000$ s.

Figure 11 shows the reconstruction images of an optical experiment result. A conjugate beam of reference beam was used to illuminate the holographic plate, and a camera used to record the reconstruction image of holographic plate. Its model is Canon EOS 5D-mark3 DS126091, the focal length of camera lens is 100 mm, the camera was set 50 cm away from holographic plate, and a real image of 3D scene was captured by camera.

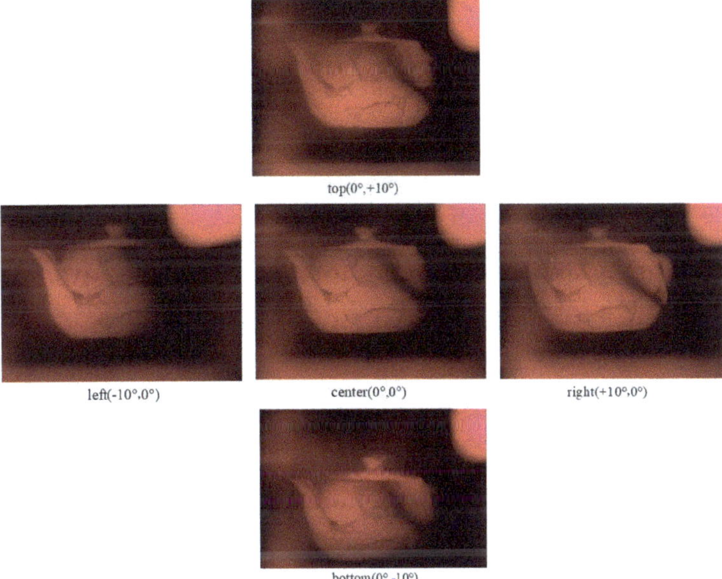

Figure 11. Optical reconstruction images in different viewing position obtained by experiments

As shown in Figure 11, by the effective display size and the distance L_2, the FOV of H_2 plate is about 30°. The details of the original scene are perfectly reproduced, and full parallax information can be presented by the holographic element-based EPISM method. When the camera position is close to the limited FOV, due to the use of simple camera sampling, the reconstructed scene is as incomplete as the original scene, with the observation area unable to achieve 30° FOV.

To confirm the position of reconstructed image, we put two rulers together with the holographic plate. A camera was put 50 cm away from holographic plate, and the result is shown in Figure 12. The position relation of the rulers and holographic plate are shown in Figure 12a; the ruler on the left side is on the same plane as the holographic plate, and the ruler on the right side is on the position of the reconstruction image, away from the holographic plate by 18.6 cm. In Figure 12b, focusing on the left ruler, both the figure on the ruler and the hogel on holographic plate are clear, and the figure on the right ruler is blurred. In Figure 12c, focusing on the right ruler, the figure on the ruler and the map on the reconstruction image are clear, and the left ruler is blurred.

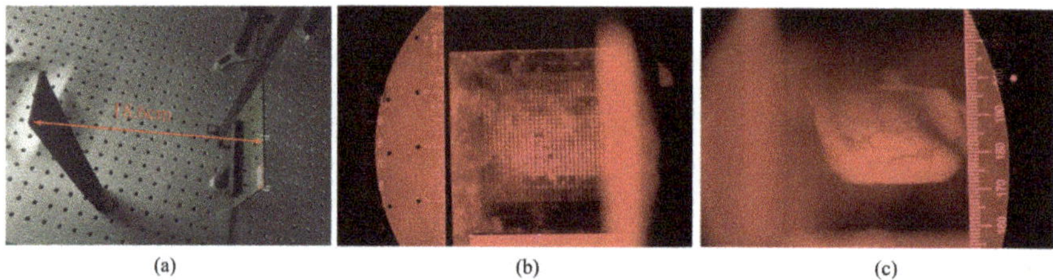

Figure 12. (**a**)The spatial position relation of two rulers and the holographic plate. (**b**) Focused on the left ruler, both rulers the holographic plate are clear. (**c**) Focused on the right ruler, both ruler and the surface of the teapot is clear.

As shown in Figure 13, the camera was also put in the position 50 cm away from the holographic plate; an EPISM method holographic stereogram was made according to the same configuration. Comparing the reconstructed images obtained by the EPISM method and the holographic element-based EPISM method, Figure 13a shows the position relation of rulers and the holographic plate; the farther ruler is placed in the position of the reconstruction image, with the ruler as far from the holographic plate as 18.6 cm, and the closer ruler is placed 1.5 cm behind the position of the farther ruler. Figure 13b,c show the reconstruction images of the EPISM method. Both the figure on the ruler and the map on the teapot are clear when focusing on the farther ruler in Figure 13b; as shown in Figure 13c, the figure on the ruler is clear and the map on the teapot is blurred when focusing on the closer ruler, and there is a mosaic misplacement on the map of the teapot. The reconstruction images of the holographic element-based EPISM method are shown in Figure 13d,e. As shown in Figure 13d, the figure on the ruler and the map of the teapot are clear when focusing on the farther ruler; the figure on the ruler is clear and the map on the teapot is only blurred when focusing on the closer ruler in Figure 13e.

Figure 14 shows the detail of Figure 13c,e, Figure 14a magnifies part of the Figure 13c, and the mosaic misplacement of the map of the teapot is revealed; Figure 14b magnifies the same part on the Figure 13e, and this part of the surface map of the teapot is blurred without mosaic misplacement. The brightness and contrast of the reconstructed image will be affected by the efficiency of the developer and the slight adjustment of the camera when shooting, but under the same conditions, the effective perspective images segments in EPISM method is smaller than in the proposed method, whether this difference will affect the quality of the reconstruction image still needs further study.

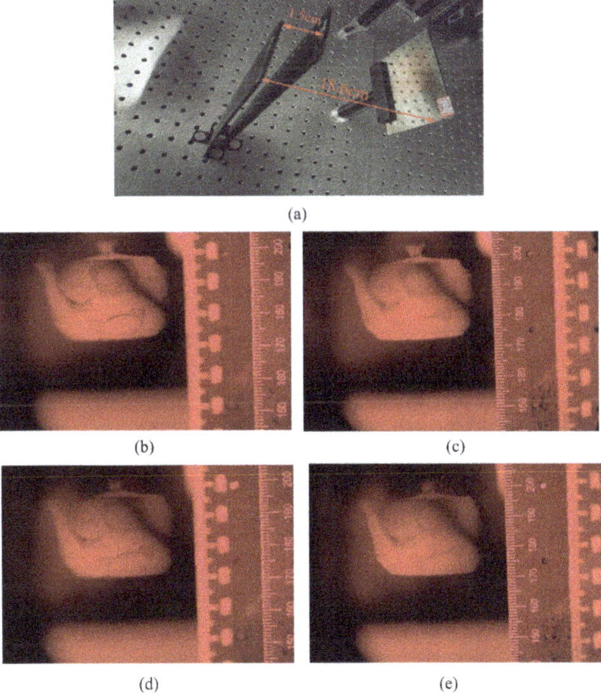

Figure 13. Comparison of reconstructed images between the EPISM method and the holographic element-based EPISM method. (b,c) are the reconstructed images of the EPISM method, (d,e) are the reconstruction images of the holographic element-based EPISM method. (a) The spatial position relation of two rulers and the holographic plate. (b) The reconstruction image of the EPISM method when focused on the farther ruler. (c) The reconstruction image of the EPISM method when focused on the closer ruler. (d) The reconstruction image of the holographic element-based EPISM method when focused on the farther ruler. (e) The reconstruction image of the holographic element-based EPISM method when focused on the closer ruler.

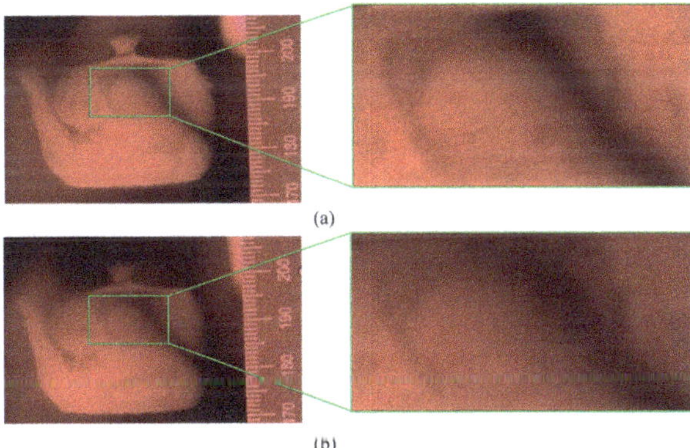

Figure 14. Comparison of details between reconstructed images of the EPISM method and reconstructed images of the holographic element-based EPISM method (a) Details of reconstructed images of the EPISM method (b) Details of reconstructed images of the holographic element-based EPISM method.

The experimental results show that the holographic element-based EPISM method can reproduce the 3D scene as well as the EPISM method, and solve the problem of mosaic misplacement in the EPISM method, but there are some restrictions on the EPISM method. The condition $l_1 = l_2$ and $L_1 = L_2$ must be satisfied so that there will be a hogel on virtual H_1 plate corresponding to the area formed by hogel correspondence between H_2 plate and LCD panel. In the future, we will consider how to generate effective perspective image segments when we relax this restriction appropriately.

5. Conclusions

In this paper, holographic element-based effective perspective image segmentation and mosaicking method was proposed for holographic stereogram printing. By using hogel correspondence to obtain the effective perspective images segments, the mosaic misplacement in the EPISM method is avoided, and high-quality reconstruction image also can be obtained as the EPISM method; however, in order to establish hogel correspondence, some restrictions are introduced in the proposed method. We will try to remove these restrictions in future work.

Author Contributions: Conceptualization, F.F. and X.J.; Methodology, F.F.; Software, J.W.; Validation, J.W.; Formal Analysis, S.C.; Writing—Original Draft Preparation, T.Z.; Writing—Review & Editing, C.H.; Supervision, X.Y.; Project Administration, X.J.; Funding Acquisition, X.Y.

Funding: This research was funded by [the National Key Research and Development Program of China] grant number [2017YFB 1104500], [National Natural Science Foundation of China] grant number [61775240], [Foundation for the Author of National Excellent Doctoral Dissertation of the People's Republic of China] (FANEDD) grant number [201432].

Conflicts of Interest: The author declares no conflict of interest.

References

1. DeBitetto, D.J. Holographic Panoramic Stereograms Synthesized from White Light Recordings. *Appl. Opt.* **1969**, *8*, 1740–1741. [CrossRef] [PubMed]

2. King, M. A new approach to computer-generated holography. *Appl. Opt.* **1970**, *9*, 471–475. [CrossRef] [PubMed]

3. Yamaguchi, M.; Ohyama, N. Holographic 3-D printer. *Proc. SPIE* **1990**, *1212*, 84–92. [CrossRef]

4. Yamaguchi, M.; Ohyama, N.; Honda, T. Holographic three-dimensional printer: new method. *Appl. Opt.* **1992**, *31*, 217–222. [CrossRef] [PubMed]

5. Yamaguchi, M.; Endoh, H.; Honda, T.; Ohyama, N. High-quality recording of a full-parallax holographic sterogram with a digital diffuser. *Opt. Lett.* **1994**, *19*, 135–137. [CrossRef] [PubMed]

6. Halle, M.W.; Benton, S.A.; Klug, M.A.; Underkoffler, J.S. Ultrgram: A generalized holographic stereogram. In Proceedings of the SPIE Practical Holography V, San Jose, CA, USA, 1–7 February 1991; Volume 1461, pp. 142–155. [CrossRef]

7. Halle, M.W. The Generalized Holographic Stereogram. Ph.D. Thesis, Massachusetts Institute of Technology, Cambridge, MA, USA, 1991.

8. Klug, M.A.; Halle, M.W.; Hubel, P.M. Full-color ultragrams. *Proc. SPIE* **1992**, *1667*, 110–119. [CrossRef]

9. Klug, M.A.; Halle, M.W.; Lucente, M.E.; Plesniak, W.J. Compact prototype one-step Ultragram printer. *Proc. SPIE* **1993**, *1914*, 15–24. [CrossRef]

10. Klug, M.A. Holographic stereograms as discrete imaging systems. *Proc. SPIE* **1994**, *2176*, 73–84. [CrossRef]

11. Klug, M.A.; Kropp, A.B. Fast computer graphics rendering for full parallax spatial displays. *Proc. SPIE* **1997**, *3011*, 105–114. [CrossRef]

12. Klug, M.A.; Klein, A.; Plesniak, W.J.; Kropp, A.B.; Chen, B.J. Optics for full-parallax holographic stereograms. *Proc. SPIE* **1997**, *3011*, 78–88. [CrossRef]

13. Brotherton-Ratcliffe, D.; Vergnes, F.; Rodin, A.; Grichine, M. Holographic Printer. U.S. Patent 6930811, 16 August 2005.

14. Brotherton-Ratcliffe, D.; Rodin, A.; Hrynkiw, L. A Method of Writing a Composite 1-Step Hologram. U.S. Patent 7333252, 19 February 2008.

15. Brotherton-Ratcliffe, D.; Nikolskij, A.; Zacharovas, S.; Pileckas, J.; Bakanas, R. Image Capture System for a Digital Holographic Printer. U.S. Patent 8154584, 10 April 2010.

16. Su, J.; Yuan, Q.; Huang, Y.Q.; Jiang, X.Y.; Yan, X.P. Method of single-step full parallax synthetic holographic stereogram printing based on effective perspective images' segmentation and mosaicking. *Opt. Express* **2017**, *25*, 23523–23544. [CrossRef] [PubMed]

17. Su, J.; Yan, X.P.; Jiang, X.Y.; Huang, Y.Q.; Chen, Y.B.; Zhang, T. Characteristic and optimization of the effective perspective images' segmentation and mosaicking (EPISM) based holographic stereogram: An optical transfer function approach. *Sci. Rep.* **2018**, *8*, 4488. [CrossRef] [PubMed]

18. Su, J.; Yan, X.P.; Huang, Y.Q.; Jiang, X.Y.; Chen, Y.B.; Zhang, T. Improvement of printing efficiency in holographic stereogram printing with the combination of a field lens and holographic diffuser. *Appl. Opt.* **2018**, *57*, 7159–7166. [CrossRef] [PubMed]

19. Su, J.; Yan, X.P.; Huang, Y.Q.; Jiang, X.Y.; Chen, Y.B.; Zhang, T. Progress in the Synthetic Holographic Stereogram Printing Technique. *Appl. Sci.* **2018**, *8*, 851. [CrossRef]

20. Hong, K.; Park, S.; Yeom, J.; Kim, J.; Chen, N.; Pyun, K.; Choi, C.; Kim, S.; An, J.; Lee, H.; et al. Resolution enhancement of holographic printer using a hogel overlapping method. *Opt. Express* **2013**, *21*, 14047–14055. [CrossRef] [PubMed]

21. Park, J.; Kang, H.; Stoykova, E.; Kim, Y.; Hong, S.; Choi, Y.; Kim, Y.; Kwon, S.; Lee, S. Numerical reconstruction of a full parallax holographic stereogram with radial distortion. *Opt. Express* **2014**, *22*, 20776–20788. [CrossRef] [PubMed]

 *applied sciences*

Article

Dual-View Integral Imaging 3D Display Based on Multiplexed Lens-Array Holographic Optical Element

Hanle Zhang [1,2], Huan Deng [2,*], Minyang He [1,2], Dahai Li [2] and Qionghua Wang [1,*]

[1] School of Instrumentation and Optoelectronic Engineering, Beihang University, Beijing 100191, China
[2] School of Electronics and Information Engineering, Sichuan University, Chengdu 610065, China
* Correspondence: huandeng@scu.edu.cn (H.D.); qionghua@buaa.edu.cn (Q.W.)

Received: 16 August 2019; Accepted: 11 September 2019; Published: 13 September 2019

Featured Application: In this paper, the dual-view integral imaging 3D display can be used in head-up display, augmented reality and so on.

Abstract: We propose a dual-view integral imaging 3D display based on a multiplexed lens-array holographic optical element (MHOE). A MHOE is a volume holographic optical element obtained by multiplexing technology, which can be used for dual-view integral imaging 3D display due to the angle selectivity of the volume HOE. In the fabrication of the MHOE, two spherical wavefront arrays with different incident angles are recorded using photopolymer material. In the reconstruction, two projectors are used to project the elemental image arrays (EIA) with corresponding angles for two viewing zones. We have developed a prototype of the dual-view integral imaging display. The experimental results demonstrate the correctness of the theory.

Keywords: dual-view display; 3D display; integral imaging; lens-array holographic optical element

1. Introduction

Dual-view display provides different images for viewers in different directions. It can meet the needs of different viewers at the same time, and it saves a lot of space and cost. It is increasingly commercialized, especially for car navigation, home entertainment and business applications. For example, when a dual-view display is used in an automobile, it can represent navigation information for the driver and entertainment information for the passenger simultaneously. Some dual-view liquid crystal displays have been proposed [1,2], but they can only display 2D images. For automobile applications, optical see-through capacity is needed because the driver needs to see the road while watching the navigation information.

Recently, 3D displays, including holography display [3–5], integral imaging display [6–9], volumetric display [10,11] and so on, have received extensive attention. Integral imaging is considered to be one of the most promising display technologies. It has the advantages of full parallax [12] and continuous viewpoint [13], and it supplies various physiological depth cues, so viewers can freely change the accommodation and convergence without feeling any visual discomfort [14,15]. Integral imaging has a small viewing angle, and the viewing zone is periodically repeated. Therefore, it is suitable for dual-view or multi-view displays. By using some kinds of multiplexing methods, it can present different 3D images in different directions, and meet the needs of different viewers. In recent years, our team proposed several ways to implement dual-view 3D displays [16–18]. We proposed a dual-view integral imaging 3D display [16], but the viewing angle is only half of the conventional one. In order to increase the viewing angle of the dual-view 3D display, we proposed a dual-view integral imaging 3D display using polarizer parallax barriers [17], but the resolution of each 3D image is reduced to half of the conventional one. In order to solve these issues, Seoul University proposed a projection-type dual-view 3D display based on integral imaging [19], but the display system is bulky.

Then we proposed a dual-view integral imaging 3D display using an orthogonal polarizer array and polarization switcher [18], but the brightness is not high.

In this paper, we propose a dual-view integral imaging 3D display system based on a MHOE. The viewing angle and resolution of 3D images in the dual viewing zones are almost same as the conventional one. The dual-view 3D display system can provide different 3D images in different zones via controlling the projection beams. In Section 2, the fabrication and reconstruction principles of the MHOE are analyzed, and the projection part of the dual-view 3D display system is designed. In Section 3, the different 3D images are displayed in different viewing zones, and the imaging characteristics of the MHOE are analyzed. The principle of the proposed system is verified by the display experiments.

2. Structure and Principle

The schematic diagram of the dual-view integral imaging 3D display system by using MHOE is shown in Figure 1, which consists of an MHOE, projector I and II. The MHOE is used as a projection screen, which records the wavefronts of a micro-lens array from two different directions by using two-step fabrication methods. Projection beams I are projected by projector I, which contain elemental image array (EIA) I, and projection beams I are projected on the MHOE with a pair of angles θ_2 and θ_3, which satisfies the Bragg condition. Then, 3D images I are reconstructed in the left viewing zone. Projection beams II with EIA II are projected by projector II, and projection beams II are incident onto the MHOE with a pair of angles θ_5 and θ_6. Then the Bragg condition is satisfied and 3D images II are reconstructed in the right viewing zone. Finally, two different 3D images are reconstructed in the dual viewing zones, respectively.

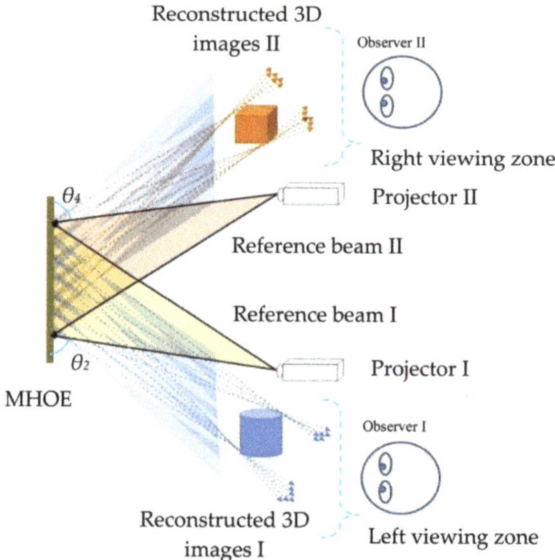

Figure 1. Schematic diagram of the dual-view 3D display system by using the MHOE.

2.1. Fabrication and Reconstruction Principles of the MHOE

Compared with the conventional holographic optical element (HOE), the MHOE is recorded on the volume hologram, which has the characteristics of angular and spatial multiplexing. HOE has angular selectivity; when the incident beam deviates from the Bragg condition, the diffraction efficiency of the reconstructed beam is reduced. When the incident beam greatly deviates from Bragg condition, the diffracted beam cannot be reconstructed. Therefore, multiple optical elements can be recorded on a single holographic plate by angular and spatial multiplexing.

The angular and spatial multiplexing method are used in the fabrication process. Two sets of micro-lens array wavefronts are recorded in a single green sensitive photopolymer material with two different angles. The fabrication process is divided into two steps, which are shown in Figure 2. In the first step, the parallel beam I is incident into the micro-lens array at an oblique angle θ_1, and then the spherical wavefront array I is generated. The reference beam I is generated by point source I. Reference beam I interferes with spherical wavefront array I with a pair of angles θ_2 and θ_3, as shown in Figure 2a. In the second step, the parallel beam II is incident into the micro-lens array at an oblique angle θ_4, and then the spherical wavefront array II is generated. The reference beam II is generated by point source II, and reference beam II interferes with spherical wave array II with a pair of angles θ_5 and θ_6, as show in Figure 2b. The parallel beams I and II are symmetric to each other. By strictly controlling the exposure time of the first and second steps in the fabrication process. Finally, the required MHOE is obtained.

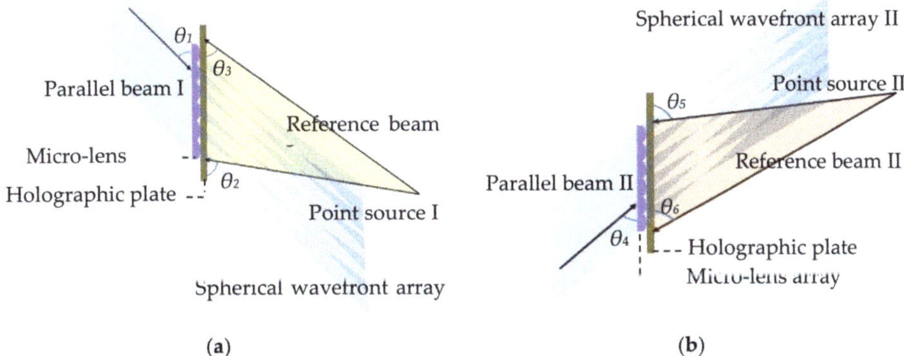

Figure 2. Fabrication principle of the MHOE, (**a**) the first step and (**b**) the second step.

The reconstruction principle of the MHOE is shown in Figure 3. Reference beam I is projected onto the MHOE with a pair of angles θ_2 and θ_3. The incident angles of the reference beam I are consistent with the angles of fabrication, and the Bragg condition is satisfied. So the spherical wavefront array I can be reconstructed. Reference beam II is also projected onto the MHOE with a pair of angles θ_5 and θ_6. The incident angles of the reference beam II are the same as the angles of fabrication, and the Bragg condition is also satisfied. Then, the spherical wavefront II is reconstructed. Therefore, the MHOE has the optical characteristics of two sets of micro-lens arrays in different directions. Reference beams I and II are generated by the point source I and II, which have various directional light components. In the reconstruction of 3D images, a Bragg-mismatch occurs when reference beams are produced by projector I and II. So, the angular variation caused by angular deviation of the reconstructed beam should be analyzed.

2.2. MHOE as Imaging Optical Element in the Dual-View 3D Display

In the fabrication process, reference beams interfere with the spherical wavefront arrays. The spherical wavefront array consists of many small spherical waves, and the interference fringes are recorded on the holographic material. Each elemental lens in the MHOE is generated by the interference of a large spherical wave (reference beam) and a small spherical wave. The grating vectors of different positions on the MHOE are different. The reference beam is considered as a series of plane beams [20], whose wave vectors ($\mathbf{K_{r1 \sim rn}}$) are shown in Figure 4a. The spherical wavefront array can be regarded as a series of small spherical waves, which is shown in Figure 2. Each small spherical wave in the spherical wavefront array can also be considered as a series of plane beams, whose wave vectors ($\mathbf{K_{s1\sim sn}}$) are shown in Figure 4a. So the change of the reconstructed beam of the MHOE could be explained by the wave vector relationship of the plane beams.

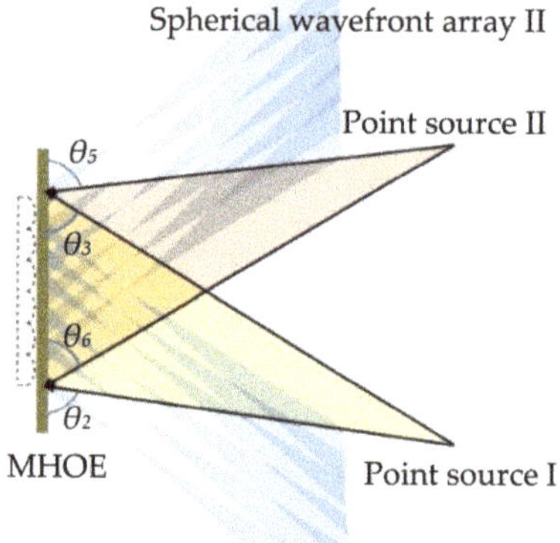

Figure 3. Reconstruction principle of the MHOE.

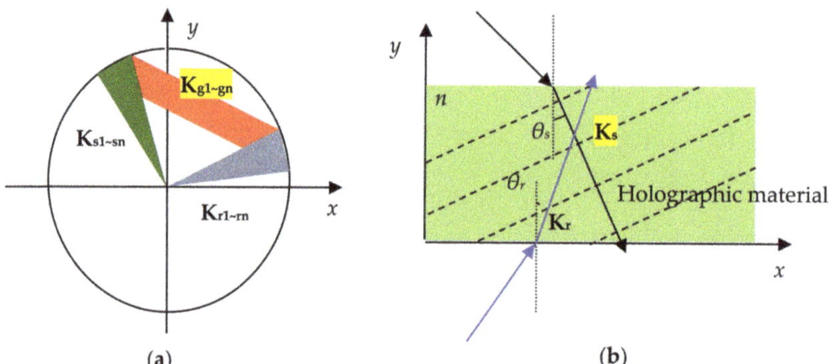

Figure 4. (**a**) Grating vector diagram of the elemental lens in the MHOE, and (**b**) wavefront recording schematic diagram of the elemental lens in the MHOE.

The recorded schematic diagram of the small spherical wave in the MHOE is shown in Figure 4b; the recorded grating vectors $\mathbf{K_g}$ can be expressed as [20]:

$$\mathbf{K_g} = \vec{x}\left[|\mathbf{K_s}|\sin[\theta_s] - |\mathbf{K_r}|\sin[\theta_r]\right] + \vec{y}\left[|\mathbf{K_s}|\cos[\theta_s] + |\mathbf{K_r}|\cos[\theta_r]\right] \tag{1}$$

where $\mathbf{K_s}$ and $\mathbf{K_r}$ are the vectors of the small spherical wave in the spherical wavefront array and reference beam, respectively. In the reconstruction process, the projection beam project onto the MHOE with the vector $\mathbf{K_p}$. The angle of the reconstructed beam can be expressed as [20]:

$$\theta_c = \sin^{-1}\left[\sin[\theta_p] + \sin[\theta_s] - \sin[\theta_r]\right] \tag{2}$$

When the projection angle θ_p of the projection beam, the incident angle θ_r of the reference beam and the incident angle θ_s of the small spherical wave incident into the MHOE are determined, the angle

θ_c of the reconstructed beam changes almost linearly, within the range of the projection angle of the projection beam. Therefore, the MHOE is suitable to be used as the imaging optical element in the dual-view 3D display.

2.3. Design of Projection Part for the Dual-View 3D Display

The projection part for the MHOE has to satisfy the Bragg conditions. Projection beam I must satisfy the projection angles θ_2 and θ_3 for reconstructed 3D images I in the left viewing zone, and projection beam II also must satisfy the projection angles θ_5 and θ_6 for reconstructed 3D image II in the right viewing zone. In this section, the projection part is realized based on two projectors, so that 3D image I and II are displayed simultaneously, as depicted in Figure 1.

The principle and design of the projection part for the dual-view 3D display are shown in Figure 5. When reference beam I is projected onto the MHOE with a pair of angles θ_2 and θ_3, spherical wavefront array I is reconstructed with the angle θ_1. When reference beam II is projected onto the MHOE with a pair of angles θ_5 and θ_6, spherical wavefront array II is reconstructed with the angle θ_4. The best view distance of the dual-view 3D display system is L, the focal length of the MHOE is f and the pitch is p. According to the geometric optics, the viewing angle of the left viewing zone can be expressed as:

$$\Omega_1 = \arctan\frac{2f\tan\theta_1}{2f - p\tan\theta_1} - \arctan\frac{2f\tan\theta_1}{2f + p\tan\theta_1} \tag{3}$$

and the viewing angle of the right viewing zone can be expressed as:

$$\Omega_2 = \arctan\frac{2f\tan\theta_4}{2f - p\tan\theta_4} - \arctan\frac{2f\tan\theta_4}{2f + p\tan\theta_4} \tag{4}$$

The projection part is designed to make sure that the left and right view zone are completely separated, and reconstructed 3D images I and II cannot cause crosstalk, which is shown in Figure 5.

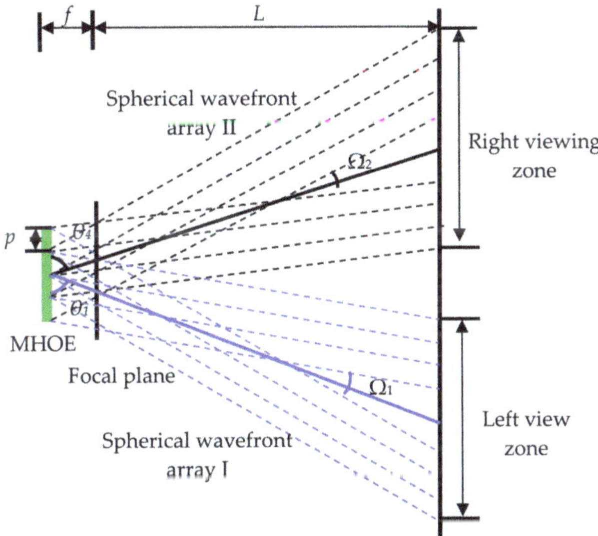

Figure 5. Principle and design of the projection part for the dual-view 3D display.

3. Experiments and Results

3.1. Fabrication of the MHOE

In the first step, the experimental setup is shown in Figure 6a. The laser's wavelength and power are 532 nm, and 400 mW, respectively. The thickness, resolution, sensitivity and average refractive index of the photopolymer material are 15 um, 12,000 line/mm, 10 mJ/cm^2 and 1.47, respectively. The size of the micro-lens array used to fabricate the MHOE is 150 mm × 150 mm, in which the pith of each lens element is 1 mm and the focal length is about 3.3 mm. The laser beam is divided into two beams by a beam splitter (BS). One of the laser beams is expanded by a spatial filter (SF) and a collimating lens (CL), then the laser beam incident into the micro-lens array with the angle $\theta_1 = 50°$ forms spherical wavefront array I. Another laser beam is reflected by mirror 1 (M1) and mirror 2 (M2), and then passes through the divergent lens (DL) to form spherical wave I. The angles of spherical wave I incident into the holographic plate is controlled to be $\theta_2 = 61°$, and $\theta_3 = 81°$. The interference occurs between the spherical waves I and spherical wavefront array I, and the interference fringes are recorded on the holographic plate.

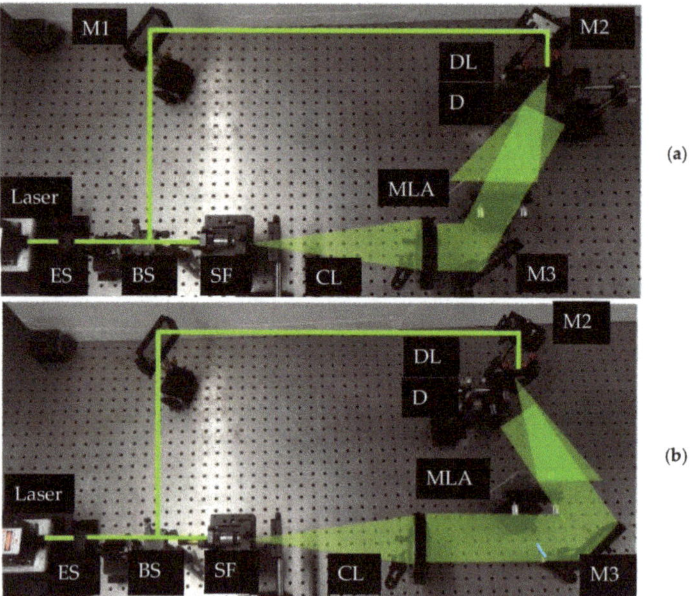

Figure 6. Setup for fabrication of the MLAHOE, (a) the first step, and (b) the second step.

In the second step, the experimental setup is shown in Figure 6b. The laser beam is divided into two beams by a BS. One of the laser beams is expanded by a SF and a CL, then the laser beam incident into the micro-lens array with the angle $\theta_4 = 50°$ forms spherical wave array II. Another laser beam passes through the DL to form spherical wave II. The angle of spherical wave II incident into the holographic material is controlled to be $\theta_5 = 61°$, and $\theta_6 = 81°$. The interference occurs between the spherical waves I and spherical wavefront array I, and the interference fringes are recorded on the holographic plate. The holographic plate is post-processed to obtain the required MHOE. The size of the MHOE is 80 mm × 80 mm, and the diffraction efficiency and transmittance of the MHOE are greater than 85% and 80%, respectively. The exposure time is controlled by the electronic shutters. The detailed parameters of the optical devices are shown in Table 1.

Table 1. Parameters of the optical devices in our experiments.

Components	Parameters	Values
Green laser	Power	400 mW
	Wavelength	532 nm
Green sensitive photopolymer material	Thickness	15 um
	Resolution	12000 line/mm
	Sensitivity	10 mJ/cm^2
	Averaged refractive index	1.47
Micro-lens array	Pitch	1 mm
	Focal length	3.3 mm
Beam splitter	Coupling ration	1:1
Projector	Model	MX518F
	Resolution	1600 × 1200
	Horizontal scanning frequency	102 KHz
	Vertical scanning frequency	102 Hz

In this paper, the MHOE is used as a projection screen, when the incident projection beam deviates greatly from the Bragg condition, the diffracted beam cannot be reconstructed and the 3D images also cannot be reconstructed. Therefore, the angular selectivity of the MHOE needs to be measured. The maximum deviated angle of the projection beam must be measured but the diffracted beam and 3D images cannot be reconstructed. The diffraction efficiency of the MHOE can be measured by the intensity of the transmitted and diffracted beam. The measurement formula of the diffraction efficiency can be expressed as [21]:

$$\eta = \frac{I_D}{I_T + I_D} \tag{5}$$

where I_D and I_T represent the intensity of the diffracted and transmitted beam, respectively. During the measurement, the incident angle of the reference beam is controlled by a rotation table, and the intensity of the diffracted beam of the MHOE is obtained by an optical power meter. The diffraction efficiency of the left and right view zones of the MHOE is measured, as shown in Figure 7. Through observation, when the incident angle of the reference beam is at 13°, the diffraction efficiency of the MHOE is less than 5%. When the incident angle of the reference beam is between −1.5° and 1.5°, the diffraction efficiency of the MHOE is higher than 50%. As shown in Figure 7, the maximum deviated angle of the MHOE is between −5° and 5° in the left and right viewing zones.

The angle between the incident angle θ_3 of projector I and the incident angle θ_5 of projector II is much larger than the maximum deviated angle of the MHOE, that is 38° > 10°. The angle between the incident angle θ_2 of projector I and the incident angle θ_6 of projector II is much larger than the maximum deviated angle of the MHOE, that is 38° > 10°. Therefore, there is no crosstalk between the left and right viewing zones.

The projector (BenQ MX518F) is used to ensure the diffraction efficiencies of the MHOE. The horizontal scanning frequency is 102 KHz, and the vertical scanning frequency is 120 Hz. The resolution of the projector is 1600 × 1200. The detailed parameters of the projector are shown in Table 1. Projection beam I is projected onto the MHOE with a pair of angles $\theta_2 = 61°$ and $\theta_3 = 81°$, which are the same as in the first step. Then, spherical wavefront array I is reconstructed in the left direction, as shown in Figure 8a. Projection beam II is projected onto the MHOE with a pair of angles $\theta_5 − 81°$ and $\theta_6 − 61°$, which are the same as in the second step. Then, spherical wavefront array II is reconstructed in the right direction, as shown in Figure 8b.

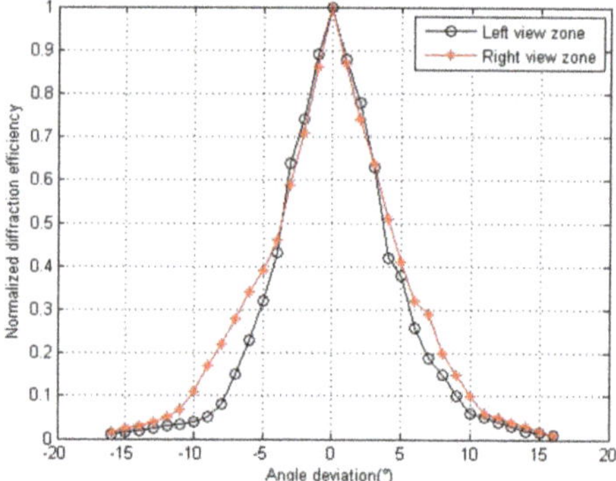

Figure 7. Diffraction efficiency of the left and right viewing zone.

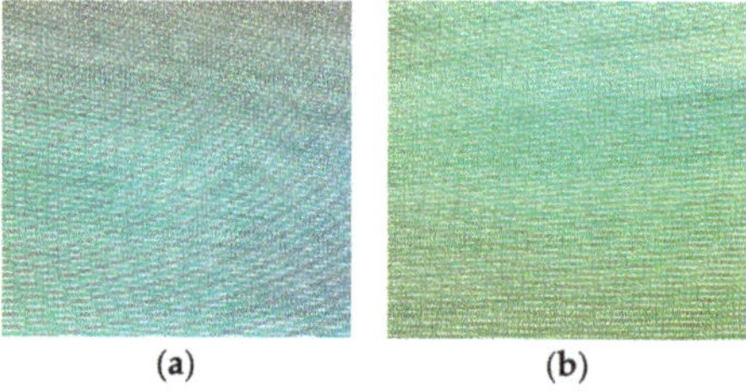

(a)   (b)

Figure 8. Reconstruction of the wavefronts of the lens array (**a**) in the left viewing zone, and (**b**) in the right viewing zone.

3.2. Experiments for Displaying Images on the Proposed Dual-View 3D Display

The projector is used in the experiment for displaying images. According to the incident angle requirement of the MHOE, projector I is placed on the left, and projection beam I is incident into the MHOE with a pair of angles $\theta_2 = 61°$ and $\theta_3 = 81°$. Projector II is placed on the right, and projection beam II is incident onto the MHOE with a pair of angles $\theta_5 = 81°$ and $\theta_6 = 61°$. Therefore, projection beams I and II are symmetrical to each other.

Figure 9 shows the experiment setup for the dual-view 3D display. In order to satisfy the incident angle about the MHOE, the two projectors are placed vertically, and the projection lens is positioned inside. The distance between the projection lenses of the projectors and the MHOE is 31 cm. Projectors I and II are 20 cm apart from each other in the horizontal direction. The projection lenses of projectors I and II are both 21 cm away from the optical table. The height of the MHOE is 19.5 cm from the optical table. The distance between the MHOE and the car model is 3 cm. The projectors are used to project the EIAs onto the MHOE. The projection beams of the projectors that meet the Bragg-match are directly projected onto the MHOE. The MHOE is used as a screen of the dual-view 3D display. Projector I projects EIA I as projection beam I, with a pair of angles $\theta_2 = 61°$ and $\theta_3 = 81°$ onto the MHOE. Then, 3D image I is reconstructed in the left viewing zone. Projector II projects EIA II as projection beam II, with a pair of angles $\theta_5 = 61°$ and $\theta_6 = 81°$ onto the MHOE. Then, 3D image II is reconstructed in the right viewing zone. In this paper, the pith of the elemental lens in the micro-lens array is 1 mm and the focal length is 3.3 mm. According to Equations (3) and (4), the viewing angles of the left and right viewing zone are $\Omega_1 = 10.2°$ and $\Omega_2 = 10.2°$, respectively. According to the parameters of the micro-lens array, the viewing angle of the conventional integral imaging is about 12°. Therefore, the viewing angles in the dual viewing zones are almost the same as the conventional one.

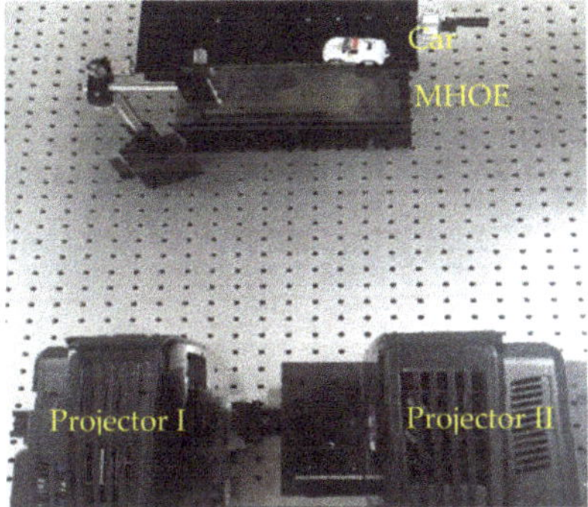

Figure 9. Experiment setup for the dual-view 3D display.

EIAs I and II are shown in Figure 10a,b, respectively. Each EIA consists of 70×70 image elements. EIA I is generated by a computer and consists of the characters "3" and "D", in which "3" is located at −15 mm, and "D" is located at +15 mm. EIA II is also generated by a computer and consists of the characters "A" and "B", in which "A" is located at −15 mm, and "B" is located at +15 mm.

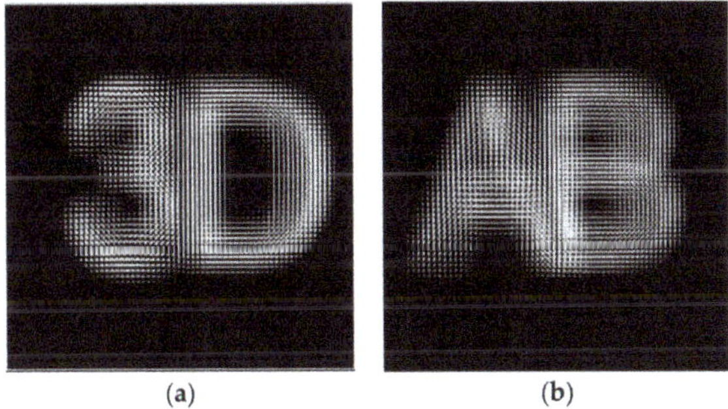

(a)   (b)

Figure 10. (a) EIA I and (b) EIA II

3D image I reconstructed in the left view zone taken from the left, right, top and bottom directions are shown in Figure 11. We can see the horizontal parallax of the "3D" images from the left and right directions, and the vertical parallax of the "3D" images from the top and bottom directions. Reconstructed 3D image I has obvious horizontal and vertical parallaxes. Visualizations 1 and 2 clearly show the horizontal and vertical parallaxes between the "3", "D" and the real object. The resolution of reconstructed 3D image I is almost the same as the conventional integral imaging 3D display, and 3D image I in the left view zone has high brightness and definition. The resolution of 3D image I is the same as the number of the image element 70×70 in EIA I.

Figure 12 shows 3D image II reconstructed in the right view zone taken from the left, right, top and bottom directions. We can see the horizontal parallax of the "AB" images from the left and right directions, and the vertical parallax of the "AB" images from the top and bottom directions. The reconstructed 3D image II has obvious horizontal and vertical parallaxes. Visualizations 2 and 3 clearly show the horizontal and vertical parallaxes between the "A", "B" and the real object. The resolution of reconstructed 3D image II is also almost the same as the conventional integral imaging 3D display. The resolution of 3D image II is also the same as the number of the image element 70×70 in EIA II. The 3D images I and II have the same resolution, according to the parameters of the micro-lens array, the resolution of the conventional integral imaging is 70×70. Therefore, the resolution in the dual viewing zones are the same as the conventional one.

It was observed that there is slight image crosstalk in the 3D image reconstructed in the left and right viewing zone. When projection beam I is projected onto the MHOE, the light produces diffuse reflection on the glass substrate of the MHOE, and the diffuse light can produce slight crosstalk with reconstructed 3D image II in the right viewing zone. Reconstructed 3D image I in the left viewing zone has slight crosstalk, which is caused by the same reason as in the right viewing zone.

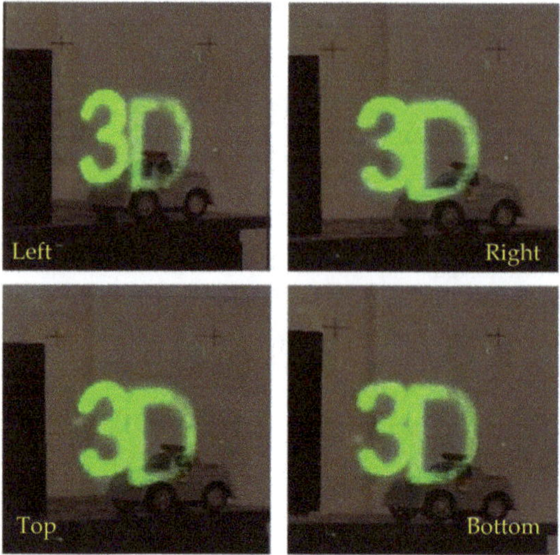

Figure 11. Experimental results in the left viewing zone (see Supplementary Materials visualization 1 and visualization 2), the parallax images in the left, right, top, and bottom four directions.

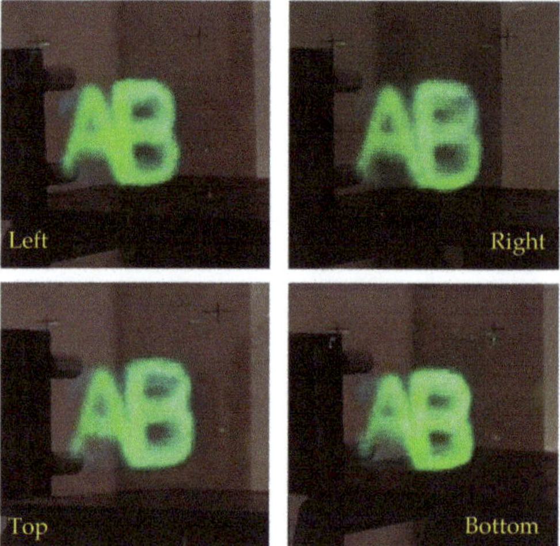

Figure 12. Experimental results in the right viewing zone (see Supplementary Materials visualization 2 and visualization 3), the parallax images in the left, right, top, and bottom four directions.

4. Conclusions

In this paper, we propose a dual-view 3D display. Two projectors are used to present different 3D images on the MHOE. Viewers can see the two different 3D images in the left and right viewing zones. The viewing angle and resolution of the reconstructed 3D images in the two viewing zones are almost the same as the conventional integral imaging 3D display. In addition, the fusion of 3D images with real objects is realized. It can be used for the car head up display and medical devices.

Supplementary Materials: The following are available online at http://www.mdpi.com/2076-3417/9/18/3852/s1.

Author Contributions: Conceptualization, Q.W. and D.L.; methodology, H.Z.; software, H.Z.; validation, Q.W., H.D.; formal analysis, H.Z.; investigation, M.H.; writing—original draft preparation, H.Z.; writing—review and editing, H.Z., Q.W., D.H., M.H., D.L.; visualization, H.Z.; funding acquisition, Q.W.

Funding: This work is supported by the National Key R&D Program of China (2017YFB1002900) and National Natural Science Foundation of China (61535007 and 61775151).

Conflicts of Interest: The authors declare no conflict of interest.

References

1. Cui, J.P.; Li, Y.; Yan, J.; Cheng, H.C.; Wang, Q.H. Time-multiplexed dual-view display using a blue phase liquid crystal. *J. Disp. Technol.* **2013**, *9*, 87–90. [CrossRef]

2. Hsieh, C.T.; Shu, J.N.; Chen, H.T.; Huang, C.Y.; Tian, C.J.; Lin, C.H. Dual-view liquid crystal display fabricated by patterned electrodes. *Opt. Express* **2012**, *20*, 8641–8648. [CrossRef] [PubMed]

3. Zhao, Y.; Cao, L.; Zhang, H.; Kong, D.; Jin, G. Accurate calculation of computer-generated holograms using angular-spectrum layer-oriented method. *Opt. Express* **2015**, *23*, 25440–25449. [CrossRef] [PubMed]

4. Zhang, H.; Zhao, Y.; Cao, L.; Jin, G. Fully computed holographic stereogram based algorithm for computer-generated holograms with accurate depth cues. *Opt. Express* **2015**, *23*, 3901–3913. [CrossRef] [PubMed]

5. Gao, Q.; Liu, J.; Han, J.; Li, X. Monocular 3D see-through head-mounted display via complex amplitude modulation. *Opt. Express* **2016**, *24*, 17372–17383. [CrossRef] [PubMed]

6. Sang, X.; Fan, F.C.; Jiang, C.C.; Choi, S.; Dou, W.; Yu, C. Demonstration of a large-size real-time full-color three-dimensional display. *Opt. Lett.* **2009**, *34*, 3803–3805. [CrossRef] [PubMed]

7. Yu, C.; Yuan, J.; Fan, F.C.; Jiang, C.C.; Choi, S.; Sang, X. The modulation function and realizing method of holographic functional screen. *Opt. Express* **2010**, *18*, 27820–27826. [CrossRef] [PubMed]

8. Lee, H.H.; Huang, P.J.; Wu, J.Y.; Hsieh, P.Y.; Huang, Y.P. A 2D/3D hybrid integral imaging display by using fast switchable hexagonal liquid crystal lens array. In Proceedings of the Three-dimensional Imaging, Visualization, & Display. International Society for Optics and Photonics, Anaheim, CA, USA, 10 May 2017; Volume 10219, pp. 1021910–1021916.

9. Oi, R.; Chou, P.Y.; Jackin, B.J.; Wakunami, K.; Ichihashi, Y.; Okui, M.; Huang, Y.P.; Yamamoto, K. Three-dimensional reflection screens fabricated by holographic wavefront printer. *Opt. Eng.* **2018**, *57*, 61605–61612.

10. Patel, S.K.; Cao, J.; Lippert, A.R. A volumetric three-dimensional digital light photoactivatable dye display. *Nat. Commun.* **2017**, *8*, 15239–15246. [CrossRef] [PubMed]

11. Kota, K.; Satoshi, H.; Yoshio, H. Volumetric bubble display. *Optica* **2017**, *4*, 298–302.

12. Wang, X.; Hua, H. Theoretical analysis for integral imaging performance based on microscanning of a microlens array. *Opt. Lett.* **2008**, *33*, 449–451. [CrossRef] [PubMed]

13. Javidi, B.; Hua, H. A 3d integral imaging optical see-through head-mounted display. *Opt. Express* **2014**, *22*, 13484–13491.

14. Zhang, H.L.; Deng, H.; Yu, W.T.; He, M.Y.; Li, D.H.; Wang, Q.H. Tabletop augmented reality 3D display system based on integral imaging. *J. Opt. Soc. Am. B* **2017**, *34*, 16. [CrossRef]

15. He, M.Y.; Zhang, H.L.; Deng, H.; Li, X.W.; Wang, Q.H. Dual-view-zone tabletop 3D display system based on integral imaging. *Appl. Opt.* **2018**, *57*, 952–958. [CrossRef] [PubMed]

16. Wu, F.; Deng, H.; Luo, C.G.; Li, D.H.; Wang, Q.H. Dual-view integral imaging three-dimensional display. *Appl. Opt.* **2013**, *52*, 4911–4914. [CrossRef] [PubMed]

17. Wu, F.; Wang, Q.H.; Luo, C.G.; Li, D.H.; Deng, H. Dual-view integral imaging 3D display using polarizer parallax barriers. *Appl. Opt.* **2014**, *53*, 2037–2039. [CrossRef] [PubMed]

18. Wang, Q.H.; Ji, C.C.; Li, L.; Deng, H. Dual-view integral imaging 3D display by using orthogonal polarizer array and polarization switcher. *Opt. Express* **2016**, *24*, 9 16. [CrossRef] [PubMed]

19. Jeong, J.; Lee, C.K.; Hong, K.; Yeom, J.; Lee, B. Projection-type dual-view three-dimensional display system based on integral imaging. *Appl. Opt.* **2014**, *53*, 12–18. [CrossRef] [PubMed]
20. Li, G.; Lee, D.; Jeong, Y.; Lee, B. Fourier holographic display for augmented reality using holographic optical element. *Disp. Technol. Int. Soc. Opt. Photonics* **2016**, *9770*, 97700D.
21. Tao, S.Q.; Wang, D.Y.; Qing, J.Z.; Yuan, Q. *Optical Hologram Storage*; Beijing University of Technology Press: Beijing, China, 1998.

Review

Volume Holographic Optical Elements as Solar Concentrators: An Overview

Maria Antonietta Ferrara [1,*], Valerio Striano [2] and Giuseppe Coppola [1]

[1] National Research Council, Institute for Microelectronics and Microsystems, Via P. Castellino 111, 80131 Naples, Italy; giuseppe.coppola@cnr.it

[2] CGS S.p.A., Via Tiengo snc, 82100 Benevento, Italy; VStriano@cgspace.it

* Correspondence: antonella.ferrara@na.imm.cnr.it; Tel.: +39-081-6132-343

Received: 10 December 2018; Accepted: 2 January 2019; Published: 7 January 2019

Featured Application: Energy harvesting is the principal application for Volume Holographic Optical Elements, thus promoting cheap solar concentrators applied for piped sunlight indoor illumination. Moreover, a further interesting potential application is in the aerospace sector where low weight holographic concentrators could be useful for recovering electricity in long missions or in space bases.

Abstract: Generally, to reduce the area of a photovoltaic cell, which is typically very expensive, solar concentrators based on a set of mirrors or mechanical structures are used. However, such solar concentrators have some drawbacks, as they need a tracking system to track the sun's position and also they suffer for the overheat due to the concentration of both light and heat on the solar cell. The fundamental advantages of volume holographic optical elements are very appealing for lightweight and cheap solar concentrators applications and can become a valuable asset that can be integrated into solar panels. In this paper, a review of volume holographic-based solar concentrators recorded on different holographic materials is presented. The physical principles and main advantages and disadvantages, such as their cool light concentration, selective wavelength concentrations and the possibility to implement passive solar tracking, are discussed. Different configurations and strategies are illustrated and the state-of-the-art is presented including commercially available systems.

Keywords: holographic solar concentrator; holographic lens; volume phase holographic optical elements

---

1. Introduction

Solar energy conversion processes provide clean, sustainable and renewable energy, thus there is a growing interest in research and development in the study of conversion systems and their cost effectiveness. An improvement in this sense needs simultaneous consideration of three issues: initial cost, durability-reliability and performance [1]. Considering the current technology, photovoltaic (PV) cells with triple-junction InGaAs are the most efficient (37%); nevertheless, their high cost makes them unattractive, even if their use for domestic applications, as well as in the aerospace industry, is desirable. This problem can be solved by using optical concentrators that allow focussing the sun's rays onto the active solar cell area, thus letting to reduce a significant amount of the expensive PV material [2]. Consequently, in the last 15 years, solar concentration technologies were explored to direct all the available light towards small solar cells.

Currently, there are two principal types of concentrators involved in conversion technology: conventional solar concentrators (e.g., lenses or mirrors) and holographic solar concentrators. Holographic optical elements (HOEs) could in part overcome the limitations of the conventional solar concentrator, such as complex designs, thermal management due to the excessively heated of the solar cell when illuminated with concentrated solar radiation and active solar tracking. Additionally, holography is much more versatile and cheaper with respect to other concentrating optical systems.

In particular, volume holographic optical elements (V-HOEs) have been proposed for use as solar concentrators [3,4] thanks to the following features: (i) potential to achieve nearly 100% efficiency for certain wavelengths and directions; (ii) very rapid and low-cost effective manufacturing and easy customizability of the recorded devices; (iii) narrowness and lightness; (iv) potential to fabricate elements with multiple optical responses (multiplexing); and (v) potential to be very inexpensive in mass production. V-HOEs applications range from spectral splitting applications to increase the conversion efficiency of PV cells [3,5,6] to simultaneous concentration and spectral splitting applications [7].

Despite all these advantages offered by V-HOEs, currently only a few holographic concentrators, patented by Prism Solar Technologies [8], are on market. They work by total internal reflection by means of multiplexed gratings [9], have low cost (around 1 $/W) and are easy to be integrated into buildings [10]. Nevertheless, Prism Solar Modules are still under test and maybe this is the main reason for which this technology is not yet so diffused.

However, with the aim of obtaining high performance of V-HOEs as solar concentrators, it is necessary to keep in mind that volume holograms have high efficiency only when the incident rays vary in a given portion of the plane (angular selectivity) and their efficiency depends on the wavelength: it is high for a bandwidth centred on a wavelength determined by both the refractive index modulation obtained in the recording material and the angle of incidence (chromatic selectivity) [4]. Thus, V-HOEs should be designed to have a high efficiency for the spectrum of the sunlight inside the PV conversion range (for multijunction PV cells, 350 ÷ 1750 nm [11]).

It is worth noting that, due to the chromatic selectivity, the V-HOE diffraction efficiency depends on the wavelength [10]. Exploiting this feature, the heating of the cell, due to the solar spectrum region associated with wavelengths above 1200 nm, can be managed. Thus, one of the main problems of conventional solar concentrators can be overcome allowing a higher conversion efficiency and so lower cost/watt [12,13].

V-HOEs can be used as solar concentrators in both earth and space (satellites) applications. Regarding the first application, several studies are reported in the literature, whereas for aerospace applications only a few works are available [4,14–16], maybe due to the hostile space environment that has to be taken into account.

To obtain an efficient V-HOE as solar concentrator, the main requirements are: (i) the recording material, (ii) the concentration ratio, (iii) the angular selectivity, (iv) the possibility to implement the passive solar tracking, (v) the efficiency of single and multiplexed elements and (vi) the possibility to split the solar spectrum. The resulting system performance depends on each of these points. However, in our knowledge, the aforementioned focusing points required to achieve an efficient V-HOE as solar concentrator are never considered all together. Thus, we think that future researches should be focused on all of these features.

The purpose of the present manuscript is to give an overview of different solar concentrator based on volume holographic devices. Thus, after a brief introduction to the theory behind it and the recording materials commonly used, we analyse the configuration and the performance of several significant results reported in the literature.

2. Theoretical Background

Kogelnik's theory [17] is widely accepted for modelling the performance of volume holograms. This theory shows that high diffraction efficiency can be related to some physical characteristics, such as thickness, spatial frequency and refractive index modulation. According to Kogelnik's theory, the diffraction efficiency (η) can be theoretically evaluated as: [18]

$$\eta = \frac{\sin^2 \sqrt{\zeta^2 + v^2}}{\left(1 + \frac{\zeta^2}{v^2}\right)}, \tag{1}$$

where the parameters ζ and v are defined as:

$$\zeta = \Delta\theta \frac{|\vec{G}|d}{2}, \tag{2}$$

$$v = \frac{\pi \Delta n d}{\lambda \cos\theta_{Bragg}}, \tag{3}$$

In Equations (2) and (3) d is the holographic film thickness, Δn is the refractive index modulation; λ is the wavelength of the reconstructing beam; θ_{Bragg} is Bragg diffraction angle; $\Delta\theta$ is the deviation from the Bragg angle and $\vec{G}$ is the grating vector, normal to the fringes with a magnitude $|\vec{G}| = 2\pi/\Lambda$, with Λ grating period:

$$\Lambda = \lambda/2 * \sin(\theta/2), \tag{4}$$

here λ is the recording wavelength and θ is the angle between the incident and diffracted angles inside the medium.

When the Bragg condition is satisfied, that is, $\Delta\theta = 0$, the volume holographic grating (VHG) diffraction efficiency η can be theoretically evaluated as [17,19]:

$$\eta = \sin^2\left(\frac{\pi \Delta n d}{\lambda \cos\theta_{Bragg}}\right), \tag{5}$$

Depending on the applications, the holographic optical element and in particular for solar applications the holographic lens, requires particular values of angular and/or wavelength selectivity. Due to the chromatic selectivity, hologram diffraction efficiency depends on wavelength; while regarding the angular selectivity, the hologram diffraction efficiency decreases very quickly when the direction of the incident radiation does not fulfil the Bragg condition in the recording plane [10]. It is important to point out that the main difference between conventional optics and holographic optics is that to determine diffracted ray direction, Snell's law is replaced by the grating equation [20].

As expected, the angle at which the diffraction intensity reaches its maximum is strictly related to the incident wavelength [21]. In particular, the angle increases as the wavelength increases. This behaviour is due to Bragg phase-matching condition for VHG, which (considering a non-slanted transmission grating, i.e., gratings with a grating period vector forming an angle of 90° with the perpendicular to the plane of incidence) is defined as:

$$\vec{k_i} - \vec{k_d} = \vec{G}, \tag{6}$$

where $|\vec{k_i}| = |\vec{k_d}| = 2\pi n/\lambda$ are the magnitude of the incident and diffracted wave vectors inside the medium and angles θ_i, θ_d are the incident and diffracted angles inside the medium. $\vec{G}$ is the grating vector defined above.

Generally, two regimes in which phase gratings operate are defined: the Raman-Nath regime, in which several diffracted waves are produced and the Bragg regime, where basically only one diffracted wave is formed and this occurs only for near Bragg incidence. It has been usual to refer to gratings that work in the Raman-Nath or the Bragg regimes as thin and thick gratings, respectively. A confirmation that the recorded hologram is a volume and not a surface hologram can be obtained by evaluating a parameter called Q factor. Indeed, the criterion for whether a hologram is thick or thin is given by the Q factor as:

$$Q = \frac{2\pi\lambda d}{n\Lambda^2},$$ (7)

where λ is the recording wavelength, d is the photosensitive layer thickness, n is the refraction index of the material and Λ is the fringe spacing. A holographic grating is considered to be thin (surface hologram) when $Q \leq 1$, thick (volume hologram) when $Q \geq 10$ [22].

Over the Q parameter, there is another parameter ρ defined as:

$$\rho = \frac{\lambda_0^2}{\Lambda^2 n_0 \Delta n},$$ (8)

where Λ and Δn are defined before, λ_0 is the vacuum wavelength of the light, n_0 is the mean refractive index. This parameter can be used to predict whether a grating is in the Raman-Nath regime or in the Bragg regime: if ρ is nearly zero, the diffraction effect will be prominent, otherwise it can be neglected in the case where ρ very large [23].

3. V-HOE Recording Process

The simplest V-HOE is a VHG which acts as a non-focusing element; therefore, it basically redirects the light [21]. Usually, a hologram can be recorded by interference between two beams, namely reference and object beam, respectively. When the two interfering beams are two collimated light beams, a VHG can be recorded. A scheme of the experimental setup typically used to record VHGs is shown in Figure 1a; the angle α between two incident beams is related to the final grating period of the recorded VHG. The interference fringe pattern leads to a photoinduced modulation of the refractive index in a photosensitive thick film.

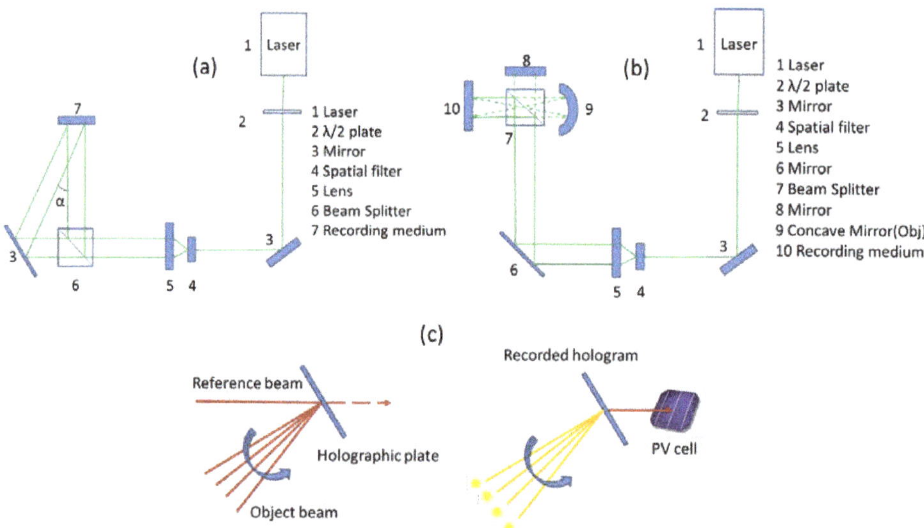

Figure 1. (**a**) Set-up for volume holographic grating recording; (**b**) Set-up for volume phase holographic lenses recording; (**c**) Schematic representation of the recording of multiplexed holographic lenses: the reference beam is fixed while the object beam incident angle is changed during the process. The multiplexed holographic lenses recorded behaves like a passive solar tracker: the light coming from the sun in different positions, that is, at different times of the day, is always focused on the same point, where a photovoltaic cell is positioned.

When in the object beam path a focusing optics is placed during the recording process, the interference between a reference (collimated) beam and an object (converging) beam is induced. Thus, an interferometric pattern that replaces the response of the focusing optical systems used as object is generated, obtaining a V-HOE that acts as a focusing element. This V-HOE shows the same effect as spherical or cylindrical lenses [4,24]. A typical recording setup is reported in Figure 1b.

Conventional concentration modules need the aid of complex mechanical systems (active solar tracking) to improve their efficiency. These aiding systems increase the complexity of the PV cell power system design, the panel volume and development costs. The possibility to reduce or even eliminate, the moving system using multiplexed holographic lenses as passive solar trackers, makes the system more competitive in the perspective of new generation solar panel development. As a result, there are fewer problems related both to the wear of moving parts and possible vibrations due to movements [16]. Multiplexed V-HOEs can be achieved by changing the angle between the two incident wavefronts during the writing process, as schematically shown in Figure 1c. In this way, a V-HOE that addresses the solar radiation incident with different angles, on a single photovoltaic cell, can be realized.

4. Holographic Materials for Solar Concentrators

The holographic recording medium should be able to resolve fully all the fringes resulting from interference between the two incident beams. These fringe spacing can range from tens of micrometres to less than one micrometre, corresponding to spatial frequencies in the range of a few hundred to several thousand cycles/mm. If the performance of the recording medium for these spatial frequencies is low, the diffraction efficiency of the hologram will be poor.

Currently, to record a V-HOE several photosensitive materials can be used, among them substrates based on silver halide emulsions, dichromatic gelatines and photopolymers are the most widely used.

4.1. Silver Halide Emulsions

Silver halide emulsion is one of the oldest recording materials for holography. It is a fine suspension of microscopic grains of silver halide (usually silver bromide, grain size in the range of tens of nanometres) in a colloid sol, usually consisting of gelatin. Typically, a layer of emulsion with a thickness in the range of 5 to 15 µm, is coated onto glass or film substrate. The recorded image is then developed by chemical post-processing, allowing multiple holograms recording. Additionally, emulsions show high sensitivity (10^{-5} to 10^{-3} mJ/cm^2) and good resolution (greater than 6000 lines/mm).

Silver halide emulsions were recently used to obtain a panchromatic holographic material for the fabrication of wavelength multiplexed holographic solar concentrators [25].

4.2. Dichromatic Gelatines

Dichromatic gelatine (DCG) is composed of ammonium or potassium dichromate, gelatin and water and needs chemical post-processing. Ammonium dichromate becomes progressively harder on exposure to light, inducing high refractive index modulation, thus allowing high diffraction efficiency, high resolution, low noise and high optical quality [26–28]. The drawback of DCG is its low exposure sensitivity and limited spectral response.

Even if DCG can be considered an ideal recording material for volume phase holograms, it is very sensitive to environmental changes, therefore it requires a cover plate to ensure environmental stability thus weighting the structure [29]. This reason makes it virtually less eligible for solar concentration applications.

4.3. Photopolymers

Photopolymers are the most studied holographic materials since the 1970s; they are based on polymerization and cross-linking reactions induced by absorption of light and they offer several advantages respect to silver halide and DCG. Indeed, photopolymerizable materials show high diffraction efficiencies, allow real-time monitoring of the recording process, do not require development processes but only a bleaching process, can be produced from raw materials at low cost and give the possibility to modulate the properties through chemical synthesis [4]. Additional, photopolymers can be mass produced allowing to reducing production costs [30,31].

Typically, a photopolymer is made from a photoinitiator system, one or more polyfunctional monomers or oligomers and a polymeric binder. When it is exposed to light, the polymerization occurs in the areas of constructive interference (high-light intensity areas) leading to increased consumption of the monomers, while the polymerization is limited or absent in the areas of destructive interference (low-light intensity areas). The difference of monomer consumption rate leads to a concentration gradient that drives monomer diffusion from dark to illuminated areas [32–34]. Thus, the polymer concentration distribution will take over the sinusoidal pattern of the light intensity, resulting in a permanent modulation of the refractive index, that is, a volume-phase hologram.

Some examples of the most commonly used photopolymers as recording material for holography are: (i) a photopolymer developed within Bayer MaterialScience based upon an orthogonal two chemistry formulation; it is capable of achieving transmission above 90% in the film samples [35]; (ii) photopolymer materials developed by DuPont for recording volume phase holograms, in particular new materials with panchromatic sensitivity, designed for multicolour holographic recording have been developed [36]; (iii) acrylamide-based photopolymers have also been extensively used for fabrication of HOEs in solar energy applications [37].

Regarding the resistance of these materials for holographic concentrators in environmental conditions, there are durability tests on some of the most widely used photopolymers in conventional Fresnel-type solar concentrators, such as acrylic polymers [38].

The fabrication of lenses for concentrators and protective layers for photovoltaic cells is often made by using polymethyl methacrylate (PMMA) [39] since it allows a good resistance to UV radiation and high transmittance (>92%). The UV sensitivity can be further reduced by introducing protective layers or by adding radical scavengers or antioxidants to the formulation of the material. Also polydimethylsiloxane (PDMS) shows very high stability against UV radiation and making it suitable for use in the space environment [40]. Additional, PDMS features greater optical transmittance compared to PMMA.

Considering all the advantages offered by photopolymeric materials, to obtain solar compliant holographic materials that should be less sensitive to thermal and photochemical degradation phenomena, the study of new photopolymers based on inorganics or hybrid organic/inorganic components instead of organic material is still in progress. Some examples in this sense are photopolymers containing nanoparticles of inorganic species such as SiO_2, ZrO_2 and TiO_2 [41], that show a lower shrinkage due to the polymerization and a higher refractive index modulation [42,43] and photopolymer containing zeolite nanocrystals as inorganic dopant, that allows to improve compatibility between inorganic particles and polymer and reduce the optical losses due to scattering [44]. Finally, the sol-gel chemistry versatility allows obtaining a high level of interpenetration between the organic and inorganic networks, as reported in Refs. [4,45–47]. Photopolymer with a very low outgassing level and a high resistance to strong thermal excursions, that can be useful for some applications, was obtained by sol-gel techniques [4].

5. V-HOE Based Solar Concentrators

5.1. Solar Concentrator and Spectral Splitting

Since the pioneering work published by Ludman in 1982 [48], V-HOEs have been proposed as solar concentrators; additionally, by using DCG as a recording material, they have been employed to obtain a spatial separation of the solar spectrum allowing the use of solar cell materials with optimized band gaps achieving high PV efficiency [3,49]. Moreover, an optimized holographic concentrating and spectral splitting systems can reduce the cooling requirements of the photovoltaic cells and, considering V-HOE design versatility, PV cells can be placed on a side of the hologram, thus avoiding shadow effects and simplifies cooling, as can be seen in Figure 2 [5,50]. In this case, the hologram not only concentrates the solar radiation but also split and directs the red and near-infrared spectrum on one PV cell and the green and blue spectrum on another one, while the far-infrared wavelengths of the solar spectrum are diffracted away from the cells, reducing the main cause of overheating.

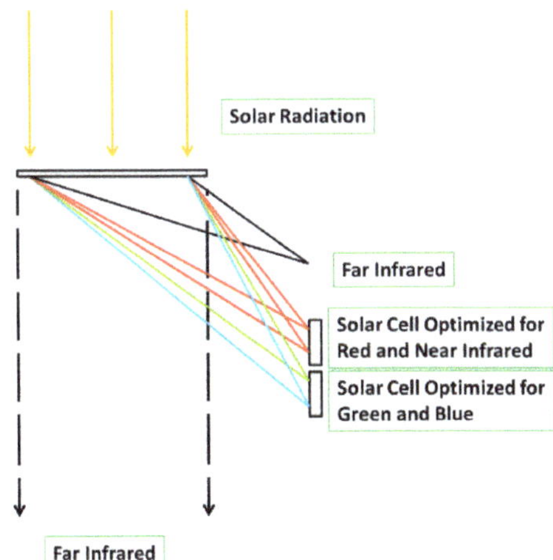

Figure 2. Holographic solar concentrator with spectral splitting systems [5,50].

The authors also demonstrated that this holographic system concentrates higher power than a Fresnel-based solar concentrator and shows a larger relative efficiency; additionally, large heat sinks are not required leading to a decrease of both the bulk and the cost.

With the aim of providing a holographic solar concentrator that reduces solar cell exposure to harmful radiations, Okorogu et al. [51] patented a structure composed of hybrid transmission and reflection VHOEs that spectrally and spatially separate the incident solar spectrum into component wavelength bands. Using internal reflection, the separated solar radiation is propagated through a waveguide far away from the incident direction. At the end of the waveguide, band selection reflecting HOEs inject the solar radiation into PV cells of appropriate band gap energies.

5.2. Multiplexed Hologram

To improve the angular selectivity, multiplexed solutions have been considered. Bainier et al. [7] reported a detailed study on superimposed transmission holograms recorded on the same holographic medium based on DCG. In particular, the authors compared a system consisting of a single holographic element as a concentrator with the maximum of reconstruction wavelength (620 nm) centred in the middle of the range of the PV cell (i.e., 500–800 nm) and a system composed of two holographic recordings with the two maximum reconstruction wavelengths (514.5 and 620 nm) designed both to overlap the operating spectrum of the PV cell (GaAs with an efficiency of 23%) and to avoid the coupling effect. Both the reflecting and transmitting version for the double hologram system were characterized, with one of the two holograms superimposed on the same holographic medium in the transmission configuration. Theoretical and experimental evaluation of the energy efficiency of the holographic systems were carried out. Values of 6 and 5% for the single holographic elements and of 11 and 9% for the double holographic element were obtained, respectively.

A newsworthy structure with a double superimposed slanted reflecting hologram tilted of 30° and recorded on DCG by using 488 and 632 nm laser sources, was proposed in 2010 [52]. This geometry gives rise to a total internal reflection at the surfaces of the holographic medium. The PV cells were placed at the edges of the holographic plate where the collected light emerges. A sketch of the experimental set-up used for the doubly slanted layer structures is reported in Figure 3a, while the diffraction pattern of normally incident white light for the doubly slanted hologram is showed in Figure 3b,c.

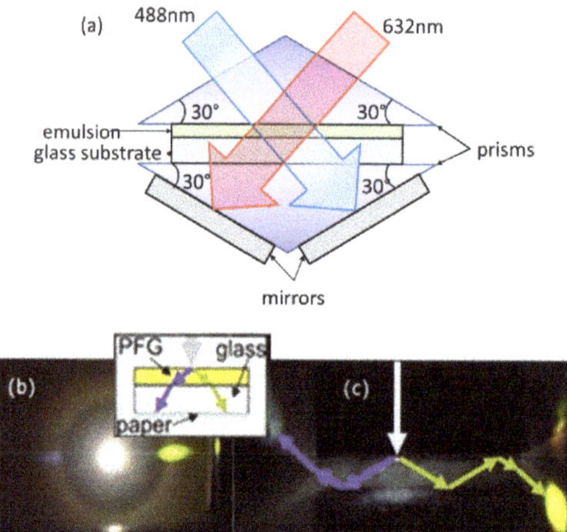

Figure 3. (**a**) A schematic representation of the experimental set-up used in Reference [52] for the doubly slanted layer structure. Red and blue arrows are the incident laser sources emitting at 632 nm and 488 nm, respectively, used to record the double superimposed slanted reflecting hologram. Diffraction patterns evaluated taken from behind (**b**) and above (**c**). Here, the arrow represents the incident light.

Lee's group [53] realized an angular multiplexed HOE recorded on a photopolymer. They optimized a suitable recording method for an angular multiplexed HOE solar concentrator and find that by using the iterative recording method, the low efficiency of multiplexed holograms due to overexposure is compensated. Moreover, the authors evaluated the performance of the HOE as a solar concentrator by introducing a new efficiency calculation method, named concentrated diffraction efficiency (CDE), that is considered as the percentage ratio between the effective concentration rate (ECR) of the HOE and the ECR of the convex lens used in the recording process. The fabricated HOE that uses the modified iterative recording method shows CDEs of 26.73%, 35.31% and 22.78% from incidence angles $-10°$, $0°$ and $+10°$, respectively. These values are considerably higher than those obtained considering multiplexed holograms recorded with equal exposure time without considering the saturation time.

Naydenova et al. recorded a focusing HOE as well as multiplexed gratings on an acrylamide-based photopolymer. High diffraction efficiency HOE consisting of a single spherical lens using thin layers and lower spatial frequency was recorded and a larger acceptance angle was obtained respect to the optical component. Regarding the multiplexed geometry, the authors calibrated the intensity of the beam and then the exposure energy to obtain high efficiency. A diffraction efficiency of three multiplexed gratings at $51.9 \pm 3.5\%$ was demonstrated [54].

5.3. Holographic Solar Deflector

In 2010, Castro et al. [9] designed and characterized a holographic grating, recorded on dichromatic emulsion, able to deflect the direct sunlight on a PV cell with the higher energy efficiency possible. A detailed study of the effects of incident spectra that vary hourly, daily and seasonally was performed, and, in order to maximize the energy collection efficiency per year, the authors proposed the structure illustrated in Figure 4a. The designed cell is composed by two cascaded holographic grating on each side of the PV cell (holograms A and B), that are conjugated (i.e., A and A′ or B and B′) to provide peak energy collection at different seasons.

The optical crosstalk of the V-HOEs was reduced by designing the two cascaded holograms to diffract light in opposite directions with the incident angles in different quadrants of the Bragg circle (Figure 4b). To guarantee that maximum of the diffracted rays of the sunlight within the solar responsivity spectrum of PV cell can reach the surface of the cell independently of the incident angle, the geometrical parameters of the system (i.e., the hologram width and the distance hologram-PV cell) have been optimized, too. Results reported an average daily energy increase of 147% due to the concentrator and approximately 50% of the total energy that reaches the hologram areas can be collected by PV cells without the need of tracking.

An interesting application of V-HOE-based solar concentrator for solar control of domestic conservatories and sunrooms was published in 2005 [55]. In particular, selective use of HOE with given working angles is predicted to maintain the daytime temperatures to an acceptable level when only 62% of glazed the area is used with HOE. Additionally, thanks to the angular selectivity of the V-HOEs, during the winter months theoretically none of the incident beam radiation is attenuated allowing to obtain a comfortable temperature.

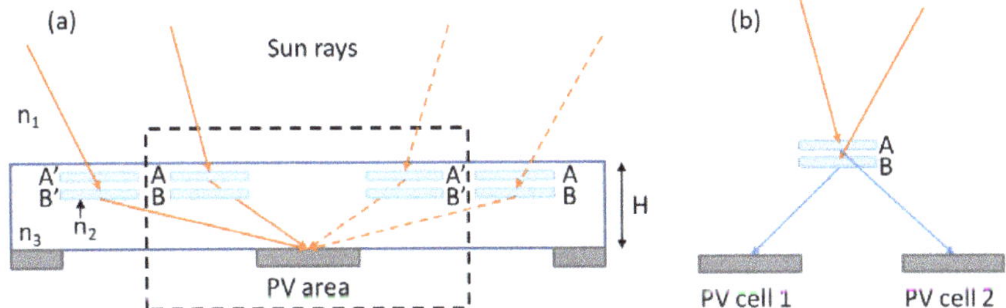

Figure 4. (**a**) Holographic solar concentrator structure realized by Castro et al. [9]. In the dashed box the unit cell is highlighted. (**b**) Holographic design to reduce the optical crosstalk.

5.4. Cylindrical Holographic Lenses

Cylindrical holographic lenses, that allow obtaining a compact and wide-angle structure, were also considered as a solar concentrator. With the aim to take into account a specific set of designed parameter, such as bandwidth, angular selectivity, PV cell size, optical polarization and so on, a proper simulation tool has been developed [10,13]. The possibility to realize a high-efficient system that only requires one-axis tracking was demonstrated.

An interesting application of cylindrical holographic lenses is reported in Figure 5a where the conceptual recording set-up is illustrated [56]. Here, the reference and object beams are an edge-lit and a cylindrical converging beam, respectively, allowing to simultaneously record a combination of a lens and a mirror. An array of V-HOEs is then recorded simply by shifting the medium and by repeating the recording process; this method allows having a large angular acceptance. In fact, the diffracted wave of the incident sunlight coming from different incident angles is guided to the edge of the recording medium where a PV cell is positioned (Figure 5b). When a 2 mm thick holographic recording material (phenanthrenequinone doped PMMA photopolymer) was used to record the proposed architecture, the collection angle from increases 0.01° to 6°.

Recently, Marín-Saez et al. [57] theoretically and experimentally explored a novel approach to overcoming the problem related to the V-HOEs chromatic selectivity by using HOEs operating in the transition regime, which led to a lower chromatic selectivity while maintaining rather high efficiencies. In particular, they developed a model that takes into account the recording material's response to evaluate the index modulation reached for different spatial frequency and exposure dosage. Three cylindrical holographic lenses with different spatial frequency ranges were recorded in Bayfol HX photopolymer to experimentally validate the method. Promising results were obtained when a system composed of two cylindrical holographic lenses with lower spatial frequencies and a mono-c Si PV cell is implemented (see Figure 6). Indeed, a total current intensity of 3.72 times respect that would be reached without the concentrator was achieved.

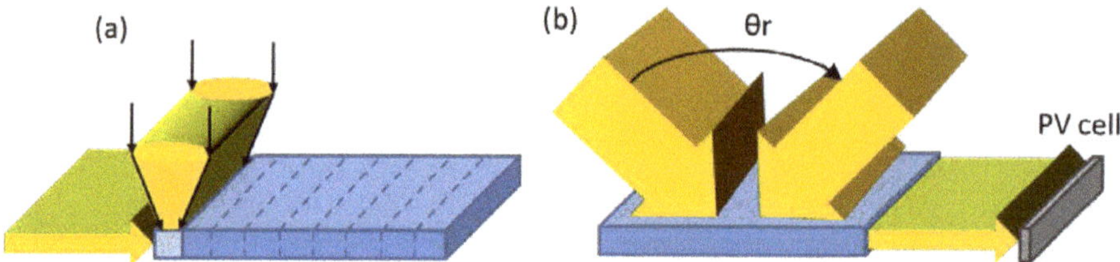

Figure 5. (**a**) Recording set-up and (**b**) configuration of volume holographic concentrator [56].

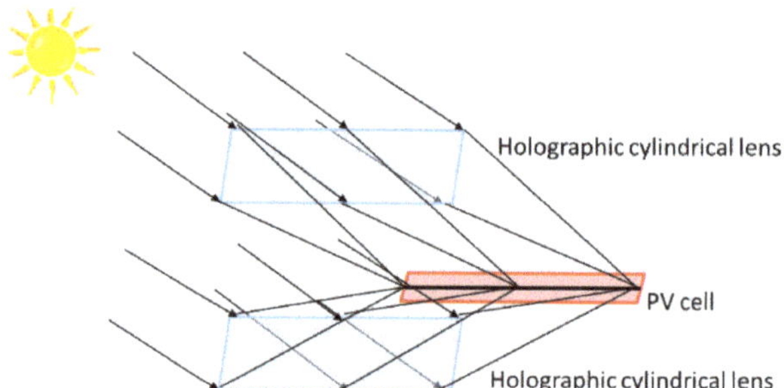

Figure 6. Schematic of the system considered: each cylindrical holographic lens redirects sun rays towards a PV cell [48].

5.5. Commercial and Space Holographic Concentrators

Regarding commercial systems that make use of holographic concentrators, the best performing solutions are offered by Prismsolar, which propose patented enhanced glass-on-glass holographic modules composed of highly efficient bifacial N-type silicon cells with holographic technology. A single module is made alternating strips of bifacial solar cells with parallel strips of holographic film. It allows to collect not only the light that hits the solar cells directly, as normally occurs but also the light that hits the holographic film, which diffracts the light and a portion of this light is guided to the cell through total internal reflection, thus increasing the total harvested sunlight [8,58].

Finally, as a possible future application, taking into account their thin, lightweight and flexible, holographic elements can be studied, designed and fabricated for space solar concentrators. In fact, the solar power conversion is the primary power source for most of the satellite and loss of power, even for a short time, can lead to catastrophic consequences [50]. Thus, an enhancement of the collected light is desirable. Of course, in the space the holographic materials must be able to withstand much more drastic conditions respect to earth applications, due to strong thermal excursions, high vacuum and the presence of high-energy gamma, electronic and protonic radiation originating from the solar wind [4,46,47].

6. Conclusions and Perspectives

The growing demand for new devices able both to improve the harvested sunlight by a PV cell and to reduce the amount of expensive PV material is the main motivation of this work.

In this field, the goal is to find and develop solar concentrators that can overcome the limit presented by conventional solar concentrators, such as active solar tracking, overheating induced on the PV cell, cost and size. With this aim, V-HOEs seem to be the best candidate, satisfying, at the same time, various technological and economic requirements. In fact, V-HOEs can be recorded very easily and allow superimposed structures permitting passive solar tracking. Moreover, a key factor for an efficient V-HOE is the recording material; currently, the most used are silver halides, dichromate gelatins and photopolymers. However, the development of new holographic recording materials is an on ongoing research field; for solar applications this study must be guided by a clear understanding of the behaviour of these new materials in given environmental conditions, that are different for earth and space applications.

In this review, the state of the art of V-HOE based solar concentrators is reported. A number of investigations regarding V-HOE recorded on different material and with different geometry have been described and discussed. However, there is a crucial point that we want to highlight: despite the literature, only a few commercial solutions are available for earth applications, while, to date no hologram solar concentrator has been used for space applications.

Considering the increasing need to obtain clean, renewable and sustainable energy for the welfare of the whole planet and the stringent requirements to reduce area, weight and footprint occupied by photovoltaic cells for space missions, V-HOEs with their low cost for mass production, planar configuration, high efficiency with self-tracking and easy installation, can play a significant role in solar energy conversion applications. Thus, we strongly encourage new research in this field.

Author Contributions: Conceptualization, M.A.F., V.S. and G.C.; writing—original draft preparation, M.A.F., V.S. and G.C.

Funding: This research received no external funding.

Conflicts of Interest: The authors declare no conflict of interest. The funders had no role in the design of the study; in the collection, analyses or interpretation of data; in the writing of the manuscript or in the decision to publish the results.

References

1. Call, P.J. Overview of solar energy conversion technologies: Quantum processes and thermal processes. *Mater. Sci. Eng.* **1982**, *53*, 7–16. [CrossRef]
2. Lyons, V.J. Power and propulsion at NASA Glenn Research Center: Historic perspective of major accomplishments. *J. Aerosp. Eng.* **2013**, *26*, 288–299. [CrossRef]
3. Bloss, W.H.; Griesinger, M.; Reinhardt, E.R. Dispersive concentrating systems based on transmission phase holograms for solar applications. *Appl. Opt.* **1982**, *21*, 3739–3742. [CrossRef] [PubMed]
4. Ferrara, M.A.; Bianco, G.; Borbone, F.; Centore, R.; Striano, V.; Coppola, G. Volume holographic optical elements as solar concentrators. In *Holographic Materials and Optical Systems*; Naydenova, I., Ed.; InTechOpen: London, UK, 2017; pp. 27–50. ISBN 978-953-51-3038-3. [CrossRef]
5. Ludman, J.; Riccobono, J.; Reinhand, N.; Semenova, I.; Martin, J.; Tai, W.; Li, X.L. Holographic solar concentrator for terrestrial photovoltaics. In Proceedings of the 1994 IEEE First World Conference on Photovoltaic Energy Conversion, Waikoloa, HI, USA, 5–9 December 1994; IEEE: Waikoloa, HI, USA, 1994; pp. 1212–1215. [CrossRef]
6. Imenes, A.G.; Mills, D.R. Spectral beam splitting technology for increased conversion efficiency in solar concentrating systems: A review. *Sol. Energy Mater. Sol. Cells* **2004**, *84*, 19–69. [CrossRef]
7. Bainier, C.; Hernandez, C.; Courjon, D. Solar concentrating systems using holographic lenses. *Sol. Wind Technol.* **1988**, *5*, 395–404. [CrossRef]
8. Prismsolar. Prism Solar Technologies. Available online: http://www.prismsolar.com (accessed on 1 December 2018).
9. Castro, J.M.; Zhang, D.; Myer, B.; Kostuk, R.K. Energy collection efficiency of holographic planar solar concentrators. *Appl. Opt.* **2010**, *49*, 858–870. [CrossRef]
10. Chemisana, D.; Collados, M.V.; Quintanilla, M.; Atencia, J. Holographic lenses for building integrated concentrating photovoltaics. *Appl. Energy* **2013**, *110*, 227–235. [CrossRef]

11. Dimroth, F.; Karam, N.H.; Ermer, J.H.; Haddad, M.; Colter, P.; Isshiki, T.; Yoon, H.; Cotal, H.L.; Joslin, D.E.; Krut, D.D.; et al. Next generation GaInP/GaInAs/Ge multi-junction space solar cells. In Proceedings of the 17th European Photovoltaic Solar Energy Conference, Munich, Germany, 22–26 October 2001; ETA: Florence, Italy; WIP: Munich, Germany, 2001.

12. Riccobono, J.R.; Ludman, J.E. Solar holography. In *Holography for the New Millennium*; Ludman, J., Caulfield, H.J., Riccobono, J., Eds.; Springer: New York, NY, USA, 2002; pp. 157–178. [CrossRef]

13. Palacios, P.B.; Álvarez-Álvarez, S.; Marín-Sáez, J.; Collados, M.V.; Chemisana, D.; Atencia, J. Broadband behavior of transmission volume holographic optical elements for solar concentration. *Opt. Express* **2015**, *23*, A671–A681. [CrossRef]

14. Renk, K.; Jacques, Y.; Felts, C.; Chovit, A. *Holographic Solar Energy Concentrators for Solar Thermal Rocket Engines*; No. NTS-6006; NTS Engineering: Long Beach, CA, USA, 1988.

15. Loicq, J.; Venancio, L.M.; Stockman, Y.; Georges, M.P. Performances of volume phase holographic grating for space applications: Study of the radiation effect. *Appl. Opt.* **2013**, *52*, 8338–8346. [CrossRef]

16. Bianco, G.; Ferrara, M.A.; Borbone, F.; Roviello, A.; Pagliarulo, V.; Grilli, S.; Ferraro, P.; Striano, V.; Coppola, G. Multiplexed holographic lenses: Realization and optical characterization. In Proceedings of the 17th Italian Conference on Photonics Technologies, Fotonica AEIT 2015, Turin, Italy, 6–8 May 2015; IET: Stevenage, UK, 2015; Volume 2015. [CrossRef]

17. Kogelnik, H. Coupled-wave theory of thick hologram gratings. *Bell Syst. Tech. J.* **1969**, *48*, 2909–2947. [CrossRef]

18. Martin, S.; Akbari, H.; Keshri, S.; Bade, D.; Naydenova, I.; Murphy, K.; Toal, V. Holographically Recorded Low Spatial Frequency Volume Bragg Gratings and Holographic Optical Elements. In *Holographic Materials and Optical Systems*; Naydenova, I., Ed.; InTechOpen: London, UK, 2017; Chapter 4; pp. 73–98. ISBN 978-953-51-3038-3. [CrossRef]

19. Goodman, J.W. *Introduction to Fourier Optics*, 2nd ed.; McGraw-Hill: New York, NY, USA, 1996; ISBN 978-0974707723.

20. Close, D.H. Holographic Optical Elements. *Opt. Eng.* **1975**, *14*, 408–419. [CrossRef]

21. Barden, S.C.; Arns, J.A.; Colburn, W.S. Volume-phase holographic gratings and their potential for astronomical applications. *Proc. SPIE* **1998**, *3355*, 866. [CrossRef]

22. Kress, B.C.; Meyureis, P. *Applied Digital Optics: From Micro-Optics to Nanophotonics*; John Wiley & Sons: Chippenham, UK, 2009; 638p, ISBN 978-0-470-02263-4.

23. Moharam, M.G.; Young, L. Criterion for Bragg and Raman-Nath diffraction regimes. *Appl. Opt.* **1978**, *17*, 1757–1759. [CrossRef] [PubMed]

24. Leutz, R.; Suzuki, A. Solar concentration in space. In *Nonimaging Fresnel Lenses: Design and Performance of Solar Concentrators*; Springer: Heidelberg, Germany, 2001; pp. 246–256. ISBN 978-3-540-45290-4.

25. Vadivelan, V. Recording of holographic solar concentrator in ultra-fine grain visible wavelength sensitive silver halide emulsion. *Am. J. Electron. Commun.* **2015**, *2*, 15–17. [CrossRef]

26. Hull, J.; Lauer, J.; Broadbent, D. Holographic solar concentrators. *Energy* **1987**, *12*, 209–215. [CrossRef]

27. Quintana, J.A.; Boj, P.G.; Crespo, J.; Pardo, M.; Satorre, M.A. Line-focusing holographic mirrors for solar ultraviolet energy concentration. *Appl. Opt.* **1997**, *36*, 3689–3693. [CrossRef]

28. Ranjan, R.; Khan, A.; Chakraborty, N.R.; Yadav, H.L. Use of holographic lenses recorded in dichromated gelatin film for PV concentrator applications to minimize solar tracking. In *Energy Problems and Environmental Engineering*; Perlovsky, L., Dionysiou, D.D., Zadeh, L.A., Kostic, M.M., Gonzalez-Concepcion, C., Jaberg, H., Sopian, K., Eds.; WSEAS Press: Athens, Greece, 2009; pp. 49–52. ISBN 978-960-474-093-2.

29. Chang, B.J. Dichromated Gelatin Holograms and Their Applications. *Opt. Eng.* **1980**, *19*, 195642. [CrossRef]

30. Bruder, F.K.; Fäcke, T.; Grote, F.; Hagen, R.; Hönel, D.; Koch, E.; Rewitz, C.; Walze, G.; Wewer, B. Mass Production of Volume Holographic Optical Elements (vHOEs) Using Bayfol® HX Photopolymer Film in a Roll-to-Roll Copy Process. In *Practical Holography XXXI: Materials and Applications, Proceedings of the SPIE, San Francisco, CA, USA, 6 April 2017*; SPIE: Washington, DC, USA, 2017; Volume 10127, pp. 101270A-1–101270A-20. [CrossRef]

31. Vather, D.; Naydenova, I.; Cody, D.; Zawadzka, M.; Martin, S.; Mihaylova, E.; Curran, S.; Duffy, P.; Portillo, J.; Connell, D.; et al. Serialized holography for brand protection and authentication. *Appl. Opt.* **2018**, *57*, E131–E137. [CrossRef]

32. Zhao, G.; Mouroulis, P. Diffusion model of hologram formation in dry photopolymer materials. *J. Mod. Opt.* **1994**, *41*, 1929–1939. [CrossRef]
33. Mackey, D.; O'Reilly, P.; Naydenova, I. Theoretical modelling of the effect of polymer chain immobilization rates on holographic recording in photopolymers. *JOSA A* **2016**, *33*, 920–929. [CrossRef]
34. Gleeson, M.R.; Sheridan, J.T.; Bruder, F.K.; Rölle, T.; Berneth, H.; Weiser, M.S.; Fäcke, T. Comparison of a new self developing photopolymer with AA/PVA based photopolymer utilizing the NPDD model. *Opt. Express* **2011**, *19*, 26325–26342. [CrossRef] [PubMed]
35. Jurbergs, D.; Bruder, F.K.; Deuber, F.; Fäcke, T.; Hagen, R.; Hönel, D.; Rölle, T.; Weiser, M.S.; Volkov, A. New recording materials for the holographic industry. *Proc. SPIE* **2001**, *7233*, 72330K-1–72330K-10. [CrossRef]
36. Stevenson, S.H. DuPont multicolor holographic recording films. *Proc. SPIE* **1997**, *3011*, 231–241. [CrossRef]
37. Akbari, H.; Naydenova, I.; Martin, S. Using Acrylamide Based Photopolymers for Fabrication of Holographic Optical Elements in Solar Energy Applications. *Appl. Opt.* **2014**, *53*, 1343–1353. [CrossRef] [PubMed]
38. Miller, D.C.; Kurtz, S.R. Durability of Fresnel lenses: A review specific to the concentrating photovoltaic application. *Sol. Energy Mater. Sol. Cells* **2011**, *95*, 2037–2068. [CrossRef]
39. Schissel, P.; Jorgensen, G.; Kennedy, C.; Goggin, R. Silvered-PMMA reflectors. *Sol. Energy Mater. Sol. Cells* **1994**, *33*, 183–197. [CrossRef]
40. Dever, J.A.; Banks, B.A.; Yan, L. Effects of vacuum ultraviolet radiation on DC93-500 silicone. *J. Spacecr. Rocket.* **2006**, *43*, 386–392. [CrossRef]
41. Vaia, R.A.; Dennis, C.L.; Natarajan, L.V.; Tondiglia, V.P.; Tomlin, D.W.; Bunning, T.J. Onestep, micrometer-scale organization of nano- and mesoparticles using holographic photopolymerization: A generic technique. *Adv. Mater.* **2001**, *13*, 1570. [CrossRef]
42. Lei, Z.; Jun-He, H.; Ruo-Ping, L.; Long-Ge, W.; Ming-Ju, H. Resisting shrinkage properties of volume holograms recorded in TiO$_2$ nanoparticle-dispersed acrylamide-based photopolymer. *Chin. Phys. B* **2013**, *22*, 124207. [CrossRef]
43. Suzuki, N.; Tomita, Y.; Ohmori, K.; Hidaka, M.; Chikama, K. Highly transparent ZrO$_2$ nanoparticle-dispersed acrylate photopolymers for volume holographic recording. *Opt. Express* **2006**, *14*, 12712–12719. [CrossRef] [PubMed]
44. Moothanchery, M.; Naydenova, I.; Mintova, S.; Toal, V. Nanozeolites doped photopolymer layers with reduced shrinkage. *Opt. Express* **2011**, *19*, 25786. [CrossRef]
45. Cheben, P.; del Monte, F.; Levy, D.; Belenguer, T.; Nuñez, A. Holographic diffraction gratings recording in organically modified silica gels. *Opt. Lett.* **1996**, *21*, 1857. [CrossRef] [PubMed]
46. Bianco, G.; Ferrara, M.A.; Borbone, F.; Roviello, A.; Pagliarulo, V.; Grilli, S.; Ferraro, P.; Striano, V.; Coppola, G. Volume Holographic Gratings: Fabrication and Characterization. *Proc. SPIE* **2015**, *9508*, 950807. [CrossRef]
47. Bianco, G.; Ferrara, M.A.; Borbone, F.; Roviello, A.; Striano, V.; Coppola, G. Photopolymer-based volume holographic optical elements: Design and possible applications. *J. Eur. Opt. Soc.* **2015**, *10*, 15057. [CrossRef]
48. Ludman, J. Holographic solar concentrator. *Appl. Opt.* **1982**, *21*, 3057–3058. [CrossRef] [PubMed]
49. Frohlich, K.; Wagemann, U.; Schulat, J.; Schutte, H.; Stojanoff, C.G. Fabrication and test of a holographic concentrator for two color PV-operation. *Proc. SPIE* **1994**, *2255*, 812–821. [CrossRef]
50. Ludman, J.E.; Riccobono, J.; Semenova, I.V.; Reinhand, N.O.; Tai, W.; Li, X.; Syphers, G.; Rallis, E.; Sliker, G.; Martín, J. The optimization of a holographic system for solar power generation. *Sol. Energy* **1997**, *60*, 1–9. [CrossRef]
51. Okorogu, A.O.; Marvin, D.C.; Liu, S.H.; Prater, A. Holographic Solar Concentrator. U.S. Patent 2010/0186818A1, 29 July 2010.
52. Hung, J.; Chan, P.S.; Sun, C.; Ho, C.W.; Tam, W.Y. Doubly slanted layer structures in holographic gelatin emulsions: Solar concentrators. *J. Opt.* **2010**, *12*, 045104. [CrossRef]
53. Lee, J.-H.; Wu, H.-Y.; Piao, M.-L.; Kim, N. Holographic Solar Energy Concentrator Using Angular Multiplexed and Iterative Recording Method. *IEEE Photonics J.* **2016**, *8*, 8400511. [CrossRef]
54. Naydenova, I.; Akbari, H.; Dalton, C.; Yahya, M.; Pang Tee Wei, C.; Toal, V.; Martin, S. Photopolymer Holographic Optical Elements for Application in Solar Energy Concentrators. In *Holography—Basic Principles and Contemporary Applications*; Mihaylova, E., Ed.; InTechOpen: London, UK, 2013; pp. 129–145. ISBN 978-953-51-1117-7. [CrossRef]
55. James, P.A.B.; Bahaj, A.S. Holographic optical elements: Various principles for solar control of conservatories and sunrooms. *Sol. Energy* **2005**, *78*, 441–454. [CrossRef]

56. Hsieh, M.; Lin, S.; Hsu, K.Y.; Burr, J.; Lin, S. An efficient solar concentrator using volume hologram. In *CLEO:2011–Laser Applications to Photonic Applications, OSA Technical Digest*; Optical Society of America: Baltimore, MD, USA, 2011; p. PDPB8. [CrossRef]
57. Marin-Saez, J.; Atencia, J.; Chemisana, D.; Collados, M.V. Full modeling and experimental validation of cylindrical holographic lenses recorded in Bayfol HX photopolymer and partly operating in the transition regime for solar concentration. *Opt. Express* **2018**, *26*, A398–A412. [CrossRef] [PubMed]
58. aSolarus. Bifacial Solar PV. Available online: http://www.asolarus.com/product-distribution/bifacial-solar-pv/#prettyPhoto (accessed on 1 December 2018).

Article

An Early Study on Imaging 3D Objects Hidden Behind Highly Scattering Media: a Round-Trip Optical Transmission Matrix Method

Bin Zhuang [1,2], Chengfang Xu [1,2], Yi Geng [1,2], Guangzhi Zhao [1,2], Hui Chen [1,2], Zhengquan He [1] and Liyong Ren [1,*]

[1] Research Department of Information Photonics, Xi'an Institute of Optics and Precision Mechanics, Chinese Academy of Sciences, Xi'an 710119, China; zhuangbin@opt.cn (B.Z.); xuchengfang@opt.cn (C.X.); gengyi2015@opt.cn (Y.G.); zhaoguangzhi2015@opt.cn (G.Z.); chenhui2016@opt.cn (H.C.); zhqhe@opt.ac.cn (Z.H.)

[2] University of Chinese Academy of Sciences, Beijing 100049, China

* Correspondence: renliy@opt.ac.cn; Tel.: +86-29-88881538

Received: 13 May 2018; Accepted: 23 June 2018; Published: 25 June 2018

Abstract: Imaging an object hidden behind a highly scattering medium is difficult since the wave has gone through a round-trip distortion: On the way in for the illumination and on the way out for the detection. Although various approaches have recently been proposed to overcome this seemingly intractable problem, they are limited to two-dimensional (2D) intensity imaging because the phase information of the object is lost. In such a case, the morphological features of the object cannot be recovered. Here, based on the round-trip optical transmission matrix of the scattering medium, we propose an imaging method to recover the complex amplitude (both the amplitude and the phase) information of the object. In this way, it is possible to achieve the three-dimensional (3D) complex amplitude imaging. To preliminarily verify the effectiveness of our method, a simple virtual complex amplitude object has been tested. The experiment results show that not only the amplitude but also the phase information of the object can be recovered directly from the distorted output optical field. Our method is effective to the thick scattering medium and does not involve scanning during the imaging process. We believe it probably has potential applications in some new fields, for example, using the scattering medium itself as an imaging sensor, instead of a barrier.

Keywords: round-trip imaging; scattering media; 3D imaging; transmission matrix

---

1. Introduction

When imaging an object through a scattering medium, the transmitted light is consisted by the ballistic light and the scattered light [1]. The ballistic light that carries the object information migrates through the scattering medium without deviating from the forward direction. The scattered light that has been scattered randomly in all directions loses the object information and undermines the imaging quality [2]. Most methods only use the ballistic light for imaging, and the scattered light is suppressed by employing a short time gate based on nonlinear phenomena [3–6]. However, these methods are only effective for the weak scattering medium. When imaging through a highly scattering medium, almost all the light has been scattered randomly. In order to solve this problem, some methods that use the scattered light for imaging have recently been proposed. These include computational ghost imaging [7,8], wavefront shaping [9,10], speckle correlation [11–13], and optical transmission matrix (TM) [14–17].

Generally speaking, imaging 3D objects that are hidden behind scattering media is a challenging work. On one hand, compared with the traditional one-way imaging through a scattering medium where the light wave has only been distorted once, imaging an object that hidden behind scattering media is more difficult since the light wave has gone through a round-trip distortion: On the way in for the illumination and on the way out for the detection. In such a case, the information of the object seemingly cannot be recovered. On the other hand, two-dimensional (2D) imaging only needs to recover the intensity (amplitude) features of the object. Three-dimensional (3D) imaging is much more difficult, as both the intensity (amplitude) and the morphological (phase) features of the object should be recovered.

Although a lot of methods have ever been proposed, they are not effective for imaging 3D objects hidden behind highly scattering media. For example, the computational ghost imaging method cannot achieve round-trip imaging through scattering media because the information of the illumination wave is lost via the on-the-way-in distortion [18]. The wavefront shaping method only can image the 2D intensity object as the phase information of the object cannot be recovered, and due to the scanning process, its imaging speed is also restricted [19]. Similarly, the speckle correlation method relies on the inherent correlations in scattered speckle patterns, which makes it only applied to thin opaque layers, as the field-of-view is inversely proportional to the scattering medium thickness [12]. Different from the above methods, the optical transmission matrix (TM) can inherently characterize the scattering medium by giving the relationship between the input optical field and the output optical field [14]. Theoretically, it has the ability to recover the information of the object without loss and achieve the 3D complex amplitude imaging. Recently, this method has shown potential in focusing [15], delivering imaging [16], controlling transmitted energy [20], multispectral control [21], and acoustically modulated light [22]. However, generally speaking, they all belong to the cases working in the transmission imaging mode where the light is only distorted once. More recently, we also show that a 2D intensity object, which is completely hidden behind a scattering medium, can be imaged directly from the distorted output optical field [23]. Nevertheless, to the best of our knowledge, using the TM method to recover the complex amplitude information of the object that is hidden behind highly scattering media has not yet been reported.

In this paper, we show in principle that based on the round-trip TM, it is possible to recover the complex amplitude information of the object that is hidden behind scattering media. The imaging method and experiment setup are introduced, and a simple virtual complex amplitude object has also been constructed to verify the effectiveness of our method. The experiment results show that not only the amplitude but also the phase information of the object can be recovered.

2. Principle

As shown in Figure 1, when a target object is hidden behind a scattering medium, an active light is used to illuminate the object. In such a case, the incident plane wave has gone through a round-trip distortion, and the optical field reflected by the object can be expressed as

$$R(\xi,\eta) = S(\xi,\eta) * O(\xi,\eta), \tag{1}$$

where the '*' operation indicates that the corresponding (ξ,η) elements are multiplied. $O(\xi,\eta)$ is the complex amplitude of the target object, and $S(\xi,\eta)$ is the complex field distribution of the illumination wave which is seriously distorted due to the on-the-way-in process via the scattering medium. Based on these definitions, the final output optical field can be expressed as

$$E(x,y) = \sum_{\xi,\eta} K_{R\to E}(x,y;\xi,\eta) * R(\xi,\eta). \tag{2}$$

The $K_{R\to E}(x,y;\xi,\eta)$ is a traditional one-way TM of scattering media, which has given the relationship between the reflected optical field $R(\xi,\eta)$ and the output optical field $E(x,y)$ [15].

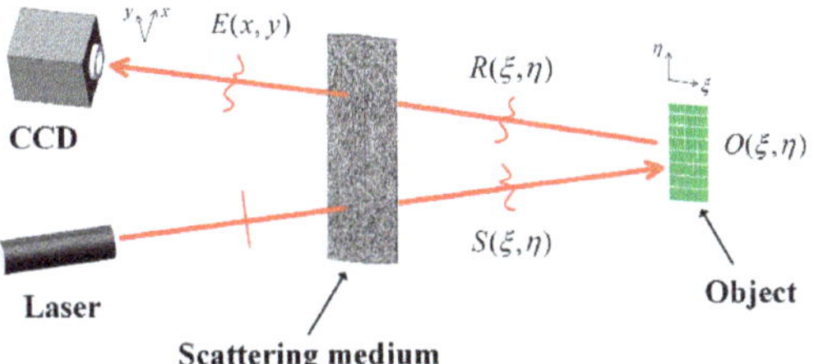

Figure 1. Schematic of imaging an object hidden behind a scattering medium.

As long as $K_{R\to E}(x,y;\xi,\eta)$ is measured in advance, the reflected optical field $R(\xi,\eta)$ can be recovered from $E(x,y)$ directly [16]. Nevertheless, the object $O(\xi,\eta)$ cannot be recovered yet since $S(\xi,\eta)$ is still unknown. To overcome this problem, specially, we construct a new TM

$$T_{O\to E}(x,y;\xi,\eta) = K_{R\to E}(x,y;\xi,\eta) * S(\xi,\eta) \tag{3}$$

as the round-trip TM of the scattering medium. It can be seen from Equation (3) that, different from the traditional one-way TM $K_{R\to E}(x,y;\xi,\eta)$, this new TM $T_{O\to E}(x,y;\xi,\eta)$ has additionally recorded the complex field distribution $S(\xi,\eta)$ of the distorted illumination wave. In this way, Equation (2) can also be expressed as

$$E(x,y) = \sum_{\xi,\eta} T_{O\to E}(x,y;\xi,\eta) * O(\xi,\eta). \tag{4}$$

Now, once the $T_{O\to E}(x,y;\xi,\eta)$ is measured, the relationship between $O(\xi,\eta)$ and $E(x,y)$ can be given directly.

Furthermore, for the convenience of calculation, the Equation (4) is converted to the matrix form as

$$E(x,y) = T_{O\to E}(x,y;\xi,\eta) \times O(\xi,\eta), \tag{5}$$

where the ' $\times$ ' represents a multiplication operation between two matrices. In this way, the object information $O(\xi,\eta)$ can be recovered directly from the distorted output optical field $E(x,y)$ by the inverse operation:

$$O_{\text{rec}}(\xi,\eta) = T_{O\to E}(x,y;\xi,\eta)^{-1} \times E(x,y), \tag{6}$$

where $T_{O\to E}(x,y;\xi,\eta)^{-1}$ is the inverse matrix (or pseudoinverse matrix for any $(\xi,\eta)/(x,y)$ segments ratio) of .

3. Experimental Study

3.1. Measure the Round-Trip TM

Before imaging the object, the round-trip TM $T_{O\to E}(x,y;\xi,\eta)$ of the scattering medium should be measured in advance.

The experimental setup is shown in Figure 2. A He-Ne laser with a wavelength of 632.8 nm is split into two by a beam splitter (BS1). The transmitted beam is reflected off by a mirror, after being reflected at a second beam splitter (BS2), the beam is distorted via the scattering medium to illuminate the DMD, then, the beam reflected by the DMD is distorted again via the scattering medium to generate the output wave. Meanwhile, the beam reflected by BS1, with the wave front being modulated by the LCVR, is used as the reference wave. At last, the output wave and the reference wave are combined to form an interference image at the CCD. In this way, the complex field distribution (both the amplitude and the phase) of the output wave can be acquired by using the phase-shifting digital holography technology [24].

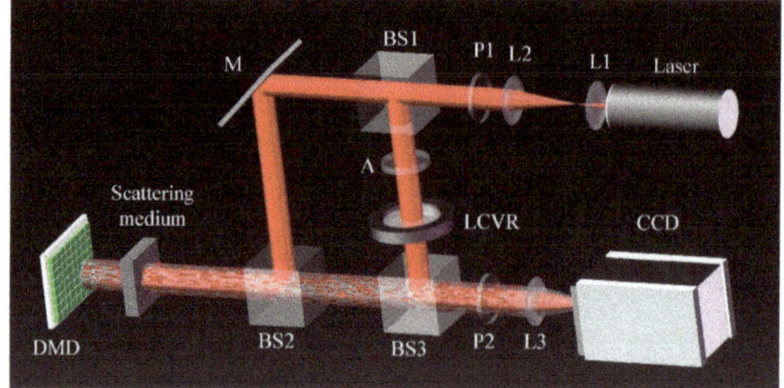

Figure 2. Experimental setup. Scattering medium (a 220-grit and a 1500-grit Thorlabs Optics ground-glass diffuser are stacked together to enhance the scattering ability), DMD: digital micromirror devices (ViALUX, V-7001VIS, 1024 × 768 pixels), LCVR: liquid crystal variable retarder (Meadowlark Optics, D5020), CCD (Hamamatsu, C13440-20CU, 2048 × 2048 segments (pixels), the central 400 × 300 segments are used for imaging), L#: lens, P#: polarizer, BS#: beam splitter, A: attenuation film, M: mirror.

To acquire the round-trip TM, we divide the DMD into 64 × 48 segments (the size of each segment: 219 μm × 219 μm) and turn on only one segment in sequence. The sketch map of the measured $T_{O \to E}(x, y; \xi, \eta)$ are shown in Figure 3a,b for the amplitude part and the phase part, respectively (Two-dimensional input DMD segments (ξ, η) and output CCD segments (x, y) are both stretched to one-dimensional to facilitate the computation, and only part of the DMD segments and CCD segments are shown). What we want to emphasize is that the measured $T_{O \to E}(x, y; \xi, \eta)$ has the ability to overcome the round-trip distortion simultaneously without information loss, which is owing to the fact that it not only has played the role of the one-way TM, but has also recorded the complex field distribution of the distorted illumination wave.

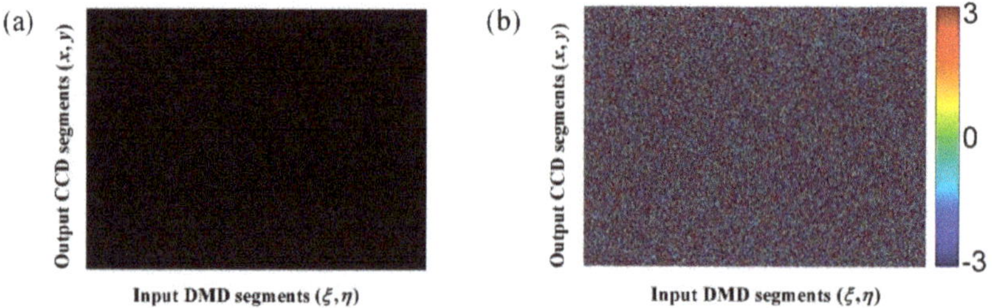

Figure 3. Sketch map of the measured round-trip TM $T_{O \to E}(x, y; \xi, \eta)$. Amplitude (**a**) and corresponding phase (**b**), respectively. Color bar: phase in radian.

3.2. Imaging the Object

After the $T_{O \to E}(x, y; \xi, \eta)$ is measured, the system is ready to image the target object. In order to test the imaging quality precisely, we don't capture the image of a natural real object directly. Instead, a binary complex amplitude object which consists of the characters part and the background part, is constructed by a phase-only-modulation spatial light modulator (SLM: Meadowlark Optics, P1920-0635-HDMI, 1920 × 1152 pixels) using two phase masks $O_{\text{SLM}}^{(1)}(\xi, \eta)$ and $O_{\text{SLM}}^{(2)}(\xi, \eta)$ [15,16]. The mask $O_{\text{SLM}}^{(1)}(\xi, \eta)$ is a plane phase whose amplitude = 1 and phase = π. The mask $O_{\text{SLM}}^{(2)}(\xi, \eta)$ is obtained by flipping the phase of $O_{\text{SLM}}^{(1)}(\xi, \eta)$ from π to $2\pi/3$ for the characters part, and from π to $-\pi/3$ for the background part. As a result, a virtual 3D complex amplitude object $O_{\text{SLM}}^{(2)}(\xi, \eta) - O_{\text{SLM}}^{(1)}(\xi, \eta)$

can be constructed, which corresponds to the characters part with amplitude $= 1$ and phase $= \pi/3$ [as shown in Figure 4a], as well as the background part with amplitude $= \sqrt{3}$ and phase $= -\pi/6$ [as shown in Figure 4b], respectively.

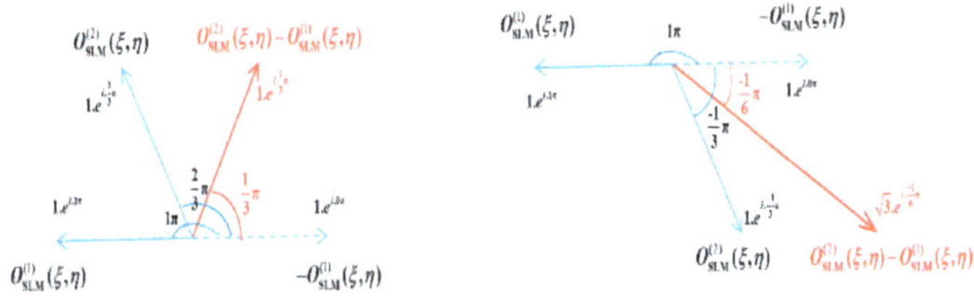

(a) Complex amplitude of the character part   (b) Complex amplitude of the background part

Figure 4. Complex amplitude object constructed by the SLM using two phase masks $O^{(1)}_{\mathrm{SLM}}(\xi, \eta)$ and $O^{(2)}_{\mathrm{SLM}}(\xi, \eta)$. (**a**) amplitude $= 1$ and phase $= \pi/3$ for the characters part, (**b**) amplitude $= \sqrt{3}$ and phase $= -\pi/6$ for the background part.

Next, we will perform the imaging, and the SLM is located at the same place after removing the DMD. To match the size of the DMD segment (219 µm × 219 µm) used for measuring the round-trip TM of the scattering medium, the SLM is divided into 80 × 48 segments (the size of each segment 220 µm × 220 µm), and the central 18 × 14 segments are used to construct the complex amplitude object $O^{(2)}_{\mathrm{SLM}}(\xi, \eta) - O^{(1)}_{\mathrm{SLM}}(\xi, \eta)$ whose amplitude and phase are shown in Figure 5a1,a2, respectively. The corresponding output optical field $F^{(2)}_{\mathrm{CCD}}(x, y) - E^{(1)}_{\mathrm{CCD}}(x, y)$ is shown in Figure 5b1,b2 for the amplitude and the phase, respectively, where $E^{(1)}_{\mathrm{CCD}}(x, y)$ and $E^{(2)}_{\mathrm{CCD}}(x, y)$ are the output optical field correspond to the $O^{(1)}_{\mathrm{SLM}}(\xi, \eta)$ and $O^{(2)}_{\mathrm{SLM}}(\xi, \eta)$, respectively. It can be seen from Figure 5b1,b2 that the information of the object $O^{(2)}_{\mathrm{SLM}}(\xi, \eta) - O^{(1)}_{\mathrm{SLM}}(\xi, \eta)$ has indeed been seriously destroyed, and it is impossible to image the object directly.

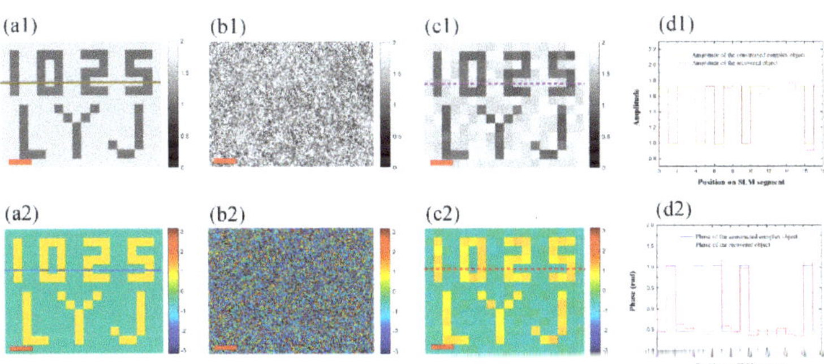

Figure 5. Imaging the complex amplitude object. (**a1,a2**) Amplitude and phase of the object $O^{(2)}_{\mathrm{SLM}}(\xi, \eta) - O^{(1)}_{\mathrm{SLM}}(\xi, \eta)$, respectively. (**b1,b2**) Amplitude and phase of the corresponding output optical filed $E^{(2)}_{\mathrm{CCD}}(x, y) - E^{(1)}_{\mathrm{CCD}}(x, y)$, respectively. (**c1,c2**) Amplitude and phase of the recovered object, respectively. (**d1**) Section profiles corresponding to the lines in (**a1,c1**); (**d2**) section profiles corresponding to the lines in (**a2,c2**). Scale bars indicate 400 µm in (**b1,b2**) and 600 µm in (**a1,a2,c1,c2**). Gray bar: amplitude. Color bar: phase in radian.

Finally, with the round-trip TM $T_{O \to E}(x, y; \xi, \eta)$ being measured previously, according to Equation (6), the object information can be recovered directly from the distorted output optical field $E_{\text{CCD}}^{(2)}(x, y) - E_{\text{CCD}}^{(1)}(x, y)$ by the inverse operation:

$$O_{\text{rec}}(\xi, \eta) = T_{O \to E}(x, y; \xi, \eta)^{-1} \times \left[E_{\text{CCD}}^{(2)}(x, y) - E_{\text{CCD}}^{(1)}(x, y) \right]. \tag{7}$$

The results are shown in Figure 5c1,c2 for the amplitude and the phase, respectively. The section profiles in Figure 5d1, corresponding to the lines in Figure 5a1,c1, show that the amplitude of the object can be recovered; Similarly, the section profiles in Figure 5d2, corresponding to the lines in Figure 5a2,c2, show that the phase of the object can also be recovered well.

3.3. Verify the Effectiveness of the Round-Trip TM

In this part, to verify the effectiveness of our round-trip TM method, we will show that the illumination wave is indeed seriously distorted (for simplicity, only the amplitude is tested, and it can be predicted that the phase will get the same result). In such a case, the traditional one-way TM is ineffective.

Firstly, we have recorded the intensity (amplitude) distribution of the illumination wave (corresponding to $S(\xi, \eta)$ in Figure 1) by placing a CCD in the place of the DMD, the result is shown in Figure 6a. It is seen that the originally uniformed incident light is seriously distorted due to the on-the-way-in process via the scattering medium. Furthermore, we multiply the amplitude of the object [shown in Figure 6b] by Figure 6a to indicate the amplitude of the reflected wave indirectly (corresponding to $R(\xi, \eta)$ in Figure 1). The result is shown in Figure 6c. It is seen that the amplitude information of the object [$O(\xi, \eta)$] is completely submerged. This means that, even adopting the traditional one-way imaging method using the one-way TM, Figure 6c might be the best recovery result instead of the Figure 5c1 which could be obtained adopting our round-trip imaging method using the round-trip TM.

Figure 6. (**a**) Intensity (amplitude) distribution of the illumination wave. (**b**) Amplitude of the object. (**c**) Amplitude of the reflected wave. Scale bar: 550 µm.

4. Discussion

The pre-processing of our method is relatively complicated, as the round-trip TM of a scattering medium should be measured prior to image the object, which will usually take a few minutes. However, this is just a one-time procedure, once calibrated, the TM is effective continuously as long as the medium has not been disturbed.

Our method is effective for imaging through the thick scattering medium, and it does not involve any scanning operations during the imaging process. Therefore, it probably has potential applications in some new fields. On one hand, one can monitor the target through an apparently opaque screen (used as a barrier), while it is incapable of being observed by it. On the other hand, instead of as a barrier which undermines the imaging ability, a scattering medium can also be used actively as an imaging sensor. For example, the single multimode optical fiber, as a scattering medium due to the mode dispersion, could potentially open up new, less invasive forms of endoscopy to perform high-resolution imaging of tissues out of reach of current conventional endoscopes [19,25,26].

At present, the preliminary result shows that our method has the ability to recover the complex amplitude information of the object. However, as an early study, no real 3D object has been tested yet. In fact, many problems should be further resolved when imaging a real 3D object. For example, the 2π ambiguity problem when there is an abrupt change in surface height of the object. This problem could can be solved if the size of the DMD segment that used to measure the TM is further reduced.

5. Conclusions

In conclusion, based on the round-trip TM, we have done some preliminary studies on imaging 3D objects hidden behind highly scattering media. A simple virtual complex amplitude object has been tested, and the results show that both the amplitude and the phase information of the object can be recovered. As an early study, this work may have potential reference value for the endoscopic imaging with more research.

Author Contributions: Funding acquisition, L.R. and Z.H.; Investigation, B.Z., C.X., Y.G., G.Z. and H.C.; Methodology, B.Z.; Supervision, L.R.; Validation, B.Z. and C.X.; Writing—original draft, B.Z.; Writing—review & editing, L.R.

Funding: This work was supported by the National Natural Science Foundation of China (Grant No. 61535015).

Conflicts of Interest: The authors declare no conflict of interest.

References

1. Das, B.B.; Yoo, K.M.; Alfano, R.R. Ultrafast time-gated imaging in thick tissues: A step toward optical mammography. *Opt. Lett.* **1993**, *18*, 1092–1094. [CrossRef] [PubMed]
2. Yoo, K.M.; Liu, F.; Alfano, R.R. Imaging through a scattering wall using absorption. *Opt. Lett.* **1991**, *16*, 1068–1070. [CrossRef] [PubMed]
3. Sappey, A.D. Optical imaging through turbid media with a degenerate four wave mixing correlation time gate. *Appl. Opt.* **1994**, *33*, 8346–8354. [CrossRef] [PubMed]
4. Ambekar, R.; Lau, T.-Y.; Walsh, M.; Bhargava, R.; Toussaint, K.C. Quantifying collagen structure in breast biopsies using second-harmonic generation imaging. *Biomed. Opt. Exp.* **2012**, *3*, 2021–2035. [CrossRef] [PubMed]
5. Paciaroni, M.; Linne, M. Single-shot, two-dimensional ballistic imaging through scattering media. *Appl. Opt.* **2004**, *43*, 5100–5109. [CrossRef] [PubMed]
6. Ren, Y.; Si, J.; Tan, W.; Zhong, Y.; Tong, J.; Hou, X. Speckle Suppression of OKG Imaging In Highly Turbid Medium Using SC-Assisted Fundamental Frequency. *IEEE Photon. Technol. Lett.* **2017**, *29*, 106–109. [CrossRef]
7. Shapiro, J.H. Computational ghost imaging. *Phys. Rev. A* **2008**, *78*, 061802. [CrossRef]
8. Bromberg, Y.; Katz, O.; Silberberg, Y. Ghost imaging with a single detector. *Phys. Rev. A* **2009**, *79*, 053840. [CrossRef]
9. Vellekoop, I.M.; Mosk, A.P. Focusing coherent light through opaque strongly scattering media. *Opt. Lett.* **2007**, *32*, 2309–2311. [CrossRef] [PubMed]
10. Horstmeyer, R.; Ruan, H.; Yang, C. Guidestar-assisted wavefront-shaping methods for focusing light into biological tissue. *Nat. Photonics* **2015**, *9*, 563–571. [CrossRef] [PubMed]
11. Bertolotti, J.; van Putten, E.G.; Blum, C.; Lagendijk, A.; Vos, W.L.; Mosk, A.P. Non-invasive imaging through opaque scattering layers. *Nature* **2012**, *491*, 232–234. [CrossRef] [PubMed]
12. Katz, O.; Heidmann, P.; Fink, M.; Gigan, S. Non-invasive real-time imaging through scattering layers and around corners via speckle correlations. *Nat. Photonics* **2014**, *8*, 784–790. [CrossRef]
13. Wu, T.; Dong, J.; Shao, X.; Gigan, S. Imaging through a thin scattering layer and jointly retrieving the point-spread-function using phase-diversity. *Opt. Exp.* **2017**, *25*, 27182–27194. [CrossRef] [PubMed]
14. Beenakker, C.W.J. Random-matrix theory of quantum transport. *Rev. Mod. Phys.* **1997**, *69*, 731–808. [CrossRef]
15. Popoff, S.M.; Lerosey, G.; Carminati, R.; Fink, M.; Boccara, A.C.; Gigan, S. Measuring the transmission matrix in optics: an approach to the study and control of light propagation in disordered media. *Phys. Rev. Lett.* **2010**, *104*, 100601. [CrossRef] [PubMed]
16. Popoff, S.; Lerosey, G.; Fink, M.; Boccara, A.C.; Gigan, S. Image transmission through an opaque material. *Nat. Commun.* **2010**, *1*, 81. [CrossRef] [PubMed]

17. Choi, Y.; Yang, T.D.; Fang-Yen, C.; Kang, P.; Lee, K.J.; Dasari, R.R.; Feld, M.S.; Choi, W. Overcoming the diffraction limit using multiple light scattering in a highly disordered medium. *Phys. Rev. Lett.* **2011**, *107*, 023902. [CrossRef] [PubMed]

18. Le, M.; Wang, G.; Zheng, H.; Liu, J.; Zhou, Y.; Xu, Z. Underwater computational ghost imaging. *Opt. Exp.* **2017**, *25*, 22859–22868. [CrossRef] [PubMed]

19. Papadopoulos, I.N.; Farahi, S.; Moser, C.; Psaltis, D. High-resolution, lensless endoscope based on digital scanning through a multimode optical fiber. *Biomed. Opt. Exp.* **2013**, *4*, 260–270. [CrossRef] [PubMed]

20. Kim, M.; Choi, Y.; Yoon, C.; Choi, W.; Kim, J.; Park, Q.-H.; Choi, W. Maximal energy transport through disordered media with the implementation of transmission eigenchannels. *Nat. Photonics* **2012**, *6*, 583–585. [CrossRef]

21. Mounaix, M.; Andreoli, D.; Defienne, H.; Volpe, G.; Katz, O.; Grésillon, S.; Gigan, S. Spatiotemporal coherent control of light through a multiply scattering medium with the Multi-Spectral Transmission Matrix. *Phys. Rev. Lett.* **2016**, *116*, 253901. [CrossRef] [PubMed]

22. Chaigne, T.; Katz, O.; Boccara, A.C.; Fink, M.; Bossy, E.; Gigan, S. Controlling light in scattering media non-invasively using the photoacoustic transmission matrix. *Nat. Photonics* **2013**, *8*, 58–64. [CrossRef]

23. Zhuang, B.; Xu, C.; Geng, Y.; Zhao, G.; Chen, H.; He, Z.; Wu, Z.; Ren, L. Round-trip imaging through scattering media based on optical transmission matrix. *Chin. Opt. Lett.* **2018**, *16*, 041102. [CrossRef]

24. Yamaguchi, I.; Zhang, T. Phase-shifting digital holography. *Opt. Lett.* **1997**, *22*, 1268–1270. [CrossRef] [PubMed]

25. Choi, Y.; Yoon, C.; Kim, M.; Yang, T.D.; Fang-Yen, C.; Dasari, R.R.; Lee, K.J.; Choi, W. Scanner-Free and Wide-Field Endoscopic Imaging by Using a Single Multimode Optical Fiber. *Phys. Rev. Lett.* **2012**, *109*, 203901. [CrossRef] [PubMed]

26. Psaltis, D.; Moser, C. Imaging with multimode fibers. *Opt. Photonics News* **2016**, *27*, 24–31. [CrossRef]

Article

An Efficient Neural Network for Shape from Focus with Weight Passing Method

Hyo-Jong Kim [1], Muhammad Tariq Mahmood [2] and Tae-Sun Choi [1,*]

[1] School of Mechanical Engineering, Gwangju Institute of Science and Technology, 123 Cheomdangwagi-ro, Buk-gu, Gwangju 61005, Korea; hyojongkim@gist.ac.kr
[2] School of Computer Science and Engineering, Korea University of Technology and Education, 1600 Chungjeolno, Byeogchunmyun, Cheonan 31253, Korea; tariq@koreatech.ac.kr
* Correspondence: tschoi@gist.ac.kr; Tel.: +82-62-715-2413

Received: 23 August 2018; Accepted: 12 September 2018; Published: 13 September 2018

Abstract: In this paper, we suggest an efficient neural network model for shape from focus along with weight passing (WP) method. The neural network model is simplified by reducing the input data dimensions and eliminating the redundancies in the conventional model. It helps for decreasing computational complexity without compromising on accuracy. In order to increase the convergence rate and efficiency, WP method is suggested. It selects appropriate initial weights for the first pixel randomly from the neighborhood of the reference depth and it chooses the initial weights for the next pixel by passing the updated weights from the present pixel. WP method not only expedites the convergence rate, but also is effective in avoiding the local minimization problem. Moreover, this proposed method may also be applied to neural networks with diverse configurations for better depth maps. The proposed system is evaluated using image sequences of synthetic and real objects. Experimental results demonstrate that the proposed model is considerably efficient and is able to improve the convergence rate significantly while the accuracy is comparable with the existing systems.

Keywords: shape from focus; neural network; weight passing

---

1. Introduction

Among the variety of three-dimensional (3D) shape recovery techniques, shape from focus (SFF) is a passive optical method that infers 3D structure of an object from its image stack taken with different focus settings. During the last decade, SFF method has gained considerable attention due to its simplicity and economical computational cost and its considerable usage in computer vision and industrial applications such as 3D cameras, manufacture of thin-film-transistor liquid-crystal display (TFT-LCD) color filters, measurement of surface roughness, medical examination, microelectronics, focus variation for 3D surface [1–8]. In SFF, a sequence of images is captured through a single charge-coupled device (CCD) camera by translating object towards the camera in limited and small steps. Next, a focus measure is applied to measure the focus quality for each pixel in the sequence then an initial depth map is obtained by maximizing the focus measure in the direction of optical axes. Finally, an approximation or a machine learning method is applied to refine the initial depth estimate. Various approximation and machine learning-based methods have been suggested to extract and refine the object shape. The refinement process, generally, provides better depth maps as compared to the focus measure aggregation; however, these are computationally expensive.

Artificial neural network models have been successfully used to estimate optimal or near-optimal solutions of complex problems using a set of input-output data examples [9]. In case of SFF, however, exemplary data set is not available. In [10], SFF problem is defined as optimization of focus measure in terms of neural network weights over a three dimensional focused image surface (FIS). An artificial neural network is trained to learn the shape of FIS (NN.FIS) in a small window by updating network weights in the direction of gradient of the focus measure. It yields considerably accurate depth estimates, but it suffers from high computational complexity and low convergence rate. The high complexity and slow convergence of the model are due to the large numbers of inputs and inappropriate initial weight. In NN.FIS, the inputs are taken by including the neighborhood frames that enlarge the size of weight vector. The initial weights are randomly set that can slow the convergence rate and lead to high computational cost. In addition, the accuracy is limited by the fact that the model may face local minimization problem.

In this paper, we propose a neural network over planar (NN.Planar) model for decreasing computational complexity. In order to increase the convergence rate and efficiency, weight passing (WP) is suggested. Our two contributions can be summarized as (1) a simplified neural network model is proposed that takes input and weight vector of smaller size; and (2) the WP method is proposed that selects proper initial weights for the first pixel randomly and it updates the initial weights for the next pixel efficiently by passing the updated weight from the present pixel. The proposed NN.Planar with WP method is not only efficient and simplified but also expedites the convergence rate and helps in avoiding the local minimization problem. Moreover, the proposed methods may also be applied to (NN.FIS) for better depth maps. The proposed method improves the convergence rate significantly and provides accurate depth maps. Experimental results demonstrate that the proposed model is significantly efficient while the accuracy is comparable with the conventional model.

2. Related Work

The goal of SFF is to detect the 3D depth of every point of an object from a camera lens. At first, an image sequence of the object is obtained by a single camera at different distances between lens and object. Principle of image sequence acquisition is illustrated in Figure 1. It shows an arrangement to estimate the shape of the object which is unknown body. The reference plane is the initial stage and the focused plane means a plane which is completely focused onto the sensor plane. The displacement of translational stage and the distance between the reference plane and the focused plane can be measured by experiment. Specifically, the translational stage is moved from reference plane in the direction of optical axes with a finite steps. At every step, an image is taken and a sequence of images at different step is obtained.

Focus measure plays an important role in SFF methods. In literature, several focus measure operators have been proposed in the spatial and the frequency domains. The performance of the focus measure operator depends on several factors including local window size, imaging device specifications, illumination conditions and complexity of object shape. The effect of window sizes on depth map recovery is studied by Malik and Choi [11] and concluded that a larger window may distort the object shape while a smaller window size may not be able to suppress noise properly. Recently, a comprehensive study by Pertuz et al. [12] demonstrates the effects of various factors on 3D shape recovery using various focus measures. The study also serves a good reference for focus measure operators in various domains. In noise free environment, second derivative-based focus measure Sum-Modified-Laplacian (SML) [13] performs better whereas in noisy environment Gray-Level-Variance (GLV) [12] provides better depth maps.

Once focus values for all pixels in the sequence are computed, various approximation and machine learning-based methods have been suggested to extract and refine the object shape. The simplest approximation method used in earlier works is to replace the central pixel with aggregated focus measures in a small window. Then depth map is obtained by maximizing focus measure along the optical direction from the refined focus volume [11]. The aggregation of focus measures is efficient to compute but it may not provide accurate depth map. On the other hand, several optimization and machine learning-based approaches have been proposed for better 3D shapes [10,14–17]. Ahmad and Choi [14] proposed the usage of dynamic programming to obtain optimal focus measure. It is difficult to apply dynamic programming on 3D volume directly so authors applied it on two-dimensional

(2D) slices that has decreased its effectiveness. In [17], authors proposed maximum a posteriori (MAP)-based framework for SFF that utilizes local learning process to build a consistency model directly from image focus volume. It improves accuracy and obtains consistent depth maps, but it needs additional computations in local learning process. Recently, in [15], authors suggested anisotropic diffusion process with regularization to obtain better depth map. The above-mentioned approaches provide better depth maps as compared to the focus measure aggregation however, these are also computationally expensive.

In machine learning-based methods, one of the important issues is to develop proper representations for complex data. In literature, various dimensionality reduction models are proposed to reduce the complexity. In [18], authors suggested global geometric framework to discover the nonlinear degrees of freedom that consists of natural multivariate data. It improves efficiency for computing a globally optimal solution than the conventional techniques such as principal component analysis (PCA) [19] and multidimensional scaling (MDS) [20]. In [21], authors developed predictors based on support vector machine (SVM) for feature extraction. They applied various dimensionality reduction models including PCA and independent component analysis (ICA). The results of this study show that SVM by feature extraction using PCA performs better than SVM without using PCA or ICA. In [22], an unsupervised learning algorithm, called locally linear embedding (LLE), is proposed. It maps its inputs into a single global coordinate without local minimization problem. In [23], authors proposed a geometrically motivated algorithm for nonlinear dimensionality reduction having locality-preserving properties. Another alternative research suggests pre-process the input data before feature extraction. In [24], authors presented an application of deep networks to learn features over multiple modalities using an extension of restricted Boltzmann machines (RBM). The multiple features provide considerably better models and reduce dimensionality. Similarly, in [25], authors proposed a method to estimate the motion based on virtual character. The motion features are pre-processed using RBM. In [26], authors proposed a data-processing pipeline based on sparse feature learning using RBM.

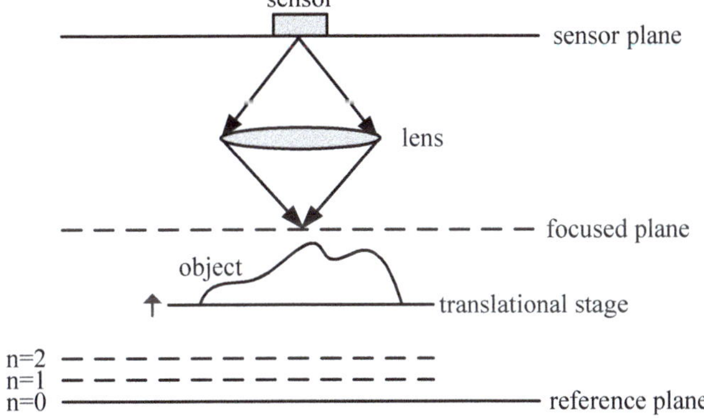

Figure 1. Principle of image sequence acquisition in shape from focus (SFF).

3. Neural Network for SFF

3.1. Neural Network Model over FIS

The neural network model over FIS (NN.FIS) proposed in [10] employs a three layers neural network model as shown in Figure 2. The input layer consists of three linear neurons and hidden layer contains sigmoidal neurons with an additional linear neuron. Whereas the third layer is composed of one sigmoidal neuron and it provides output in the range [0, 1].

At an iteration (n), the local depth map $z_{(x,y)}^{(n)}(j,k)$ for a small window of size $M \times M$ centered at global index (x,y) is described as:

$$z_{(x,y)}^{(n)}(j,k) = \phi\left(\left(\mathbf{v}^{(n)}\right)^{T} \cdot \phi\left(\mathbf{U}^{(n)} \cdot \mathbf{i}\right) + w_{b}^{(n)} \right) \tag{1}$$

where (j,k) is spatial representation of the local index values in the window, $\mathbf{U}^{(n)} \in \mathbb{R}^{H \times 3}$ and $\mathbf{v}^{(n)} \in \mathbb{R}^{H}$ are the network weight matrices between the layers, H is the number of sigmoidal neurons. $w_{b}^{(n)}$ represents the bias, $\mathbf{i} = [j\ k\ 1]^{T}$ is the input vector, $(\cdot)^{T}$ indicates transpose function of the matrix function, and $\phi(\cdot)$ is the sigmoidal function.

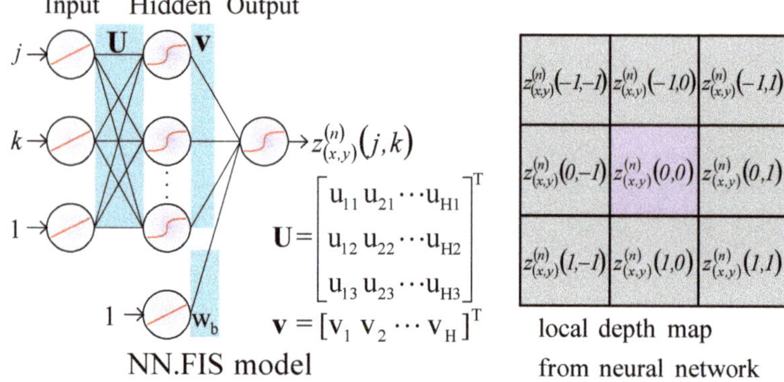

Figure 2. Multilayer feedforward neural network structure for shape recovery of neural network model over FIS and local depth map from the neural network model. **Left:** NN.FIS model; **Right:** local depth map from NN.FIS model. The local depth map consists of 9 neighboring pixels centered around (x,y), when small window size is 3×3. In addition, each local depth is produced by network with changing the local index j and k.

The focus measure F is calculated from $z_{(x,y)}^{(n)}(j,k)$ by GLV as follows:

$$F = \sum_{\substack{-m \leq j \leq m \\ -m \leq k \leq m}} \left[I\left(x+j, y+k, z_{(x,y)}^{(n)}(j,k)\right) - \mu \right]^{2} \tag{2}$$

where $I\left(x+j, y+k, z_{(x,y)}^{(n)}(j,k)\right)$ is the gray level at $(x+j, y+k)$ in the frame number $z_{(x,y)}^{(n)}(j,k)$, $m = (M-1)/2$ the local index bound, and μ represents the average gray level of the pixels on the window surface defined by $z_{(x,y)}^{(n)}(j,k)$ as:

$$\mu = \frac{1}{M^{2}} \sum_{\substack{-m \leq j \leq m \\ -m \leq k \leq m}} I\left(x+j, y+k, z_{(x,y)}^{(n)}(j,k)\right) \tag{3}$$

It is important to note that F and μ are the functions of $z_{(x,y)}^{(n)}(j,k)$. Accordingly, these functions not only depend on global index (x,y) but also depend on local index (j,k). It means, they can only be calculated during the network learning. Consequently, the complexity rises sharply.

Generally, feedforward neural networks are utilized for learning an input-output relationship from the given example data set. In case of neural networks for SFF, there is no such an example data set available. Therefore, it is a distinctive application of neural networks where a focus measure, which is role of a performance function, is to be optimized. The network weights are updated according to the gradient decent rule as follows:

$$\Delta \mathbf{W}^{(n)} = -\beta \left(\frac{\partial E}{\partial \mathbf{W}^{(n)}} \right)$$
$$= \beta \left(\frac{\partial F}{\partial \mathbf{W}^{(n)}} \right) \tag{4}$$

where $\Delta \mathbf{W}^{(n)}$ means the net change of the weight matrix $\mathbf{W}^{(n)}$, which includes $\mathbf{U}^{(n)}$, $\mathbf{v}^{(n)}$, and $w_b^{(n)}$, whereas β is the learning rate. In addition, $E = -F$ is the minimization criterion. At each iteration weights are updated and consequently $z_{(x,y)}^{(n)}(j,k)$ are also updated. At the last iteration (N), the final depth at (x,y) is obtained as:

$$d(x,y) = z_{(x,y)}^{(N)}(0,0) \tag{5}$$

and the final restored depth map from the NN.FIS model is represented as:

$$F_{NN.FIS} = \{d(x,y) | x \in [1,\ldots,I_W], y \in [1,\ldots,I_H]\} \tag{6}$$

where I_W and I_H are the width and the height of each image in the sequence, respectively.

3.2. Proposed Model

In NN.FIS a focus value is computed from a current frame $(z_{(x,y)}^{(n)}(0,0))$ and its neighboring frames $(z_{(x,y)}^{(n)}(j,k))$ that increases the size of the weight matrix $\mathbf{W}$ and over all complexity for shape recovery become high. We have observed through the experiments that the focus measure computed from the current frame not only provides smaller weight matrix but also reduces the time complexity remarkably without compromising on accuracy. Figure 3 shows the proposed neural network model over planar surface (NN.Planar) for shape from focus. In SFF, a planar surface is considered in 3D space, however, in this paper "planar" means only on the same frame i.e., 2D surface. From the figure, it can be observed that the number of neurons at the input layer are decreased from three to one. In the NN.Planar, the depth $z_{(x,y)}^{(n)}$ at iteration (n) for the central pixel (x,y) of the window can be expressed as:

$$z_{(x,y)}^{(n)} = \phi \left(\left(\mathbf{v}^{(n)} \right)^T \cdot \phi \left(\mathbf{u}^{(n)} \right) + w_b^{(n)} \right) \tag{7}$$

where $\mathbf{u}^{(n)} \in \mathbb{R}^H$, $\mathbf{v}^{(n)} \in \mathbb{R}^H$, and $w_b^{(n)} \in \mathbb{R}$, and $\phi(\cdot)$ is sigmoidal function defined as:

$$\phi(\eta) = \frac{1}{1 + e^{-\eta}} \tag{8}$$

where η is a dummy variable.

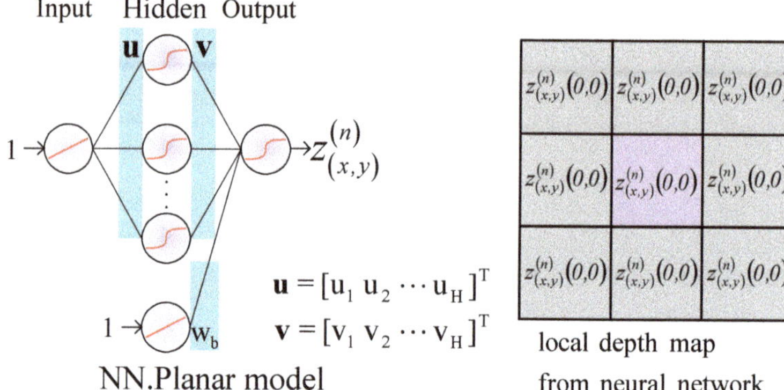

Figure 3. Proposed neural network structure for shape recovery of neural network model over Planar and local depth map from the neural network model. **Left**: NN.Planar model; **Right**: local depth map from NN. Planar model. The local depth map consists of 9 neighboring pixels centered around (x,y), when small window size is 3×3. In addition, all local depths are same to the depth at center, so it can reduce the dimension of input layer from 3 to 1.

Accordingly, in the NN.Planar, the focus measure F is calculated as

$$F = \sum_{\substack{-m \leq j \leq m \\ -m \leq k \leq m}} \left[I\left(x+j, y+k, z^{(n)}_{(x,y)}\right) - \mu \right]^2 \tag{9}$$

where $I\left(x+j, y+k, z^{(n)}_{(x,y)}\right)$ is the gray level at the pixel position of $(x+j, y+k)$ in the frame number $z^{(n)}_{(x,y)}$, and μ represents the mean of gray levels within the window and it is expressed as

$$\mu = \frac{1}{M^2} \sum_{\substack{-m \leq j \leq m \\ -m \leq k \leq m}} I\left(x+j, y+k, z^{(n)}_{(x,y)}\right) \tag{10}$$

In order to update weights, we use gradient descent rule. Note that, in NN.Planar, a scalar depth is computed instead of a matrix (as in the case of NN.FIS model) therefore the computation of gradient becomes simplified. Moreover, it becomes more efficient due to the reduced dimensions of $\mathbf{u}^{(n)}$. By using chain rule, Equation (4) can be written as

$$
\begin{aligned}
\Delta \mathbf{W}^{(n)} &= \beta \left(\frac{\partial F}{\partial \mathbf{W}^{(n)}} \right) \\
&= \beta \left(\frac{\partial F}{\partial z^{(n)}_{(x,y)}} \right) \cdot \left(\frac{\partial z^{(n)}_{(x,y)}}{\partial \mathbf{W}^{(n)}} \right) \\
&= 2\beta \sum_{\substack{-m \leq j \leq m \\ -m \leq k \leq m}} \left[I\left(x+j, y+k, z^{(n)}_{(x,y)}\right) - \mu \right] \\
&\quad \cdot \left(\frac{\partial \left[I\left(x+j, y+k, z^{(n)}_{(x,y)}\right) - \mu \right]}{\partial z^{(n)}_{(x,y)}} \right) \\
&\quad \cdot \left(\frac{\partial z^{(n)}_{(x,y)}}{\partial \mathbf{W}^{(n)}} \right)
\end{aligned}
\tag{11}
$$

Here μ is constant, thus Equation (11) can be written as

$$
\begin{aligned}
\Delta \mathbf{W}^{(n)} \;=\; & 2\beta \sum_{\substack{-m \leq j \leq m \\ -m \leq k \leq m}} \left[I\left(x+j, y+k, z^{(n)}_{(x,y)}\right) - \mu \right] \\
& \cdot \left(\frac{\partial I\left(x+j, y+k, z^{(n)}_{(x,y)}\right)}{\partial z^{(n)}_{(x,y)}} \right) \\
& \cdot \left(\frac{\partial z^{(n)}_{(x,y)}}{\partial \mathbf{W}^{(n)}} \right)
\end{aligned}
\tag{12}
$$

where $I\left(x+j, y+k, z^{(n)}_{(x,y)}\right)$ represents gray level values in the window, and the term $\left(\partial I\left(x+j, y+k, z^{(n)}_{(x,y)}\right) / \partial z^{(n)}_{(x,y)} \right)$ is the derivative of the image intensity of the window surface with respect to the image frame number $z^{(n)}_{(x,y)}$. The derivative can be implemented as

$$
\frac{\partial I\left(x+j, y+k, z^{(n)}_{(x,y)}\right)}{\partial z^{(n)}_{(x,y)}} = \frac{I_+ - I_-}{2h}
\tag{13}
$$

where h is a step size, $I_+ = I\left(x+j, y+k, z^{(n)}_{(x,y)}+h\right)$, and $I_- = I\left(x+j, y+k, z^{(n)}_{(x,y)}-h\right)$. The remaining term $\left(\partial z^{(n)}_{(x,y)} / \partial \mathbf{W}^{(n)} \right)$, which is surface gradient, can be acquired using backpropagation algorithm [9]. Here, the weight $\mathbf{W}^{(n)}$ includes $\mathbf{u}^{(n)}$, $\mathbf{v}^{(n)}$, and $w_b^{(n)}$, and their derivatives are expressed as

$$
\begin{aligned}
\frac{\partial z^{(n)}_{(x,y)}}{\partial \mathbf{u}^{(n)}} = \; & z^{(n)}_{(x,y)} \cdot \left(1 - z^{(n)}_{(x,y)}\right) \cdot \mathbf{v}^{(n)} \\
& * \left[\phi\left(\mathbf{u}^{(n)}\right) * \left(1 - \phi\left(\mathbf{u}^{(n)}\right)\right) \right]
\end{aligned}
\tag{14}
$$

$$
\frac{\partial z^{(n)}_{(x,y)}}{\partial \mathbf{v}^{(n)}} = z^{(n)}_{(x,y)} \cdot \left(1 - z^{(n)}_{(x,y)}\right) \cdot \phi\left(\mathbf{u}^{(n)}\right)
\tag{15}
$$

$$
\frac{\partial z^{(n)}_{(x,y)}}{\partial w_b^{(n)}} = z^{(n)}_{(x,y)} \cdot \left(1 - z^{(n)}_{(x,y)}\right)
\tag{16}
$$

where $*$ denotes the element-by-element multiplication of vectors. Using Equations (12)–(16), the weights are updated as

$$
\mathbf{u}^{(n+1)} = \mathbf{u}^{(n)} + \beta \cdot G^{(n)} \cdot \frac{\partial z^{(n)}_{(x,y)}}{\partial \mathbf{u}^{(n)}}
\tag{17}
$$

$$
\mathbf{v}^{(n+1)} = \mathbf{v}^{(n)} + \beta \cdot G^{(n)} \cdot \frac{\partial z^{(n)}_{(x,y)}}{\partial \mathbf{v}^{(n)}}
\tag{18}
$$

$$
w_b^{(n+1)} = w_b^{(n)} + \beta \cdot G^{(n)} \cdot \frac{\partial z^{(n)}_{(x,y)}}{\partial w_b^{(n)}}
\tag{19}
$$

where $G^{(n)}$ is represented as

$$
\begin{aligned}
G^{(n)} = \; & 2 \sum_{\substack{-m \leq j \leq m \\ -m \leq k \leq m}} \left[I\left(x+j, y+k, z^{(n)}_{(x,y)}\right) - \mu \right] \\
& \cdot \left(\frac{\partial I\left(x+j, y+k, z^{(n)}_{(x,y)}\right)}{\partial z^{(n)}_{(x,y)}} \right)
\end{aligned}
\tag{20}
$$

Note that $G^{(n)}$ is pre-calculable value before network learning. Essentially, it is dependent on global indexes (x, y) and the network output $z_{(x,y)}^{(n)}$, and thus it is dependent on the network learning iteration (n). However, a possible set of $G^{(n)}$ can be calculated in advance of network learning, because the number of elements of the possible set is same to the total frame number I_D of image sequence. Therefore, it gives additional time efficiency. After updating $\mathbf{W}^{(n+1)}$ as Equations (17)–(19), the present iteration (n) is ended, and it follows the next iteration $(n + 1)$. In this manner, all $z_{(x,y)}^{(n)}$ at each iteration are updated and the final depth map for (x, y) can be acquired as

$$d(x, y) = z_{(x,y)}^{(N)} \tag{21}$$

And the final depth map of NN.Planar at all points is described as

$$F_{NN.Planar} = \{d(x,y) | x \in [1, \ldots, I_W], y \in [1, \ldots, I_H]\} \tag{22}$$

The proposed NN.Planar looks an instance of the NN.FIS as proposed in [10]. Consider a particular instance of NN.FIS by restricting the input vector $\mathbf{i} = [j\ k\ 1]^T$ with $j = 0, k = 0$. In this case, it needs to update not only weights connected from the bias 1 but also weights connected from j and k. Weights update, related to j and k, for focus measure over a planar surface is redundant in NN.FIS. Therefore, the proposed NN.Planar is not an instance of the NN.FIS. Moreover, in the NN.FIS, neighboring points are taken into account to compute the complex shape of FIS. Whereas the proposed NN.Planar is based on the assumption of planar surface. In other words, the surface within the local window is assumed to be a planar surface. It may seem to be similar to the piece-wise constant approximation. However, it is not constant as it is modelled through the functions (7) and (9) to provide accurate depth.

4. Initial Weight Setting

In neural network-based SFF methods, initial network weights play important role in acquiring accurate depth map, network convergence, and time complexity. If the initial weights for a pixel are not set properly, more likely an incorrect initial depth map is obtained. Consequently, the convergence of the network is affected by these inaccuracies in depth map and improper initial weights. Therefore, in NN.FIS and NN.Planar networks, it is important to select suitable initial weights for rapid convergence of the networks and for obtaining accurate final depth maps.

Usually, initial weights are selected randomly. However, random weights may not provide rapid convergence and accurate depth map. In order to overcome these problems, we propose WP method for setting proper initial weights. Using WP method, the network converges rapidly, provides more precise depth map and reduces the time complexity. Furthermore, it helps in avoiding the local minimization problem effectively. Before explaining the proposed WP method, it would be worthy to describe the random setting (RS).

4.1. RS Initial Weight Setting Method

In [10], the initial weights for each pixel are determined randomly. If the initial depth map significantly deviate from the desired depth map generated from the initial weight, it may lead to local minimization problem and network may fail in obtaining a precise depth map. Furthermore, this method may lead to slow convergence of the network as it needs larger numbers of iterations for 3D shape learning. In other words, it fails to estimate a depth precisely with a high probability if the initial weights for a pixel are determined randomly. Therefore, in [27], initial weights are taken randomly around a reference depth map $z_r(x, y)$ to complement the problem. In this paper, we call it RS method. The RS method is summarized in Algorithm 1. At first, a test set of initial weights ($\mathbf{u}$, $\mathbf{v}$, and w_b) are randomly generated in the range $[0, 1]$. Next, $z_{(x,y)}^{(1)}$ is calculated using Equation (7). Then, the distance between initial depth $z_{(x,y)}^{(1)}$ and the reference depth $z_r(x, y)$ is compared with the threshold T_1. If the distance is less than or equal to T_1, the test set of initial weights are regarded as the proper initial weights for (x, y) and are used for further computation of $(\mathbf{u}^{(1)}, \mathbf{v}^{(1)}, \text{and } w_b^{(1)})$. On the other side, if the distance is greater than to T_1 then again initial weights are generated randomly and the distance is compared with T_1. This process is repeated for a certain times; say $R = 1000$. If the proper weights are not found within R iterations then an inaccurate depth for (x, y) may be computed. Once weights are obtained, the depth for (x, y) is estimated through NN.Planar (or NN.FIS) model using Equations (7)–(21) (or Equations (1)–(5)) with $\mathbf{u}^{(1)}$, $\mathbf{v}^{(1)}$, and $w_b^{(1)}$. In this way, the entire depth map is obtained by computing depth for all pixel positions.

Algorithm 1 Random Setting method.

1: **procedure** RANDOM SETTING
2:     **for** $x = 1$ to I_W **do**
3:         **for** $y = 1$ to I_H **do**
4:             **for** $i = 1$ to R **do**
5:                 Generate initial weights $\mathbf{u}, \mathbf{v}, w_b$ randomly.
6:                 $z_{(x,y)}^{(1)} \leftarrow \phi\left((\mathbf{v})^T \cdot \phi(\mathbf{u}) + w_b\right)$
7:                 **if** $|z_{(x,y)}^{(1)} - z_r(x, y)| \leq T_1$ **then**
8:                     $\mathbf{u}^{(1)} \leftarrow \mathbf{u}$
9:                     $\mathbf{v}^{(1)} \leftarrow \mathbf{v}$
10:                     $w_b^{(1)} \leftarrow w_b$
11:                     **break**
12:                 **end if**
13:             **end for**
14:             $d(x, y) \leftarrow$ estimated depth at (x, y) by using Equations (7)–(21) (or Equations (1)–(5))
15:         **end for**
16:     **end for**
17: **end procedure**

In summary, this method has two main problems: (1) the time complexity is significantly increased by determining the initial weights for every pixel position; and (2) If the initial depth map is close to the reference depth map, RS method may avoid the local minimization problem and can provide reasonable initial weights however, when the reference depth map is near to the minimum or maximum value of depth range, it is hard to get proper initial weights and accurate depth map. This phenomena is highlighted in Figure 4. It shows the histogram of $z_{(x,y)}^{(1)}$ and represents the distribution of numerical data of $z_{(x,y)}^{(1)}$ from 10,000 trials of Equation (7) by using random weights. From Figure 4, we can find that the frequency is very low near to the minimum or maximum depth values. Consequently, it is hard to meet the requirements of $|z_{(x,y)}^{(1)} - z_r(x, y)| \leq T_1$ within R trials, and thus it is hard to set the proper initial weights in RS method. For example, if $z_r(x, y) = 0.05$ and $T_1 = 0.02$, and then $z_{(x,y)}^{(1)}$ should be in $[0.03, 0.07]$. However, there is low probability that $z_{(x,y)}^{(1)}$ be within $[0.03, 0.07]$ after R trials, thus there is high probability of not setting proper initial weights.

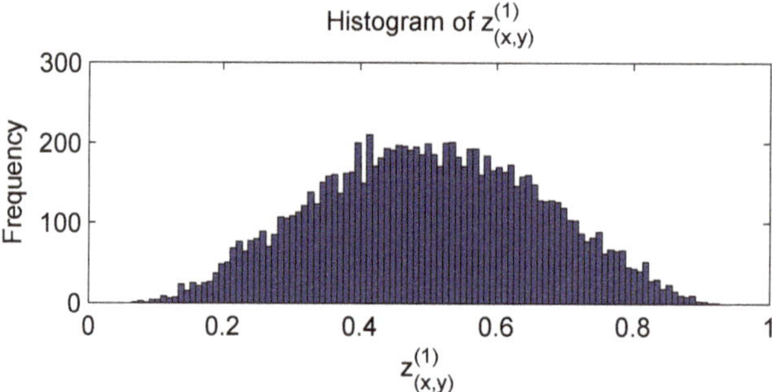

Figure 4. Histogram of $z^{(1)}_{(x,y)}$ values. It graphically represents the distribution of numerical data of $z^{(1)}_{(x,y)}$. In addition, the bins are same size of 0.01 and the total frequency is 10,000. Note that it is similar to the Gaussian distribution, and thus the frequency number is very low at a range of $z^{(1)}_{(x,y)} \leq 0.1$ or $z^{(1)}_{(x,y)} \geq 0.9$. Consequently, in the range, it is hard to meet the requirements of $|z^{(1)}_{(x,y)} - z_r(x,y)| \leq T_1$ within R, and thus it fails to set the proper initial weights in RS method.

4.2. Proposed WP Method

Figure 4 highlights the distribution of initial depth z_i. From the figure, it can be observed that the initial depth values near the extremities may affect the procedure of choosing the proper initial weights. Particularly, the RS method may produce an erroneous initial depth values near the extremities as theses depth values may considerably deviate from the precise depth. This phenomena can also be understandable as there is high probability that $G^{(1)}$ may be near to zero at these condition. Therefore, weights will become stagnant as this fact can be observed from Equations (17)–(19). It can be concluded that the initial depth needs to be close to the precise depth for avoiding local minimization problem. Therefore, WP method is proposed to relieve the local minimization regardless of a depth range of a reference depth map.

The proposed WP method is described in Algorithm 2. In the proposed algorithm, in first step, a 2-dimensional reference depth map $z_r(x,y)$ patch is converted into 1-dimensional reference depth vector $z_r(s)$. In the second step, a new vector $\hat{z}_r(t)$ is obtained by sorting $z_r(s)$ with respect to depth in ascending order. In third step, an index vector $\hat{s}(t)$ is acquired by rearranging the indices of $z_r(s)$ corresponding to the depth at $\hat{z}_r(t)$. In fourth step, 2-dimentional indices $(\hat{x}(t), \hat{y}(t))$ are acquired by rearranging the index of 2-dimentional patch $z_r(x,y)$ corresponding to the depth at $\hat{z}_r(t)$. In order to describe these four steps, an example is shown in Figure 5. In fifth step, the proper initial weights for the first pixel $z_r(\hat{x}(1), \hat{y}(1))$ are obtained by comparing the initial depth with the reference depth. In sixth step, a depth for $(\hat{x}(t), \hat{y}(t))$ is estimated through the NN.Planar model as using Equations (7)–(21) with $\mathbf{u}^{(1)}$, $\mathbf{v}^{(1)}$, and $\mathbf{w_b}^{(1)}$. In next step, initial weights for the next pixel are set by passing updated weights from the present pixel. In this way, the entire depth map is estimated by repeating the steps 6–7 for all pixel positions.

Algorithm 2 Weight Passing method.

1: **procedure** WEIGHT PASSING
2:     $z_r(s) \leftarrow z_r(x,y)$ ▷ Change 2-dimensional $z_r(x,y)$ into 1-dimensional vector form $z_r(s)$
3:     $\hat{z}_r(t) \leftarrow$ sorted $z_r(s)$ into depth order
4:     $\hat{s}(t) \leftarrow$ rearranged the index of $z_r(s)$ corresponding to the depth of $\hat{z}_r(t)$
5:     $(\hat{x}(t), \hat{y}(t)) \leftarrow$ rearranged the index of $z_r(x,y)$ corresponding to the depth of $\hat{z}_r(t)$
6:     **for** $i = 1$ to R **do**
7:         Generate initial weights $\mathbf{u}, \mathbf{v}, w_b$ randomly.
8:         $z_{(x,y)}^{(1)} \leftarrow \phi\left((\mathbf{v})^T \cdot \phi(\mathbf{u}) + w_b\right)$
9:         **if** $|z_{(x,y)}^{(1)} - z_r(\hat{x}(1), \hat{y}(1))| \leq T_1$ **then**
10:            $\mathbf{u}^{(1)} \leftarrow \mathbf{u}$
11:            $\mathbf{v}^{(1)} \leftarrow \mathbf{v}$
12:            $w_b^{(1)} \leftarrow w_b$
13:            **break**
14:        **end if**
15:    **end for**
16:    **for** $t = 1$ to $(I_W \cdot I_H)$ **do**
17:        $d(\hat{x}(t), \hat{y}(t)) \leftarrow$ estimated depth at $(\hat{x}(t), \hat{y}(t))$ pixel position by using Equations (7)–(21)
    (or Equations (1)–(5))
18:            $\mathbf{u}^{(1)} \leftarrow$ updated weight $\mathbf{u}^{(N)}$
19:            $\mathbf{v}^{(1)} \leftarrow$ updated weight $\mathbf{v}^{(N)}$
20:            $w_b^{(1)} \leftarrow$ updated weight $w_b^{(N)}$
21:    **end for**
22: **end procedure**

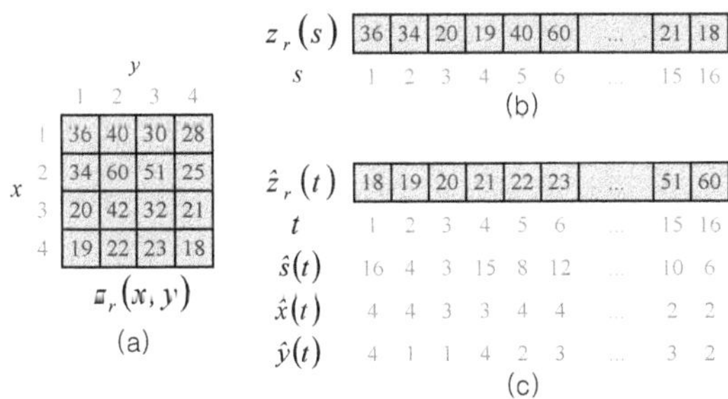

Figure 5. The proposed WP initial weight setting method. (a) $z_r(x,y)$ is a reference depth map estimated by Sum-Modified-Laplacian (SML) or Gray-Level-Variance (GLV); (b) $z_r(s)$ is 1-dimensional vector form of the $z_r(x,y)$ and s is the index of $z_r(s)$; (c) $\hat{z}_r(t)$ is the sorted reference depth map into depth order, t is the index of $\hat{z}_r(t)$, $\hat{s}(t)$ is the index of $z_r(s)$ corresponding to the depth of $\hat{z}_r(t)$, and $(\hat{x}(t), \hat{y}(t))$ is the index of $z_r(x,y)$ corresponding to the depth of $\hat{z}_r(t)$.

Note that the initial depth of $(\hat{x}(i+1), \hat{y}(i+1))$ is close to $d(\hat{x}(i), \hat{y}(i))$, because $\mathbf{W}^{(1)}$ for $(\hat{x}(i+1), \hat{y}(i+1))$ is allocated from $\mathbf{W}^{(N)}$ for $(\hat{x}(i), \hat{y}(i))$. In addition, there is high probability that the resultant depths for $d(\hat{x}(i+1), \hat{y}(i+1))$ and $d(\hat{x}(i), \hat{y}(i))$ may be very close to each other. Accordingly, z_i satisfies the condition $|z_{(x,y)}^{(1)} - z_r(x(i), y(i))| \leq T_1$ when $i \geq 2$. In summary, WP is advantageous in (1) reducing the chances of local minimization problem; (2) decreasing the total numbers of iterations for the stable solution; and (3) decreasing computational complexity for setting initial weights for all pixels.

5. Experiments

5.1. Experimental Setup

The performance of the proposed system consisting of NN.Planar model, WP initial weight setting method is evaluated using image sequences of synthetic and real objects. Synthetic objects plane, sinusoidal, cone, and wave are virtually created by simulation tool. For each object, synthetic image sequence is generated by using a defocus simulation tool box [28] with varying focus settings. Each synthetic image in the sequences is resultant of convolution of actual texture image and Point Spread Function (PSF). Thus, in the synthetic image sequence, the Gaussian function is taken as PSF, camera parameters and the true depth map of the synthetic object are used. For more details, refer to [12,29]. Image sequences for each synthetic object consists of 80 images (or frames) each with size 360×360 pixels.

Figure 6a represents sample frames from image sequences of plane, sinusoidal, cone, and wave objects. From the figure, we can observe that some parts of the objects are well focused and others are out of focus with a degree of blur. And for evaluating the proposed algorithms in real world, we have performed experiments using real objects of real cone, engraved letter I, coin, and TFT-LCD color filter. An image sequence from real cone is obtained by using a CCD camera system with varying focus levels [3], and it consists of 97 images of 200×200 pixels. Other image sequences from microscopic objects of engraved letter I, coin, and TFT-LCD color filter are acquired by using a microscopic control system [2]. In the case of coin object, the image sequence consists of 68 frames of 300×300 pixels. And each image sequence from engraved letter I and TFT-LCD color filter consists of 60 images of 300×300 pixels. Sample frames from each image sequence are represented in Figure 6b.

In order to produce comparative analysis, first, depth maps have been obtained from image sequences by applying conventional methods and the proposed method. Then, the resultant depth maps are compared qualitatively and quantitatively. We performed experiments for networks NN.FIS and NN.Planar with combinations of RS initial weight setting method [10] and WP initial weight setting method. In the experiments, the numbers of hidden layer H, the total iteration number T, the window size $M \times M$ are set as 20, 50, 7×7, respectively. Moreover, the learning rate βs: 19.92, 2.35, 6.15, 7.23, and 1.25 are taken for synthetic objects, real cone, engraved letter I, coin, and TFT-LCD color filter, respectively. There is not any specific method to determine the size of the hidden layer H. If H is too large it suffers from expensive computation. On the other hand, if H is too small the network may not provide desired results. Therefore, H is determined empirically through experiments.

5.2. Quantitative Analysis

In order to evaluate the performance of the proposed system quantitatively, two quantitative metrics: Root-Mean-Square-Error (RMSE) and correlation (Corr). In the case of experiments for synthetic objects, it is possible to calculate RMSE and correlation as their true depth maps are available. Whereas it is impracticable to compute RMSE and correlation metrics for real objects as there true depth maps are not available. RMSE measures the distortion between the true depth map and the estimated depth map. It can be calculated as:

$$\text{RMSE} = \sqrt{\frac{1}{I_W \cdot I_H} \sum_{\substack{1 \leq x \leq I_W \\ 1 \leq y \leq I_H}} (z_t(x,y) - z_e(x,y))^2} \tag{23}$$

where $z_t(x,y)$ and $z_e(x,y)$ are the true depth map and the estimated depth map respectively, I_W and I_H are the width and the height of the true and the estimated depth maps. correlation measures the similarity between the true depth map and the estimated depth map. It can be calculated as follows:

$$\text{Corr} = \frac{\sum_{\substack{1 \leq x \leq I_W \\ 1 \leq y \leq I_H}} (z_t(x,y) - \overline{z_t(x,y)})(z_e(x,y) - \overline{z_e(x,y)})}{\sum_{\substack{1 \leq x \leq I_W \\ 1 \leq y \leq I_H}} (z_t(x,y) - \overline{z_t(x,y)})^2 (z_e(x,y) - \overline{z_e(x,y)})^2} \tag{24}$$

where $\overline{z_t(x,y)}$ and $\overline{z_e(x,y)}$ are the mean of the true and the estimated depth map, respectively.

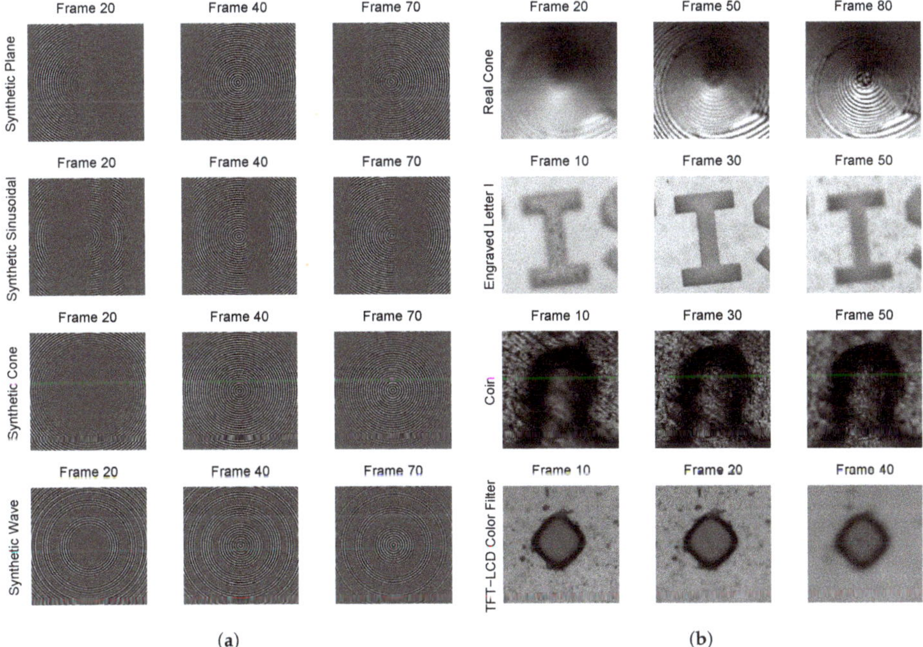

Figure 6. (a) Sample frames from image sequences of synthetic objects: plane object (first row), sinusoidal object (second row), cone object (third row), and wave object (fourth row). In addition, each column shows frame number 20, frame number 40, and frame number 70. In a sample frame, the texture clearly distinguishes only at the well focused area; (b) Sample frames from image sequences of real objects: real cone (first row), engraved letter I (second row), coin (third row), and TFT-LCD color filter (fourth row). In a sample frame, the texture clearly distinguishes only at the well focused area.

The performance comparison between conventional and proposed methods in terms of RMSE and correlation is shown in Table 1. The quantitative measures are calculated by using estimated depth maps through SML, GLV, NN.FIS with RS, NN.Planar with RS, NN.FIS with WP, and NN.Planar with WP. From Table 1, two meaningful results are (1) The performance is similar to each other between NN.FIS and NN.Planar with same initial weight setting method (NN.FIS with RS vs. NN.Planar with RS or NN.FIS with WP vs. NN.Planar with WP), and (2) the proposed WP generates more accurate estimated depth map than RS initial weight setting method (NN.FIS with RS vs. NN.FIS with WP or NN.Planar with RS vs. NN.Planar with WP). In the same manner, Table 2 shows the experiment time for various SFF methods using various synthetic objects. There are also two notable results (1) NN.Planar model is faster than NN.FIS model with same initial weight setting method; and (2) WP is faster than RS method. Specifically, NN.Planar with WP is about 90 times faster than NN.FIS with

WP, and NN.Planar with WP is about 100 to 140 times faster than NN.FIS with RS. Note that RS method gains additional complexity because of setting initial weights for all pixels of an object. In addition, the additional complexity varies considerably with the reference depth of the object. As shown in Algorithm 1, if z_r at a point is considerably high near to 1, then it fails to find the proper initial weight for $z_{(x,y)}^{(1)} < |z_r(x,y) - \sigma|$. For this reason, Sinusoidal and Plane objects, which include many pixels having the reference depth near to 1, provide higher additional complexity than others. Moreover, WP method gives more stable performance than RS as shown in Figure 7. It shows box plots of the normalized RMSE from 30 experiments for Planar model with WP and RS method using various objects. Each box plot displays variation in the normalized RMSE results out of 30 experiments, and it shows the degree of spread. From Figure 7, we can observe that the variable width of WP is less than that of RS.

Table 1. Performance Comparison for Various SFF methods using Various Synthetic Objects.

Object	SML RMSE (Corr)	GLV RMSE (Corr)	FIS with RS RMSE (Corr)	Planar with RS RMSE (Corr)	FIS with WP RMSE (Corr)	Planar with WP RMSE (Corr)
Plane	4.589 (0.970)	3.805 (0.979)	4.196 (0.972)	4.073 (0.974)	3.217 (0.985)	3.215 (0.985)
Sinusoidal	4.649 (0.978)	3.782 (0.985)	4.630 (0.977)	4.505 (0.978)	3.092 (0.990)	3.128 (0.990)
Cone	8.548 (0.957)	8.462 (0.971)	8.567 (0.967)	8.503 (0.970)	8.402 (0.978)	8.395 (0.978)
Wave	2.824 (0.979)	2.004 (0.989)	2.581 (0.981)	2.366 (0.984)	1.644 (0.992)	1.661 (0.992)

Table 2. Experiment Time for Various SFF methods using Various Synthetic Objects.

Object	SML	GLV	FIS with RS	Planar with RS	FIS with WP	Planar with WP
Plane	17.8	24.6	15,347.3	2837.5	12,377.9	130.9
Sinusoidal	17.9	24.6	18,169.5	5336.5	12,504.5	129.9
Cone	17.8	24.8	13,183.0	928.4	12,922.5	135.9
Wave	17.8	24.6	13,829.0	977.9	12,595.6	130.4

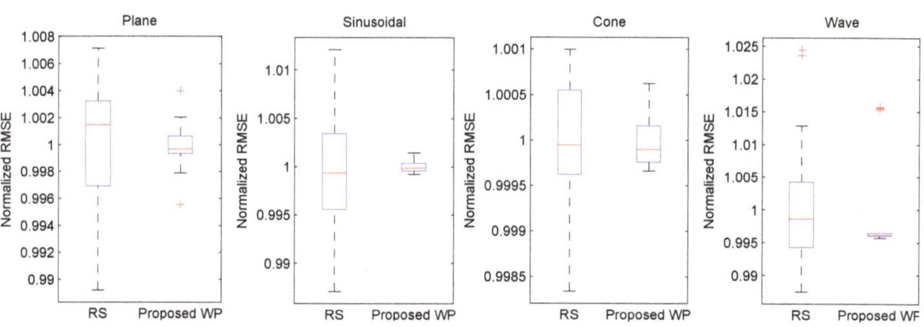

Figure 7. Box plot of the normalized Root-Mean-Square-Error (RMSE) from 30 experiments for Planar with WP and Planar with RS using various objects.

5.3. Qualitative Analysis

Figure 8 shows restored 3D depth maps for synthetic objects with different learning iterations. At first or second column, when total iteration is 1 or 5, the network produces inaccurate results at pixel positions having high reference depth. The reason is that it needs sufficient iteration to converge the depth at the previous pixel of inaccuracy. Therefore, it needs a sufficient total iteration number, over 50 iterations in our simulation, to ensure depth convergence for all pixels. In addition, note that we can roughly acquire the restored depth map, appearing to be the corresponding object, even 1 iteration. That is the strength of WP method. Specifically, when the total iteration is not enough, depths at the first few pixels (its pixel positions having low reference depth) are considerably precise; however, depths return to be distorted after some pixels (its pixel positions having high reference depth). In addition, overall depth map can be precisely restored at 50 iterations, which is also smaller iterations than other usual neural networks.

Figure 8. Restored three-dimensional (3D) depth maps for synthetic objects with different learning iterations, first row: plane object, second row: sinusoidal object, third row: cone object, and fourth row: wave object. In addition, first column: Iteration 1, second column: Iteration 5, and third column: Iteration 50. In this experiment, the Planar model with WP method is utilized for measuring depth map.

Resultant depth maps for various SFF methods using synthetic objects are show in Figure 9. From the figure, it is clear that (1) the precision of restored depth map by using NN.FIS or NN.Planar is almost similar to each other with a same initial weight setting. It means that the NN.Planar model is more efficient than the NN.FIS model because it gives similar performance with less complexity; And (2) WP initial setting method facilitates more precise restoring than RS when using a same neural network model. Specifically, the Planer and Sinusoidal objects include pixel positions having high depth near to the maximum depth. In the pixel positions, it is hard to estimate a depth precisely with RS method because of short of iteration, thus it affects the precision of entire restored depth map. However, WP relieves the problem regardless of the histogram of depth. In addition, Figure 10 shows the restored 3D depth map by using real objects. Similarly, NN.FIS and NN.Planar with a same initial weight setting method give almost same restored depth. In addition, WP gives more precise restored depth map than RS in a same neural network model. The reference depth map is utilized as a reference for generating an initial depth map. In the case of coin object, the restored depth map by SML is used as the reference map for coin object. In addition, other objects utilize the restored depth map by GLV as the reference map for the objects. Note that the reference depth map of each object contains several errors, but the errors are reduced by using Planar with WP. In addition the errors increase slightly or heavily with an object by using Planar with RS, because the RS method needs enough iteration larger than 50. In order to solve the errors in RS, the total iteration number should be increased enough but it gives time complexity.

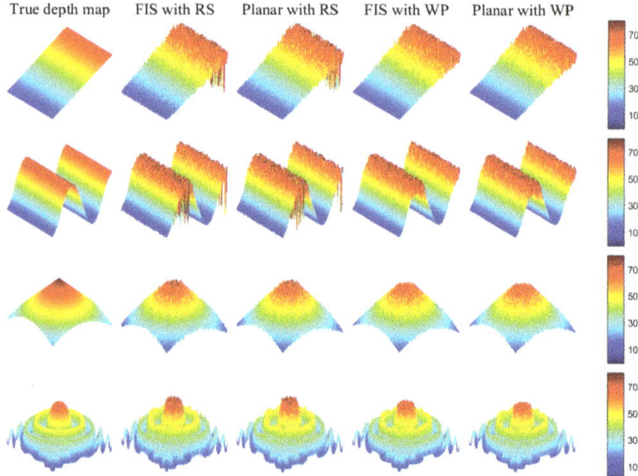

Figure 9. Restored 3D depth map for various SFF methods using various synthetic objects, first row: plane object, second row: sinusoidal object, third row: cone object, and fourth row: wave object. In addition, first column: True depth map, second column: NN.FIS with RS, third column: NN.Planar with RS, fourth column: NN.FIS with WP, and fifth column: NN.Planar with WP.

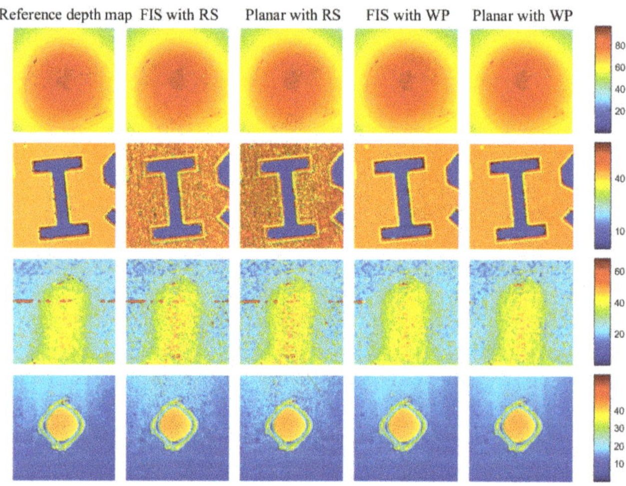

Figure 10. Restored 3D depth map for various SFF methods using various real objects, first row: real cone object, second row: engraved I object, third row: coin object, and fourth row: TFT-LCD filter object. In addition, first column: True depth map, second column: NN.FIS with RS, third column: NN.Planar with RS, fourth column: NN.FIS with WP, and fifth column: NN.Planar with WP.

6. Discussion

The presented method is one of the paradigms of 3D shape recovery techniques known as SFF. Three-dimensional shapes have possible potentials for various applications including motion reconstruction [30–32], 3D object retrieval [33], 3D cameras, manufacture of TFT-LCD color filter, and measurement of surface roughness, medical examinations, microelectronics, and focus variation for 3D surface. The proposed method is a simplified version of the previous well-known methods based on neural network. From the experimental results, it is clear that the proposed model has considerably reduced the complexity without a loss in accuracy. It is expected that if the PCA is used to reduce the dimensionality to select the surface locally, it can give marginally better accurate depth than the NN.Planar. However, the additional computational complexity by PCA will rise proportional to the total iteration numbers. It would be interesting to do a separate study that can make fair comparisons among the methods based on different dimensionality reduction techniques for three shape refinement through the SFF in term of complexity and accuracy.

In literature, a number of machine learning techniques such as convolutional neural network (CNN) [34,35], deep convolutional neural network (DCNN) [36,37] recurrent neural network (RNN) [38], graphical model [39], have been proposed. Among these, CNN includes the fully connected layers which connect each neuron in a layer to all neurons in the next layer. It is similar to the feedforward artificial neural network. Therefore, WP method can be applied with the same settings by using CNN for 3D imaging applications to reduce local minimization. Similarly, the other well-known techniques are also applicable with modifications in the weight update procedures.

In machine learning techniques, the restricted Boltzmann machine [24–26] are considered powerful methods to pre-process the input data to expedite the learning process. In our future work, the restricted Boltzmann machines will be used, instead of reducing the dimensionality of the data, in refining the depth maps in SFF. It is expected that RBM will provide depth maps efficiently with higher accuracy. In addition, it would be interesting to do a separate study for comparing the performances of different machine learning-based methods in 3D shape recovery in SFF in terms of efficacy and accuracy.

7. Conclusions

In this paper, a simplified neural network over planar model is proposed for SFF. In addition, for rapid convergence and efficiency of the network, WP method has been introduced. The simplified neural network model takes input and weight vector of smaller size, and the WP method selects proper initial weights for the first pixel randomly and it updates the initial weights for the next pixel efficiently by passing the updated weight from the present pixel. The proposed method significantly has reduced the computational complexity while the accuracy is comparable with the conventional model.

8. Patents

There is a patent [40] resulting from the work reported in this manuscript.

Author Contributions: Funding Acquisition, T.-S.C. and M.T.M.; Conceptualization and Validation, H.-J.K.; Writing-Original Draft Preparation, H.-J.K.; Writing-Review & Editing, M.T.M.; Supervision, T.-S.C.

Funding: This research was funded by [Ministry of Science and ICT—Korea] grant number (NRF-2015R1A2A2A01002734 and NRF-2015R1A2A1A15053854) and [Ministry of Education—Korea] grant number (NRF 2016R1D1A1B03933860).

Acknowledgments: We thank Jun-seop Lee for his assistance with debugging the conventional neural network model, Ji-Seok Yoon and Hoon-Seok Jang for their assistance with useful discussion, and the anonymous reviewers for their precious comments.

Conflicts of Interest: The authors declare no conflict of interest.

References

1. Lee, I.H.; Mahmood, M.T.; Choi, T.S. Robust Depth Estimation and Image Fusion Based on Optimal Area Selection. *Sensors* **2013**, *13*, 11636–11652. [CrossRef] [PubMed]
2. Ahmad, M.; Choi, T.S. Application of Three Dimensional Shape from Image Focus in LCD/TFT Displays Manufacturing. *IEEE Trans. Consum. Electron.* **2007**, *53*, 1–4. [CrossRef]
3. Mahmood, M. MRT letter: Guided filtering of image focus volume for 3D shape recovery of microscopic objects. *Microsc. Res. Tech.* **2014**, *77*, 959–963. [CrossRef] [PubMed]
4. Mahmood, M.; Choi, T.S. Nonlinear Approach for Enhancement of Image Focus Volume in Shape from Focus. *IEEE Trans. Image Process.* **2012**, *21*, 2866–2873. [CrossRef] [PubMed]
5. Thelen, A.; Frey, S.; Hirsch, S.; Hering, P. Improvements in Shape-From-Focus for Holographic Reconstructions with Regard to Focus Operators, Neighborhood-Size, and Height Value Interpolation. *IEEE Trans. Image Process.* **2009**, *18*, 151–157. [CrossRef] [PubMed]
6. Tang, H.; Cohen, S.; Price, B.; Schiller, S.; Kutulakos, K.N. Depth from Defocus in the Wild. In Proceedings of the 2017 IEEE Conference on Computer Vision and Pattern Recognition (CVPR), Honolulu, HI, USA, 21–26 July 2017; pp. 4773–4781. [CrossRef]
7. Frommer, Y.; Ben-Ari, R.; Kiryati, N. Shape from Focus with Adaptive Focus Measure and High Order Derivatives. In Proceedings of the British Machine Vision Conference (BMVC), Swansea, UK, 7–10 September 2015; pp. 134.1–134.12. [CrossRef]
8. Suwajanakorn, S.; Hernandez, C.; Seitz, S.M. Depth from focus with your mobile phone. In Proceedings of the 2015 IEEE Conference on Computer Vision and Pattern Recognition (CVPR), Boston, MA, USA, 7–12 June 2015; pp. 3497–3506, [CrossRef]
9. Bishop, C.M. *Pattern Recognition and Machine Learning*; Springer: New York, NY, USA, 2006. Available online: https://www.springer.com/us/book/9780387310732 (accessed on 13 September 2018).
10. Asif, M.; Choi, T.S. Shape from focus using multilayer feedforward neural networks. *IEEE Trans. Image Process.* **2001**, *10*, 1670–1675. [CrossRef] [PubMed]
11. Malik, A.S.; Choi, T.S. Consideration of illumination effects and optimization of window size for accurate calculation of depth map for 3D shape recovery. *Pattern Recognit.* **2007**, *40*, 154–170. [CrossRef]
12. Pertuz, S.; Puig, D.; Garcia, M.A. Analysis of focus measure operators for shape-from-focus. *Pattern Recognit.* **2013**, *46*, 1415–1432. [CrossRef]
13. Huang, W.; Jing, Z. Evaluation of focus measures in multi-focus image fusion. *Pattern Recognit. Lett.* **2007**, *28*, 493–500. [CrossRef]
14. Ahmad, M.; Choi, T.S. A heuristic approach for finding best focused shape. *IEEE Trans. Circuits Syst. Video Technol.* **2005**, *15*, 566–574. [CrossRef]
15. Boshtayeva, M.; Hafner, D.; Weickert, J. A focus fusion framework with anisotropic depth map smoothing. *Pattern Recognit.* **2015**, *48*, 3310–3323. [CrossRef]
16. Hariharan, R.; Rajagopalan, A. Shape-From-Focus by Tensor Voting. *IEEE Trans. Image Process.* **2012**, *21*, 3323–3328. [CrossRef] [PubMed]
17. Tseng, C.Y.; Wang, S.J. Shape-From-Focus Depth Reconstruction with a Spatial Consistency Model. *IEEE Trans. Circuits Syst. Video Technol.* **2014**, *24*, 2063–2076. [CrossRef]
18. Tenenbaum, J.B.; de Silva, V.; Langford, J.C. A Global Geometric Framework for Nonlinear Dimensionality Reduction. *Science* **2000**, *290*, 2319–2323. [CrossRef] [PubMed]
19. Shlens, J. A Tutorial on Principal Component Analysis. *arXiv* **2014**, arXiv:1404.1100.
20. Borg, I.; Groenen, P.J.; Mair, P. *Applied Multidimensional Scaling and Unfolding*, 2nd ed.; Springer: New York, NY, USA, 2018. Available online: https://www.springer.com/gb/book/9783319734705 (accessed on 13 September 2018).
21. Cao, L.; Chua, K.; Chong, W.; Lee, H.; Gu, Q. A comparison of PCA, KPCA and ICA for dimensionality reduction in support vector machine. *Neurocomputing* **2003**, *55*, 321–336. [CrossRef]
22. Roweis, S.T.; Saul, L.K. Nonlinear Dimensionality Reduction by Locally Linear Embedding. *Science* **2000**, *290*, 2323–2326. [CrossRef] [PubMed]
23. Belkin, M.; Niyogi, P. Laplacian Eigenmaps for Dimensionality Reduction and Data Representation. *Neural Comput.* **2003**, *15*, 1373–1396. [CrossRef]

24. Ngiam, J.; Khosla, A.; Kim, M.; Nam, J.; Lee, H.; Ng, A.Y. Multimodal Deep Learning. In Proceedings of the 28th International Conference on International Conference on Machine Learning (ICML'11), Bellevue, WA, USA, 28 June–2 July 2011; Omnipress: Madison, WI, USA, 2011; pp. 689–696.

25. Mousas, C.; Anagnostopoulos, C.N. Learning Motion Features for Example-Based Finger Motion Estimation for Virtual Characters. *3D Res.* **2017**, *8*, 25. [CrossRef]

26. Nam, J.; Herrera, J.; Slaney, M.; Smith, J. Learning Sparse Feature Representations for Music Annotation and Retrieval. In Proceedings of the 2012 International Society for Music Information Retrieval (ISMIR), Porto, Portugal, 8–12 October 2012; pp. 565–570.

27. Asif, M. Shape from Focus Using Multilayer Feedforward Neural Networks. Master's Thesis, Gwangju Institute of Science and Technology, Gwangju, Korea, 1999.

28. Pertuz, S. Defocus Simulation. Available online: https://kr.mathworks.com/matlabcentral/fileexchange/55095-defocus-simulation (accessed on 12 September 2018).

29. Favaro, P.; Soatto, S.; Burger, M.; Osher, S.J. Shape from Defocus via Diffusion. *IEEE Trans. Pattern Anal. Mach. Intell.* **2008**, *30*, 518–531. [CrossRef] [PubMed]

30. Holden, D.; Komura, T.; Saito, J. Phase-functioned Neural Networks for Character Control. *ACM Trans. Graph.* **2017**, *36*, 42. [CrossRef]

31. Mousas, C.; Newbury, P.; Anagnostopoulos, C.N. Evaluating the Covariance Matrix Constraints for Data-driven Statistical Human Motion Reconstruction. In Proceedings of the 30th Spring Conference on Computer Graphics (SCCG'14), Smolenice, Slovakia, 28–30 May 2014; ACM: New York, NY, USA, 2014; pp. 99–106. [CrossRef]

32. Mousas, C.; Newbury, P.; Anagnostopoulos, C.N. Data-Driven Motion Reconstruction Using Local Regression Models. In *Artificial Intelligence Applications and Innovations, Proceedings of the 10th IFIP International Conference on Artificial Intelligence Applications and Innovations (AIAI), Rhodes, Greece, September 2014*; Iliadis, L., Maglogiannis, I., Papadopoulos, H., Eds.; Part 8: Image–Video Processing; Springer: Berlin/Heidelberg, Germany, 2014; Volume AICT-436, pp. 364–374. Available online: https://hal.inria.fr/IFIP-AICT-436/hal-01391338 (accessed on 13 September 2018). [CrossRef]

33. Krizhevsky, A.; Sutskever, I.; Hinton, G.E. Imagenet classification with deep convolutional neural networks. In Proceedings of the 25th International Conference on Neural Information Processing Systems, Lake Tahoe, Nevada, USA, 3–6 December 2012; pp. 1097–1105.

34. Cheron, G.; Laptev, I.; Schmid, C. P-CNN: Pose-based CNN Features for Action Recognition. *arXiv* **2015**, arXiv:1506.03607.

35. Abdel-Hamid, O.; Mohamed, A.R.; Jiang, H.; Penn, G. Applying convolutional neural networks concepts to hybrid NN-HMM model for speech recognition. In Proceedings of the 2012 IEEE International Conference on Acoustics, Speech and Signal Processing (ICASSP), Kyoto, Japan, 25–30 March 2012; pp. 4277–4280.

36. Saito, S.; Wei, L.; Hu, L.; Nagano, K.; Li, H. Photorealistic Facial Texture Inference Using Deep Neural Networks. *arXiv* **2016**, arXiv:1612.00523.

37. Li, R.; Si, D.; Zeng, T.; Ji, S.; He, J. Deep Convolutional Neural Networks for Detecting Secondary Structures in Protein Density Maps from Cryo-Electron Microscopy. In Proceedings of the IEEE International Conference on Bioinformatics and Biomedicine, Shenzhen, China, 15–18 December 2016; pp. 41–46.

38. Li, Z.; Zhou, Y.; Xiao, S.; He, C.; Li, H. Auto-Conditioned LSTM Network for Extended Complex Human Motion Synthesis. *arXiv* **2017** arXiv:1707.05363.

39. Bilmes, J.; Bartels, C. Graphical model architectures for speech recognition. *IEEE Signal Process Mag.* **2005**, *22*, 89–100. [CrossRef]

40. Kim, H.J.; Mahmood, M.T.; Choi, T.S. A Method for Reconstruction 3-D Shapes Using Neural Netwok. Korea Patent 1,018,166,630,000, 2018. Available online: https://doi.org/10.8080/1020160136767 (accessed on 9 January 2018).

Article

Absolute Phase Retrieval Using One Coded Pattern and Geometric Constraints of Fringe Projection System

Xu Yang [1], Chunnian Zeng [1], Jie Luo [1], Yu Lei [1], Bo Tao [2] and Xiangcheng Chen [1,*]

[1] School of Automation, Wuhan University of Technology, Wuhan 430070, China; yx_auto@whut.edu.cn (X.Y.); zengchn@whut.edu.cn (C.Z.); luo_jie@whut.edu.cn (J.L.); leiyu9087@whut.edu.cn (Y.L.)

[2] Key Laboratory of Metallurgical Equipment and Control Technology, Ministry of Education, Wuhan University of Science and Technology, Wuhan 430081, China; taoboq@wust.edu.cn

* Correspondence: chenxgcg@ustc.edu; Tel.: +86-139-66733394

Received: 30 October 2018; Accepted: 14 December 2018; Published: 18 December 2018

Abstract: Fringe projection technologies have been widely used for three-dimensional (3D) shape measurement. One of the critical issues is absolute phase recovery, especially for measuring multiple isolated objects. This paper proposes a method for absolute phase retrieval using only one coded pattern. A total of four patterns including one coded pattern and three phase-shift patterns are projected, captured, and processed. The wrapped phase, as well as average intensity and intensity modulation, are calculated from three phase-shift patterns. A code word encrypted into the coded pattern can be calculated using the average intensity and intensity modulation. Based on geometric constraints of fringe projection system, the minimum fringe order map can be created, upon which the fringe order can be calculated from the code word. Compared with the conventional method, the measurement depth range is significantly improved. Finally, the wrapped phase can be unwrapped for absolute phase map. Since only four patterns are required, the proposed method is suitable for real-time measurement. Simulations and experiments have been conducted, and their results have verified the proposed method.

Keywords: absolute phase retrieval; phase-shift; fringe order; geometric constraints

1. Introduction

Optical 3D measurement plays a pivotal role in all aspects of our lives, such as industrial production, biological medicine, and consumer entertainment [1–5]. Many optical technologies including structured light, stereo vision, and digital fringe projection (DFP) have been exploited to achieve high-density and full-field 3D measurement [6]. Among those technologies, DFP has become the most popular one because of its speed, accuracy, and flexibility [7]. Fourier transform and phase-shift are two main methods applied in the DFP system [8]. The former method only uses one pattern for computing phase map, but the measured surfaces must be rather simple to avoid a spectral overlapping problem. On the other hand, the phase-shift method exploits at least three patterns to compute the phase map pixel-by-pixel, which can achieve higher accuracy and stronger robustness, especially for complex surfaces. However, those two methods can only work out wrapped phases which need to be unwrapped for absolute phase maps.

Ideally, when referring to the neighboring pixels, the wrapped phase can be unwrapped by adding integral multiple of 2π at each pixel. In reality, however, local shadows, random noises, and isolated objects are very usual occurrences that make the unwrapping phase difficult [9]. Thus, many absolute phase retrieval algorithms have been proposed, which can be divided into two major classes: spatial algorithms and temporal algorithms [7]. The spatial algorithms are generally used for smooth surfaces, while the temporal algorithms are more suitable for complex surfaces and attract more attention [10]. Research conducted in this field brings forth several typical examples. Chen et al. [11,12] first proposed two-wavelength phase-shift interferometry, and then developed multi-wavelength phase-shift interferometry to enhance the measurement capability. Sansoni et al. [13] combined phase-shift and gray-code into the 3D vision system, which greatly improved the measurement performance. Wang et al. [14] put forward an effective and robust phase-coding method. Zheng et al. [15] improved the phase-coding method for a large number of code words. Chen et al. [16,17] successively developed a quantized phase-coding method and a modified gray-level coding method, which achieved good results when measuring isolated objects. Nevertheless, all the aforementioned methods require three or more extra patterns, which will limit the speed of measurement. To reduce the number of patterns, some researchers have utilized color patterns for 3D measurement [18–20]. However, these methods have always failed for colorful objects. Other researchers have employed more cameras to capture the patterns from different perspectives, such that the multi-view geometric constraints can be used for absolute phase calculation [21–23]. However, the measurement field reduces because of the multiple perspectives, and the cost and complexity of the system increase due to additional cameras [24].

To realize high-speed measurement, An et al. [25] recently proposed a pixel-wise phase unwrapping method with no additional pattern. Based on the geometric constraints of fringe projection system, an artificial phase map Φ_{min} at the closest depth plane z_{min} is generated, and then the phase unwrapping can be executed by referring to Φ_{min}. Subsequently, a number of algorithms were developed for phase unwrapping based on An's method [26–29]. However, the maximum depth range this method can handle is within 2π in phase domain. When the object points far away from depth plane z_{min} brings more than 2π changes, this method is no longer applicable.

Inspired by An's method, this paper presents an absolute phase retrieval method using only one additional coded pattern to improve the measurement depth range. Firstly, the wrapped phase is calculated from three phase-shift patterns, and the code word is extracted from the coded pattern. Secondly, an artificial fringe order map k_{min} of depth plane z_{min} is generated, and then the code word is mapped to the fringe order by referring to the fringe order map k_{min}. Finally, the wrapped phase is unwrapped for the absolute phase map. Simulations and experiments have been conducted to verify the proposed method.

2. Principle

2.1. Fringe Projection System

The setup of a typical fringe projection system is shown in Figure 1. This system mainly includes a projector, a camera, and measured objects. The patterns are projected by the projector onto the measured objects from one direction, modulated by the objects' surfaces, and then captured by the camera from another direction. In Figure 1, Points O_c and O_p respectively denote the optical centers of the camera and the projector. The optical axes of the projector and the camera intersect at point O on the reference plane. Note that line O_cO_p is parallel to the reference plane, so points O_c and O_p have the same distance L from the reference surface. Based on the triangulation principle, the height of the measured objects can be computed as [30]:

$$h = \frac{L * \Delta\phi}{2\pi f_0 d + \Delta\phi} \tag{1}$$

where $\Delta\phi$ denotes the phase difference between the point P on the object and the point B on the reference plane, f_0 denotes the frequency of the fringe on the reference plane. For a specific system, parameters L, d_0 and f_0 are fixed, which can be obtained by calibration [31].

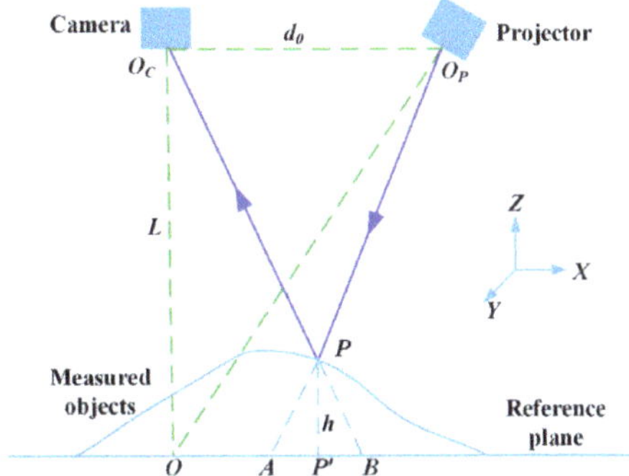

Figure 1. Fringe projection system.

2.2. Phase-Shift and Coded Patterns

Phase-shift methods have been widely used for optical measurement because of their measurement accuracy, spatial resolution, and data density [8]. The three-step phase-shift method requires the fewest number of patterns among various phase-shift methods, thus it is desirable for high-speed applications. Three-step phase-shift patterns can be described as:

$$\begin{cases} I_1(x,y) = A(x,y) + B(x,y)\cos[\phi(x,y) - 2\pi/3] \\ I_2(x,y) = A(x,y) + B(x,y)\cos[\phi(x,y)] \\ I_3(x,y) = A(x,y) + B(x,y)\cos[\phi(x,y) + 2\pi/3] \end{cases} \tag{2}$$

where $A(x,y)$ denotes the average intensity, $B(x,y)$ denotes the intensity modulation, and $\phi(x,y)$ denotes the phase to be solved for. Figure 2a–c shows three phase-shift patterns generated using the above equations, and same rows of the three patterns are shown in Figure 3a. Solving the above equations, the three variables can be calculated as:

$$\begin{cases} A(x,y) = (I_1 + I_2 + I_3)/3 \\ B(x,y) = [(I_1 - I_3)^2/3 + (2I_2 - I_1 - I_3)^2/9]^{1/2} \\ \phi(x,y) = \tan^{-1}[\sqrt{3}(I_1 - I_3)/(2I_2 - I_1 - I_3)] \end{cases} \tag{3}$$

Because of the arctangent operation, the wrapped phase $\phi(x,y)$ is limited in range of $[-\pi, \pi]$. Thus, phase unwrapping should be carried out to recover the absolute phase. If the fringe order $k(x,y)$ is determined, the absolute phase $\Phi(x,y)$ can be calculated as:

$$\Phi(x,y) = \phi(x,y) + 2\pi * k(x,y) \tag{4}$$

To determine the fringe order, we designed one coded pattern. Figure 2d shows the coded pattern, and one row of this pattern is shown in Figure 3b. The coded pattern can be described as:

$$I_M(x,y) = A(x,y) + B(x,y) * M(x,y) = A(x,y) + B(x,y) * [2 * \mod(\lceil x/P \rceil, N)/N - 1] \tag{5}$$

where P represents the fringe period, the truncated integer $k = \lceil x/P \rceil$ represents the fringe order, and the remainder $C = \mod(k, N)$ represents the code word; note that it is a periodic function with a period of N. Once these four patterns are captured, the coded coefficient $M(x,y)$ ranging from -1 to 1 can be calculated as:

$$M(x,y) = \cos^{-1}[(I_m - A)/B] \tag{6}$$

Then the code word $C(x,y)$ can be computed as:

$$C(x,y) = round[(M+1) * N/2] \tag{7}$$

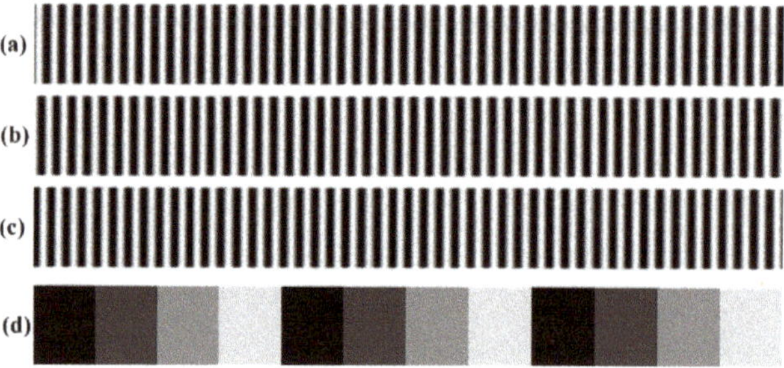

Figure 2. Projected patterns. (**a–c**) phase-shift patterns; (**d**) coded pattern.

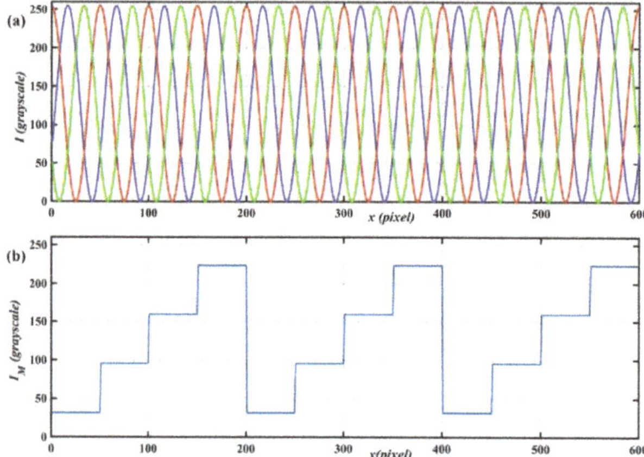

Figure 3. Same rows as in Figure 2. (**a**) Phase-shift patterns; (**b**) coded pattern.

2.3. Geometric Constraints for Phase Unwrapping

An et al. [25] have recently proposed a pixel-wise phase unwrapping method based on geometric constraints of the fringe projection system. The main idea is to create the minimum phase map Φ_{min} at the closest depth plane z_{min} of the measured volume. Then phase unwrapping can be performed with reference to minimum phase map Φ_{min}. The details of this method have been described in [25]. The following briefly introduces the main idea of this method.

Figure 4 illustrates the phase unwrapping method using the minimum phase map Φ_{min}. If the wrapped phase ϕ is less than Φ_{min}, we need to add k times of 2π to the wrapped phase ϕ to obtain the absolute phase Φ. The fringe order k can be computed as:

$$k(x,y) = ceil\left(\frac{\Phi_{min} - \phi}{2\pi}\right) \tag{8}$$

where function *ceil()* returns the closest upper integer value. It should be noted that the above equation must satisfy the following condition:

$$0 \leq \Phi - \Phi_{min} < 2\pi \tag{9}$$

Its physics signification is that the measured objects should be close to the depth plane z_{min} and within 2π in phase domain. In other words, the maximum depth range should be less than 2π changes which will limit the applications of this method. For example, at point A, $\Phi - \Phi_{min} < 2\pi$, and wrapped phase ϕ is correctly unwrapped for the absolute phase $\Phi' = \Phi$; at point B, $\Phi - \Phi_{min} > 2\pi$, but wrapped phase ϕ is wrongly unwrapped for the absolute phase $\Phi' \neq \Phi$.

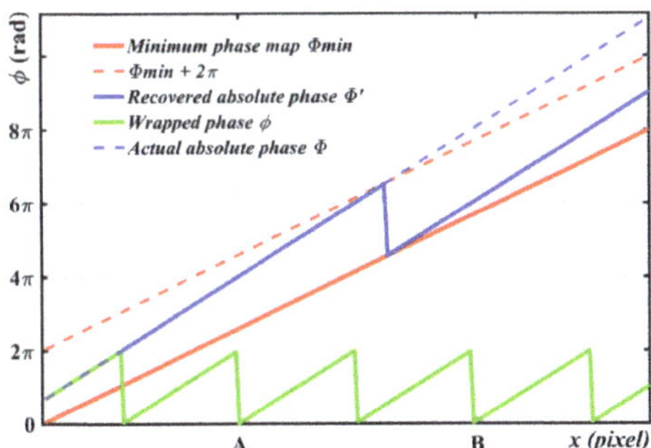

Figure 4. Phase unwrapping using the minimum phase map Φ_{min}.

2.4. Phase Unwrapping with One Coded Pattern

To improve the measurement depth range, we utilized an additional coded pattern to provide more information for fringe order determination. Assume that the camera captures an object placed at the depth plane z_{min}, there exists a one-to-one mapping between the camera sensor and the projector sensor, and the minimum fringe order k_{min} can be uniquely defined on the projector sensor. Figure 5 illustrates the main idea to determine the fringe order k, in which line k_{min} plots the minimum fringe order, the line C plots the code word at depth plane z, and line k plots the corresponding fringe order. The relationship between the three variables can be described as:

$$k = C + N * ceil\left(\frac{k_{min} - C}{N}\right) \tag{10}$$

For example, at point D, $k_{min} - C < 0$, thus $k = C$; at point E, $0 < k_{min} - C < N$, thus $k = C + N$; at point F, $N < k_{min} - C < 2 * N$, thus $k = C + 2 * N$. Similarly, the above equation must satisfy the following condition:

$$0 \leq k - k_{min} < N \tag{11}$$

In other words, the measured objects should be close to the depth plane z_{min} within $2\pi N$ in phase domain. Through the above analysis, the proposed method raises the measurement depth range by N times compared with the traditional method.

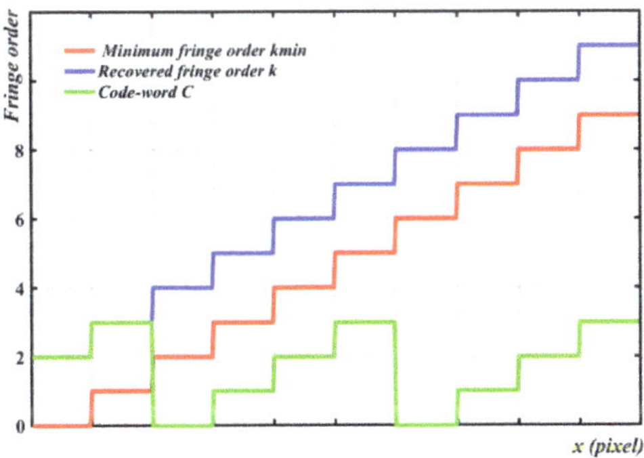

Figure 5. Fringe order determination using the minimum fringe order k_{min}.

3. Simulation

To test the performance of the proposed method, some simulations were carried out. Figure 6 shows the simulation of the closet depth plane z_{min}. Specifically, Figure 6a–c shows three phase-shift patterns with eight periods; Figure 6d shows the corresponding wrapped phase ranging from $-\pi$ to π; Figure 6e shows the fringe order map regarded as k_{min}; and Figure 6f shows the absolute phase map regarded as Φ_{min}.

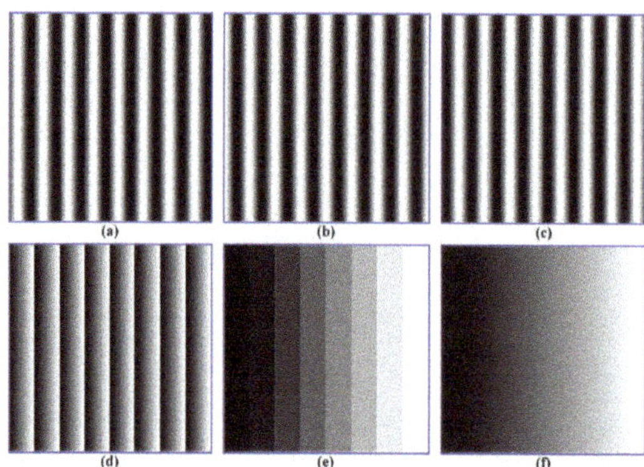

Figure 6. Simulation of depth plane z_{min}. (**a–c**) Phase-shift patterns; (**d**) wrapped phase map; (**e**) minimum fringe order map k_{min}; (**f**) minimum phase map Φ_{min}.

Then, a hemisphere was selected as the measure object and simulated, as shown in Figure 7. Specifically, Figure 7a–c shows the three phase-shift patterns; Figure 7d shows the coded pattern with $N = 4$; Figure 7e shows the fringe order determined by the proposed method; Figure 7f shows the fringe order map determined by An's method for comparison; Figure 7g shows the absolute phase map recovered by the proposed method; Figure 7h shows the absolute phase map recovered by An's method. Obviously, the fringe order and the absolute phase map are correctly determined by the proposed method. However, An's method fails in contrast. The 3D reconstruction results of the hemisphere using the two methods are shown in Figure 8. As we can see, the proposed method can accurately recover the whole surface of the hemisphere, but An's method fails to measure the overall hemisphere surface. The maximum depth range of the proposed method can deal with is $2\pi N$, which is four times that of An's method.

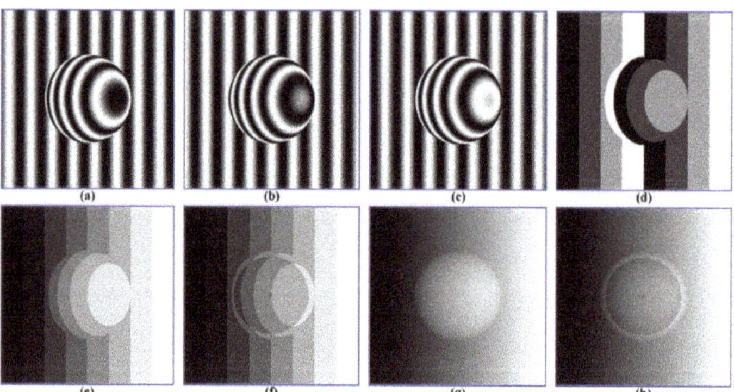

Figure 7. Simulation of a hemisphere. (**a–c**) Phase-shift patterns; (**d**) coded pattern; (**e**) fringe order map using the proposed method; (**f**) fringe order map using An's method; (**g**) absolute phase map using the proposed method; (**h**) absolute phase map using An's method.

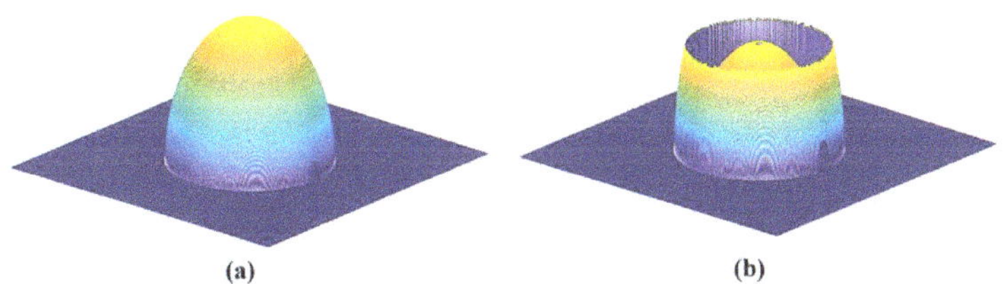

Figure 8. A 3D reconstruction of the hemisphere. (**a**) The proposed method; (**b**) An's method.

4. Experimental Setup

To test the proposed method in real condition, a fringe projection system was set up. The system consisted of a projector (Light Crafter 4500) with resolution of 912×1140 pixels, and a camera (IOI Flare 2M360-CL) with resolution of 1280×1024 pixels. A flat board was placed at the closest depth plane of the measured volume, and used as the reference plane. Two isolated objects were selected as the measured objects. Total four patterns, including three phase-shift patterns and one coded pattern, were projected onto the reference plane and the measured objects by the projector, and sequentially captured by the camera.

Figure 9a–c shows three phase-shift patterns projected onto the reference plane, respectively. Figure 9d shows the corresponding wrapped phase. Figure 9e shows the fringe order, also regarded as the minimum fringe order map k_{min}. Figure 9f shows the absolute phase map also regarded as the minimum phase map Φ_{min}. Similarly, Figure 10a–c shows the images of three phase-shift patterns projected onto the measured objects, respectively. Figure 10d shows the corresponding wrapped phase map calculated from the three phase-shift patterns. Meanwhile, the average intensity and intensity modulation were calculated. Figure 10e shows the coded pattern with $N = 4$, and Figure 10f shows the corresponding code word map.

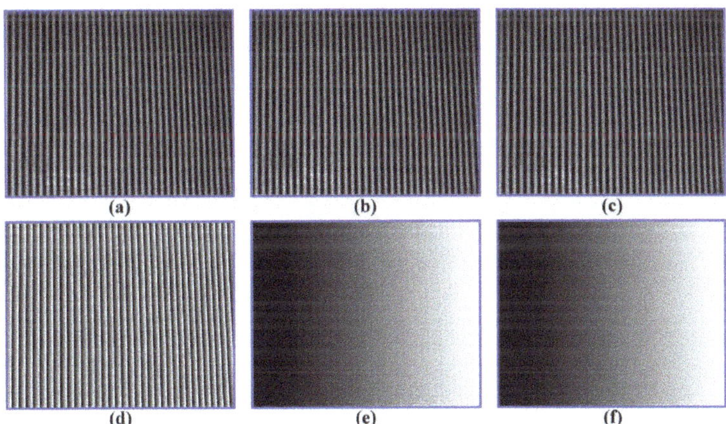

Figure 9. Images of the reference plane. (**a–c**) Phase-shift patterns; (**d**) wrapped phase map; (**e**) minimum fringe order map k_{min}; (**f**) minimum phase map Φ_{min}.

In order to compare the proposed method and An's method, Equations (12) and (14) were both used for computing fringe order. Figure 11a,b shows the fringe order maps recovered by the two methods. As we can see, the proposed method recovered the fringe order map Φ correctly; however, An's method led to the wrong fringe order map Φ' at some areas. There are obvious differences between the two fringe order maps within the two circular areas plotted in Figure 11. The pixels of the same stripe had the same fringe order k in Figure 11a. However, the pixels of the same stripe had a different fringe order k' in Figure 11b.

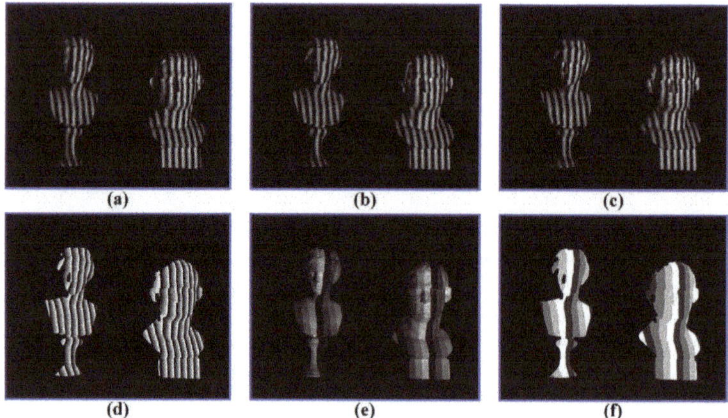

Figure 10. Images of the measured objects. (**a**–**c**) Phase-shift patterns; (**d**) wrapped phase map; (**e**) coded pattern; (**f**) code-word map.

For better illustration, Figure 12a,b shows the 600th rows of the two fringe order maps and absolute phase maps. Clearly, $\Phi - \Phi_{min} < 8\pi$ and $\Phi' - \Phi_{min} < 2\pi$. This indicates that the maximum depth range of the proposed method is up to 8π, and that of An's method is only 2π. Therefore, the proposed method can obtain much larger depth range than An's method. Finally, we reconstructed the 3D shapes of the two isolated objects, as shown in Figure 13.

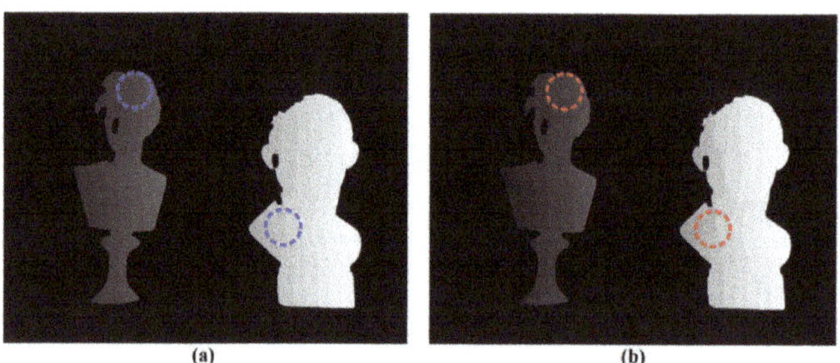

Figure 11. Fringe order maps. (**a**) The proposed method; (**b**) An's method.

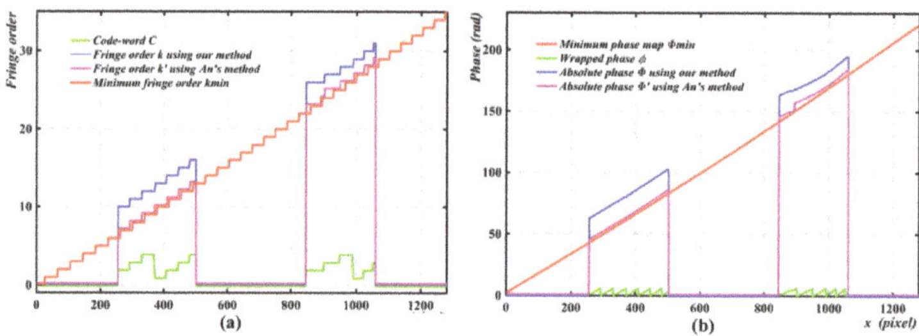

Figure 12. The 600th rows. (**a**) Fringe order maps; (**b**) absolute phase maps.

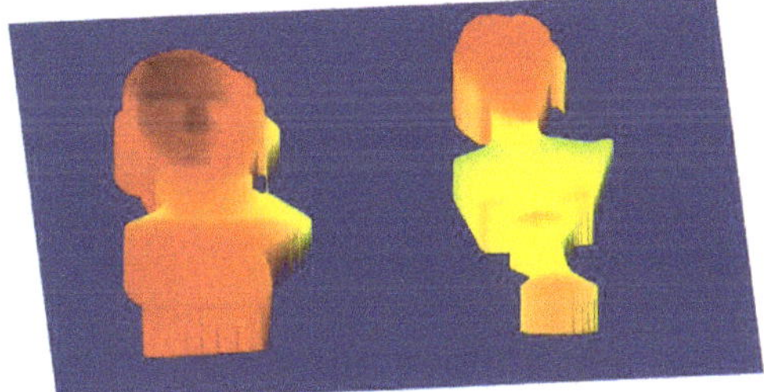

Figure 13. Measurement result of two isolated objects.

In order to further verify our method, two separate planes were also measured using the proposed method. Figure 14a–c shows three phase-shift patterns projected onto the two planes, respectively. Figure 14d shows the corresponding wrapped phase map. Figure 14e shows the coded pattern, and Figure 14f shows the corresponding code-word map. Then the fringe order was calculated, as shown in Figure 15a. Using Equation (4), the absolute phase map was recovered, as shown in Figure 15b. Finally, the 3D shapes of two planes were reconstructed, as shown in Figure 16. There are no obvious mistakes in the measurement results. The experimental results illustrate the performance of the proposed method.

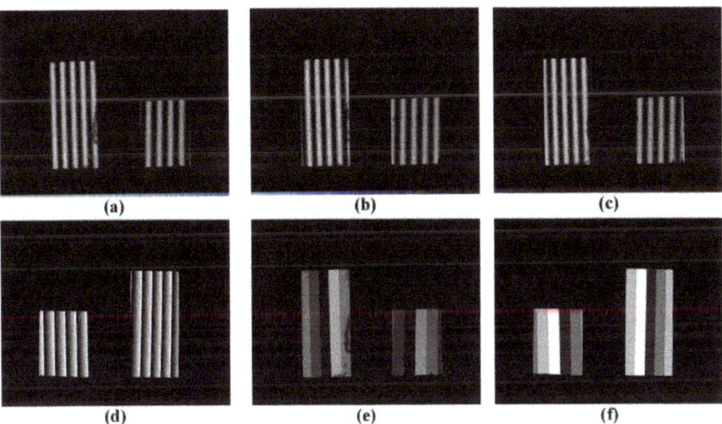

Figure 14. Images of two planes. (**a–c**) Phase-shift patterns; (**d**) wrapped phase map; (**e**) coded pattern; (**f**) code-word map.

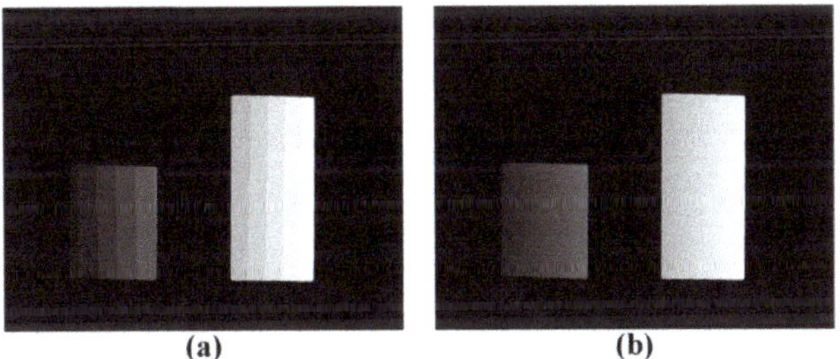

Figure 15. (**a**) Fringe order map; (**b**) absolute phase map.

Figure 16. Measurement result of two planes.

5. Conclusions

This paper has presented an absolute phase retrieval method using only one coded pattern. A total of four patterns are used for 3D shape measurement, which is suitable for high-speed applications. The code words are encoded into the coded pattern, which can be correctly recovered using the average intensity and intensity modulation of phase-shift patterns. Based on the geometric constraints of fringe projection system, the minimum fringe order map is generated, then the code word can be easily converted into fringe order. Compared with the conventional method, the proposed method can significantly enhance the measurement depth range.

Author Contributions: X.C. and B.T. conceived and designed the experiments; X.Y. and J.L. performed the experiments; X.C. and C.Z. analyzed the data; X.Y. and Y.L. wrote the paper.

Funding: This research was funded by National Natural Science Foundation of China (51605130), Hubei Provincial Natural Science Foundation of China (2018CFB656), Fundamental Research Funds for the Central Universities (WUT: 2017IVA059), Open Fund of the Key Laboratory for Metallurgical Equipment and Control of Ministry of Education in Wuhan University of Science and Technology (2018B03).

Conflicts of Interest: The authors declare no conflicts of interest.

References

1. Geng, J. Structured-light 3D surface imaging: A tutorial. *Adv. Opt. Photonics* **2011**, *3*, 128–160. [CrossRef]
2. Jung, K.; Kim, S.; Im, S.; Choi, T.; Chang, M. A Photometric Stereo Using Re-Projected Images for Active Stereo Vision System. *Appl. Sci.* **2017**, *7*, 1058. [CrossRef]
3. Nguyen, H.; Nguyen, D.; Wang, Z.; Kieu, H.; Le, M. Real-time, high-accuracy 3D imaging and shape measurement. *Appl. Opt.* **2015**, *54*, A9–A17. [CrossRef] [PubMed]
4. Salvi, J.; Fernandez, S.; Pribanic, T.; Llado, X. A state of the art in structured light patterns for surface profilometry. *Pattern Recognit.* **2010**, *43*, 2666–2680. [CrossRef]
5. Chen, S.Y.; Li, Y.F.; Zhang, J. Vision processing for realtime 3-D data acquisition based on coded structured light. *IEEE Trans. Image Process.* **2008**, *17*, 167–176. [CrossRef] [PubMed]
6. Zhang, S. Recent progresses on real-time 3D shape measurement using digital fringe projection techniques. *Opt. Lasers Eng.* **2010**, *48*, 149–158. [CrossRef]
7. Zuo, C.; Huang, L.; Zhang, M.; Chen, Q.; Asundi, A. Temporal phase unwrapping algorithms for fringe projection profilometry: A comparative review. *Opt. Lasers Eng.* **2016**, *85*, 84–103. [CrossRef]
8. Zuo, C.; Feng, S.; Huang, L.; Tao, T.; Yin, W.; Chen, Q. Phase shifting algorithms for fringe projection profilometry: A review. *Opt. Lasers Eng.* **2018**, *109*, 23–59. [CrossRef]
9. Su, X.; Chen, W. Reliability-guided phase unwrapping algorithm: A review. *Opt. Lasers Eng.* **2004**, *42*, 245–261. [CrossRef]
10. Zhang, S. Absolute phase retrieval methods for digital fringe projection profilometry: A review. *Opt. Lasers Eng.* **2018**, *107*, 28–37. [CrossRef]
11. Cheng, Y.Y.; Wyant, J.C. Multiple-wavelength phase-shifting interferometry. *Appl. Opt.* **1985**, *24*, 804–807. [CrossRef] [PubMed]

12. Cheng, Y.Y.; Wyant, J.C. Two-wavelength phase shifting interferometry. *Appl. Opt.* **1984**, *23*, 4539–4543. [CrossRef] [PubMed]

13. Sansoni, G.; Carocci, M.; Rodella, R. Three-dimensional vision based on a combination of gray-code and phase-shift light projection: Analysis and compensation of the systematic errors. *Appl. Opt.* **1999**, *38*, 6565–6573. [CrossRef] [PubMed]

14. Wang, Y.; Zhang, S. Novel phase-coding method for absolute phase retrieval. *Opt. Lett.* **2012**, *37*, 2067–2069. [CrossRef] [PubMed]

15. Zheng, D.; Da, F. Phase coding method for absolute phase retrieval with a large number of codewords. *Opt. Express* **2012**, *20*, 24139–24150. [CrossRef] [PubMed]

16. Chen, X.; Chen, S.; Luo, J.; Ma, M.; Wang, Y.; Wang, Y.; Chen, L. Modified Gray-Level Coding Method for Absolute Phase Retrieval. *Sensors* **2017**, *17*, 2383. [CrossRef] [PubMed]

17. Chen, X.; Wang, Y.; Wang, Y.; Ma, M.; Zeng, C. Quantized phase coding and connected region labeling for absolute phase retrieval. *Opt. Express* **2016**, *24*, 28613–28624. [CrossRef] [PubMed]

18. Su, W.H. Color-encoded fringe projection for 3D shape measurements. *Opt. Express* **2007**, *15*, 13167–13181. [CrossRef] [PubMed]

19. Yee, C.K.; Yen, K.S. Single frame profilometry with rapid phase demodulation on colour-coded fringes. *Opt. Commun.* **2017**, *397*, 44–50. [CrossRef]

20. Rao, L.; Da, F. Neural network based color decoupling technique for color fringe profilometry. *Opt. Laser Technol.* **2015**, *70*, 17–25. [CrossRef]

21. Li, Z.; Zhong, K.; Li, Y.F.; Zhou, X.; Shi, Y. Multiview phase shifting: A full-resolution and high-speed 3D measurement framework for arbitrary shape dynamic objects. *Opt. Lett.* **2013**, *38*, 1389–1391. [CrossRef] [PubMed]

22. Garcia, R.R.; Zakhor, A. Consistent stereo-assisted absolute phase unwrapping methods for structured light systems. *IEEE J. Sel. Top. Signal Process.* **2012**, *6*, 411–424. [CrossRef]

23. Wang, M.; Yin, Y.; Deng, D.; Meng, X.; Liu, X.; Peng, X. Improved performance of multi-view fringe projection 3D microscopy. *Opt. Express* **2017**, *25*, 19408–19421. [CrossRef] [PubMed]

24. Dai, J.; An, Y.; Zhang, S. Absolute three-dimensional shape measurement with a known object. *Opt. Express* **2017**, *25*, 10384–10396. [CrossRef] [PubMed]

25. An, Y.; Hyun, J.S.; Zhang, S. Pixel-wise absolute phase unwrapping using geometric constraints of structured light system. *Opt. Express* **2016**, *24*, 18445–18459. [CrossRef] [PubMed]

26. Yun, H.; Li, B.; Zhang, S. Pixel-by-pixel absolute three-dimensional shape measurement with modified Fourier transform profilometry. *Appl. Opt.* **2017**, *56*, 1472–1480. [CrossRef]

27. Jiang, C.; Li, B.; Zhang, S. Pixel-by-pixel absolute phase retrieval using three phase-shifted fringe patterns without markers. *Opt. Lasers Eng.* **2017**, *91*, 232–241. [CrossRef]

28. Li, B.; An, Y.; Zhang, S. Single-shot absolute 3D shape measurement with Fourier transform profilometry. *Appl. Opt.* **2016**, *55*, 5219–5225. [CrossRef] [PubMed]

29. Hyun, J.S.; Zhang, S. Enhanced two-frequency phase-shifting method. *Appl. Opt.* **2016**, *55*, 4395–4401. [CrossRef] [PubMed]

30. Zeng, Z.; Li, B.; Fu, Y.; Chai, M. Stair phase-coding fringe plus phase-shifting used in 3D measuring profilometry. *J. Eur. Opt. Soc. Rapid Publ.* **2016**, *12*, 9. [CrossRef]

31. Zhang, S.; Huang, P.S. Novel method for structured light system calibration. *Opt. Eng.* **2006**, *45*, 083601.

 *applied sciences*

Article

Sampling Based on Kalman Filter for Shape from Focus in the Presence of Noise

Hoon-Seok Jang [1,†], Mannan Saeed Muhammad [2,†], Guhnoo Yun [3] and Dong Hwan Kim [3,*]

[1]   Center for Imaging Media Research, Korea Institute of Science and Technology, Seoul 02792, Korea
[2]   College of Information and Communication, Natural Sciences Campus, Sungkyunkwan University, Suwon 16419, Korea
[3]   Center for Intelligent and Interactive Robotics, Korea Institute of Science and Technology, Seoul 02792, Korea
*   Correspondence: gregorykim@kist.re.kr; Tel.: +82-2-958-6719
†   These authors contributed equally as first authors to this work.

Received: 10 July 2019; Accepted: 7 August 2019; Published: 9 August 2019

Abstract: Recovering three-dimensional (3D) shape of an object from two-dimensional (2D) information is one of the major domains of computer vision applications. Shape from Focus (SFF) is a passive optical technique that reconstructs 3D shape of an object using 2D images with different focus settings. When a 2D image sequence is obtained with constant step size in SFF, mechanical vibrations, referred as jitter noise, occur in each step. Since the jitter noise changes the focus values of 2D images, it causes erroneous recovery of 3D shape. In this paper, a new filtering method for estimating optimal image positions is proposed. First, jitter noise is modeled as Gaussian or speckle function, secondly, the focus curves acquired by one of the focus measure operators are modeled as a quadratic function for application of the filter. Finally, Kalman filter as the proposed method is designed and applied for removing jitter noise. The proposed method is experimented by using image sequences of synthetic and real objects. The performance is evaluated through various metrics to show the effectiveness of the proposed method in terms of reconstruction accuracy and computational complexity. Root Mean Square Error (RMSE), correlation, Peak Signal-to-Noise Ratio (PSNR), and computational time of the proposed method are improved on average by about 48%, 11%, 15%, and 5691%, respectively, compared with conventional filtering methods.

Keywords: shape from focus (SFF); jitter noise; focus curve; Kalman filter

1. Introduction

Inferring three-dimensional (3D) shape of an object from two-dimensional (2D) images is a fundamental problem in computer vision applications. Many 3D shape recovery techniques have been proposed in literature [1–5]. The methods can be categorized into two categories based on the optical reflective model. The first one includes active techniques which use projected light rays. The second category consists of passive techniques which utilize reflected light rays without projection. The passive methods can further be classified into Shape from X, where X denotes the cue used to reconstruct the 3D shape as *Stereo* [6], *Texture* [7], *Motion* [8], *Defocus* [9], and *Focus* [10]. Shape from Focus (SFF) is a passive optical method that utilizes a series of 2D images with different focus levels for estimating 3D information of an object [11]. For SFF, a focus measure is applied to each pixel of the image sequence, to evaluate the focus quantity at every point. The best focused position is acquired by maximizing the focus measure values along the optical axis.

Many focus measures have been reported in literature [12–16]. Initial depth map, obtained through any of the focus measure operators, has the problem of information loss between consecutive frames due to the discreteness of predetermined sampling step size. To solve this problem, refined depth map is acquired using approximation techniques, as reported in literature [17–22]. As an important issue of SFF, when images are obtained by translating the object plane with constant step size, mechanical vibration, referred to as jitter noise, occurs in each step, as shown in Figure 1 [21].

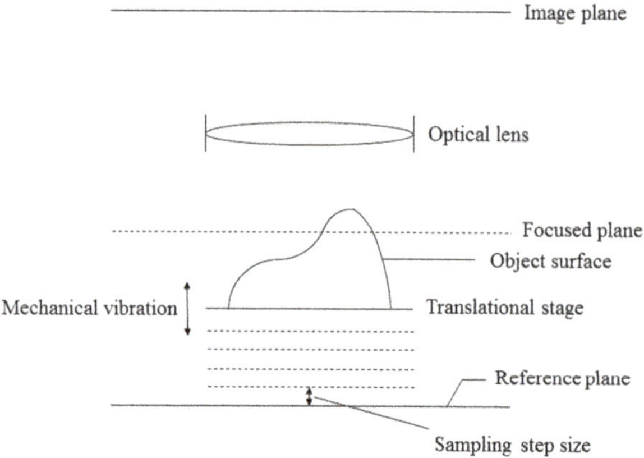

Figure 1. Image acquisition for Shape from Focus.

Since this noise changes the focus values of images by oscillating along the optical axis, accuracy of 3D shape recovery is considerably degraded. Unlike any image noise [23,24], this noise is not detectable by simply observing images.

Many filtering methods for removing the jitter noise have been reported [25–27]. In [25,26], Kalman and Bayes filtering methods for removing Gaussian jitter noise have been proposed, respectively. In [27], a modified Kalman filtering method for removing Lévy noise has been presented.

In this paper, a new filtering method for removing the jitter noise is proposed as an extended version of [25]. First, jitter noise is modeled as Gaussian or speckle function to reflect more types of noise that can occur in SFF. At the second stage, the focus curves acquired by one of the focus measure operators are modeled as Gaussian function for application of the filter and a clearer performance comparison of various filters. Finally, Kalman filter as the proposed method is designed and applied. Kalman filter is a recursive filter that tracks the state of a linear dynamic system containing noise, and is used in many fields such as computer vision, robotics, radar, etc. [28–33]. In many cases, this algorithm is based on measurements made over time. More precise results can be expected than from using only measurements at that moment. As the filter recursively processes input data, including noise, optimal statistical prediction for the current state can be performed. The proposed method is experimented by using image sequences of synthetic and real objects. The performance of the proposed method is analyzed through various metrics to show its effectiveness in terms of reconstruction accuracy and computational complexity. Root Mean Square Error (RMSE), correlation, Peak Signal-to-Noise Ratio (PSNR), and computational time of the proposed method are improved by an average of about 48%, 11%, 15%, and 5691%, respectively, compared with conventional filtering methods. In the remainder of this paper, Section 2 presents the concept of SFF and a summary of previously proposed focus measures as background. Sections 3 and 4 provide the modeling of jitter noise and focus curves, respectively. Section 5 explains the Kalman filter as the proposed method in detail. Experimental results and discussion are presented in Section 6. Finally, Section 7 concludes this paper.

2. Related Work

2.1. Shape from Focus

In SFF methods, images with different focus levels (such that some parts are well focused and the rest of the parts are defocused with some blur) are obtained by translating the object plane at a predetermined step size along the optical axis [11]. By applying a focus measure, the best focused frame for each object point is acquired to find the depth of the object with an unknown surface. The distance of the corresponding object point is computed by using the camera parameters for the frame, and utilizing the lens formula as follows:

$$\frac{1}{f} = \frac{1}{u} + \frac{1}{v} \tag{1}$$

where, f is the focal length, u and v are the distances of object and image from the lens, respectively. Figure 2 shows the image formation in the optical lens. The object point at the distance u is focused to the image point at the distance v.

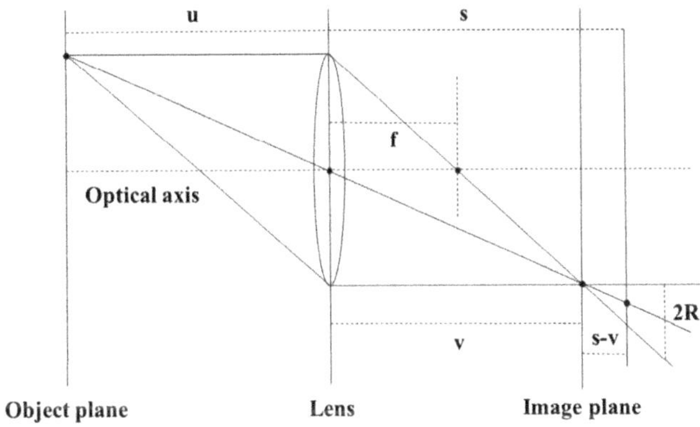

Figure 2. Image formation in optical lens.

2.2. Focus Measures

A focus measure operator calculates the focus quality of each pixel in the image sequence, and is evaluated locally. As the image sharpness increases, the value of the focus measure increases. When the image sharpness is maximum, the best focused image is attained. Some of the popular gradient-based, statistical-based, and Laplacian-based operators are briefly given in [12].

First, there are Modified Laplacian (ML) and Sum of Modified Laplacian (SML) as Laplacian-based operators. When Laplacian is used in textured images, x and y components of the Laplacian operator may cancel out and provide no response. ML is calculated by adding the squared second derivatives for each pixel of the image I as:

$$F_{ML}(x,y) = \left(\frac{\partial^2 I(x,y)}{\partial x^2}\right)^2 + \left(\frac{\partial^2 I(x,y)}{\partial y^2}\right)^2 \tag{2}$$

If the image has rich textures with high variability at each pixel, focus measure can be evaluated for each pixel. In order to improve robustness for weak-textured images, SML is computed by adding the ML values in a $W \times W$ window as:

$$F_{SML}(i,j) = \sum_{x \in W} \sum_{y \in W} \left\{ \left(\frac{\partial^2 I(x,y)}{\partial x^2}\right)^2 + \left(\frac{\partial^2 I(x,y)}{\partial y^2}\right)^2 \right\} \tag{3}$$

where, i and j are the x and y coordinates of center pixel in a $W \times W$ window, respectively.

Next, there is Tenenbaum (TEN) as a gradient-based operator. TEN is calculated by adding the squared responses of horizontal and vertical Sobel operators. For robustness, it is also computed by adding the TEN values in a $W \times W$ window as:

$$F_{TEN}(i,j) = \sum_{x \in W} \sum_{y \in W} \left\{ (G_x(x,y))^2 + (G_y(x,y))^2 \right\}$$ (4)

where, $G_x(x,y)$ and $G_y(x,y)$ are images acquired through convolution with the horizontal and vertical Sobel operators, respectively.

Finally, there is Gray-Level Variance (GLV) as a statistics-based operator. It has been proposed on the basis of the idea that the variance of gray level in a sharp image is higher than in a blurred image. GLV for a central pixel in a $W \times W$ window is calculated as:

$$F_{GLV}(i,j) = \frac{1}{N^2} \sum_{x \in W} \sum_{y \in W} \left\{ (I(x,y) - \mu)^2 \right\}$$ (5)

where, μ is the mean of the gray values in a $W \times W$ window.

3. Noise Modeling

When a sequence of 2D images is obtained by translating the object at a constant step size along the optical axis, mechanical vibrations, referred as jitter noise, occur in each step. In this manuscript, two probability density functions are used for modeling the jitter noise. At first, the jitter noise is modeled as Gaussian function with mean μ_n and standard deviation σ_n, as shown in Figure 3.

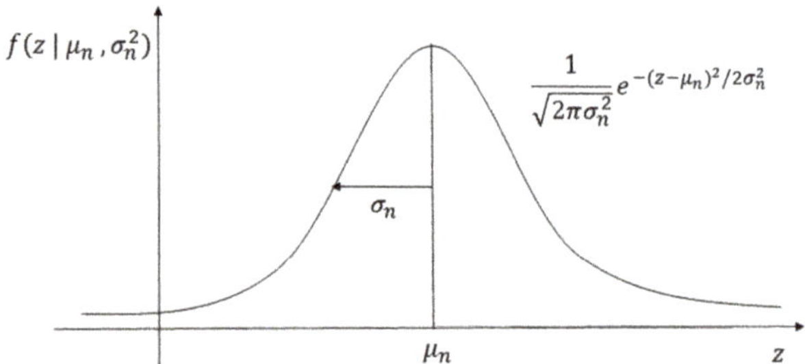

Figure 3. Noise modeling through Gaussian function.

μ_n represents the position of each image frame without the jitter noise, and σ_n represents the amount of jitter noise occurred in each image frame. σ_n is determined by checking depth of field and corresponding image position. The depth of field is affected by magnification and different factors. σ_n is selected as $\sigma_n \le 10$ µm through repeated experiments with real objects used in this manuscript. Second, the jitter noise is modeled as speckle function as follows [34,35]:

$$f(\zeta) = \frac{1}{2\sigma_n^2} \times e^{\frac{-\zeta}{2\sigma_n^2}}$$ (6)

where, ζ is the amount of jitter noise before or after filtering.

4. Focus Curve Modeling

In order to filter out jitter noise, the focus curve obtained by one of the focus measure operators is modeled by Gaussian approximation with mean z_f and standard deviation σ_f [11]. This focus curve modeling is shown in Figure 4.

prediction can be made through the values from the present state to the observed measurement. The prediction of the state and its variance is represented as follows:

$$S = T \times S + C \times U \tag{17}$$

$$V = T \times V \times T' + N_p \tag{18}$$

where, S is "estimate of the system state", T is "transition coefficient of the state", U is "input"; C is "control coefficient of the input", V is "variance of the state estimate", and N_p is "variance of the process noise". In the SFF system, S is represented as the position of each image frame in the 2D image sequence estimated by the Kalman filter, and C, U, and N_p are all set as 0, since there is no control input in the SFF system, and only jitter noise, as the measurement noise, is considered in this manuscript. Next, the computation of the Kalman gain is given for updating the predicted state as follows:

$$G = V \times A' \times inv(A \times V \times A' + N_m) \tag{19}$$

where, G is "Kalman gain", A is "observation coefficient", and N_m is "variance of the measurement noise". In an SFF system, N_m is defined as the variance of the previously modeled jitter noise. Finally, the update of the predicted state on the basis of the observed measurement is provided by:

$$S = S + G \times (O - A \times S) \tag{20}$$

$$V = V - G \times A \times V \tag{21}$$

where, O is "observed measurement". In an SFF system, O is represented as the position of each image frame in the image sequence before filtering. Parameters that are not set to values, T and A, are all set to 1 for simplicity. For the start of the algorithm, S and V are initialized as:

$$S = inv(A) \times O \tag{22}$$

$$V = inv(A) \times N_m \times inv(A') \tag{23}$$

Through the Kalman filter algorithm, the optimal position S of each image frame in the image sequence is estimated. The pseudo code for the Kalman filter algorithm is shown in Algorithm 1.

Algorithm 1 Computing optimal position of each image frame and remaining jitter noise

---

1: **procedure** Optimal position S & remaining jitter noise ζ
2: $S \leftarrow O$ ▷ Set initial position of image frame with observed position
3: $V \leftarrow N_m$ ▷ Initialize variance of position of image frame to variance of jitter noise
4. **for** $i = 1 \rightarrow N$ **do**   ▷ Total number of iterations of Kalman filter
5:    $G \leftarrow V \cdot (V + N_m)^{-1}$ ▷ Compute Kalman gain
6:    $V \leftarrow V - G \cdot V$ ▷ Correct variance of position of image frame
7:    $S \leftarrow S + G \cdot (O - S)$ ▷ Update position of image frame
8:    $\zeta \leftarrow |S - \mu_n|$ ▷ Compute remaining jitter noise
9:   **end for**
10: **end procedure**

The difference between the true position μ_n, which is the position of each image frame without the jitter noise, and optimal position S, is put to ζ. This algorithm is repeated for all image frames in the image sequence. After acquiring a filtered image sequence, a depth map is obtained by maximizing the focus measure obtained by using the previously modeled focus curve, for each pixel in the image sequence. A list of frequently used symbols and notations is shown in Table 1.

Table 1. List of frequently used symbols and notation.

Notation	Description
μ_n	Position of each image frame without jitter noise
σ_n	Standard deviation of jitter noise
z_f	Best focused position through Gaussian approximation in each object point
σ_f	Standard deviation of Gaussian focus curve
ζ	The amount of jitter noise before or after filtering
S	Position of each image frame after Kalman filtering
O	Position of each image frame before Kalman filtering
N	Total number of iterations of filters

6. Results and Discussion

6.1. Image Acquisition and Parameter Setting

For experiments, four objects were used, as shown in Figure 6, consisting of one simulated and three real objects.

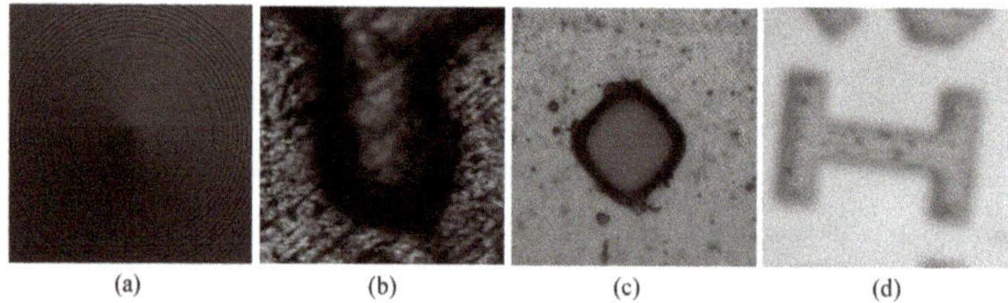

(a)   (b)   (c)   (d)

Figure 6. 10th frame of experimented objects: (**a**) Simulated cone, (**b**) Coin, (**c**) Liquid Crystal Display-Thin Film Transistor (LCD-TFT) filter, (**d**) Letter-I.

First, a simulated cone image sequence consisting of 97 images, with dimensions of 360×360 pixels, was acquired. These images were generated using camera simulation software [40].

The real objects used for experiments were: coin, Liquid Crystal Display-Thin Film Transistor (LCD-TFT) filter, and letter-I. The coin images were magnified images of Lincoln's head from the back of the US penny. The coin sequence consisted of 80 images, with dimensions of 300×300 pixels. The LCD-TFT filter images consisted of microscopic images of an LCD color filter. The image sequence of LCD-TFT filter had 60 images, with the dimensions of 300×300 pixels each. The third image sequence consisted of letter-I, engraved on the metallic surface. It consisted of 60 images, with dimensions of 300×300 pixels each. The real objects were acquired through a microscopic control system (MCS) [18]. The system consists of a personal computer integrated with a frame grabber board (Matrox Meteor-II) and a CCD camera (SAMSUNG CAMERA SCC-341) mounted on a microscope (NIKON OPTIPHOT-100S). Computer software obtains images by translating the object plane through a stepper motor driver (MAC 5000), possessing a 2.5 nm minimum step size. The coin and letter-I images were obtained under 10× magnification, while the LCD-TFT filter images were acquired under 50× magnification.

In parameter setting, the standard deviation of the jitter noise for each object was assumed to be ten times the sampling step size of each image sequence, i.e., 254 mm, 6.191 μm, 1.059 m, and 1.529 μm for simulated cone, coin, LCD-TFT filter, and letter-I, respectively. For comparison of 3D shape recovery results, a local window 7×7 for focus measure operators was used. The total number of iterations N of the Kalman filter was set as 100.

For performance comparison, Bayes filter and particle filter were employed [41–46]. The depth estimation through the Bayes filter is presented in Figure 7.

In Figure 7, z_0 is defined as a total number of 2D images obtained for SFF, and $p_j(i)$ is presented as follows:

$$p_j(i) = \frac{1}{\sqrt{2\pi\sigma_n^2}}e^{\frac{-(z(j)-r(i))^2}{2\sigma_n^2}}, \quad 1 \le i \le M, \; 1 \le j \le N \tag{24}$$

where, $p_j(i)$ is Gaussian probability density function, $z(j)$ is the position of each image frame changed by the jitter noise, $r(i)$ is the possible positions of each image frame in the presence of the jitter noise, σ_n is standard deviation of previously modeled jitter noise, M is the total length of $r(i)$ with intervals of 0.01, and N is the total number of iterations of Bayes filter. The reason why $3\sigma_n$ is set in the range of r, is because $3\sigma_n$ makes $z(j)$ be in the range of r with the probability of 99.7% due to the Gaussian probability density function. The recursive Bayesian estimation was applied to all 2D image frames obtained for SFF. After the filtered image sequence was acquired, an optimal depth map was obtained using the previously modeled focus curve.

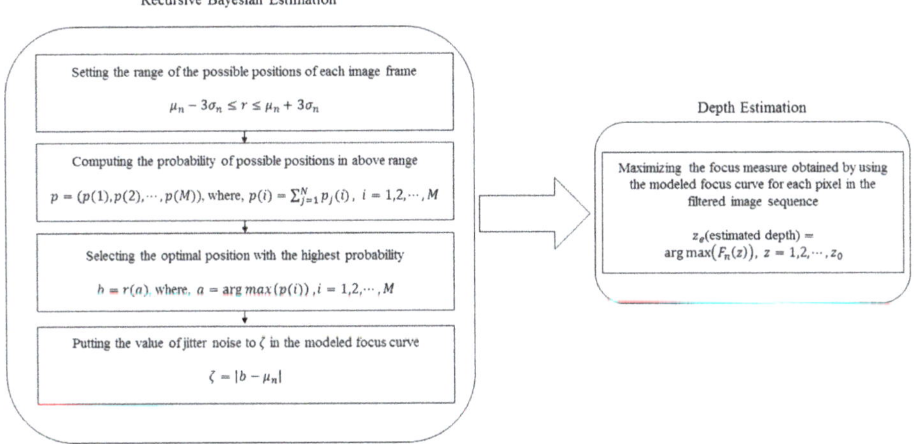

Figure 7. Depth estimation through Bayes filter.

Particle filter algorithm is mainly divided into two steps: Generating the weights for each of particles and resampling for acquiring new estimated particles. In the first step, the weights are based on the probability of the given observation for each of the particles as:

$$p_w(i) = \frac{1}{\sqrt{2\pi\sigma_n^2}}e^{\frac{-(z-z_p(i))^2}{2\sigma_n^2}}, \quad 1 \le i \le P \tag{25}$$

where, $p_w(i)$ is Gaussian probability density function, z is the observed position of each image frame changed by the jitter noise, $z_p(i)$ is vector of particles, and P is the number of particles the SFF system generates. In this manuscript, $z_p(i)$ is initialized by randomly selecting the values on the x-axis from the previously modeled jitter noise, and P is set as 1000. After the weights are normalized, resampling, as the second step, is needed for acquiring new estimated particles. The new estimated particles are obtained by sampling the cumulative distribution of the normalized $p_w(i)$, randomly and uniformly. Through this sampling, the particles with the higher weights are selected. This particle filter algorithm is repeated N times, as the total number of iterations of the particle filter. The optimal position of each image frame is the mean of the final estimated particles $z_p(i)$ obtained through resampling in iteration N. After the filtered image sequence is acquired through application of the particle filter to all 2D image frames, optimal depth map is obtained in the same way as the depth estimation in Figure 7.

6.2. Experimental Results

Figure 8 presents the performance comparison of the filters in the 97th frame of the simulated cone using various iterations in the presence of Gaussian jitter noise.

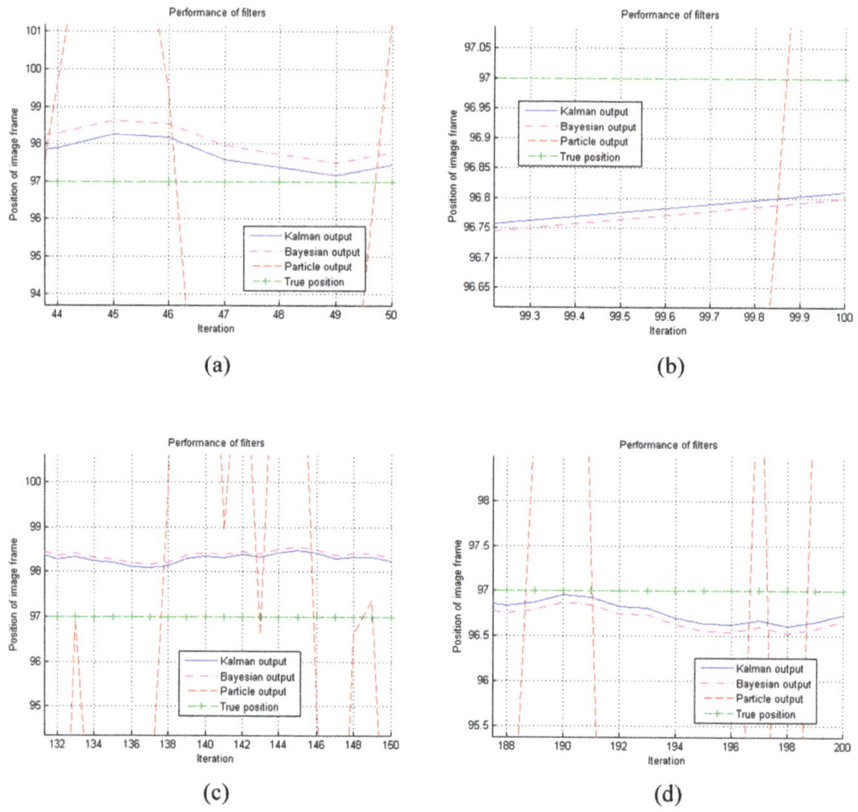

Figure 8. Performance of the filters: (**a**) Iteration—50, (**b**) Iteration—100, (**c**) Iteration—150, (**d**) Iteration—200.

Figure 9 provides performance comparison of the filters in the 100th iteration using various frames of the simulated cone in the presence of Gaussian jitter noise.

These Figures are intensively enlarged versions of the last iteration. "Kalman output" is the estimated position through Kalman filter, "Bayesian output" is the estimated position through Bayes filter, "Particle output" is the estimated position through particle filter, and "True position" is the position without the jitter noise. It is clear from these Figures that Kalman output converged better to True position than Bayesian output and Particle output. It means that Kalman filter outperformed the other filters compared for experiments.

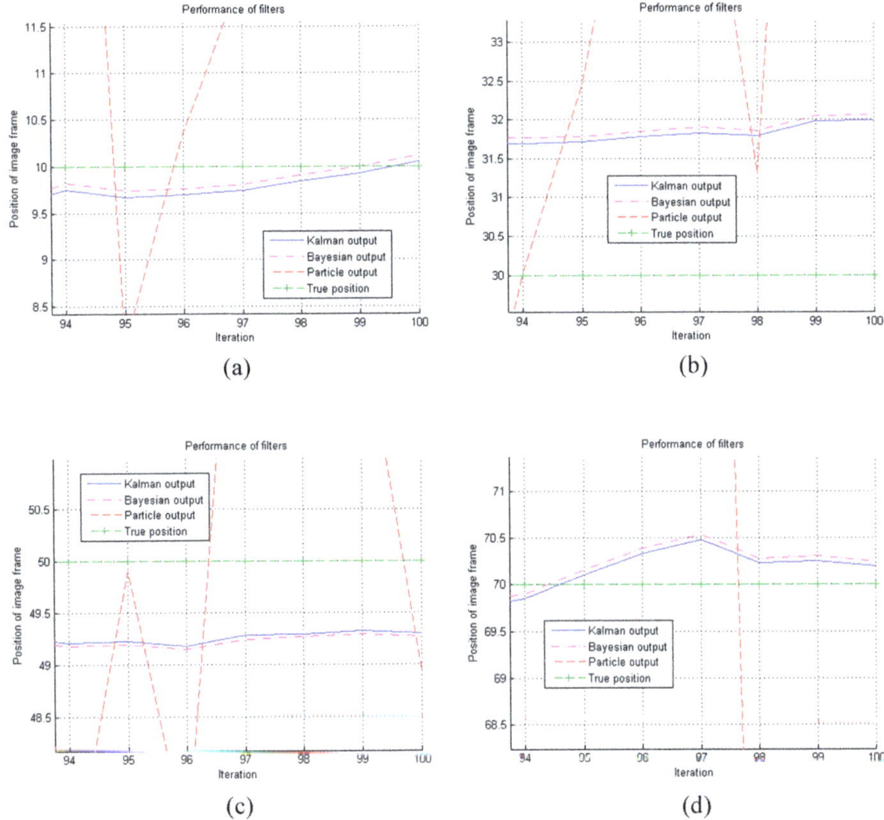

Figure 9. Performance of the filters: (**a**) Frame number—10, (**b**) Frame number—30, (**c**) Frame number—50, (**d**) Frame number—70.

Figure 10 shows the Gaussian approximation of the focus curves using experimented objects in the presence of Gaussian jitter noise.

"Without Noise" is the Gaussian approximation of the focus curve without the jitter noise, "After Kalman Filtering" is the Gaussian approximation of the focus curve after Kalman filtering, "After Bayesian Filtering" is the Gaussian approximation of the focus curve after Bayes filtering, and "After Particle Filtering" is the Gaussian approximation of the focus curve after particle filtering. It is clear from Figure 10 that the optimal position with the highest focus value in After Kalman Filtering is closer to the optimal position in Without Noise than the optimal positions in the focus curves obtained after using other filtering techniques.

For performance evaluation of 3D shape recovery, three metrics were used in case of simulated cone, since the synthetic object had an actual depth map, as in Figure 11 [47].

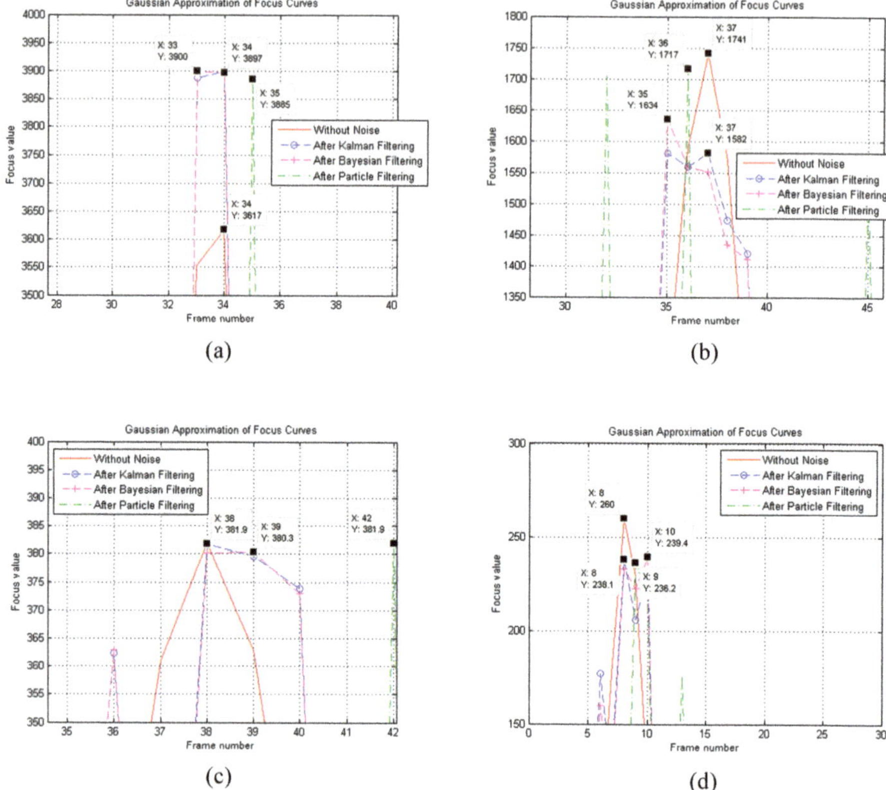

Figure 10. Gaussian approximation of focus curves: (**a**) Simulated cone (60, 60), (**b**) Coin (120, 120), (**c**) LCD-TFT filter (180, 180), (**d**) Letter-I (240, 240).

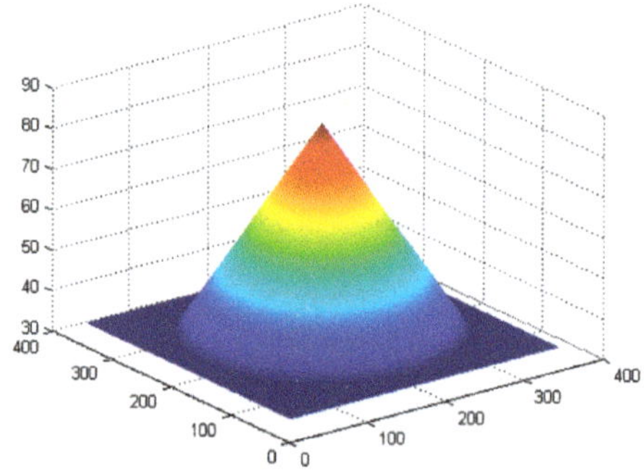

Figure 11. Actual depth map of simulated cone.

The first one is Root Mean Square Error (RMSE), which is a commonly used measure when dealing with the difference between estimated and actual value, as follows:

$$RMSE = \sqrt{\frac{1}{XY} \sum_{x=0}^{X-1} \sum_{y}^{Y-1} \left(d(x,y) - \hat{d}(x,y) \right)^2} \tag{26}$$

where, $d(x,y)$ and $\hat{d}(x,y)$ are the actual and estimated depth map, respectively, and X and Y are width and height of 2D images, which are used for SFF, respectively.

The second one is correlation, which shows the linear relationship and strength between two variables as:

$$Correlation = \frac{\sum_{x=0}^{X-1}\sum_{y=0}^{Y-1}\left(d(x,\ y)-\overline{d}(x,y)\right)\left(\hat{d}(x,y)-\overline{\hat{d}(x,y)}\right)}{\sqrt{\left(\sum_{x=0}^{X-1}\sum_{y=0}^{Y-1}\left(d(x,y)-\overline{d}(x,y)\right)^2\right)\left(\sum_{x=0}^{X-1}\sum_{y=0}^{Y-1}\left(\hat{d}(x,y)-\overline{\hat{d}(x,y)}\right)^2\right)}} \tag{27}$$

where, $\overline{d(x,y)}$ and $\overline{\hat{d}(x,y)}$ are the means of the actual and estimated depth map, respectively.

The third one is Peak Signal-to-Noise Ratio (PSNR), which is the power of noise over the maximum power a signal can have. It is usually represented in terms of the logarithmic decibel scale as:

$$PSNR = 10log_{10}(\frac{d_{max}^2}{MSE}) \tag{28}$$

where, d_{max} is maximum depth value in the depth map and MSE is the Mean Square Error, which is the square of the RMSE. The lower the RMSE, the higher the correlation, and the higher the PSNR, the higher the accuracy of 3D shape reconstruction.

Tables 2–4 provide the quantitative performance of 3D shape recovery of the simulated cone using three focus measures, SML, GLV, and TEN, before and after filtering in the presence of Gaussian jitter noise.

Table 2. Comparison of focus measure operators with proposed method for simulated cone in the presence of Gaussian noise by using RMSE (Root Mean Square Error). SML: Sum of Modified Laplacian; GLV: Gray-Level Variance; TEN: Tenenbaum.

Focus Measure Operators	SML	GLV	TEN
Before Filtering	9.2629	12.4038	15.2304
After Particle Filtering	9.1993	10.9459	11.2293
After Bayesian Filtering	7.3260	8.2659	8.4961
After Kalman Filtering	7.3169	8.1400	8.3652

Table 3. Comparison of focus measure operators with proposed method for simulated cone in the presence of Gaussian noise by using correlation.

Focus Measure Operators	SML	GLV	TEN
Before Filtering	0.7831	0.7430	0.7121
After Particle Filtering	0.7925	0.8157	0.7914
After Bayesian Filtering	0.9536	0.9427	0.9200
After Kalman Filtering	0.9541	0.9438	0.9206

Table 4. Comparison of focus measure operators with proposed method for simulated cone in the presence of Gaussian noise by using PSNR.

Focus Measure Operators	SML	GLV	TEN
Before Filtering	20.1276	17.8644	16.0812
After Particle Filtering	20.4604	18.9504	18.7284
After Bayesian Filtering	22.3481	21.3897	21.1510
After Kalman Filtering	22.3588	21.5229	21.2859

The order of the general performance of the focus measures is that SML is the best, then the GLV, and finally the TEN. In Before Filtering, it is difficult to distinguish the performance of the focus measures due to the jitter noise. However, in After Bayesian Filtering and After Kalman Filtering, it is shown in Tables 2–4 that the performance order of the focus measures is almost correct, as described above. The particle filter suitable for nonlinear systems does not remove jitter noise well in a linear SFF system. It is seen in Tables 2–4 that the performance order of the focus measures in After Particle Filtering is slightly different from the one presented above. Tables 5–7 provide the quantitative performance of 3D shape recovery of the simulated cone using three focus measures, SML, GLV, and TEN, before and after filtering in the presence of speckle noise. The performance order of the focus measures for each filtering technique is almost the same as that of focus measures when Gaussian jitter noise is present. However, in the presence of speckle noise, After Kalman Filtering and After Bayesian Filtering have poor performance in terms of RMSE and PSNR. This is because these two filters estimate the position of each 2D image after assuming the jitter noise to be Gaussian function. It is evident from Tables 2–7 that the best overall performance is that of the Kalman filter as the proposed method, which provides the optimal estimation results in a linear system. The Bayes filter comes second, and finally the particle filter, which estimates the optimal value in a nonlinear system.

Table 5. Comparison of focus measure operators with proposed method for simulated cone in the presence of speckle noise by using RMSE.

Focus Measure Operators	SML	GLV	TEN
Before Filtering	21.6257	19.5639	19.1356
After Particle Filtering	18.2074	18.1522	18.3674
After Bayesian Filtering	16.9866	17.4630	17.9747
After Kalman Filtering	16.9458	17.4064	17.9344

Table 6. Comparison of focus measure operators with proposed method for simulated cone in the presence of speckle noise by using correlation.

Focus Measure Operators	SML	GLV	TEN
Before Filtering	0.8133	0.8661	0.8570
After Particle Filtering	0.8941	0.9083	0.8873
After Bayesian Filtering	0.9518	0.9496	0.9316
After Kalman Filtering	0.9527	0.9504	0.9325

Table 7. Comparison of focus measure operators with proposed method for simulated cone in the presence of speckle noise by using PSNR (Peak Signal-to-Noise Ratio).

Focus Measure Operators	SML	GLV	TEN
Before Filtering	12.2884	13.3517	13.3074
After Particle Filtering	13.2806	13.8092	13.5967
After Bayesian Filtering	14.0878	13.8476	13.6162
After Kalman Filtering	14.1087	13.8758	13.6389

Tables 8 and 9 present the time taken to estimate the position of one image frame by using the filters for the experimented objects in the presence of Gaussian and speckle noise, respectively.

Table 8. Computation time of filters for the experimented objects in the presence of Gaussian noise.

Experimented Objects	Particle Filter	Bayes Filter	Kalman Filter
Simulated cone	0.651821	0.112929	0.006780
Coin	0.699162	0.125152	0.009283
LCD-TFT filter	0.624884	0.112957	0.007769
Letter-I	0.688404	0.112758	0.007259

Table 9. Computation time of filters for the experimented objects in the presence of speckle noise.

Experimented Objects	Particle Filter	Bayes Filter	Kalman Filter
Simulated cone	0.637766	0.113022	0.008112
Coin	0.677874	0.124471	0.009683
LCD-TFT filter	0.625544	0.116263	0.007738
Letter-I	0.636709	0.112466	0.009179

The computation time in Tables 8 and 9 is expressed in seconds. It is evident that the computation time of the Kalman filter was about 14 times better than the Bayes filter and about 80 times better than the particle filter. Figures 12–14 show the qualitative performance of 3D shape reconstruction of the experimented objects using three focus measures, SML, GLV, and TEN, before and after filtering in the presence of Gaussian noise. Figures 15 and 16 provide the qualitative performance of 3D shape reconstruction of the experimented objects using three focus measures, SML, GLV, and TEN, before and after filtering in the presence of speckle noise. In Before Filtering and After Particle Filtering, it can be seen that the performance of 3D shape reconstruction was very poor due to unremoved or poorly removed jitter noise. However, in After Bayesian Filtering and After Kalman Filtering, it is evident that the performance of the 3D shape recovery was greatly improved due to the elimination of most of the jitter noise. It is proved from these experimental results that filtering the jitter noise using the Kalman filter improves the 3D shape reconstruction faster and more accurately.

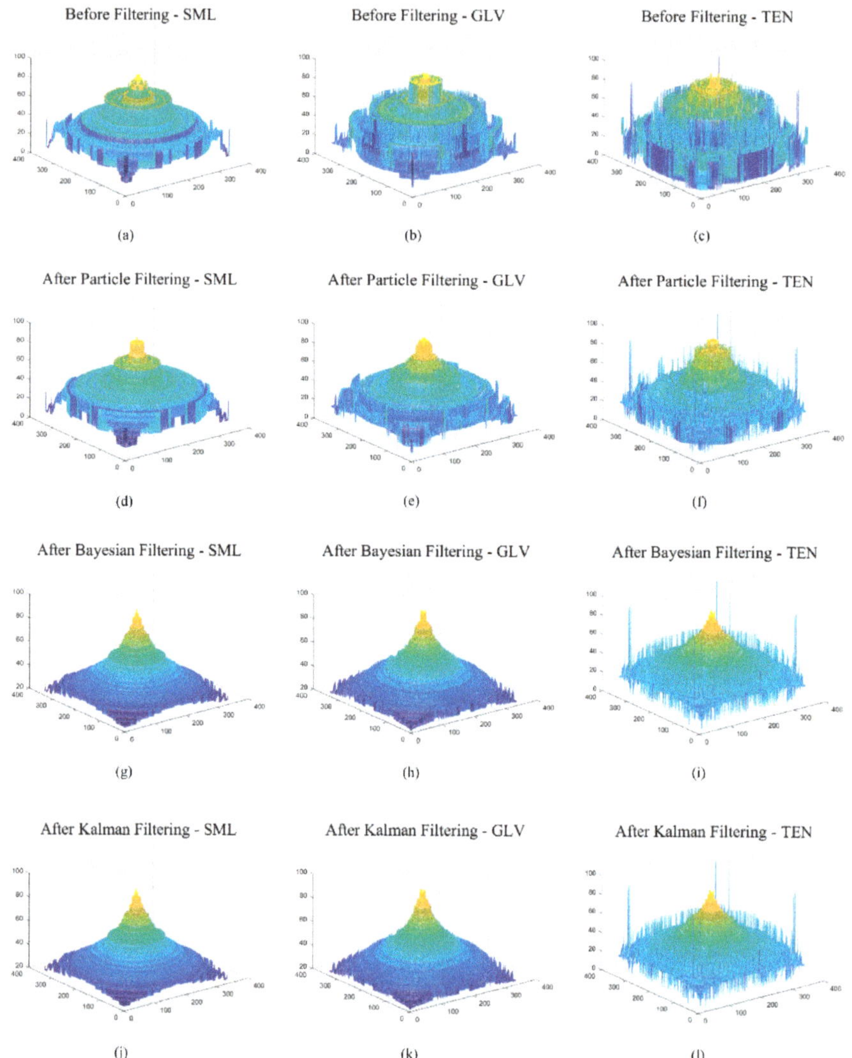

Figure 12. 3D shape recovery of simulated cone, before and after filtering, using SML, GLV, and TEN in the presence of Gaussian noise. (**a**) Before filtering for SML; (**b**) Before filtering for GLV; (**c**) Before filtering for TEN; (**d**) After particle filtering for SML; (**e**) After particle filtering for GLV; (**f**) After particle filtering for TEN; (**g**) After Bayesian filtering for SML; (**h**) After Bayesian filtering for GLV; (**i**) After Bayesian filtering for TEN; (**j**) After Kalman filtering for SML; (**k**) After Kalman filtering for GLV; (**l**) After Kalman filtering for TEN.

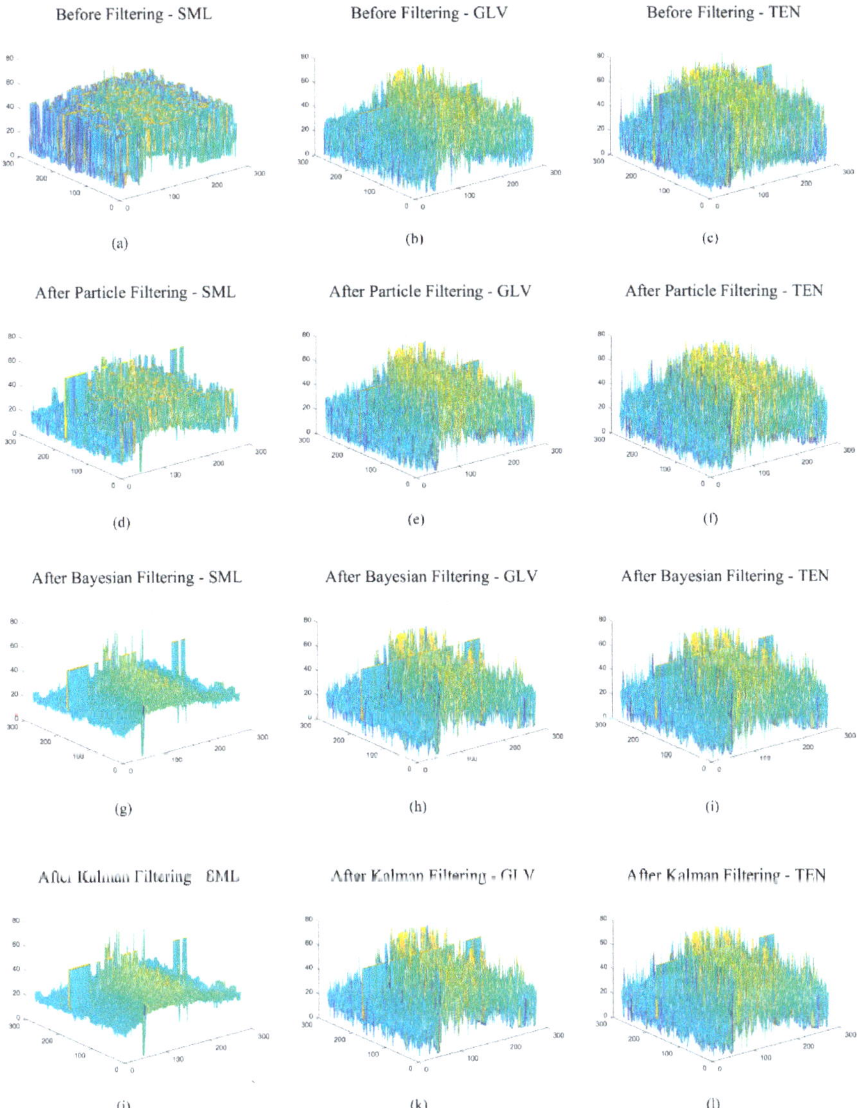

Figure 13. 3D shape recovery of coin, before and after filtering, using SML, GLV, and TEN in the presence of Gaussian noise. (**a**) Before filtering for SML; (**b**) Before filtering for GLV; (**c**) Before filtering for TEN; (**d**) After particle filtering for SML; (**e**) After particle filtering for GLV; (**f**) After particle filtering for TEN; (**g**) After Bayesian filtering for SML; (**h**) After Bayesian filtering for GLV; (**i**) After Bayesian filtering for TEN; (**j**) After Kalman filtering for SML; (**k**) After Kalman filtering for GLV; (**l**) After Kalman filtering for TEN.

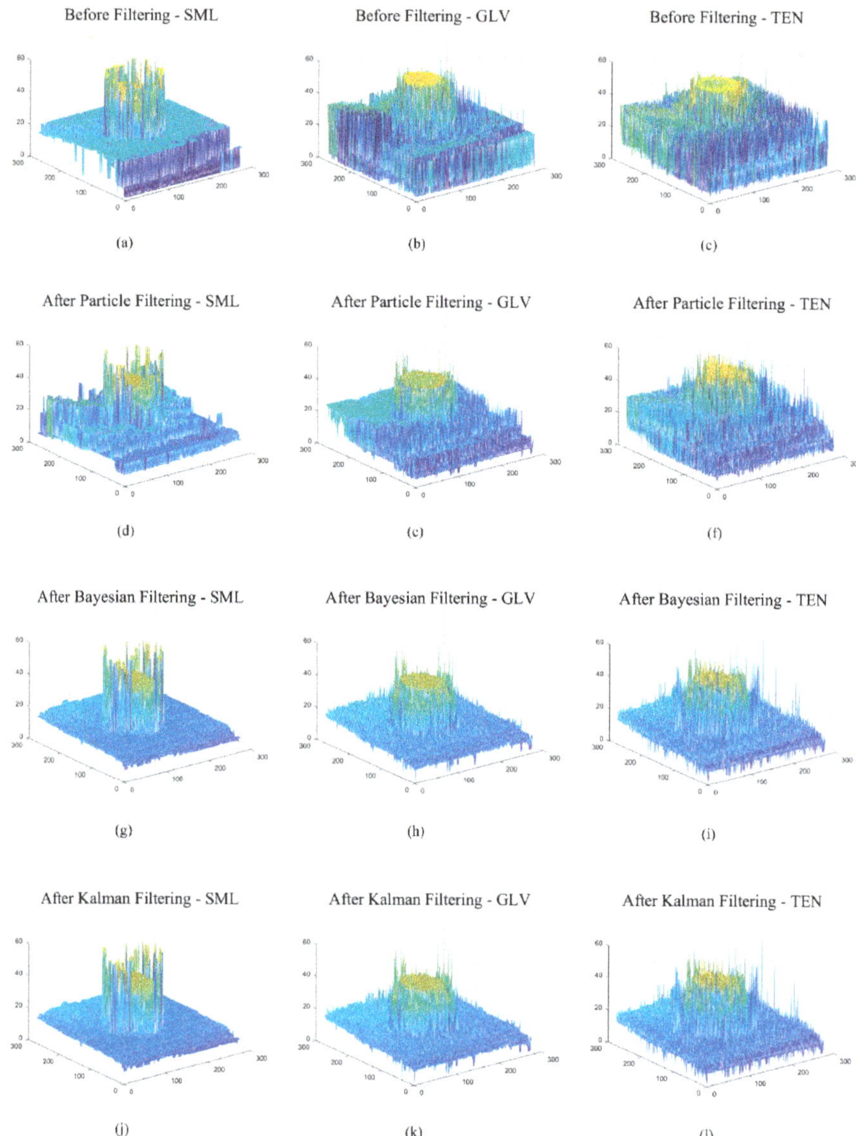

Figure 14. 3D shape recovery of LCD-TFT filter, before and after filtering, using SML, GLV, and TEN in the presence of Gaussian noise. (**a**) Before filtering for SML; (**b**) Before filtering for GLV; (**c**) Before filtering for TEN; (**d**) After particle filtering for SML; (**e**) After particle filtering for GLV; (**f**) After particle filtering for TEN; (**g**) After Bayesian filtering for SML; (**h**) After Bayesian filtering for GLV; (**i**) After Bayesian filtering for TEN; (**j**) After Kalman filtering for SML; (**k**) After Kalman filtering for GLV; (**l**) After Kalman filtering for TEN.

Figure 15. 3D shape recovery of simulated cone, before and after filtering, using SML, GLV, and TEN in the presence of speckle noise. (**a**) Before filtering for SML; (**b**) Before filtering for GLV; (**c**) Before filtering for TEN; (**d**) After particle filtering for SML; (**e**) After particle filtering for GLV; (**f**) After particle filtering for TEN; (**g**) After Bayesian filtering for SML; (**h**) After Bayesian filtering for GLV; (**i**) After Bayesian filtering for TEN; (**j**) After Kalman filtering for SML; (**k**) After Kalman filtering for GLV; (**l**) After Kalman filtering for TEN.

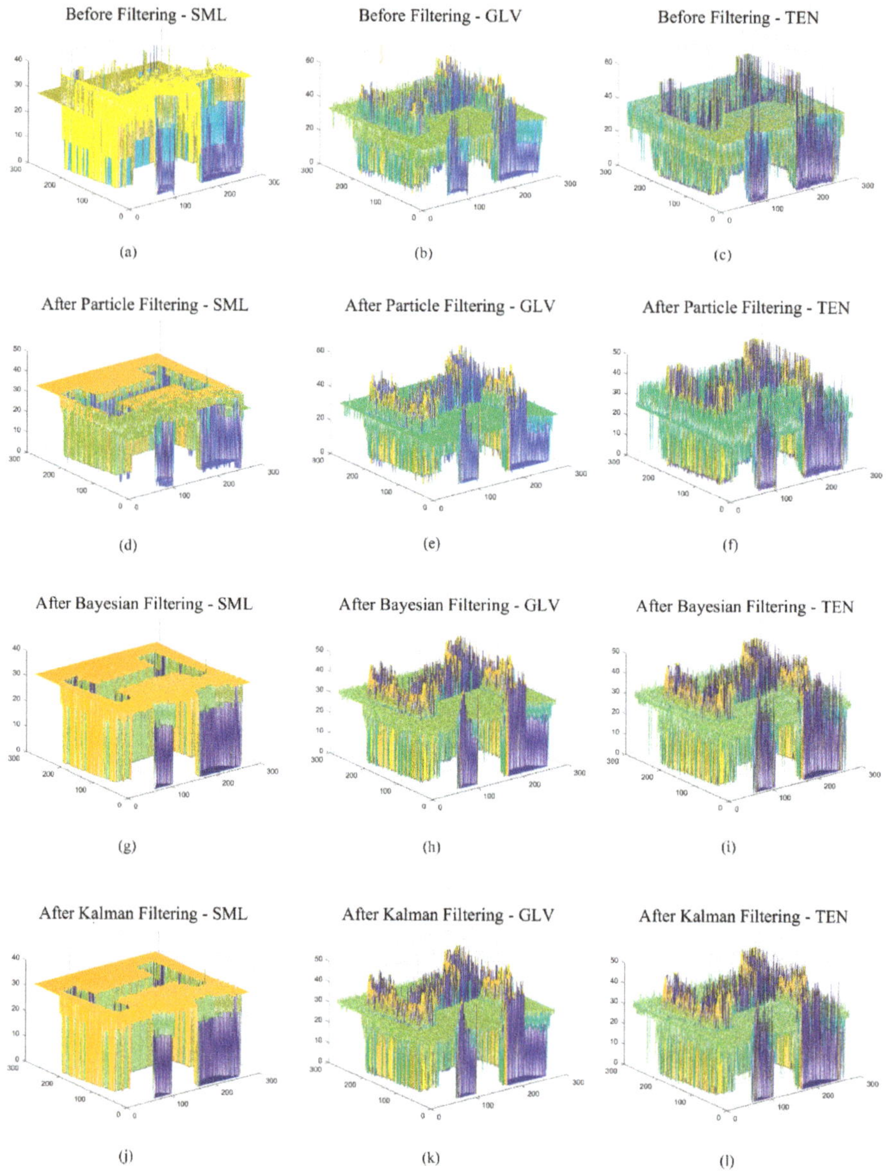

Figure 16. 3D shape recovery of letter-I, before and after filtering, using SML, GLV, and TEN in the presence of speckle noise. (**a**) Before filtering for SML; (**b**) Before filtering for GLV; (**c**) Before filtering for TEN; (**d**) After particle filtering for SML; (**e**) After particle filtering for GLV; (**f**) After particle filtering for TEN; (**g**) After Bayesian filtering for SML; (**h**) After Bayesian filtering for GLV; (**i**) After Bayesian filtering for TEN; (**j**) After Kalman filtering for SML; (**k**) After Kalman filtering for GLV; (**l**) After Kalman filtering for TEN.

7. Conclusions

For SFF, an object is translated at a constant step size along the optical axis. When an image of the object is captured in each step, mechanical vibrations occur, which are referred as jitter noise.

In this manuscript, jitter noise is modeled as Gaussian function with mean z_n and standard deviation σ_n for simplicity. Then, the focus curves obtained by one of the focus measure operators are also modeled as Gaussian function, with mean z_f and standard deviation σ_f, for the application of the proposed method. Finally, a new filter is proposed to provide optimal estimation results in a linear SFF system with the jitter noise, utilizing Kalman filter to eliminate jitter noise in the modeled focus curves. Through experimental results, it was found that the Kalman filter provided significantly improved 3D reconstruction of the experimented objects compared with before filtering, and that the 3D shapes of the experimented objects were recovered with more accurate and faster performance than with other existing filters, such as the Bayes filter and the particle filter.

Author Contributions: Conceptualization, H.-S.J. and M.S.M.; Methodology, H.-S.J. and M.S.M.; Software, H.-S.J. and G.Y.; Validation, H.-S.J.; Writing—original draft preparation, H.-S.J.; Writing—review and editing, M.S.M.; Supervision, D.H.K.; Funding acquisition, D.H.K.

Funding: This work was supported by Institute for Information & communications Technology Promotion (IITP) grant funded by the Korea government (MSIT) (No.2018-0-00677, Development of Robot Hand Manipulation Intelligence to Learn Methods and Procedures for Handling Various Objects with Tactile Robot Hands).

Acknowledgments: We thank Tae-Sun Choi for his assistance with useful discussion.

Conflicts of Interest: The authors declare no conflict of interest.

References

1. Lin, J.; Ji, X.; Xu, W.; Dai, Q. Absolute Depth Estimation from a Single Defocused Image. *IEEE Trans. Image Process.* **2013**, *22*, 4545–4550. [PubMed]
2. Humayun, J.; Malik, A.S. Real-time processing for shape-from-focus techniques. *J. Real Time Image Process.* **2016**, *11*, 49–62. [CrossRef]
3. Lafarge, F.; Keriven, R.; Brédif, M.; Vu, H.H. A Hybrid Multiview Stereo Algorithm for Modeling Urban Scenes. *IEEE Trans. Pattern Anal. Mach. Intell.* **2013**, *35*, 5–17. [CrossRef] [PubMed]
4. Ciaccio, E.J.; Tennyson, C.A.; Bhagat, G.; Lewis, S.K.; Green, P.H. Use of shape-from-shading to estimate three-dimensional architecture in the small intestinal lumen of celiac and control patients. *Comput. Methods Programs Biomed.* **2013**, *111*, 676–684. [CrossRef] [PubMed]
5. Barron, J.T.; Malik, J. Shape, Illumination, and Reflectance from Shading. *IEEE Trans. Pattern Anal. Mach. Intell.* **2015**, *37*, 1670–1687. [CrossRef]
6. De Vries, S.C.; Kappers, A.M.L.; Koenderink, J.J.; Vries, S.C. Shape from stereo: A systematic approach using quadratic surfaces. *Percept. Psychophys.* **1993**, *53*, 71–80. [CrossRef] [PubMed]
7. Super, B.J.; Bovik, A.C. Shape from Texture Using Local Spectral Moments. *IEEE Trans. Pattern Anal. Mach. Intell.* **1995**, *17*, 333–343. [CrossRef]
8. Parashar, S.; Pizarro, D.; Bartoli, A. Isometric Non-Rigid Shape-From-Motion in Linear Time. In Proceedings of the IEEE Conference on Computer Vision and Pattern Recognition (CVPR), Las Vegas, NV, USA, 27–30 June 2016; pp. 4679–4687.
9. Favaro, P.; Soatto, S.; Burger, M.; Osher, S.J. Shape from Defocus via Diffusion. *IEEE Trans. Pattern Anal. Mach. Intell.* **2008**, *30*, 518–531. [CrossRef] [PubMed]
10. Tiantian, F.; Hongbin, Y. A novel shape from focus method based on 3D steerable filters for improved performance on treating textureless region. *Opt. Commun.* **2018**, *410*, 254–261.
11. Nayar, S.K.; Nakagawa, Y. Shape from Focus. *IEEE Trans. Pattern Anal. Mach. Intell.* **1994**, *16*, 824–831. [CrossRef]
12. Pertuz, S.; Puig, D.; García, M.A. Analysis of focus measure operators for shape-from-focus. *Pattern Recognit.* **2013**, *46*, 1415–1432. [CrossRef]
13. Rusiñol, M.; Chazalon, J.; Ogier, J.M. Combining Focus Measure Operators to Predict OCR Accuracy in Mobile-Captured Document Images. In Proceedings of the 2014 11th IAPR International Workshop on Document Analysis Systems, Tours, France, 7–10 April 2014; pp. 181–185.
14. Xie, H.; Rong, W.; Sun, L. Construction and evaluation of a wavelet-based focus measure for microscopy imaging. *Microsc. Res. Tech.* **2007**, *70*, 987–995. [CrossRef] [PubMed]
15. Choi, T.S.; Malik, A.S. *Vision and Shape—3D Recovery Using Focus*; Sejong Publishing: Busan, Korea, 2008.

16. Lee, I.-H.; Mahmood, M.T.; Choi, T.-S. Robust Focus Measure Operator Using Adaptive Log-Polar Mapping for Three-Dimensional Shape Recovery. *Microsc. Microanal.* **2015**, *21*, 442–458. [CrossRef] [PubMed]
17. Asif, M.; Choi, T.-S. Shape from focus using multilayer feedforward neural networks. *IEEE Trans. Image Process.* **2001**, *10*, 1670–1675. [CrossRef] [PubMed]
18. Mahmood, M.T.; Choi, T.-S.; Choi, W.-J.; Choi, W. PCA-based method for 3D shape recovery of microscopic objects from image focus using discrete cosine transform. *Microsc. Res. Tech.* **2008**, *71*, 897–907.
19. Ahmad, M.; Choi, T.-S. A heuristic approach for finding best focused shape. *IEEE Trans. Circuits Syst. Video Technol.* **2005**, *15*, 566–574. [CrossRef]
20. Muhammad, M.S.; Choi, T.S. A Novel Method for Shape from Focus in Microscopy using Bezier Surface Approximation. *Microsc. Res. Tech.* **2010**, *73*, 140–151. [CrossRef] [PubMed]
21. Muhammad, M.; Choi, T.-S. Sampling for Shape from Focus in Optical Microscopy. *IEEE Trans. Pattern Anal. Mach. Intell.* **2012**, *34*, 564–573. [CrossRef]
22. Lee, I.H.; Mahmood, M.T.; Shim, S.O.; Choi, T.S. Optimizing image focus for 3D shape recovery through genetic algorithm. *Multimed. Tools Appl.* **2014**, *71*, 247–262. [CrossRef]
23. Malik, A.S.; Choi, T.-S. Consideration of illumination effects and optimization of window size for accurate calculation of depth map for 3D shape recovery. *Pattern Recognit.* **2007**, *40*, 154–170. [CrossRef]
24. Malik, A.S.; Choi, T.-S. A novel algorithm for estimation of depth map using image focus for 3D shape recovery in the presence of noise. *Pattern Recognit.* **2008**, *41*, 2200–2225. [CrossRef]
25. Jang, H.-S.; Muhammad, M.S.; Choi, T.-S. Removal of jitter noise in 3D shape recovery from image focus by using Kalman filter. *Microsc. Res. Tech.* **2017**, *81*, 207–213.
26. Jang, H.-S.; Muhammad, M.S.; Choi, T.-S. Bayes Filter based Jitter Noise Removal in Shape Recovery from Image Focus. *J. Imaging Sci. Technol.* **2019**, *63*, 1–12. [CrossRef]
27. Jang, H.-S.; Muhammad, M.S.; Choi, T.-S. Optimal depth estimation using modified Kalman filter in the presence of non-Gaussian jitter noise. *Microsc. Res. Tech.* **2018**, *82*, 224–231. [PubMed]
28. Vasebi, A.; Bathaee, S.; Partovibakhsh, M. Predicting state of charge of lead-acid batteries for hybrid electric vehicles by extended Kalman filter. *Energy Convers. Manag.* **2008**, *49*, 75–82. [CrossRef]
29. Frühwirth, R. Application of Kalman filtering to track and vertex fitting. *Nucl. Instrum. Methods Phys. Res. A* **1987**, *262*, 444–450. [CrossRef]
30. Harvey, A.C. *Applications of the Kalman Filter in Econometrics*; Cambridge University Press: Cambridge, UK, 1987.
31. Boulfelfel, D.; Rangayyan, R.; Hahn, L.; Kloiber, R.; Kuduvalli, G. Two-dimensional restoration of single photon emission computed tomography images using the Kalman filter. *IEEE Trans. Med. Imaging* **1994**, *13*, 102–109. [CrossRef] [PubMed]
32. Bock, Y.; Crowell, B.W.; Webb, F.H.; Kedar, S.; Clayton, R.; Miyahara, B. Fusion of High-Rate GPS and Seismic Data: Applications to Early Warning Systems for Mitigation of Geological Hazards. In Proceedings of the American Geophysical Union Fall Meeting, San Francisco, CA, USA, 15–19 December 2008.
33. Miall, R.C.; Wolpert, D. Forward Models for Physiological Motor Control. *Neural Netw.* **1996**, *9*, 1265–1279. [CrossRef]
34. Frieden, B.R. *Probability, Statistical Optics, and Data Testing*; Springer: Berlin, Germany, 2001.
35. Goodman, J.W. *Speckle Phenomena in Optics*; Roberts and Company Publishers: Englewood, CO, USA, 2007.
36. Akselsen, B. *Kalman Filter Recent Advances and Applications*; Scitus Academics: New York, NY, USA, 2016.
37. Pan, J.; Yang, X.; Cai, H.; Mu, B. Image noise smoothing using a modified Kalman filter. *Neurocomputing* **2016**, *173*, 1625–1629. [CrossRef]
38. Ribeiro, M.I. *Kalman and Extended Kalman Filters: Concept, Derivation and Properties*; Institute for Systems and Robotics: Lisbon, Portugal, 2004.
39. Welch, G.; Bishop, G. *An Introduction to the Kalman Filter*; University of North Carolina at Chapel Hill: Chapel Hill, NC, USA, 1995.
40. Subbarao, M.; Lu, M.-C. Image sensing model and computer simulation for CCD camera systems. *Mach. Vis. Appl.* **1994**, *7*, 277–289.
41. Tagade, P.; Hariharan, K.S.; Gambhire, P.; Kolake, S.M.; Song, T.; Oh, D.; Yeo, T.; Doo, S. Recursive Bayesian filtering framework for lithium-ion cell state estimation. *J. Power Sources* **2016**, *306*, 274–288. [CrossRef]
42. Hajimolahoseini, H.; Amirfattahi, R.; Gazor, S.; Soltanian-Zadeh, H. Robust Estimation and Tracking of Pitch Period Using an Efficient Bayesian Filter. *IEEE/ACM Trans. Audio Speech Lang. Process.* **2016**, *24*, 1. [CrossRef]

43. Li, T.; Corchado, J.M.; Bajo, J.; Sun, S.; De Paz, J.F. Effectiveness of Bayesian filters: An information fusion perspective. *Inf. Sci.* **2016**, *329*, 670–689. [CrossRef]
44. Yang, T.; Laugesen, R.S.; Mehta, P.G.; Meyn, S.P. Multivariable feedback particle filter. *Automatica* **2016**, *71*, 10–23. [CrossRef]
45. Zhou, H.; Deng, Z.; Xia, Y.; Fu, M. A new sampling method in particle filter based on Pearson correlation coefficient. *Neurocomputing* **2016**, *216*, 208–215. [CrossRef]
46. Wang, D.; Yang, F.; Tsui, K.-L.; Zhou, Q.; Bae, S.J. Remaining Useful Life Prediction of Lithium-Ion Batteries Based on Spherical Cubature Particle Filter. *IEEE Trans. Instrum. Meas.* **2016**, *65*, 1282–1291. [CrossRef]
47. Mahmood, M.T.; Majid, A.; Choi, T.-S. Optimal depth estimation by combining focus measures using genetic programming. *Inf. Sci.* **2011**, *181*, 1249–1263. [CrossRef]

Review

Tomographic Diffractive Microscopy: A Review of Methods and Recent Developments

Ting Zhang [1], Kan Li [2], Charankumar Godavarthi [3] and Yi Ruan [2,*]

[1] Zhejiang Provincial Key Laboratory of Information Processing, Communication and Networking (IPCN), College of Information Science & Electronic Engineering, Zhejiang University, Hangzhou 310027, China; zhang_ting@zju.edu.cn

[2] Collaborative Innovation Center for Bio-Med Physics Information Technology, College of Science, Zhejiang University of Technology, Hangzhou 310014, China; kanli@zjut.edu.cn

[3] Rue de l'hostellerie-Ville active, 30900 Nimes, France; gjcharankumar@gmail.com

* Correspondence: yiruan@zjut.edu.cn

Received: 16 August 2019; Accepted: 10 September 2019; Published: 12 September 2019

Abstract: Tomographic diffractive microscopy (TDM) is a label-free, far-field, super-resolution microscope. The significant difference between TDM and wide-field microscopy is that in TDM the sample is illuminated from various directions with a coherent collimated beam and the complex diffracted field is collected from many scattered angles. By utilizing inversion procedures, the permittivity/refractive index of investigated samples can be retrieved from the measured diffracted field to reconstruct the geometrical parameters of a sample. TDM opens up new opportunities to study biological samples and nano-structures and devices, which require resolution beyond the Rayleigh limit. In this review, we describe the principles and recent advancements of TDM and also give the perspectives of this fantastic microscopy technique.

Keywords: tomographic diffractive microscopy (TDM); diffracted field; holographic interferometry; inverse scattering

1. Introduction

Microscopy has revolutionized biological research and promoted the development of human health. Despite the number of microscopes that have been created and applied for different research objectives, such as the electron microscope [1,2], the atomic force microscope [3,4], etc., the optical microscope is still the most used tool in biological research and life science due to its non-invasive nature [5]. However, the spatial resolution of conventional optical microscopes is limited by Rayleigh criterion, which is the smallest separation distance between two point sources that can be resolved, $0.61\lambda/\mathrm{NA}$ (λ is the wavelength of light and NA is the numerical aperture of the objective). For the range of visible light used in optical microscopes, this limitation is −250 nm. Attracted by the non-contact mechanical property of optical imaging, enormous efforts have been made to find the way to overcome this diffraction barrier, collectively termed as optical super-resolution microscopes [5,6].

By placing the light source of an optical probe near the sample at a distance shorter than the wavelength, the diffraction limit can be bypassed by exploiting the properties of evanescent waves. This has led to the so-called near-field super-resolution microscopes, such as scanning near-field optical microscopy (SNOM), whose resolution is limited by the aperture size of the probe (or source), typically of −25 nm [7]. However, SNOM is operated at near-field distance and in special conditions; its application is thus limited.

Another common and powerful optical imaging technique is far-field super-resolution microscopes, including the label (fluorescence) [8,9] and label-tree microscopies [10,11], such as the stimulated emission depletion microscope [12], the stochastic optical reconstruction microscope [13], the photo-activated localization microscope [14], and non-label far-field diffractive microscopes [15]. They have successfully overcome the diffraction limit to the level of tens of nanometers resolution. Moreover, compared to near-field microscopes, these far-field ones greatly simplify the experimental setup and increase numerous possibilities of practical use in biomedical research. However, in principle, fluorescence-based methods rely on prior knowledge of the investigated sample, such as molecules and the availability of specific fluorophores; recognition of antibodies against the specific molecules. In the bio-medical usage of non-linear microscopies, such as multi-photon excitation (MPE) fluorescence microscopy [16] and second-harmonic generation microscopy (SHG) [17], the labeling procedure may also restrict the mobility of the molecules, affecting their functions. Moreover, photobleaching and phototoxicity of fluorescent microscopy play limiting roles in biology [18]. Optical coherence topography (OCT) has the ability to visualize the anatomic structures in three-dimensions and in high resolution [19], but the lateral resolution is not well developed [20]. The coherent anti-stokes Raman spectroscopy microscope (CARS) is a new imaging technique without sample labeling, but it requires laser sources with excellent intensity stabilization [21]. Holography microscopes are able to view the contrast difference, but the quantitative information of intrinsic properties of the sample, like index of refraction or permittivity, is difficult to retrieve [22,23]. Consequently, tomographic diffractive microscopy (TDM) is a super-resolution microscopy technique, it is label-free and far-field and works in a close to physiological environment, one that is non-label, and principally at normal pressure and room temperature. By illuminating the sample from various directions with coherent collimated light and detecting the complex diffracted field from many scattered angles [15,24–26], together with numerical inversion procedures, TDM has emerged to provide quantitative reconstruction of the opto-geometrical characteristics of the sample [27–32].

The purpose of this review is to give an overview of the tomographic diffractive microscopy technique. First, following a brief introduction of TDM theory, the accessible spatial frequencies are analyzed under different TDM configurations. Second, we present the recent development of TDM, including the optimization of optical setup and the improvement of inversion methods. Finally, the problems remaining and the perspectives of TDM are discussed.

2. Theoretical Background

In conventional optical microscopes [33], the object is illuminated simultaneously by a sum of plane waves spatially incoherent with each other. Each plane wave propagates with a different illumination angle, so that the object is globally illuminated simultaneously with all possible angles, within a given numerical aperture noted NA_{inc}, which is the sine of the maximum illumination angle with respect to the optical axis of the microscope. For the detection, the most commonly used architecture consists in placing the object near the object focal plane of an objective lens; the diffraction field is detected by the detector, such as a CCD camera, which is placed at the image focal plane of the objective lens. However, limited by the NA_{inc} of the objective lens, only parts of object diffracted field information could be collected by the objective lens; this leads to the well-known Rayleigh criterion, shown in Figure 1a.

To demonstrate the link between NA_{inc} of the objective lens and its achievable spatial resolution, we introduce the point-spread function (PSF) to describe the response of the imaging system to a point source, which bridges the original object O and resultant image G as, $G = O * PSF$. The Fourier transform of PSF is defined as the optical transfer function (OTF), corresponding to a particular object in the Fourier transform domain. The direction of light propagation could be defined by the wave vector $\mathbf{k} = k_x\mathbf{x} + k_y\mathbf{y} + k_z\mathbf{z}$. We define $\mathbf{k}_\parallel = k_x\mathbf{x} + k_y\mathbf{y}$, where $\mathbf{k}$ propagates on the (x, y) plane. As not all the light propagated through the sample is detectable in conventional microscopes, only the plane waves emerging from the sample with $|\mathbf{k}_\parallel| \leq k_0 NA$ are collected, shown in Figure 2, where NA is the numerical aperture of the microscope objective; Rayleigh criterion is thus introduced limiting the resolution as $\Delta r = 0.61\lambda/NA$. Moreover, by using conventional optical microscopes, one is unable to quantitatively obtain the opto-geometrical characteristics of the sample.

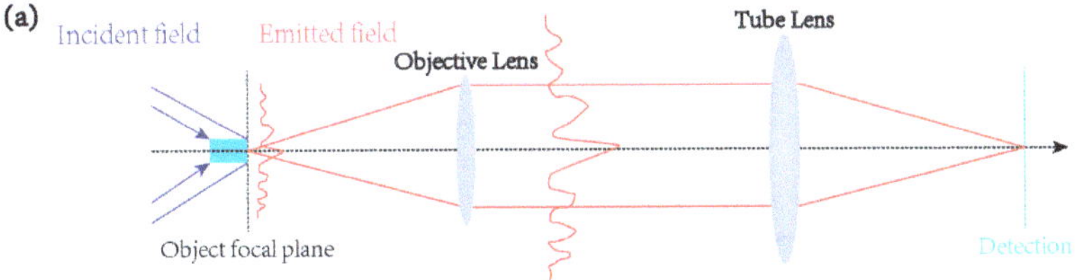

Wide-field microscopy - A general representation

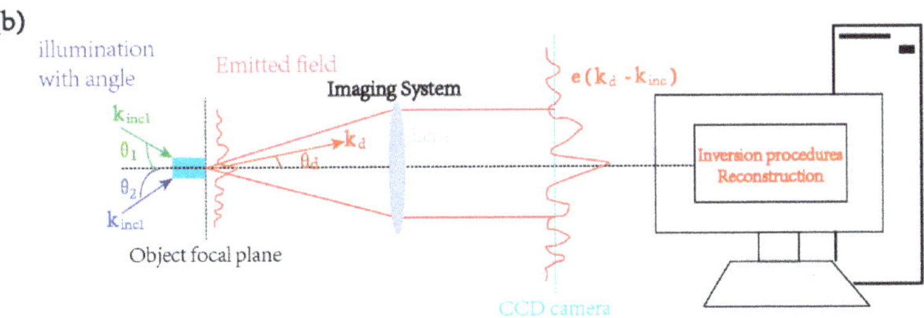

Tomographic diffractive microscopy

Figure 1. Schemas of the wide-field microscopy and tomographic diffractive microscopy. (**a**) Wide-field microscopy. (**b**) Tomographic diffractive microscopy.

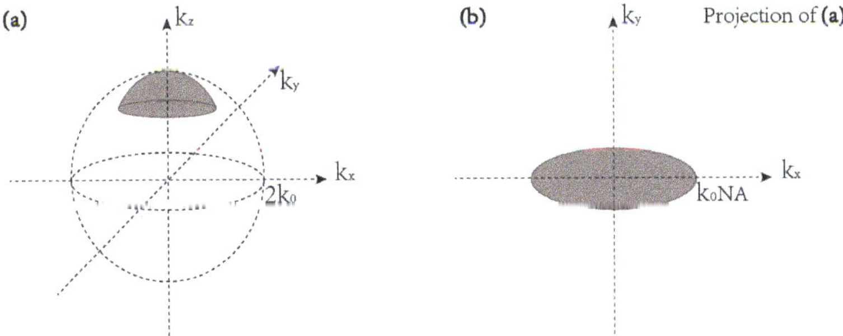

Figure 2. A description of the accessible frequency domain of conventional microscopy. (**a**) The sample is illuminated in normal incidence and detected in the same direction. (**b**) The projection of the measurable diffracted wave vector **k** onto the transverse (xOy) in conventional microscopy.

Thus, two questions arise: Can we get the three-dimensional image of the sample and is it possible to investigate the material properties of the sample? The answer is tomographic diffractive microscopy, which was introduced by E. Wolf in the year of 1969 [34]. Upon recording complex diffracted fields (amplitude and phase) by coherently illuminating the sample, the index of refraction or the permittivity of the sample are retrievable by numerical inversion procedures, shown in Figure 1b [31,35–38].

2.1. The Principle of Tomographic Diffractive Microscopy

An incident electromagnetic wave of $[\mathbf{E}_{inc}, \mathbf{H}_{inc}]$ interacts with a sample in vacuum that occupies a bounded region **V** in three-dimensional space and a relative permittivity $\varepsilon(\mathbf{r})$ for $\mathbf{r} \in \mathbf{V}$, $\varepsilon(\mathbf{r}) = 1$ for $\mathbf{r} \notin \mathbf{V}$. By deducing the Maxwell equations, the total scalar field satisfies:

$$\nabla \times \nabla \times \mathbf{E}(\mathbf{r}) - \varepsilon(\mathbf{r})k_0^2\mathbf{E}(\mathbf{r}) = 0 \tag{1}$$

where $k_0 = 2\pi/\lambda$ is the wave number, where λ is the wavelength in vacuum. We transform Equation (1) as:

$$\nabla \times \nabla \times \mathbf{E}(\mathbf{r}) - k_0^2 \mathbf{E}(\mathbf{r}) = k_0^2 \chi(\mathbf{r}) \mathbf{E}(\mathbf{r}) \tag{2}$$

where $\chi(\mathbf{r}) = \varepsilon(\mathbf{r}) - 1$ is the contrast of permittivity. The method for solving Equation (2) is to find the Green's function, i.e., to find the solution of the corresponding differential equation with a Dirac delta inhomogeneity:

$$\nabla \times \nabla \times \mathbf{G}(\mathbf{r}, \mathbf{r}') - k_0^2 \mathbf{G}(\mathbf{r}, \mathbf{r}') = \mathbf{I}\delta(\mathbf{r} - \mathbf{r}') \tag{3}$$

where $\mathbf{I}$ is the identity matrix. This Green's function is known as the free-space dyadic Green's function:

$$\mathbf{G}(\mathbf{r}, \mathbf{r}') = \left[\mathbf{I} + \frac{1}{k_0^2}\nabla\nabla\right]\frac{e^{ik_0|\mathbf{r}-\mathbf{r}'|}}{4\pi|\mathbf{r}-\mathbf{r}'|} \tag{4}$$

By solving Equation (2), we obtain the integral equation as:

$$\mathbf{E}(\mathbf{r}) = \mathbf{E}_{\text{ref}}(\mathbf{r}) + k_0^2 \int_V \mathbf{G}(\mathbf{r}, \mathbf{r}')\chi(\mathbf{r}')\mathbf{E}(\mathbf{r}')dV \tag{5}$$

where the reference field $\mathbf{E}_{\text{ref}}(\mathbf{r})$ consists of the field without a research object and is a special solution to the homogeneous equation obtained by setting $\varepsilon(\mathbf{r})$ to the homogeneous background, i.e., $\varepsilon(\mathbf{r}) - 1$. The scattered field $\mathbf{E}_{\text{sca}}(\mathbf{r})$ is the difference between the total field and the reference field:

$$\mathbf{E}_{\text{sca}}(\mathbf{r}) = k_0^2 \int_V \mathbf{G}(\mathbf{r}, \mathbf{r}')\chi(\mathbf{r}')\mathbf{E}(\mathbf{r}')dV \tag{6}$$

Notice that the conventional tomographic diffractive microscopy approach neglects the polarization effects induced by the object and the setup, so that the scalar approximation used for the field and Equation (6) can be rewritten as a scalar propagation equation in an inhomogeneous medium.

2.2. Born Approximation with Linear Inversion

In far-field cases, if the observation position $\mathbf{r}$ is sufficiently far away from the sample, or if the sample is weakly scattered enough, typically the sample with small permittivity contrasts, i.e., $\Delta\varepsilon < 0.1$; the amplitude of the scattered field is tiny compared to that of the reference field; an approximation termed Born approximation was commonly used in TDM [35,39–43]. The scalar Green function that approximates the direction given by the wave vector $\mathbf{k}$ in far field is:

$$\mathbf{G}(\mathbf{r}, \mathbf{r}') = \frac{e^{ik_0 r}}{4\pi r}e^{i\mathbf{k}\cdot\mathbf{r}} \tag{7}$$

where $\mathbf{k} = k_0\frac{\mathbf{r}}{r}$. Defining $\mathbf{E}_{\text{ref}}(\mathbf{r})$ as a plane wave with incident wave vector $\mathbf{k}_{\text{inc}}$, we obtain the scattered field as:

$$\mathbf{E}_{\text{sca}}(\mathbf{r}) = k_0^2\frac{e^{ik_0 r}}{4\pi r}\int_V \chi(\mathbf{r}')e^{-i(\mathbf{k}-\mathbf{k}_{\text{inc}})\cdot\mathbf{r}'}dV \tag{8}$$

From Equation (8), in far field and the Born approximation conditions, the field scattered $\mathbf{E}_{\text{sca}}(\mathbf{r})$ along the wave vector $\mathbf{k}$ for an illuminating wave vector $\mathbf{k}_{\text{inc}}$ and the 3D Fourier transform of χ taken at $\mathbf{k} - \mathbf{k}_{\text{inc}}$ are proportional:

$$\mathbf{E}_{\text{sca}}(\mathbf{k} - \mathbf{k}_{\text{inc}}) \propto \widetilde{\chi}(\mathbf{k} - \mathbf{k}_{\text{inc}}) \tag{9}$$

Hence, the dielectric constant contrast map of the object can be retrieved by a simple linear inverse Fourier transform of the scattered field recorded in the far field. Simon et al. reported several successful 3D biological sample reconstructions by TDM, shown in Figures 3 and 4 [41,44].

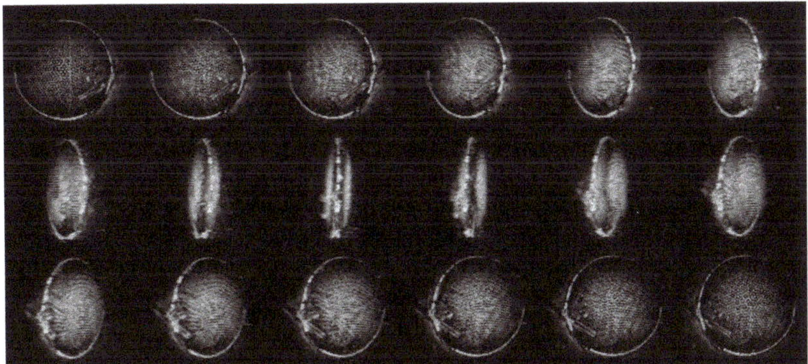

Figure 3. A 3D view of a *Coscinodiscus* sp. diatome using tomographic diffractive microscopy (TDM) with Born approximation. The pictures depict an 18-step, 0°–180° rotation of the specimen [41].

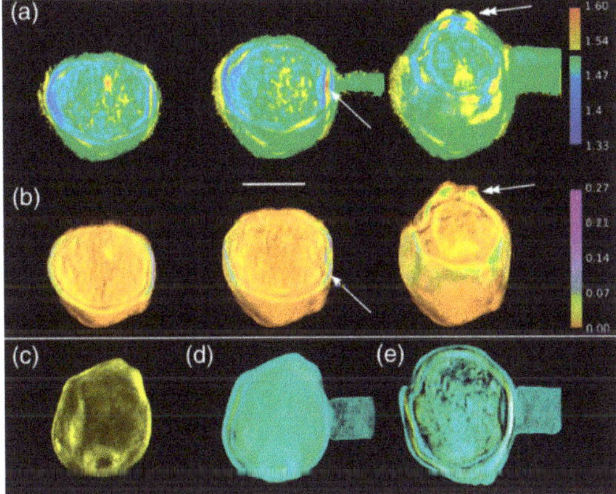

Figure 4. *Betula* pollen grain image reconstructed by tomographic diffractive microscopy (TDM) under Born approximation. (**a**,**b**) Volumetric cuts of a refraction image and an absorption image, respectively. Scale bar: 10 μm. (**c**) Outer view of the pollen: Image of the absorption component. (**d**) Outer view of the pollen: Image of the complex index of refraction. (**e**) (*x*-*y*) cut through the pollen [44].

The resolution of TDM under the Born approximation is determined by the accessible Fourier domain, which depends on the configuration of the illumination and detection [45], which we will discuss in the following section.

2.3. Rigorous Case with Non-linear Inversion

Born approximation is a scalar approximation restricted to the weakly scattered sample, while, in the case of high permittivity contrast samples or the need for high quality sample reconstruction resolution, a more sophisticated inversion procedure is required. The aim of the non-linear inversion procedure is stated as finding the permittivity $\Delta\varepsilon$ of high contrast samples, in which the multiple scattering inside the sample cannot be neglected [36,46–49]. In rigorous cases, the total field inside sample $\mathbf{E}(\mathbf{r})$, see Equation (6), cannot be simply replaced by $\mathbf{E}_{ref}(\mathbf{r})$. We rewrite Equations (5) and (6) as:

$$\mathbf{E} = \mathbf{E}_{ref} + \mathbf{A}\chi\mathbf{E}$$
$$\mathbf{E}_{sca} = \mathbf{B}\chi\mathbf{E}$$

(10)

where $\mathbf{A}$ denotes a square matrix of size (3N × 3N), contains all the tensors $\mathbf{G}(\mathbf{r}_i, \mathbf{r}_j)$; i and j present a point in the discretized sample bounded investigation domain, $i, j = 1, \cdots, N$; $\mathbf{B}$ is a matrix of size (3N × 3M) and contains the tensors $\mathbf{G}(\mathbf{r}_i, \mathbf{r}_k)$; $k = 1, \cdots, M$ is an observation point in the observation domain. Iterative methods are traditionally used for solving Equation (9) [49–51]. Assume the unknown sample is restricted in a three-dimensional box Ω, the observations are at a far-field surface of Γ, the measured field is f_l, n is the iteration number, and L is the illuminations. The cost function of iterative procedure can be written as:

$$F_n(\chi_n, \mathbf{E}_{l,n}) = W_\Gamma \sum_{l=1}^{L} \|h_{l,n}^{(1)}\|_\Gamma^2 + W_\Omega \sum_{l=1}^{L} \|h_{l,n}^{(2)}\|_\Omega^2 \tag{11}$$

where $h_{l,n}^{(1)}$ and W_Γ are the residual error and the weighting coefficient in the far-field, the lower part of Equation (10):

$$h_{l,n}^{(1)} = f_l - \mathbf{B}\chi\mathbf{E} \qquad W_\Gamma = \left(\sum_{l=1}^{L} \|f_l\|_\Gamma^2 \right)^{-1} \tag{12}$$

and $h_{l,n}^{(2)}$ and W_Ω are the residual error and the weighting coefficient in the near-field, the upper part of Equation (10):

$$h_{l,n}^{(2)} = \mathbf{E}_{l,ref} - \mathbf{E}_{l,n} + \mathbf{A}\chi_n \mathbf{E}_{l,n} \qquad W_\Omega = \left(\sum_{l=1}^{L} \|\chi_{n-1}\mathbf{E}_{l,ref}\|_\Omega^2 \right)^{-1} \tag{13}$$

A simulated scattered field could be obtained from the estimate of the relative permittivity, like estimation from Born approximation [38]. By minimizing the discrepancy between the measured scattered field and the simulated one, iterative methods are able to retrieve the distribution of sample permittivity χ and the total field $\mathbf{E}$ in the bounded investigation domain [52]. Traditionally, non-linear inversion methods are employed to investigate the arbitrarily shaped, anisotropic, and inhomogeneous samples; however, there are some disadvantages, such as: (i) It is time-consuming, the computational time is greatly increased with the enlargement of the investigation domain; and (ii) the computational accuracy is mainly dependent on the number of discretions for the sample. Thus, these efficient methods are suggested: Compromising the quality of reconstruction and computational time [53], or incorporating the prior information of the sample [38,53]. For example, prior information of sample size is helpful to minimize the investigation domain [53], or prior information of sample permittivity/refractive index is useful to speed up the inversion procedure and also to improve the resolution of the reconstruction. To the best of our knowledge, the reported resolution of TDM achieves $\lambda/10$ using approximated knowledge of the sample permittivity, see Figure 5 [10].

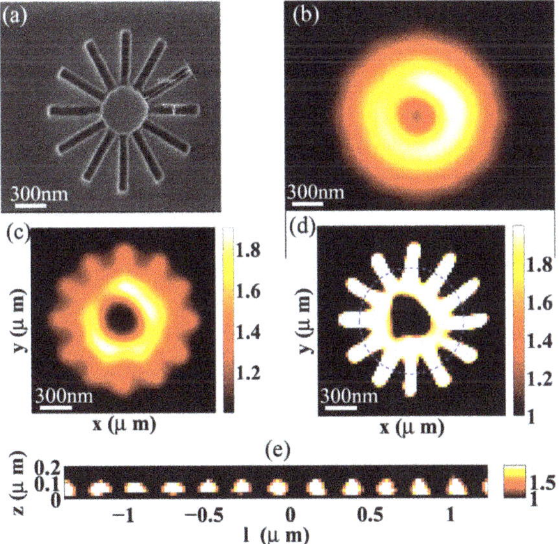

Figure 5. Images of a resin star sample, sample information: 97 nm wide rods of length 520 nm and height 140 nm on a Si substrate. (**a**) Image of the sample obtained by SEM. (**b**) Image of the sample obtained by Darkfield microscopy numerical aperture (NA) of objective 0.95. (**c**) Reconstruction obtained with a hybrid inversion method from tomographic diffraction microscopy data with NA equals to 0.95. (**d**) Permittivity reconstructed with a bounded inversion method using the prior knowledge of the resin permittivity from the same data as (**c**). (**e**) Permittivity distribution along the z direction versus the curvilinear abscissa of the dashed circle in (**d**) [10].

3. The Resolution of Tomographic Diffractive Microscopy

The aim of TDM is to obtain a 3D reconstruction of the investigated sample. The link between the measured scattered field and the 3D relative permittivity is given in Equation (9); by merging the measured components as synthetic aperture generation, the permittivity contrast of the sample is reconstructed through an inverse Fourier transform of the detected field under Born approximation [54–56]. In principle, for a given angle of illumination (e.g., normal incidence) with wave vector $\mathbf{k}_{inc}$, the Fourier components of the sample permittivity contrast were on a cap of sphere of radius $2k_0$, truncated by the numerical aperture (NA) of the objective. The sphere was centered on the extremity of wave vector $-\mathbf{k}_{inc}$, as in Figure 2. In order to increase the amount of detectable Fourier components, and thus improve the resolution of the object reconstruction, a variety of illumination angles must be used. Since the accessible frequency domain for each illumination depends on $\mathbf{k} - \mathbf{k}_{inc}$, we explain the resolution of TDM in three cases, see Figure 6.

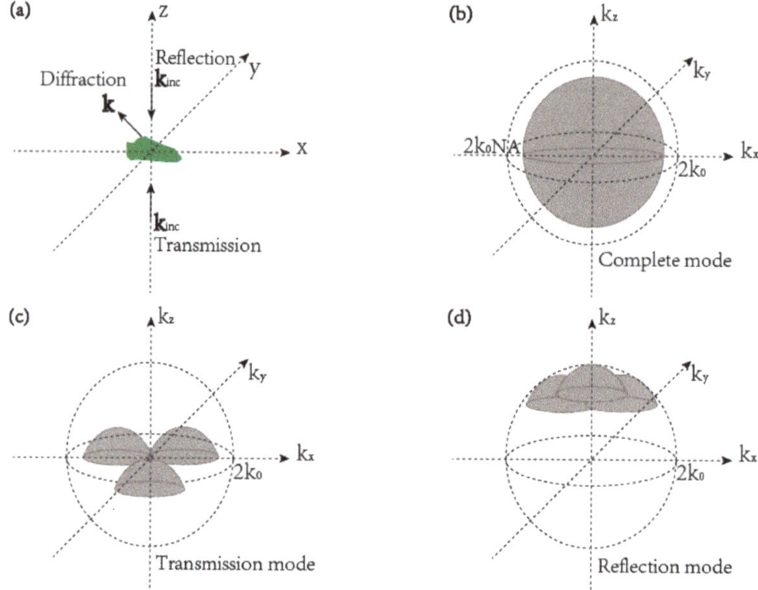

Figure 6. A description of different configurations of tomographic diffractive microscopy (TDM) and the corresponding accessible frequency domains. (**a**) The different configurations of TDM. (**b**) The accessible frequency domain for complete configuration. (**c**) The accessible frequency domain for transmission configuration. (**d**) The accessible frequency domain for reflection configuration.

3.1. TDM in Complete Configuration

Ideally, samples of all directions within 4π radians are illuminated and these directions are detected to obtain the maximum amount of Fourier components. Therefore, for a given illumination direction, the accessible Fourier component is a sphere with a radius of $2k_0NA$, as in Figure 6b. With this complete configuration, all the spatial frequencies given by $\mathbf{k} - \mathbf{k}_{inc}$ for any wave vectors $\mathbf{k}$ and $\mathbf{k}_{inc}$ are accessible. Such an OTF provides an isotropic resolution $\Delta r = 0.61\lambda/(NA + NA_{inc})$, which is nearly twice the Rayleigh criterion $\Delta r = 0.61\lambda/NA_{inc}$. In this case, we simulated the OTF, a sphere of radius $2k_0$ filled with one. The corresponding PSF is isotropic in all directions and gives the same resolution in 3D dimensions, see Figure 7.

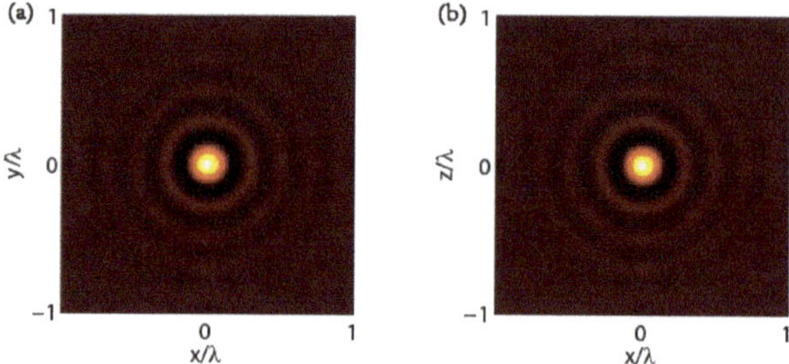

Figure 7. Point-spread function (PSF) image of TDM in complete configuration (**a**) Transverse cut of the PSF at z = 0. (**b**) Longitudinal cut of the PSF at y = 0.

3.2. TDM in Transmission Configuration

In the case of the transmission, the optical axis is irradiated along the side of the sample and detected on the other side. The reachable frequency range is a torus whose axis is symmetrical with the z-axis, and its cross-section in the longitudinal plane consisting of two circles of radius $k_0 NA_{inc}$, see Figure 6c. It is worth noting that since the accessible frequency domain of the transmission configuration along the z direction is smaller compared to the transverse direction x-y plane, the resolution of this configuration is anisotropic and the axial resolution is about three times worse than the transverse resolution. In this case, we simulated the OTF, a torus filled with one; the corresponding PSF is anisotropic, especially along the z direction, shown in Figure 8b. This means the axial resolution will deteriorate worse than the transverse resolution $\Delta r = 0.61\lambda/(NA + NA_{inc})$.

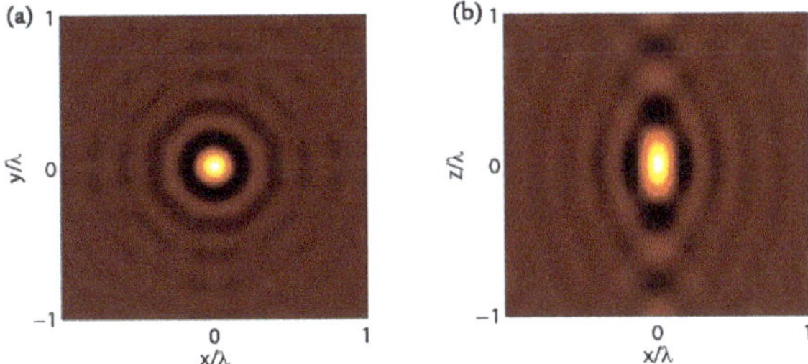

Figure 8. PSF image of TDM in transmission configuration. (**a**) Transverse cut of the PSF at z = 0. (**b**) Longitudinal cut of the PSF at y = 0.

3.3. TDM in Reflection Configuration

For illumination and detection on the same side of the sample, there is a reflective configuration. Upon varying the illumination angles, the accessible frequency domain occupies part of the complete sphere, see Figure 6d. Such a reflection configuration can significantly improve the axial resolution; however, the inverse Fourier transform of this accessible frequency domain yields a complex PSF; as a result, if the sample under investigation has a complex permittivity, the complex PSF will mix the real and imaginary part of permittivity in the reconstruction. In this case, we simulated the OTF, a half of the complete sphere of radius $2k_0$ filled with one. The corresponding PSF is isotropic. However, compared to the above two configurations, the PSF of the reflection configuration becomes a complex function, see the longitudinal cut of the imaginary part of the PSF at y = 0 in Figure 9b. This implies that the reconstructed real part and imaginary part of the permittivity of the sample will mingle in an unpredictable way [57].

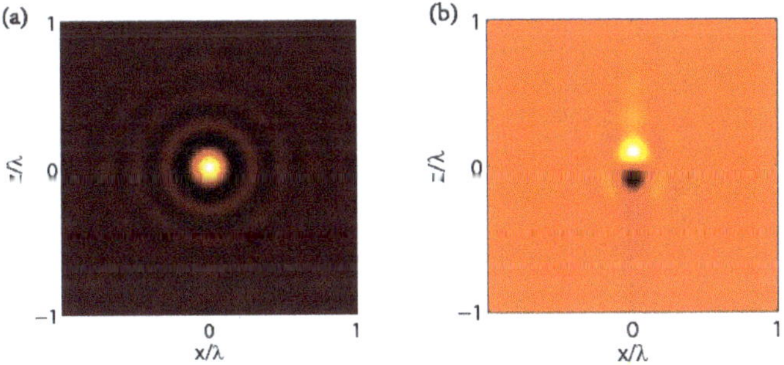

Figure 9. PSF image of TDM in reflection configuration. (**a**) Longitudinal cut of the real part of the PSF at y = 0. (**b**) Longitudinal cut of the imaginary part of the PSF at y = 0.

4. The Optical Setup of Tomographic Diffractive Microscopy

The TDM can be implemented in either transmission configuration or reflection configuration. The need of quantitative phase measurement is the significant difference between TDM and conventional wide-field microscopes [15]. The TDM setup consists of illuminating the sample with a collimated beam from controlled angles of incidence and recording the field diffracted by the sample in phase and in amplitude. Varying the illumination of TDM: In principle, two methods are available for varying illumination angles, rotation of the collimated beam [37] and rotation of the sample [58]; the former is easier to realize. Keeping the sample static is easy to manipulate for a TDM user. A rotating mirror permits control of the deflection of the collimated beam [37,59]; however, the missing parts of non-captured frequencies present a strong anisotropic resolution along the optical axis, especially in transmission configuration of TDM. To obtain an improved and isotropic resolution, Mudry et al. used a mirror-assisted setup combining the transmission and reflection configuration together [57] or used an ellipsoidal mirror for expanding the angular coverage of diffracted field detection [60]. Another way for varying the illumination is rotation of the sample and keeping incident light static. In TDM under transmission configuration, fixing the setup, the sample along the x-axis was successfully rotated to give a quasi-isotropic resolution [61]. Using optical tweezers is another promising method [62–66]. To achieve complete configuration, a combination of specimen rotation and illumination rotation simultaneously is also feasible by adopting an integrated setup [67]. Phase and amplitude detection of TDM: An interferometric arrangement of setup—phase shift interferometry—is commonly used in TDM for the detection of complex diffracted fields, see Figure 10a [38,68]; it involves shifting the phase relative to the reference beam in steps, recording of the interference between the diffracted field and reference beam for each phase step, and calculation of the phase from four or more detected intensities on a CCD camera. For example, the intensity measured on the CCD camera can be written as S, adding the phase shift by phase modulator, see Figure 10a; we could detect four successive intensities on the camera as:

$$S_1 = I_{sample} + I_{reference} + 2\sqrt{I_{sample}I_{reference}}\cos\left(\varphi + \tfrac{\pi}{2}\right)$$
$$S_2 = I_{sample} + I_{reference} + 2\sqrt{I_{sample}I_{reference}}\cos(\varphi)$$
$$S_3 = I_{sample} + I_{reference} + 2\sqrt{I_{sample}I_{reference}}\cos\left(\varphi - \tfrac{\pi}{2}\right) \tag{14}$$
$$S_4 = I_{sample} + I_{reference} + 2\sqrt{I_{sample}I_{reference}}\cos(\varphi - \pi)$$

where φ is the original phase differences between the diffracted field and the reference. The diffracted field of the sample I_{sample} could be retrieved as $E_{sample} = \frac{(S_2 - S_4) + i((S_3 - S_1))}{4\sqrt{I_{reference}}}$. Such a configuration is usually called an on-axis (or inline) system; since the phase information is obtained by making consecutive image subtractions, it provides a precise measurement of the phase of the diffracted field [69,70]. However, performing several phase steps is time consuming, especially if a large number of illumination angles have to be applied successively in TDM. Phase fluctuations due to thermal and/or mechanical drift during acquisition can interfere with the measurement [71,72]. An off-axis setup is a simple configuration for measuring the complex diffracted field from a single hologram and can avoid the image conjugation and enhance the imaging quality, but it is more sophisticated to calibrate compared to the on-axis system, see Figure 10b [23,73,74]. The main difference of the setup between on-axis and off-axis is the removal of the phase modulator. In off-axis, the references are propagated with a carefully chosen angle. For the sake of simplicity, considering the interference in one dimension, i.e., in x-axis, the intensity on the charge coupled device (CCD) camera could be written as:

$$S = I_{sample} + I_{reference} + e^{-i\alpha x}E_{sample} + e^{i\alpha x}E_{sample}{}^* \tag{15}$$

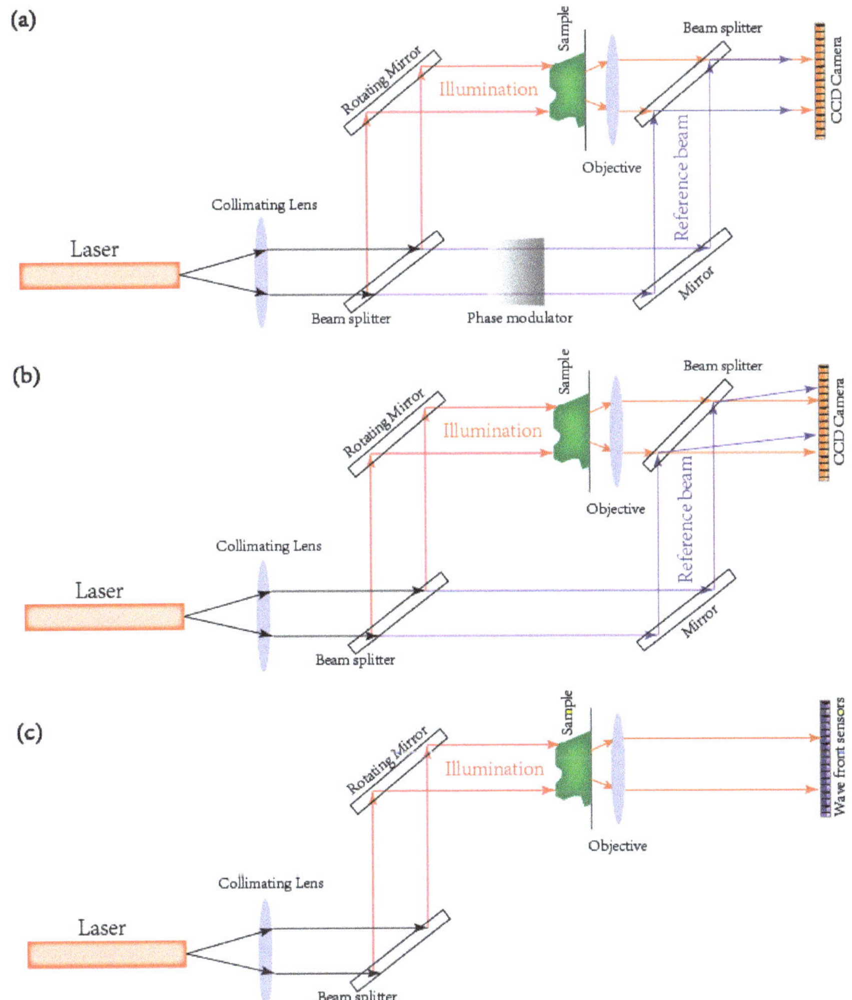

Figure 10. Schematic images of three different methods for complex diffracted field detection. (**a**) Phase shifting interferometry setup. (**b**) Off-axis interferometry setup. Compared to (**a**), the beam splitter at the right upper corner in (**b**) is rotated with a small angle so that the reference (blue) does not superimpose on the diffracted field (red). (**c**) Quadri-wave lateral shearing interferometry, wave front sensors setup.

By applying 2D Fourier transform of S, the Dirac function $\delta_{\pm\alpha}$ introduced by $e^{\pm i\alpha x}$ helps us to retrieve the complex field of the sample. Moreover, the benefit from the nature of single-shot measurement in the off-axis method is that it could significantly reduce the setup sensitivity to external fluctuations. However, it is worth noting that single-shot characteristics are measured at the cost of the camera's available pixels; a minimal off-axis angle to separate the different interference terms in the Fourier space and a maximal angle to successfully distinguish the interference fringes must be satisfied simultaneously [75]. To the best of our knowledge, the highest resolution of TDM reported as one-tenth wavelength of the illumination was using off-axis to yield the phase and amplitude of the diffracted field [10]. Besides the interferometric setup, a Shack-Hartmann wavefront sensor can be used to measure the complex diffracted field without the reference beam, see Figure 10c, to eliminate the external influences and minimize the measurement period [76–79]. In this sensor, a modified Hartmann mask (MHM) was closely set in front of the detector to create replicas of the incident wavefront in several identical but tilted waterfronts; their mutual interference patterns were then recorded on the detector. The phase recovered by applying a Fourier transform to the interferogram. Primot et al. reported the principle of recovering the intensity and phase of a field with multi-wave interferometry [80]. A successful employment of a wavefront sensor based on quadri-wave lateral shearing interferometry (QWLSI) in TDM measurement has also been presented [37], but the disadvantage of this sensors is the limitation of resolution compared to charge coupled device (CCD) or CMOS) cameras [37,81]. Notice that most tomographic diffractive microscopy (TDM) setups have been used with a laser beam that was polarized in one direction—vertically or horizontally—for both the illumination and the reference wave if they are present; however, for the resolution of the measured sample, with respect to the TMD system far beyond the Rayleigh criterion, the diffracted field close to the edge of the NA of the objective is no longer parallel to the polarization of the illumination, a modified setup developed to retrieve the full vectorial diffracted field; the resolution was thus significantly improved [82,83].

To conclude the development of the TDM setup, we present the recent remarkable progress of TDM setups in Table 1.

Table 1. Recent progress of tomographic diffractive microscopy TDM setups.

Optical Schema/Configuration	Illumination Wavelength	Measured Sample	Lateral Resolution	Inversion	Reference
Off-axis/Reflection	475 nm	nanofabricated objects	50 nm	Non-linear	[10]
Off-axis/Completeness	475 nm	*Bellis perennis* pollen grain *Betula* pollen grain	200 nm	Linear	[25]
On-axis/Reflection	633 nm	nanofabricated objects	400 nm	Non-linear	[38]
On-axis/Completeness	633 nm	prolate spheroid	~300 nm	Linear	[60]
Wavefront sensors/Reflection	633 nm	nanofabricated objects	1 μm	Non-linear	[37]

5. The Advancements of Tomographic Diffractive Microscopy

The purposes of the developments of TDM setup are to collect the diffracted field as much as possible. In the best case, the complex diffracted field, including the phase and amplitude from all possible angles of illumination and observation should be recorded for coverage of all frequency domains in Fourier space to achieve an isotropic resolution. Limited by the size and the precision, moving objectives and cameras are the challenges in a typical optical setup. Hence, illumination and sample are the only two moving components in TDM. However, varying illumination while keeping sample statics, whether in a transmission configuration or a reflection configuration, we only obtain the sample information from one side, and thus cannot get tomographic images of the sample with isotropic resolution. Rotation of the sample by optical tweezers seems promising; however, this approach is limited to certain types of samples and its controllability and measurability should be further proved [84]. A mirror-assisted method is thus a practical approach that detects the diffracted field from both sides of the sample simultaneously [57]. It is important to note that it has the similar principle of placing two opposing objectives as in a typical 4 pi microscope setup, but it provides much simpler practical implementation [85].

The purpose of the inversion algorithm of TDM is to reconstruct the nature and the three-dimensional geometry of the investigated sample from the detected complex diffracted field and its corresponding illuminations. Most TDM applications are applied under Born approximation; the linear relationship between the sample permittivity and the diffracted field permits us to reconstruct the sample opto-geometry by using a simple inverse Fourier transform. However, to image high contrast sample, multiple scattering cannot be neglected and non-linear numerical inversion procedures are necessary. However, the main bottleneck of the non-linear inversion method is computation time. To overcome this disadvantage, a combination of linear and non-linear methods is helpful; for example, a fast Born approximation may provide noisy sample localization and reconstruction [10] that are useful for minimizing the investigation domain of the non-linear inversion procedure. One reported integrating DORT (décomposition de l'opérateur de retournement temporal) and SVD (single value decomposition) methods to quickly localize the sample from a noisy environment. This not only improves the resolution but also ameliorates the reconstruction speed [53,86]. Furthermore, if prior information of a sample, such as an estimation or range of sample permittivity, is available, the resolution can be improved [10].

The advantages of TDM are remarkable; while microscopes like the digital holographic microscope are able to provide a 3D topography of the samples [23,87,88], the ability of quantitative reconstruction of the sample permittivity/refractive index using TDM is unique; moreover, the lateral resolution in TDM has been well improved far beyond the Rayleigh criterion [15]. The studies of TDM have been increasing rapidly in recent years. To our knowledge, the first commercial products of TDM were released by Nanolive in Switzerland in 2015 [89], and Tomocube KAIST in Korea at 2017 [90]. These work with transmission configuration and provide 3D reconstructions of cells with transverse resolutions of 200 nm and axial resolutions of 400 nm [91]. Hence, future developments may focus on the implementation of the reflection configuration and the use of multiple wavelengths of illumination to improve the axial resolution. Moreover, integrating TDM with other super-resolution techniques can provide multiple-modal image methods; for example, a combination of TDM with single molecule fluorescent microscopes is able to perform the structural and functional analysis of a sample simultaneously; a combination of TDM with scanning probe microscopy permits the structural, topographic, and chemical (tip-enhanced Raman spectroscopy, TERS) information of a sample. To the best of our knowledge, TDM has only been applied to investigate the individual structures, such as isolated cells and nano-fabricated structures, but three-dimensional imaging of unmarked raw tissues is also possible using TDM.

However, every coin has two sides; it is worth noting here that the main drawbacks of TDM are: (i) TDM under Born/Rytov approximation is only valid for samples with low permittivity contrast $\Delta\varepsilon < 0.1$; (ii) TDM under Born approximation has a resolution limit which is twice as good as the Rayleigh criterion resolution; (iii) the non-linear inversion procedures improve the resolution but are time-consuming. Several hours are at least necessary to reconstruct the investigated sample, thus it is difficult to realize real-time analysis for the samples using non-linear inversion methods. (iv) An isotropic resolution still remains a challenge, despite several studies reporting TDM with isotropic resolution [44,57,67], and a combination of sample rotation with illumination rotation not only needs accurate correction of the angles, but also greatly increases the amount of data collected [44,67].

6. Conclusions

Compared to other microscopies, TDM permits us to study the sample in a label-free condition It provides 3D quantitative reconstruction of the investigated sample and breaks the Rayleigh limit. With the developments of digital holographic techniques [92,93], the phase detection in optics becomes more and more simple and reliable, the off-axis methods and wave front sensors are able to retrieve the complex diffracted field from single-shot measurements, and thus provide high-speed detections, we believe they are the futuristic techniques in favor of TDM applications. Micro-manipulation tools are promising methods for enlarging the accessible frequency domain, improving the reconstruction resolution. Like using optical tweezers [94], the sample can be rotated without any mechanical contact. Similarly, using electric fields to rotate the sample is another way to achieve micro-rotation of the sample [95]. Inversion procedures are mandatory in TDM; for low contrast samples, e.g., biological cells, Born approximation [96] or Rytov approximation [97,98] is sufficient to provide a real-time reconstruction [97]; however, for high contrast sample, e.g., nano-structural materials, sophisticated non-linear inversion methods are necessary. To the best of our knowledge, it is still time-consuming to characterize the sample by iterative methods. Using a priori information of the sample, e.g., estimation of sample location and estimation of sample permittivity, is a promising approach to greatly reduce the reconstruction time.

In conclusion, we present this review to give an overview of TDM, including the principle, the setup, and the inversion methods for different applications; the limitations of TDM are also addressed. From a sample imaging point of view, the quantitative measurement of the permittivity/refractive index of a sample using TDM could provide complementary and useful information in association with other microscopies. As TDM can be used to image both low contrast samples and high contrast ones, it will play a key role not only in the exploration of biological cells but also in the investigation of nano-structural devices.

Author Contributions: T.Z., K.L., Y.R.; writing—original draft preparation, C.G. writing—review and editing.

Funding: This work was funded by National Nature Science Foundation of China Grants NSFC [61801423] (T.Z.), NSFC [61805213] (Y.R.), NSFC [61605171] (K.L.). The Fundamental Research Funds for the Central Universities 2018FZA5006 (T.Z.).

Conflicts of Interest: The authors declare no conflict of interest.

References

1. Henderson, R.; Unwin, P.N. Three-dimensional model of purple membrane obtained by electron microscopy. *Nature* **1975**, *257*, 28–32. [CrossRef] [PubMed]
2. Karnovsky, M.J.; Roots, L. A formaldehyde-glutaraldehyde fixative of high osmolarity for use in electron microscopy. *J. Cell Biol.* **1965**, *27*, A137.
3. Meyer, G.; Amer, N.M. Novel optical approach to atomic force microscopy. *Appl. Phys. Lett.* **1988**, *53*, 1045–1047. [CrossRef]
4. Giessibl, F.J. Advances in atomic force microscopy. *Rev. Mod. Phys.* **2003**, *75*, 949–983. [CrossRef]
5. Betzig, E.; Trautman, J.K.; Harris, T.D.; Weiner, J.S.; Kostelak, R.L. Breaking the diffraction barrier: Optical microscopy on a nanometric scale. *Science* **1991**, *251*, 1468–1470. [CrossRef]
6. Webb, R.H. Confocal optical microscopy. *Rep. Prog. Phys.* **1996**, *59*, 427–471. [CrossRef]
7. Hecht, B.; Sick, B.; Wild, U.P.; Deckert, V.; Zenobi, R.; Martin, O.J.F.; Pohl, D.W. Scanning near-field optical microscopy with aperture probes: Fundamentals and applications. *J. Chem. Phys.* **2000**, *112*, 7761–7774. [CrossRef]
8. Webb, D.J.; Brown, C.M. Epi-fluorescence microscopy. *Methods Mol. Biol. (Clifton, N.J.)* **2013**, *931*, 29–59. [CrossRef]
9. Denk, W.; Strickler, J.H.; Webb, W.W. Two-photon laser scanning fluorescence microscopy. *Science* **1990**, *248*, 73–76. [CrossRef]
10. Zhang, T.; Godavarthi, C.; Chaumet, P.C.; Maire, G.; Giovannini, H.; Talneau, A.; Allain, M.; Belkebir, K.; Sentenac, A. Far-field diffraction microscopy at $\lambda/10$ resolution. *Optica* **2016**, *3*, 609. [CrossRef]
11. Evanko, D. Label-free microscopy. *Nat. Methods* **2009**, *7*, 36. [CrossRef]
12. Hell, S.W.; Wichmann, J. Breaking the diffraction resolution limit by stimulated emission: Stimulated-emission-depletion fluorescence microscopy. *Opt. Lett.* **1994**, *19*, 780–782. [CrossRef] [PubMed]
13. Rust, M.J.; Bates, M.; Zhuang, X. Sub-diffraction-limit imaging by stochastic optical reconstruction microscopy (STORM). *Nat. Methods* **2006**, *3*, 793–795. [CrossRef] [PubMed]
14. Manley, S.; Gillette, J.M.; Patterson, G.H.; Shroff, H.; Hess, H.F.; Betzig, E.; Lippincott-Schwartz, J. High-density mapping of single-molecule trajectories with photoactivated localization microscopy. *Nat. Methods* **2008**, *5*, 155–157. [CrossRef] [PubMed]

15. Lauer, V. New approach to optical diffraction tomography yielding a vector equation of diffraction tomography and a novel tomographic microscope. *J. Microsc.* **2002**, *205*, 165–176. [CrossRef] [PubMed]
16. Mazza, D.; Bianchini, P.; Caorsi, V.; Cella, F.; Mondal, P.P.; Ronzitti, E.; Testa, I.; Vicidomini, G.; Diaspro, A. Non-Linear Microscopy. In *Biophotonics*; Pavesi, L., Fauchet, P.M., Eds.; Springer: Berlin Germany, 2008.
17. Zipfel, W.R.; Williams, R.M.; Christie, R.; Nikitin, A.Y.; Hyman, B.T.; Webb, W.W. Live tissue intrinsic emission microscopy using multiphoton-excited native fluorescence and second harmonic generation. *Proc. Natl. Acad. Sci. USA* **2003**, *100*, 7075–7080. [CrossRef] [PubMed]
18. Hoebe, R.A.; Van Oven, C.H.; Gadella, T.W.J.; Dhonukshe, P.B.; Van Noorden, C.J.F.; Manders, E.M.M. Controlled light-exposure microscopy reduces photobleaching and phototoxicity in fluorescence live-cell imaging. *Nat. Biotechnol.* **2007**, *25*, 249–253. [CrossRef]
19. Wachulak, P.; Bartnik, A.; Fiedorowicz, H. Optical coherence tomography (OCT) with 2 nm axial resolution using a compact laser plasma soft X-ray source. *Sci. Rep.* **2018**, *8*, 8494. [CrossRef]
20. Bousi, E.; Pitris, C. Lateral resolution improvement in Optical Coherence Tomography (OCT) images. In Proceedings of the 2012 IEEE 12th International Conference on Bioinformatics & Bioengineering (BIBE), Larnaca, Cyprus, 11–13 November 2012; pp. 598–601.
21. Cheng, J.X.; Xie, X.S. Coherent Anti-Stokes Raman Scattering Microscopy: Instrumentation, Theory, and Applications. *J. Phys. Chem. B* **2004**, *108*, 827–840. [CrossRef]
22. Mann, C.; Yu, L.; Lo, C.M.; Kim, M. High-resolution quantitative phase-contrast microscopy by digital holography. *Opt. Express* **2005**, *13*, 8693–8698. [CrossRef]
23. Zhang, H.L.; Monroy-Ramirez, F.A.; Lizana, A.; Iemmi, C.; Bennis, N.; Morawiak, P.; Piecek, W.; Campos, J. Wavefront imaging by using an inline holographic microscopy system based on a double-sideband filter. *Opt. Laser. Eng.* **2019**, *113*, 71–76. [CrossRef]
24. Bailleul, J.; Simon, B.; Debailleul, M.; Haeberlé, O. An Introduction to Tomographic Diffractive Microscopy. In *Micro- and Nanophotonic Technologies*; Meyrueis, P., van de Voorde, M., Sakoda, K., Eds.; Wiley-VCH: Weinheim, Germany, 2017; pp. 425–442.
25. Haeberlé, O. Tomographic diffractive microscopy: Principles and applications. In Proceedings of the SPIE Photonics Europe, Strasbourg, France, 24 May 2018.
26. Dändliker, R.; Weiss, K. Reconstruction of the three-dimensional refractive index from scattered waves. *Opt. Commun.* **1970**, *1*, 323–328. [CrossRef]
27. Kawata, S.; Nakamura, O.; Minami, S. Optical microscope tomography I Support constraint. *J. Opt. Soc. Am. A* **1987**, *4*, 292. [CrossRef]
28. Vögeler, M. Reconstruction of the three dimensional refractive index in electromagnetic scattering by using a propagation backpropagation method. *Inverse Probl.* **2003**, *19*, 739–753. [CrossRef]
29. Park, Y.; Diez-Silva, M.; Popescu, G.; Lykotrafitis, G.; Choi, W.; Feld, M.S.; Suresh, S. Refractive index maps and membrane dynamics of human red blood cells parasitized by Plasmodium falciparum. *Proc. Natl. Acad. Sci. USA* **2008**, *105*, 13730–13735. [CrossRef]
30. Fiolka, R.; Wicker, K.; Heintzmann, R.; Stemmer, A. Simplified approach to diffraction tomography in optical microscopy. *Opt. Express* **2009**, *17*, 12407–12417. [CrossRef]
31. Haeberlé, O.; Belkebir, K.; Giovaninni, H.; Sentenac, A. Tomographic diffractive microscopy: Basics, techniques and perspectives. *J. Mod. Optic.* **2010**, *57*, 686–699. [CrossRef]
32. Sentenac, A.; Mertz, J. Unified description of three-dimensional optical diffraction microscopy: From transmission microscopy to optical coherence tomography: Tutorial. *J. Opt. Soc. Am. A Opt. Image Sci. Vis.* **2018**, *35*, 748–754. [CrossRef]
33. Cole, E.S. Conventional Light Microscopy. *Curr. Protoc. Essent. Lab. Tech.* **2016**, *12*, 9.1.1–9.1.29. [CrossRef]
34. Wolf, E. Three-dimensional structure determination of semi-transparent objects from holographic data. *Opt. Commun.* **1969**, *1*, 153–156. [CrossRef]
35. Debailleul, M.; Simon, B.; Georges, V.; Haeberle, O.; Lauer, V. Holographic microscopy and diffractive microtomography of transparent samples. *Meas. Sci. Technol.* **2008**, *19*, 074009. [CrossRef]
36. Maire, G.; Girard, J.; Drsek, F.; Giovannini, H.; Talneau, A.; Belkebir, K.; Chaumet, P.C.; Sentenac, A. Experimental inversion of optical diffraction tomography data with a nonlinear algorithm in the multiple scattering regime. *J. Mod. Optic.* **2010**, *57*, 746–755. [CrossRef]

37. Ruan, Y.; Bon, P.; Mudry, E.; Maire, G.; Chaumet, P.C.; Giovannini, H.; Belkebir, K.; Talneau, A.; Wattellier, B.; Monneret, S.; et al. Tomographic diffractive microscopy with a wavefront sensor. *Opt. Lett.* **2012**, *37*, 1631–1633. [CrossRef] [PubMed]

38. Maire, G.; Ruan, Y.; Zhang, T.; Chaumet, P.C.; Giovannini, H.; Sentenac, D.; Talneau, A.; Belkebir, K.; Sentenac, A. High-resolution tomographic diffractive microscopy in reflection configuration. *J. Opt. Soc. Am. A Opt. Image Sci. Vis.* **2013**, *30*, 2133–2139. [CrossRef] [PubMed]

39. Debailleul, M.; Georges, V.; Simon, B.; Morin, R.; Haeberle, O. High-resolution three-dimensional tomographic diffractive microscopy of transparent inorganic and biological samples. *Opt. Lett.* **2009**, *34*, 79–81. [CrossRef] [PubMed]

40. Simon, B.; Debailleul, M.; Beghin, A.; Tourneur, Y.; Haeberle, O. High-resolution tomographic diffractive microscopy of biological samples. *J. Biophotonics* **2010**, *3*, 462–467. [CrossRef] [PubMed]

41. Simon, B.; Debailleul, M.; Georges, V.; Lauer, V.; Haeberle, O. Tomographic diffractive microscopy of transparent samples. *Eur. Phys. J. Appl. Phys.* **2008**, *44*, 29–35. [CrossRef]

42. Sung, Y.; Choi, W.; Fang-Yen, C.; Badizadegan, K.; Dasari, R.R.; Feld, M.S. Optical diffraction tomography for high resolution live cell imaging. *Opt. Express* **2009**, *17*, 266–277. [CrossRef]

43. Devaney, A.J. Inversion formula for inverse scattering within the Born approximation. *Opt. Lett.* **1982**, *7*, 111–112. [CrossRef]

44. Simon, B.; Debailleul, M.; Houkal, M.; Ecoffet, C.; Bailleul, J.; Lambert, J.; Spangenberg, A.; Liu, H.; Soppera, O.; Haeberle, O. Tomographic diffractive microscopy with isotropic resolution. *Optica* **2017**, *4*, 460–463. [CrossRef]

45. Charriere, F.; Pavillon, N.; Colomb, T.; Depeursinge, C.; Heger, T.J.; Mitchell, E.A.; Marquet, P.; Rappaz, B. Living specimen tomography by digital holographic microscopy: Morphometry of testate amoeba. *Opt. Express* **2006**, *14*, 7005–7013. [CrossRef] [PubMed]

46. Chaumet, P.C.; Sentenac, A.; Rahmani, A. Coupled dipole method for scatterers with large permittivity. *Phys. Rev. E* **2004**, *70*, 036606. [CrossRef] [PubMed]

47. Belkebir, K.; C Chaumet, P.; Sentenac, A. Influence of multiple scattering on three-dimensional imagig with optical diffraction tomography. *J. Opt. Soc. Am. A Opt. Image Sci. Vis.* **2006**, *23*, 586–595. [CrossRef] [PubMed]

48. Charriere, F.; Marian, A.; Montfort, F.; Kuehn, J.; Colomb, T.; Cuche, E.; Marquet, P.; Depeursinge, C. Cell refractive index tomography by digital holographic microscopy. *Opt. Lett.* **2006**, *31*, 178–180. [CrossRef] [PubMed]

49. Girard, J.; Maire, G.; Giovannini, H.; Talneau, A.; Belkebir, K.; Chaumet, P.C.; Sentenac, A. Nanometric resolution using far-field optical tomographic microscopy in the multiple scattering regime. *Phys. Rev. A* **2010**, *82*, 061801. [CrossRef]

50. Chaumet, P.C.; Belkebir, K. Three-dimensional reconstruction from real data using a conjugate gradient-coupled dipole method. *Inverse Probl.* **2009**, *25*, 024003. [CrossRef]

51. Mudry, E.; Chaumet, P.C.; Belkebir, K.; Sentenac, A. Electromagnetic wave imaging of three-dimensional targets using a hybrid iterative inversion method. *Inverse Probl.* **2012**, *28*, 065007. [CrossRef]

52. Maire, G.; Drsek, F.; Girard, J.; Giovannini, H.; Talneau, A.; Konan, D.; Belkebir, K.; Chaumet, P.C.; Sentenac, A. Experimental demonstration of quantitative imaging beyond Abbe's limit with optical diffraction tomography. *Phys. Rev. Lett.* **2009**, *102*, 213905. [CrossRef]

53. Zhang, T.; Chaumet, P.C.; Sentenac, A.; Belkebir, K. Improving three-dimensional target reconstruction in the multiple scattering regime using the decomposition of the time-reversal operator. *J. Appl. Phys.* **2016**, *120*, 243101. [CrossRef]

54. Alexandrov, S.A.; Hillman, T.R.; Gutzler, T.; Sampson, D.D. Synthetic aperture fourier holographic optical microscopy. *Phys. Rev. Lett.* **2006**, *97*, 168102. [CrossRef]

55. Kim, M.; Choi, Y.; Fang-Yen, C.; Sung, Y.; Dasari, R.R.; Feld, M.S.; Choi, W. High-speed synthetic aperture microscopy for live cell imaging. *Opt. Lett.* **2011**, *36*, 148–150. [CrossRef] [PubMed]

56. Kim, M.; Choi, Y.; Fang-Yen, C.; Sung, Y.; Kim, K.; Dasari, R.R.; Feld, M.S.; Choi, W. Three-dimensional differential interference contrast microscopy using synthetic aperture imaging. *J. Biomed. Opt.* **2012**, *17*, 026003. [CrossRef] [PubMed]

57. Mudry, E.; Chaumet, P.C.; Belkebir, K.; Maire, G.; Sentenac, A. Mirror-assisted tomographic diffractive microscopy with isotropic resolution. *Opt. Lett.* **2010**, *35*, 1857–1859. [CrossRef] [PubMed]

58. Vertu, S.; Delaunay, J.-J.; Yamada, I.; Haeberlé, O. Diffraction microtomography with sample rotation: Influence of a missing apple core in the recorded frequency space. *Cent. Eur. J. Phys.* **2009**, *7*, 22–31. [CrossRef]

59. Wei-Chen, H.; Jing-Wei, S.; Te-Yu, T.; Kung-Bin, S. Tomographic diffractive microscopy of living cells based on a common-path configuration. *Opt. Lett.* **2014**, *39*, 2210. [CrossRef]

60. Ding, C.; Tan, Z. Improved longitudinal resolution in tomographic diffractive microscopy with an ellipsoidal mirror. *J. Microsc.* **2016**, *262*, 33–39. [CrossRef] [PubMed]

61. Sullivan, A.C.; McLeod, R.R. Tomographic reconstruction of weak, replicated index structures embedded in a volume. *Opt. Express* **2007**, *15*, 14202–14212. [CrossRef] [PubMed]

62. Berndt, F.; Shah, G.; Power, R.M.; Brugues, J.; Huisken, J. Dynamic and non-contact 3D sample rotation for microscopy. *Nat. Commun.* **2018**, *9*, 5025. [CrossRef]

63. Leite, I.T.; Turtaev, S.; Jiang, X.; Siler, M.; Cuschieri, A.; Russell, P.S.; Cizmar, T. Three-dimensional holographic optical manipulation through a high-numerical-aperture soft-glass multimode fibre. *Nat. Photonics* **2018**, *12*, 33. [CrossRef]

64. Kreysing, M.K.; Kiessling, T.; Fritsch, A.; Dietrich, C.; Guck, J.R.; Kas, J.A. The optical cell rotator. *Opt. Express* **2008**, *16*, 16984–16992. [CrossRef]

65. Zhang, H.; Lizana, A.; Van Eeckhout, A.; Turpin, A.; Ramirez, C.; Iemmi, C.; Campos, J. Microparticle Manipulation and Imaging through a Self-Calibrated Liquid Crystal on Silicon Display. *Appl. Sci.* **2018**, *8*, 2310. [CrossRef]

66. Lizana, A.; Zhang, H.; Turpin, A.; Van Eeckhout, A.; Torres-Ruiz, F.A.; Vargas, A.; Ramirez, C.; Pi, F.; Campos, J. Generation of reconfigurable optical traps for microparticles spatial manipulation through dynamic split lens inspired light structures. *Sci. Rep.* **2018**, *8*, 11263. [CrossRef] [PubMed]

67. Vertu, S.; Flugge, J.; Delaunay, J.J.; Haeberle, O. Improved and isotropic resolution in tomographic diffractive microscopy combining sample and illumination rotation. *Cent. Eur. J. Phys.* **2011**, *9*, 969–974. [CrossRef]

68. Sweeney, D.W.; Vest, C.M. Reconstruction of three-dimensional refractive index fields by holographic interferometry. *Appl. Opt.* **1972**, *11*, 205 207. [CrossRef] [PubMed]

69. Yamaguchi, I.; Zhang, T. Phase-shifting digital holography. *Opt. Lett.* **1997**, *22*, 1268–1270. [CrossRef] [PubMed]

70. Awatsuji, Y.; Fujii, A.; Kubota, T.; Matoba, O. Parallel three-step phase-shifting digital holography. *Appl. Opt.* **2006**, *45*, 2995–3002. [CrossRef] [PubMed]

71. Voigt, D.; Ellis, J.D.; Verlaan, A.L.; Bergmans, R.H.; Spronck, J.W.; Schmidt, R.H.M. Toward interferometry for dimensional drift measurements with nanometer uncertainty. *Meas. Sci. Technol.* **2011**, *22*. [CrossRef]

72. Lazar, J.; Číp, O.; Čížek, M.; Hrabina, J.; Buchta, Z. Suppression of Air Refractive Index Variations in High-Resolution Interferometry. *Sensors* **2011**, *11*, 7644. [CrossRef]

73. Massig, J.H. Digital off-axis holography with a synthetic aperture. *Opt. Lett.* **2002**, *27*, 2179–2181. [CrossRef]

74. Herrera Ramírez, J.A.; Garcia-Sucerquia, J. Digital off-axis holography without zero-order diffraction via phase manipulation. *Opt. Commun.* **2007**, *277*, 259–263. [CrossRef]

75. Mico, V.; Zalevsky, Z.; Garcia-Martinez, P.; Garcia, J. Synthetic aperture superresolution with multiple off-axis holograms. *J. Opt. Soc. Am. A Opt. Image Sci. Vis.* **2006**, *23*, 3162–3170. [CrossRef] [PubMed]

76. Primot, J.; Guerineau, N. Extended hartmann test based on the pseudoguiding property of a hartmann mask completed by a phase chessboard. *Appl. Opt.* **2000**, *39*, 5715–5720. [CrossRef] [PubMed]

77. Primot, J. Theoretical description of Shack-Hartmann wave-front sensor. *Opt. Commun.* **2003**, *222*, 81–92. [CrossRef]

78. Zhang, H.L.; Lizana, A.; Iemmi, C.; Monroy-Ramirez, F.A.; Marquez, A.; Moreno, I.; Campos, J. LCoS display phase self-calibration method based on diffractive lens schemes. *Opt. Laser. Eng.* **2018**, *106*, 147–154. [CrossRef]

79. Zhang, H.L.; Lizana, A.; Iemmi, C.; Monroy-Ramirez, F.A.; Marquez, A.; Moreno, I.; Campos, J. Self-addressed diffractive lens schemes for the characterization of LCoS displays. In Proceedings of the Emerging Liquid Crystal Technologies Xiii, San Francisco, CA, USA, 8 February 2018.

80. Primot, J.; Sogno, L. Achromatic three-wave (or more) lateral shearing interferometer. *J. Opt. Soc. Am. A* **1995**, *12*, 2679–2685. [CrossRef]

81. Bon, P.; Maucort, G.; Wattellier, B.; Monneret, S. Quadriwave lateral shearing interferometry for quantitative phase microscopy of living cells. *Opt. Express* **2009**, *17*, 13080–13094. [CrossRef]

82. Zhang, T.; Ruan, Y.; Maire, G.; Sentenac, D.; Talneau, A.; Belkebir, K.; Chaumet, P.C.; Sentenac, A. Full-polarized tomographic diffraction microscopy achieves a resolution about one-fourth of the wavelength. *Phys. Rev. Lett.* **2013**, *111*, 243904. [CrossRef]

83. Godavarthi, C.; Zhang, T.; Maire, G.; Chaumet, P.C.; Giovannini, H.; Talneau, A.; Belkebir, K.; Sentenac, A. Superresolution with full-polarized tomographic diffractive microscopy. *J. Opt. Soc. Am. A Opt. Image Sci. Vis.* **2015**, *32*, 287–292. [CrossRef]

84. Habaza, M.; Gilboa, B.; Roichman, Y.; Shaked, N.T. Tomographic phase microscopy with 180 degrees rotation of live cells in suspension by holographic optical tweezers. *Opt. Lett.* **2015**, *40*, 1881–1884. [CrossRef]

85. Schmidt, R.; Engelhardt, J.; Lang, M. 4Pi Microscopy. *Methods Mol. Biol.* **2013**, *950*, 27–41. [CrossRef]

86. Zhang, T.; Godavarthi, C.; Chaumet, P.C.; Maire, G.; Giovannini, H.; Talneau, A.; Prada, C.; Sentenac, A.; Belkebir, K. Tomographic diffractive microscopy with agile illuminations for imaging targets in a noisy background. *Opt. Lett.* **2015**, *40*, 573–576. [CrossRef] [PubMed]

87. Castaneda, R.; Garcia-Sucerquia, J. Single-shot 3D topography of reflective samples with digital holographic microscopy. *Appl. Opt.* **2018**, *57*, A12–A18. [CrossRef] [PubMed]

88. Nicola, S.D.; Ferraro, P.; Finizio, A.; Grilli, S.; Coppola, G.; Iodice, M.; Natale, P.D.; Chiarini, M. Surface topography of microstructures in lithium niobate by digital holographic microscopy. *Meas. Sci. Technol.* **2004**, *15*, 961–968. [CrossRef]

89. Nanolive Company. Available online: www.nanolive.ch (accessed on 8 September 2019).

90. Tomocube Company. Available online: www.tomocube.com (accessed on 8 September 2019).

91. Pollaro, L.; Equis, S.; Dalla Piazza, B.; Cotte, Y. Stain-free 3D Nanoscopy of Living Cells. *Optik Photonik* **2016**, *11*, 38–42. [CrossRef]

92. Pedrini, G.; Osten, W.; Gusev, M.E. High-speed digital holographic interferometry for vibration measurement. *Appl. Opt.* **2006**, *45*, 3456–3462. [CrossRef] [PubMed]

93. Mahajan, S.; Trivedi, V.; Vora, P.; Chhaniwal, V.; Javidi, B.; Anand, A. Highly stable digital holographic microscope using Sagnac interferometer. *Opt. Lett.* **2015**, *40*, 3743–3746. [CrossRef] [PubMed]

94. Dasgupta, R.; Mohanty, S.K.; Gupta, P.K. Controlled rotation of biological microscopic objects using optical line tweezers. *Biotechnol. Lett.* **2003**, *25*, 1625–1628. [CrossRef] [PubMed]

95. Le Saux, B.; Chalmond, B.; Yu, Y.; Trouve, A.; Renaud, O.; Shorte, S.L. Isotropic high-resolution three-dimensional confocal micro-rotation imaging for non-adherent living cells. *J. Microsc.* **2009**, *233*, 404–416. [CrossRef]

96. Trattner, S.; Feigin, M.; Greenspan, H.; Sochen, N. Validity criterion for the Born approximation convergence in microscopy imaging: Reply to comment. *J. Opt. Soc. Am. A Opt. Image Sci. Vis.* **2011**, *28*, 665–666. [CrossRef]

97. Chen, B.; Stamnes, J.J. Validity of diffraction tomography based on the first born and the first rytov approximations. *Appl. Opt.* **1998**, *37*, 2996–3006. [CrossRef]

98. Maruyama, Y.; Iwata, K.; Nagata, R. Measurement of the refractive index distribution in the interior of a solid object from multi-directional interferograms. *Jpn. J. Appl. Phys.* **1977**, *16*, 1171–1176. [CrossRef]

Article

Characterization of Spatial Light Modulator Based on the Phase in Fourier Domain of the Hologram and Its Applications in Coherent Imaging

Huaying Wang [1], Zhao Dong [1,*], Feng Fan [1], Yunpeng Feng [2], Yuli Lou [3] and Xianan Jiang [4]

[1]   School of Mathematics and Physics, Hebei University of Engineering, Handan 056038, China;
      pbxsyingzi@126.com (H.W.); fanfeng@hebeu.edu.cn (F.F.)
[2]   School of Optics and Photonics, Beijing Institute of Technology, Beijing 100081, China; feng_yp@126.com
[3]   Faculty of Science, Kunming University of Science and Technology, Kunming 650500, China;
      louyuli1@163.com
[4]   Hebei Boxia Photoelectric Information Technology Co., Ltd., Handan 056000, China;
      tianshixingchen@126.com
*   Correspondence: handandong@163.com; Tel.: +86-1551-105-1296

Received: 1 June 2018; Accepted: 11 July 2018; Published: 14 July 2018

Featured Application: This work can be used for fast characterizing the phase modulation of the spatial light modulator (SLM) and shows the applications in coherent imaging by using the SLM.

Abstract: Although digital holography is used widely at present, the information contained in the digital hologram is still underutilized. For example, the phase values of the Fourier spectra of the hologram are seldom used directly. In this paper, we take full advantage of them for characterizing the phase modulation of a spatial light modulator (SLM). Incident plane light beam is divided into two beams, one of which passes the SLM and interferes with the other one. If an image with a single grey scale loads on the SLM, theoretical analysis proves that the phase of the Fourier spectra of the obtained hologram contains the added phase and a constant part relative to the optical distance. By subtracting the phase for the image with the grey scale of 0 from that for the image with other grey scales, the phase modulation can be characterized. Simulative and experimental results validate that the method is effective. The SLM after characterization is successfully used for coherent imaging, which reconfirms that this method is exact in practice. When compared to the traditional method, the new method is much faster and more convenient.

Keywords: spatial light modulator; digital holography; phase modulation; Fourier spectra; fast characterization; coherent imaging

---

1. Introduction

Digital holography (DH) is a technology that permits the numerical reconstruction of optical wave fields in both amplitude and phase from a digitally-recorded hologram [1–7]. By processing the information contained in a hologram digitally, this technology has been widely used in many fields, including particle field measurements [1,2], structure testing [3,4], and quantitative analysis of cells [5–7]. However, the digital hologram is still not utilized sufficiently. For example, the phase of the Fourier spectra is often processed together with the intensity to get final reconstructed images, and it can also be used directly in some applications without reconstruction. Dohet-Eraly et al. used the phase in the Fourier domain to achieve numerical autofocusing in DH [8]. However, to the best of our knowledge, direct usages of the phase value still need further exploitation. In this paper, we utilize the phase directly for fast characterization of the phase modulation of a spatial light modulator (SLM).

It is well known that, based on the liquid-crystal birefractive property, SLM has been widely applied in many disciplines, such as optical information processing [9], holographic and/or color display [10,11], pattern recognition [12], and vector beams [13–15]. The modulation is usually achieved by displaying a gray value map to an SLM. Different gray scale values correspond to the different directions of the liquid-crystal optical axis. As the optical axis is changed, the phase and/or intensity of the light beam passing through the SLM is modulated [16]. However, as the birefractive property can be affected by the wavelength of the incident light and the environment, it is necessary to characterize the phase and/or intensity modulation, i.e., to get the relationships between the changes of phase and/or intensity and the gray values [17,18]. The intensity modulation can be easily characterized by an optical power meter, whereas the phase modulation may not be acquired directly. Usually, the phase modulation is characterized using interferometry, such as the Mach-Zehnder [18,19], the Michelson [20], or even the simple double-beam interference [21]. Plane wave is always used as the incident beam; it is divided into two beams, one of which passes through the SLM (named as object wave) and interferes with the other one (named as reference wave) to form the patterns. Usually, two-part gray-scale maps are loaded on the SLM in the phase measurement. The gray scale in one part of the maps keeps the same (usually it is zero), and that in another part varies during the measurement. As the gray value changes, the phase added to the object wave is also changed, which bring on the relative shifts in interference patterns. Then, the phase variation can be obtained by measuring the shift and comparing it with the period of the fringes [16–18,20,21]. This traditional method is convenient, but sometimes it is time consuming to compare the intensity of the patterns to determine the real shifts. It may not be suitable for fast characterization of the phase modulation. Martín-Badosa et al. performed a one-dimensional correlation product to determine fringe displacement, as well as period [19]. This method is more convenient than the traditional method, but the procedure still seems a slightly complicated. Zhang et al. acquired the relationship of the phase shift and the gray value by writing a Ronchi grating pattern for the SLM, but the accuracy was limited by the quantization levels of the image board [22]. Recently, ptychography has been used to characterize SLM with excellent phase accuracy; however, owing to the mechanical shift in the experiment and the iterative algorithm used in the data process, this method is not suitable for fast characterization [23]. Digital holography is also applied for characterization [24]; however, a numerical autofocus process should be implemented, and the tilt distortion may also be considered in the traditional treatment for the hologram. Here, we utilize the phase in the Fourier domain to characterize the phase modulation of SLM. When compared to the traditional method, the new method is much faster and more convenient. SLM, after characterization with this method, is also successfully used for coherent imaging, confirming that the characterization is exact.

2. Analysis and Methods

Our method is based on digital holography. Figure 1 shows a sketch of the phase measurement setup. Transmissive SLM is used here as an example. Plane waves are used as the light beam. The wavefunction of the beam received at the monitor plane can be written as:

$$U = U_O + U_R \tag{1}$$

Here, the subscripts O and R represent the object and reference beam, respectively. The intensities of the two beams are adjusted to be the same and set to 1 in the analyses. The intensity of the obtained interferogram can be written as:

$$I = (U_O + U_R)(U_O^* + U_R^*) = |U_O|^2 + |U_R|^2 + U_O U_R^* + U_R U_O^* \tag{2}$$

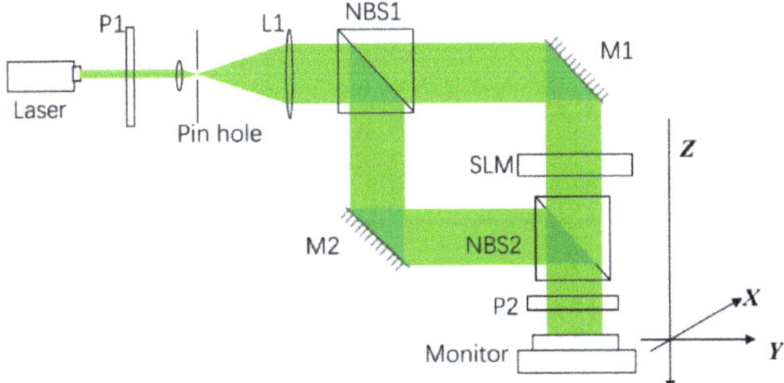

Figure 1. Sketch of the phase measurement setup. L1: collimation lens; NBS: nonpolarized beam splitter; P: Polarizer; M: Mirror.

The interferogram can be considered as the hologram of SLM after diffraction. By applying Fourier transforms to the hologram, there will be three spots in the Fourier domain, namely the 0th, +1st and −1st order spots. They correspond to $F\{|U_O|^2 + |U_R|^2\}$, $F\{U_O U_R^*\}$ and $F\{U_R U_O^*\}$, respectively. $F\{\}$ is the operator of the Fourier transform.

Set (α, β, γ) as the direction cosine of the beams, Z as the axis perpendicular to the monitor plane, X and Y as the axes parallel to it, (x, y, z) as the coordinates in the X-Y-Z coordinate system, and θ as the phase added by SLM in the object beam, then:

$$U_O = F^{-1}\left\{ F\{W \exp(j\theta)\} \exp\left(jkz\sqrt{1 - (\lambda f_x)^2 - (\lambda f_y)^2} \right) \right\} \tag{3}$$

We chose $z = 0$ as the position of the SLM; and the modulation of SLM is considered to be in the pure phase mode. W is the transmitting zone of SLM. It can be treated as a two-dimensional (2D) rectangular function, i.e.,

$$W = \mathrm{rect}\left(\frac{x}{a}, \frac{y}{b}\right) = \mathrm{rect}\left(\frac{x}{a}\right)\mathrm{rect}\left(\frac{y}{b}\right) \tag{4}$$

Here, a and b are the size of the width and length of the transmitting zone. The wave function of U_R can be expressed as:

$$U_R = \exp\left[j\vec{k}_R \cdot \vec{R} \right] = \exp[jk(\alpha_R x + \beta_R y + \gamma_R z)] \tag{5}$$

where $\vec{k} = k\left(\alpha \vec{e}_x + \beta \vec{e}_y + \gamma \vec{e}_z \right)$, and $\vec{R} = x\vec{e}_x + y\vec{e}_y + z\vec{e}_z$. Here, $\vec{k}$ is the wave vector of the beam. $\vec{e}_x$, $\vec{e}_y$ and $\vec{e}_z$ denote the unit vector of the axis X, Y, and Z, respectively. The +1 order in the Fourier domain can then be written as:

$$F\{U_O U_R^*\} = F\left\{ F^{-1}\left\{ F\{W \exp(j\theta)\} \exp\left(jkz\sqrt{1 - (\lambda f_x)^2 - (\lambda f_y)^2} \right) \right\} \times \exp[-jk(\alpha_R x + \beta_R y + \gamma_R z)] \right\} \tag{6}$$

According to the properties of the Fourier transform [25] and considering that θ and z are constant, Equation (6) can be calculated as:

$$\begin{aligned}
F\{U_O U_R^*\} = \;& |ab|\mathrm{sinc}\left(a\left(f_x + \tfrac{\alpha_R}{\lambda}\right)\right)\mathrm{sinc}\left(b\left(f_y + \tfrac{\alpha_R}{\lambda}\right)\right) \times \\
& \exp\left(jkz\sqrt{1 - \lambda^2\left(f_x + \tfrac{\alpha_R}{\lambda}\right)^2 - \lambda^2\left(f_x + \tfrac{\beta_R}{\lambda}\right)^2} \right) \exp(j\theta)\exp(-jk\gamma_R z)
\end{aligned} \tag{7}$$

The maximum value of the function sinc(x) is at $x = 0$ [25]. If we choose the origin as the center of the Fourier domain of the hologram, the coordinate of the point with the highest intensity in the +1st order spot is $\left(-\frac{\alpha_R}{\lambda}, -\frac{\beta_R}{\lambda}\right)$. It is independent of θ, and the phase of the point can then be written as:

$$\varphi = (1 - \gamma_R)kz + \theta \tag{8}$$

Usually, θ is set to be zero when the gray scale of the image loading on the SLM is zero. As $(1 - \gamma_R)kz$ is the same for all the gray values, the variation θ can then be obtained by subtracting φ_0. The subscript 0 means that the gray scale of the image loading on the SLM is zero. The phase modulation is then characterized.

In practice, the experimental setup is sometimes affected by the airflow in the environment. Then, optical path z may vary. In most situations, the direction of the beam is unchanged as the optical devices are fixed during measurement. This means that the positions of the $\pm$1st order of the spectra continues to be the same, but their phases change. An unknown phase is added to the objective beam. Considering the cross of the beam is very small, the unknown phase is thought to be the same for every part of the beam. Then, in the practice, we can get the relationship of the phase added by SLM and the gray scale of the image loading on it using the following steps:

- Step 1: Divide the image loading on the SLM into two parts. One part (named the reference part) has a constant gray scale of 0. The other one (named the variable part) has a gray scale varying from 0 to 255 (for the 8-bit-gray-scale SLM) during the experiment.
- Step 2: There are then two sets of interfered patterns in the holograms accordingly. An area with the same size is selected from each for Fourier transform. In the practice, the areas are far from the borders of the two patterns to avoid the influence of the diffraction in the border.
- Step 3: Subtract phase φ_r of the point with the highest intensity in +1st order spot for the reference part from φ_v for the variable part. Then difference $\Delta\varphi = \varphi_v - \varphi_r$ is phase variation θ.

Sometimes, the monitor may be tilted to the light beam. In this condition, the optical path z is a function of x and y. It means that $\Delta\varphi$ contains the phase difference related to Δz, which is the optical path difference between the two selected parts. As Δz is constant during the experiment, its influence will be removed by subtracting $\Delta\varphi_0$. Here, subscript 0 represents that the gray scale of the reference part being zero.

3. Results

3.1. Simulation Results

A simulation was applied to test the method shown above. The wave function of the reference beam was $U_R = \exp\{j[k(\beta y + \gamma z) + \theta_{un}]\}$. Here, θ_{un} was set randomly to simulate the unknown phase from the vibration of the environment. For simplicity, the phase was added to the reference beam only. As the reference wave was tilted to the monitor, we supposed that optical distance z_R varied from 14 to 16 mm along the x axis at the monitor plane. β was set to be 0.1 and $\gamma = \sqrt{(1 - \beta^2)} = 0.9950$. A 512 pixel $\times$ 512 pixel matrix was used to simulate the transmitting area of SLM. Its length and width were set to 9.216 mm, and the wavelength was set to 532 nm.

The object beam passes through the SLM, where the gray scale in the variable part varied from 0 to 255, and the corresponding phase shift θ_{set} was set as a variable value from 0 to $\frac{255}{256} \times 2\pi$ in the step of $\frac{\pi}{128}$. The optical distance z_o from SLM to the monitor plane was set to be 10 mm. The wavefunction U_O on the monitor plane was calculated using the angular spectrum theory. The processes were implemented in the MATLAB 2017a environment with an Intel Core i7-6500U CPU at 2.50 GHz and 8 Gb RAM.

The interferograms with different added phases of θ were then obtained. A typical one is shown in Figure 2a. θ was calculated following the steps shown in Section 2. The areas surrounded by the red lines were selected for fast Fourier transform (FFT). Their positions were fixed for different interferograms. Figure 2b shows a typical spectra pattern for the variable part; it was like that for the reference part as well. The central point in the +1 order with the highest intensity was chosen to acquire its phase. As discussed above, its position was fixed for all spectra patterns. The calculated phase shift θ_{cal} matches the set phase variation θ_{set} very well, as shown in Figure 2c.: The relative error was introduced to quantitatively evaluate the characterization. It was defined as:

$$E_r = \begin{cases} |\theta_{cal} - \theta_{set}|/\theta_{set}\,(\theta_{set} \neq 0) \\ |\theta_{cal} - \theta_{set}|\,(\theta_{set} = 0) \end{cases} \tag{9}$$

From Figure 2d, for all the θ_{cal}, the relative errors are less than 0.01%. The results show that the new method is very exact.

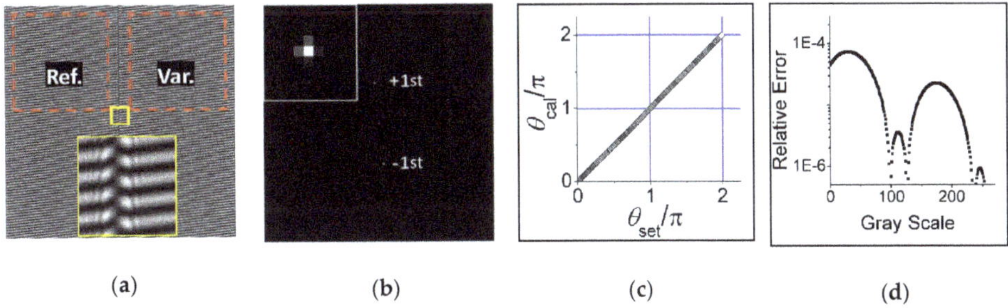

(a)   (b)   (c)   (d)

Figure 2. (**a**) The simulated interferogram with θ_{set} of π. Areas around the red line were selected for Fourier transform; the inset is the enlarged patterns surrounded by yellow lines. (**b**) the frequency spectrum of the selected variable areas; the inset is the enlarged image of the +1st order; (**c**) the relationship of θ_{set} and θ_{cal}; (**d**) the relationship of the relative error and the gray scales.

3.2. Experimental Methods and Results

This method was applied to characterize a transmissive SLM (GCS-SLM-T2, Beijing Daheng Corp, China) in the experiments. The size of the pixel of the SLM was 18 µm × 18 µm. The experimental setup can also be seen in Figure 1. A laser with a wavelength of 532 nm was used here. Polarizers P1 and P2 were adjusted to make sure that the SLM worked in the pure phase mode. Two-part gray-scale images were displayed on the SLM during the experiment, just like in the simulation. Figure 3a is a typical image, and the gray scale in its variable part is 100. The corresponding hologram is shown in Figure 3b. There is a shift in the fringes between the reference and the variable parts. Areas with the same size were selected in both parts for FFT, as mentioned in Section 2. Figure 3c shows the spectra of the selected area in the variable part after FFT. The point with the highest intensity in the +1st order of the spectra was chosen to get phase φ_v. Its position continued to be the same for all the gray scales. A similar process was also implemented on the reference part to get φ_r. Following the same steps as those described in Section 2, the relationships between the gray scale and the added phase could be obtained, as shown in Figure 3d.

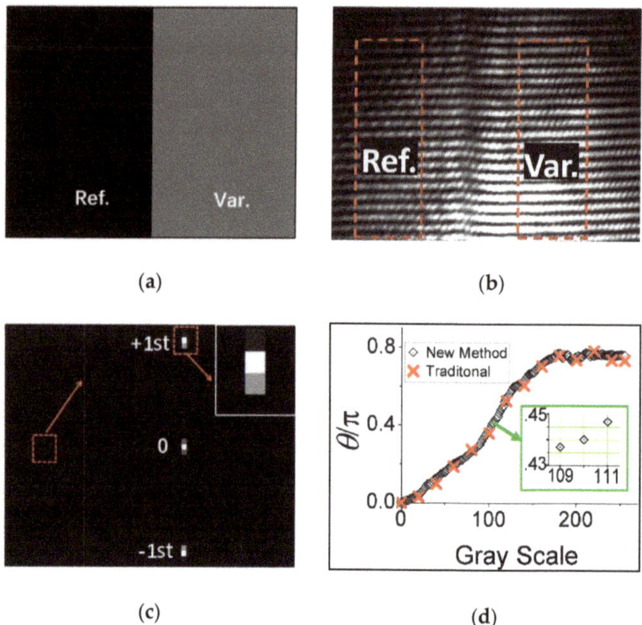

Figure 3. (a) Sketch of a typical image displayed on the spatial light modulator (SLM) with two equal parts as the reference and the variable; the gray scale in the variable part is 100; (b) the related interferogram received by the monitor; (c) the frequency spectrum of the variable part around by the red dashes in (b), left: the whole patterns; right: the enlarged image of the part around by the dashes in the left; inset: The enlarged image of +1st order; (d) the relationship between the phase variation and the gray scale of the image loaded on the SLM.

From Figure 3d, the phases could be adjusted nonlinearly by the SLM. The adjustable range was between 0 and about 0.8π. If the gray scale is below 20, the phase shift increases slowly with the increase in the gray scale value, while it is maintained almost constant with a value of about 0.8π if the gray scale value is beyond 180. For gray scale values between 80 and 130, the phase shift increases nearly linearly with the increasing gray scale value.

The traditional method like that in Reference [18] was also used to verify the new method. Figure 4 presents the main process. At first, the interference patterns were rotated parallel to one edge of the image, as closely as possible. The x axis was set to be parallel to this edge. Taking Figure 3b as an example, by properly cutting and slightly rotating it, Figure 4a could be obtained. It seems that the two figures had the same line orientation, because the angle of the rotation was quite small (about 0.05 rad). A dashed line parallel to the y axis was chosen in each part of the image (see in Figure 4a). Here, the y axis is perpendicular to the x axis. The relationships between the intensities along the lines and the y coordinates are shown in Figure 4b. Both curves varied periodically and there was a relative shift between them. By measuring the distance of the adjacent peaks or troughs, the period could be obtained. Then it was used to compare the relative shift between the two curves to get the phase shift. Here, as shown in Figure 4b, the period was about 45 pixels, and the relative shift was about 8.1 pixels. The phase shift here could then be calculated as $\frac{8.1}{45} \times 2\pi$. This means that the relative added phase was about 0.36π when the gray scale was 100. Repeating the same process, the relationships between the gray scale and the added phase could also be obtained. The results are also referred to in Figure 3d. They matched the results obtained by the new method very well, indicating that the new method was effective. The processing time of the two methods were different. For the new method, the total time consumed was about 1.694 s for characterizing all 256 gray scales in the MATLAB 2017a environment with an Intel Core i7-6500U CPU at 2.50 GHz and 8 Gb RAM. However, for the traditional method, as the comparisons of the shifts were implemented manually to avoid the influence of noises around the peaks or troughs in the curves, the processing time was long; about 30 min for about 12 gray scales. With a simpler procedure, the new method can be implemented more conveniently than the traditional one. As all the steps can be carried out automatically; it can be used to quickly characterize SLM.

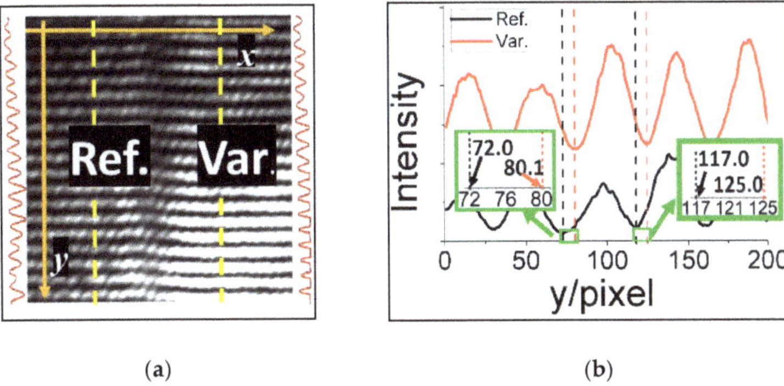

(a)   (b)

Figure 4. Sketch of the process of the traditional method. (**a**) Figure 3b after being rotated slightly and properly cut; the curves beside the pattern represent the intensity along the dashed lines at the left and right parts, respectively; (**b**) comparison of the troughs of the light intensities of the dashed lines.

3.3. Applications in Coherent Imaging

The characterization with this new method can also be tested using the SLM in coherent imaging. Zhang et al. showed that arbitrary complex-valued fields could be retrieved through aperture-plane modulation [26,27]. In Reference [26], a phase plate was used as the wave-front modulator. By shifting the plate, the authors recorded a set of diffractograms with different modulations and used them to retrieve the complex-valued fields by using an iterative retrieval technique.

Here, we used SLM as the modulator, and different modulations were produced by loading different images on the SLM. The experimental setup is referred to in Reference [26]. For simplicity, only a pure modulus-type specimen was used in the experiment as a prototype. A map with a transparent character J surrounded by a translucent background was treated as the specimen, and its size was about 3 × 3 mm (Figure 5a). The specimen is fixed on a glass by the lighttight taps. The monitor used here was a COMS camera (MER-131-210U3M, Beijing Daheng Corp) with a pixel size of 4.8 μm × 4.8 μm. In the experiment, the distance between the SLM and specimen and that between the specimen and the digital receiver were about 7 and 207 mm, respectively.

Because the phase modulation of the SLM may be affected by the environment, the phase added might have had a slight deviation from what we expected. To reduce the effect, we use random grid-like masks (Figure 5b). Only two gray scales are in the masks; one was 0 and the other was 110, corresponding to 0.440π according to our characterization. We took six different masks to adjust the diffractograms. A typical diffractogram can be seen in Figure 5c. The specimen is then retrieved using the iterative method, the details of which are referred to in Reference [26]. This process spends about 6 s for about 100 iterations. Figure 5d shows the retrieved modulus of the specimen. From the figure, not only character J, but also the background with poor transmittance and the lighttight area around the specimen were retrieved. This confirms that the characterization was available in practice.

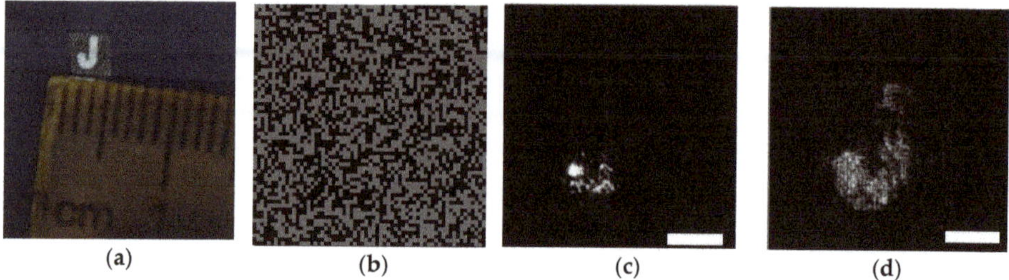

Figure 5. (**a**) A photograph of the specimen and the lighttight blue taps around it. (**b**) a typical grid-like mask sent to the SLM; (**c**) a diffractogram received by the camera; (**d**) the retrieved modulus of the specimen. The scale bar here is 1 mm.

It is possible that a value of θ in a range around 0.44π can lead to successful retrieval. During the retrieval process, we set different values of θ to investigate how they affected the retrieval. One-hundred iterations were taken for all the θ. Figure 6 shows part of the results. When θ is set to be 0.439π, it seems that the modulus is retrieved with little loss of the quality (Figure 6a). The specimen cannot be retrieved when θ equals 0.437π or 0.443π, as shown in Figure 6b,c. To quantitatively evaluate the retrieval, we introduced parameter E_I. It is defined as:

$$E_I = \frac{\sum\limits_{i} \sqrt{\Sigma(|I_{cal,i}| - |I_{exp,i}|)^2}}{N_{mask} N_{pixel}} \tag{10}$$

Here, $I_{cal,i}$ and $I_{exp,i}$ are the intensities of the calculated and the experimental diffractograms for the ith mask, respectively. N_{mask} and N_{pixel} are the total number of masks and the total pixels in a diffractogram, respectively. Considering that a pure amplitude sample was used in the experiment, $I_{cal,i}$ was calculated by the retrieved modulus instead of the complex amplitude. Obviously, a smaller E_I means a more accurate retrieval. The relationship of the E_I and θ values can be seen in Figure 6d. For Figure 6a–c, the E_I values are 0.0117, 0.0133 and 0.0245, respectively. As the θ value is farther away from 0.440π, the corresponding E_I should be larger. From the above results, it can be determined that, in this experiment, the retrieval can tolerate an error of θ less than 0.006π. From our characterization (see in the inset of Figure 3d), θ is 0.437π when the gray scale of the image loaded on the SLM is 109, and 0.447π when it is 111. They both cannot lead to a successful retrieval. This indicates that by direct use of the phase value in the Fourier domain, the phase modulation of the SLM can be precisely characterized in the practice.

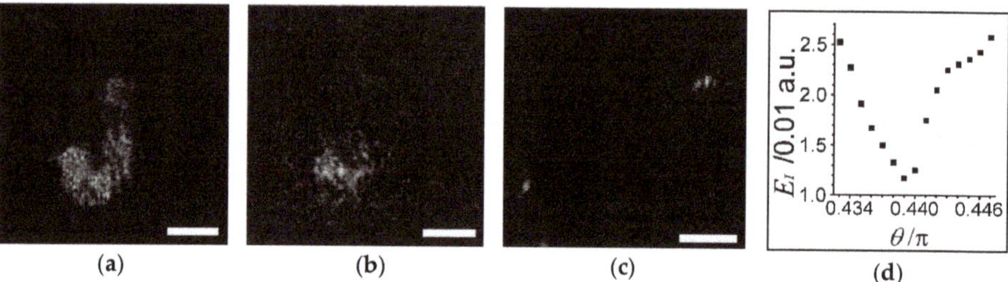

Figure 6. Results of the retrieved modulus of the specimen for different θ values. (**a**) $\theta = 0.439\pi$; (**b**) $\theta = 0.437\pi$; (**c**) $\theta = 0.443\pi$; (**d**) the relationship between E_I and θ. The scale bar is 1 mm.

4. Discussion and Conclusions

Simulative and experimental results can prove that the characterization of phase modulation of the SLM, based on the phase in the Fourier domain of the hologram, was effective. This means that, for holography, not only the reconstructed wave field from a hologram, but also the phase in the Fourier domain can be directly utilized in practice. Some processes, such as digital refocusing and/or phase compensation, are not needed in the new method and it will be faster than characterization based on the traditional treatment of a hologram [24]. In this method, all steps can be carried out automatically; it can be implemented more conveniently than the traditional method. The SLM characterized by the new method has also been applied for coherent imaging. The results show that, with characterization, the specimen was reconstructed successfully. This indicates that this characterizing method for phase modulation can be precise in practice.

Although the information of the specimen can be successfully retrieved in the coherent imaging system with SLM, the robustness of the system is not very strong. The tolerance of the error of the phases added by the SLM is very small in this system. If the environment varies, the phase modulation may need to be recharacterized. Moreover, variations can also bring pixel misalignments into the experimental setup and lead to a failure in retrieving the information. If the SLM is used effectively in coherent imaging, unexpected relative lateral shifts between it and the monitor, as well as the specimen, must be avoided, and the environment should be kept stable.

5. Patents

Related patents are showed as follows:

A ptychographic imaging setup (CN 201720808955.9; CN 201710544508.1).

A holographic imaging technology based on phase ptychography (CN 201610028448.3).

Author Contributions: Conceptualization, Z.D. and H.W.; Methodology, Z.D.; Software, Y.F. and X.J.; Validation, F.F. and Y.L.; Formal Analysis, Z.D. and Y.F.; Investigation, Z.D., H.W. and F.F.; Resources, H.W. and Y.L.; Data Curation, H.W., Z.D. and Y.F.; Writing-Original Draft Preparation, Z.D.; Writing-Review and Editing, H.W. and Y.L.; Project Administration, H.W.; Funding Acquisition, Z.D., F.F., H.W. and Y.L.

Funding: This work was supported by the Science and Technology Plan Project of Hebei Province (16273901D, 15275602D); the National Natural Science Foundation of China (61465005, 11504078); the International Science and Technology Cooperation (2014DFE00200); and the Natural Science Foundation of Hebei Province (F2018402285).

Acknowledgments: The authors are grateful to the anonymous reviewers for their constructive suggestions. We thank Xiufa Song and Sixing Xi for their help in the experiment, and we thank Eddie Wang for his help in the English writing.

Conflicts of Interest: The authors declare no conflicts of interest.

References

1. Ren, Z.; Chen, N.; Lam, E.Y. Extended focused imaging and depth map reconstruction in optical scanning holography. *Appl. Opt.* **2016**, *55*, 1040–1047. [CrossRef] [PubMed]
2. Qu, X.; Song, Y.; Jin, Y.; Li, Z.; Wang, X.; Guo, Z.; Ji, Y.; He, A. 3D SAPIV particle field reconstruction method based on adaptive threshold. *Appl. Opt.* **2018**, *57*, 1622–1633. [CrossRef] [PubMed]
3. Pourvais, Y.; Asgari, P.; Abdollahi, P.; Khamedi, R.; Moradi, A.-R. Microstructural surface characterization of stainless and plain carbon steel using digital holographic microscopy. *J. Opt. Soc. Am. B* **2017**, *34*. [CrossRef]
4. Wang, H.; Dong, Z.; Wang, S.; Lou, Y. Comparison of the refocus criteria for the phase, amplitude, and mixed objects in digital holography. *Opt. Eng.* **2018**, *57*, 054111. [CrossRef]
5. McReynolds, N.; Cooke, F.G.M.; Chen, M.; Powis, S.J.; Dholakia, K. Multimodal discrimination of immune cells using a combination of Raman spectroscopy and digital holographic microscopy. *Sci. Rep.* **2017**, *7*, 43631. [CrossRef] [PubMed]
6. Yi, F.; Moon, I.; Javidi, B. Automate red blood cells extraction from holographic images using fully convolutional neural networks. *Biomed. Opt. Express* **2017**, *8*, 4466–4479. [CrossRef] [PubMed]

7.	Merola, F.; Memmolo, P.; Miccio, L.; Savoia, R.; Mugnano, M.; Fontana, A.; D'Ippolito, G.; Sardo, A.; Iolascon, A.; Gambale, A.; et al. Tomographic flow cytometry by digital holography. *Light Sci. Appl.* **2017**, *6*, e16241. [CrossRef]

8.	Dohet-Eraly, J.; Yourassowsky, C.; Dubois, F. Fast numerical autofocus of multispectral complex fields in digital holographic microscopy with a criterion based on the phase in the Fourier domain. *Opt. Lett.* **2016**, *41*, 4071–4074. [CrossRef] [PubMed]

9.	Wang, C.; Su, Y.; Wang, J.; Zhang, C.; Zhang, Z.; Li, J. Method for holographic femtosecond laser parallel processing using digital blazed grating and the divergent spherical wave. *Opt. Eng.* **2015**, *54*, 016109. [CrossRef]

10.	Collings, N.; Christmas, J.L.; Masiyano, D.; Crossland, W.A. Real-time phase-only spatial light modulator for 2D holographic display. *J. Disp. Technol.* **2015**, *11*, 278–284. [CrossRef]

11.	Harm, W.; Jesacher, A.; Thalhammer, G.; Bernet, S.; Ritsch-Marte, M. How to use a phase-only spatial light modulator as a color display. *Opt. Lett.* **2015**, *40*, 581–584. [CrossRef] [PubMed]

12.	Xu, P.; Hong, C.; Sun, Z.; Han, F.; Cheng, G. Integrated zigzag Vander Lugt correlators incorporating an optimal trade-off synthetic discriminant filter for invariant pattern recognition. *Opt. Commun.* **2014**, *315*, 97–102. [CrossRef]

13.	Zhan, Q. Cylindrical vector beams: From mathematical concepts to applications. *Adv. Opt. Photonics* **2009**, *1*, 1–57. [CrossRef]

14.	Maluenda, D.; Juvells, I.; Martinez-Herrero, R.; Carnicer, A. Reconfigurable beams with arbitrary polarization and shape distributions at a given plane. *Opt. Express* **2013**, *21*, 5424–5431. [CrossRef] [PubMed]

15.	Maluenda, D.; Carnicer, A.; Martinez-Herrero, R.; Juvells, I.; Javidi, B. Optical encryption using photon-counting polarimetric imaging. *Opt. Express* **2015**, *23*, 655–666. [CrossRef] [PubMed]

16.	Bergeron, A.; Gauvin, J.; Gagnon, F.; Gingras, D.; Arsenault, H.H.; Doucet, M. Phase calibration and applications of a liquid-crystal spatial light modulator. *Appl. Opt.* **1995**, *34*, 5133–5139. [CrossRef] [PubMed]

17.	Dou, R.; Giles, M.K. Simple technique for measuring the phase property of a twisted nematic liquid crystal television. *Opt. Eng.* **1996**, *35*, 808–812. [CrossRef]

18.	Reichelt, S. Spatially resolved phase-response calibration of liquid-crystal-based spatial light modulators. *Appl. Opt.* **2013**, *52*, 2610–2618. [CrossRef] [PubMed]

19.	Matín-Badosa, E.; Carnier, A.; Juvells, I.; Vallmitjana, S. Complex modulation characterization of ligquid crystal devices by interferometric data correlation. *Meas. Sci. Technol.* **1997**, *8*, 764–772. [CrossRef]

20.	Otón, J.; Ambs, P.; Millán, M.S.; Pérez-Cabré, E. Multipoint phase calibration for improved compensation of inherent wavefront distortion in parallel aligned liquid crystal on silicon displays. *Appl. Opt.* **2007**, *46*, 5667–5679. [CrossRef]

21.	Bondareva, A.P.; Cheremkhin, P.A.; Evtikhiev, N.N.; Krasnov, V.V.; Starikov, R.S.; Starikov, S.N. Measurement of characteristics and phase modulation accuracy increase of LC SLM "HoloEye PLUTO VIS". *J. Phys. Conf. Ser.* **2014**, *536*, 012011. [CrossRef]

22.	Zhang, Z.; Lu, G.; Yu, F.T.S. Simple method for measuring phase modulation in liquid crystal televisions. *Opt. Eng.* **1994**, *33*, 3018–3022. [CrossRef]

23.	McDermott, S.; Li, P.; Williams, G.; Maiden, A. Characterizing a spatial light modulator using ptychography. *Opt. Lett.* **2017**, *42*, 371–374. [CrossRef] [PubMed]

24.	Dev, K.; Singh, V.R.; Asundi, A. Full-field phase modulation characterization of liquid-crystal spatial light modulator using digital holography. *Appl. Opt.* **2015**, *50*, 1593–1600. [CrossRef] [PubMed]

25.	Lv, N. *Fourier Optics*, 2nd ed.; China Machine Press: Beijing, China, 2006; pp. 31–39, ISBN 978-7-111-18480-5.

26.	Zhang, F.; Pedrini, G.; Osten, W. Phase retrieval of arbitrary complex-valued fields through aperture-plane modulation. *Phys. Rev. A* **2007**, *75*, 043805. [CrossRef]

27.	Kohler, C.; Zhang, F.; Osten, W. Characterization of a spatial light modulator and its application in phase retrieval. *Appl. Opt.* **2009**, *48*, 4003–4008. [CrossRef] [PubMed]

 *applied*
sciences

Article

Static Structures in Leaky Mode Waveguides †

Daniel Pettingill, Daniel Kurtz and Daniel Smalley *

Department of Electrical and Computer Engineering, Brigham Young University, Provo, UT 84602, USA;
dptres16@gmail.com (D.P.); kurtz117@yahoo.com (D.K.)
* Correspondence: smalley@byu.edu; Tel.: +1-1-801-422-4343
† It is an invited paper for the special issue.

Received: 20 December 2018; Accepted: 7 January 2019; Published: 11 January 2019

Featured Application: The overarching contextual objective of this research is to create transparent, near-eye devices for holographic video display.

Abstract: In this work, we suggest a new method of expanding the field of view in bottom-exit, leaky mode devices for transparent, monolithic, holographic, near-eye display. In this approach, we propose the use of static, laser-induced, grating structures within the device substrate to break the leaky mode light into diffracted orders. We then propose to use carefully timed illumination pulses to select which diffracted order is visible to the eye at every display refresh interval (up to 100 kHz). Each of these orders becomes a view for a different image point. To describe this new method, we use K-vector analysis. We give the relevant equations and a list of parameters which lead to a near-eye geometry with little or no overlap in higher-order view zones. We conclude that it should be possible to increase the field of view of our bottom-exit, leaky mode devices by as much as one order of magnitude by simply adding a laser-induced grating structure to the substrate and by carefully timing the device illumination. If successful, this method would make possible a transparent, holographic, near eye display that is simple to fabricate, relative to pixelated approaches, and which has a wide field-of-view relative to our current bottom-exit displays.

Keywords: near-eye display; leaky mode; lithium niobate; holographic video; augmented reality; 3D display; acousto-optic modulator; laser-induced structures

---

1. Introduction

This study is important to the use of leaky mode modulators as near-eye displays because it seeks to increase the device's field of view without making the fabrication of the device significantly more complex. Currently, near-eye displays are almost exclusively driven by pixelated spatial light modulators [1]. However, leaky mode modulators have a number advantages over pixelated spatial light modulators (SLMs) for the purposes of near-eye holographic display. They require only two mask steps to fabricate. They are transparent. They produce no zero order, no higher orders, and no conjugate images. They do not suffer from quantization error from pixilation, and they require no backplane. Additionally, they have the ability to rotate the polarization of the diffracted signal light for easy noise filtering. Finally, they can also multiplex color in the frequency domain, allowing all colors to be modulated in the same waveguide. However, these advantages notwithstanding, bottom-exit devices (the configuration most useful for near-eye display) suffer from a reduced field of view. In side-exit leaky mode devices, we observe a large multiplication of diffracted light angle because the device is operating in a regime where the grating equation is nonlinear [2,3]. In a bottom-exit device, leaky mode light passes through a high-spatial-frequency output grating that operates in opposition to the leaky mode diffraction [4]. This second interaction effectively erases the angle-multiplication advantage gained from illuminating the acoustic holographic pattern at a glancing angle. The result of the reduction is that the leaky mode device effectively has a smaller achievable field of view (defined in this work as the visible extent at a given depth) or view zone (defined here as the angular extent over which a point is visible). Note that these concepts are duals in this context and will be treated interchangeably in the text. Our previous work identified the narrowing of view angle in bottom-exit devices as a challenge and stated that a solution would be described in a future publication [4]. The main aim of this work is to present that solution to restore or increase the leaky mode device view zone/field of view (see Figure 1) and/or view zone by temporally stitching together multiple diffracted orders of light created by a static internal grating.

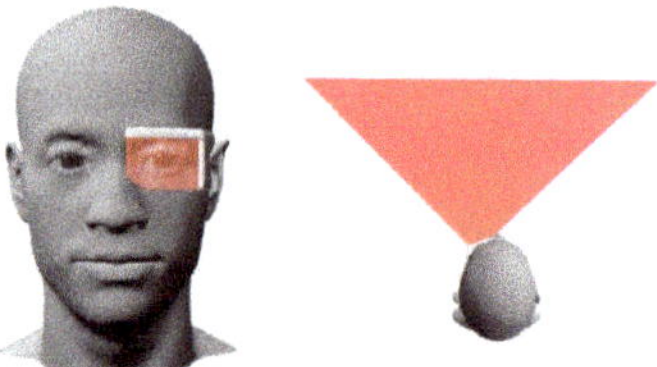

Figure 1. Low-profile near-eye display concept with a goal of large field of view.

Other researchers have used higher-order images to increase the display view zone, field of view, and space bandwidth product [5,6]. This work differs from previous efforts in several ways. First, our holographic pattern is a rapidly moving analog signal and not a pixel pattern with fixed locations. Also, leaky mode devices in near-eye applications can have an extremely high refresh rate—potentially exceeding 100 kHz for a 1 cm free-running surface acoustic wave (SAW) aperture (a leaky mode device can write a one millimeter aperture in approximately one microsecond). This refresh rate is at least two orders of magnitude higher than that of commercial near-eye displays and it allows us the ability to contemplate a line-rate approach to increasing field of view. Additionally, our leaky mode devices are made on a highly transparent lithium niobate substrate which is desirable for near-eye applications. Furthermore, using femtosecond laser pulses, we can create high-contrast, laser-induced gratings in the bulk of the substrate [7,8] with a number of diffracted orders of visible power. Finally, in addition to spatial light modulation, lithium niobate is an excellent electro-optic substrate which is frequently used to make fast (up to picosecond-rate) phase and amplitude modulators [9]. These unique attributes of a moving pattern, high refresh rate, high-contrast internal gratings, and integrated illumination pulsing are important to our proposed solution for increasing the field of view in bottom-exit leaky mode devices.

Leaky mode devices comprise a transparent slab of a piezoelectric substrate, such as lithium niobate, with a waveguide indiffused on the surface which sits adjacent to an interdigital transducer. When an RF signal excites the transducer, a surface acoustic wave is generated that travels across the surface of the waveguide. Light trapped in the waveguide, traveling contra-linearly or collinearly with the SAW pattern, may be mode-coupled from a guided mode to a leaky mode which can be steered and shaped by the SAW pattern to form a holographic image. The use of such a device in a near-eye application is shown in Figure 2. A chirped SAW pattern is generated by the transducer which is designed to mode couple light so that it appears to have originated from a point in space. The leaky mode light travels at a shallow angle through the substrate. This angle is usually very shallow, typically between 1° and 12° from parallel to the substrate surface. However, this angle can be modified by passing the leaky mode light from one substrate to another with another index (or the same index, but canted at an angle) thereby increasing the angle by an angular bias if desired (will assume an angular bias of 10° for the analysis in this paper). The lateral translation of the light ends at the bottom surface of the substrate where the light is out-coupled by a high-spatial frequency out-coupling grating usually patterned by interference lithography and etched into the substrate. This grating has only one diffracted order into the air (all others are evanescent) and toward the viewer's eye. The light rays that reach the viewer's eye can be back cast to a distant image point. As the surface acoustic wave moves across the device's SAW aperture, the image point moves across the viewer's field of view. The illumination light in the waveguide pulses in coordination with this movement to draw or emit points as the image point 'cursor' travels. When it has completed its scan, a new SAW chirp pattern is generated with a different depth and the field of view that is scanned again for as many depth planes as desired (we recently demonstrated a device with arbitrary focus from 4 ft to 10 ft [4]). Several leaky mode waveguide/transducer 'channels' can be placed adjacent to one another to create a vertically multiplexed array. Each channel becomes responsible for the one line of the display. In this way, images can be created at multiple depth planes to form a 3D image.

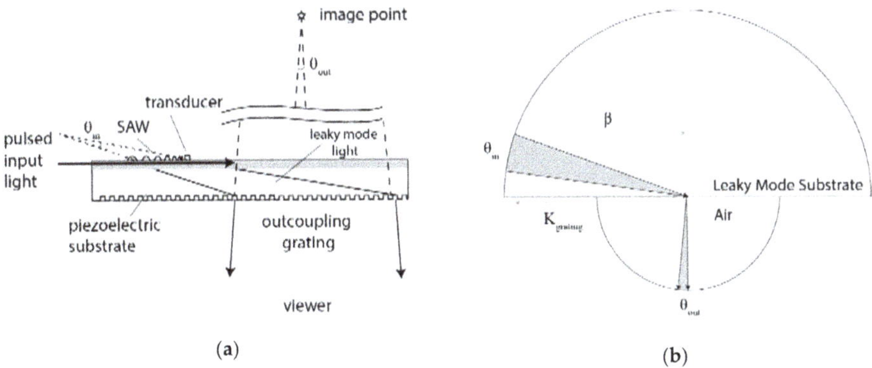

Figure 2. Reduction of view angel for bottom-exit leaky mode devices: (**a**) light diffracting from a surface acoustic wave (SAW) pattern is out-coupled with a grating; (**b**) The K-vector analysis shows a significant reduction in θ_{out} vs. θ_{in}. The former of these angles dictates the display view zone and field of view.

In this paper, we propose a simple modification to the leaky mode device by adding a laser-induced, static grating in the substrate of the leaky mode device. Femtosecond lasers can induce phase changes and 'catastrophic' lattice changes in the bulk of lithium niobate. The local nature of the change is aided by two photon up-conversion in the substrate. The result is that relatively high contrast gratings can be formed in the substrate with critical dimensions down below a micron and write depth of several tens of microns (see Figure 3). When coupled with high-accuracy, large-travel stages, whole wafers can be written with sub-surface gratings [10–12]. These gratings can be uniform, chirped, Raman–Nath, or Bragg grating. In our application, we explore a uniform thin grating underneath our leaky mode channels. It would be possible to have different periods underneath each channel, but for this analysis we will assume that the induced grating has only one period everywhere.

Figure 3. A laser-induced grating in x-cut lithium niobate and the resulting diffraction pattern with several high-order modes.

The purpose of the grating is to break the leaky mode light into different orders, each propagating at a different angle. These angles are determined by the laser-induced grating period. Depending on laser parameters, we have generated gratings with half a dozen or more orders. Our analysis assumes five such orders. Each of these diffracted orders will form a high-order image point that appears at a different location and is visible from a different angle than the original image point. We propose that these 'copies' of the original image point and others of different depths, can be superimposed over time to form a single image point, as illustrated in Figure 4a. Instead of scanning with just one point to form an image, now we scan with multiple points. We pulse only when one of these points is in a location that corresponds to a point location and a view in the target image (Figure 4b). Over time, as these spatial and angular ray bundles accumulate, a wide view angle for every point in the image is achieved (Figure 4c).

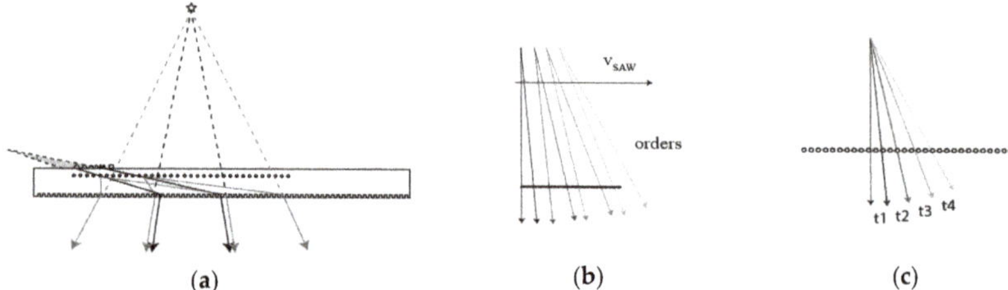

Figure 4. The multi-order leaky mode concept: (**a**) multiple orders combined to create a large view zone (or, alternatively, field of view). (**b**) Orders will travel across the display in front of the viewer. (**c**) The illumination is pulsed as orders pass by the viewer so multiple views are stitched into a wide view zone.

The success of this approach depends upon several of the unique features of leaky mode devices: high refresh rate, smoothly varying chirp functions, fast moving surface acoustic waves and translating leaky mode light. However, these attributes are not sufficient to make the approach successful if the view zones for each diffracted order cannot be adequately separated at the viewer's eye. If, for example, the viewer saw two overlapped view zones, they would see two points simultaneously and the display designer could not 'write' one point without also writing the other. Figure 5a shows a multi-order, HPO, hologram rendered in grayscale resist and reproduced with 633 nm light. Here, multiple images are created, one for each order of diffraction and each with a corresponding view zone (Figure 5b). Notice the overlap between the second- and third-order view zones. If the images appeared at different depths (which they do in this case), the viewer would see two images in the overlap position instead of one. Therefore, for our proposed multi-order method we must be able to eliminate as much as possible the overlap between adjacent view zone orders and achieve what we will call a 'no-overlap condition'. Below, we will use vector analysis to identify parameters for a 'no overlap condition', for a near-eye geometry.

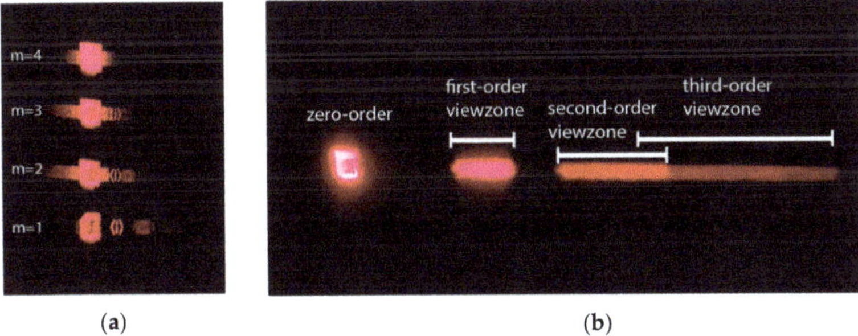

Figure 5. High-order holographic images and view zones: (**a**) first through a fourth-order image for a horizontal parallax only (HPO) hologram. This hologram, created in grayscale photoresist, is meant to simulate the output of a leaky mode modulator passing through thin grating. (**b**) Here, we see how the high-order holographic view zones separate or overlap.

2. Supplies and Methods

The geometries for order overlap are shown in Figure 6a. We will use K-vector analysis to show the relationship between points P and Q. After choosing parameters for a near-eye geometry including an internal grating period, Λ_D, and propagation distance, d_3, we will plot the high-order rays to confirm a 'no-overlap' condition.

We start with the back-cast chirp focus point, $P = (x_0, z_0)$. This parameter is a function of the leaky mode drop angle, θ, whose value will change along the chirp. We can use the angles, θ_0 and θ_1 of light rays exiting both ends of the SAW chirp to define two vectors:

$$\overline{k_0} = k_{sub} \sin(\theta_0)\hat{x} + k_{sub} \cos(\theta_0)\hat{z} \text{ and} \tag{1}$$

$$\overline{k_1} = k_{sub} \sin(\theta_1)\hat{x} + k_{sub} \cos(\theta_1)\hat{z}, \tag{2}$$

$$\text{where } k_{sub} = \frac{2\pi n_{ord}}{\lambda_0}. \tag{3}$$

We take the substrate index of refraction as $n_{ord} = 2.2864$ for $\lambda_0 = 633$ nm [13]. Conservation of transverse momentum preserves the value of the vector components parallel to the device interfaces. Therefore, in our analysis we keep track of the propagation constants:

$$\beta_0 = k_{sub} \sin \theta_0 \tag{4}$$

$$\beta_1 = k_{sub} \sin \theta_1 \tag{5}$$

The analysis can be thought of as having multiple layers separated in depth. They are as follows: d_1, the absolute distance from the chirp focus point to the SAW plane; d_2, the absolute distance from the SAW plane to the laser-induced internal grating; d_3, the absolute distance from the internal grating to the output grating; and d_4, the absolute distance traveled by light through the air to the viewer's eye (see Figure 6a).

The vectors $\overline{k_0}$ and $\overline{k_1}$ bracket a ray bundle that propagates through each device layer toward the eye. The cross section of the ray bundle at each of these layers is:

$$\Delta x^i = x_1^i - x_0^i \text{ where } x_1^i = d_1 \tan \theta_1, \text{ and } x_0^i = d_1 \tan \theta_0 \tag{6}$$

$$\Delta x^{ii} = x_1^{ii} - x_0^{ii}, \ x_1^{ii} = (d_1 + d_2) \tan \theta_1, \text{ and } x_0^{ii} = (d_1 + d_2) \tan \theta_0 \tag{7}$$

$$\Delta x^{iii} = x_1^{iii} - x_0^{iii}, \ x_1^{iii} = (d_1 + d_2 + d_3) \tan \theta_1, \text{ and } x_0^{iii} = (d_1 + d_2 + d_3) \tan \theta_0 \tag{8}$$

$$\Delta x^{iv} = x_1^{iv} - x_0^{iv}, \ x_1^{iv} = (d_1 + d_2 + d_3 + d_4) \tan \theta_1, \text{ and } x_0^{iv} = (d_1 + d_2 + d_3 + d_4) \tan \theta_0 \tag{9}$$

In order to calculate the final orientation of $\overline{k_0}$ and $\overline{k_1}$ after they have passed through both the internal grating and the output coupling grating for every order N, we simply subtract K_G once (every exiting ray must be diffracted by the output grating) and subtract (for negative orders) K_D, N times (see Figure 6b). The magnitudes of the parallel components for the resulting k-vectors are,

$$k_{0,||} = \beta_0 + mNK_D - K_G, \text{ and} \tag{10}$$

$$k_{1,||} = \beta_1 + mNK_D - K_G, \tag{11}$$

where m determines the sign of the diffracted orders; $m = -1$ in this paper. Using coordinate geometry, we determine that u_N and v_N can be expressed as:

$$u_N = \frac{x^{iii}_{0,N} M_{0,N} - x^{iii}_{1,N} M_{1,N}}{M_{0,N} - M_{1,N}} \text{ and} \tag{12}$$

$$v_N = (u_0 - x^{iii}_{1,N})M_{1,N} - (d_1 + d_2 + d_3) \text{ where,} \tag{13}$$

$$M_{0,N} = \frac{k_{0,N,\perp}}{k_{0,N,||}} = \frac{\sqrt{k^2_{air} - (\beta_0 + mNK_D - K_G)^2}}{(\beta_0 + mNK_D - K_G)} \text{ and} \tag{14}$$

$$M_{1,N} = \frac{k_{1,N,\perp}}{k_{1,N,||}} = \frac{\sqrt{k^2_{air} - (\beta_1 + mNK_D - K_G)^2}}{(\beta_1 + mNK_D - K_G)} \tag{15}$$

We define the non-overlap condition for order when the view zones are $x^{iv}_{1,1} < x^{iv}_0$ (i.e., the first $m = -1$ diffraction of $\overline{k_1}$ is below the $m = 0$ order of $\overline{k_0}$ at the viewing distance). For the non-overlap condition to be satisfied at any distance, K_D must be greater than $\Delta\beta = \beta_1 - \beta_0$.

To determine the distance, d_4, at which the non-overlap condition is satisfied, we must find the intersection of two lines. We compose the first ray from the slope of the un-diffracted $\overline{k_0}$ vector and the coordinates of the point, $[x^{iii}_0, -(d_1, d_2, d_3)]$. We compose the second ray from the slope of the negative first-order diffracted, $\overline{k_1}$ vector and the coordinates of the point, $[x^{iii}_{1,1}, -(d_1, d_2, d_3)]$. We solve for the intersection of these two lines to obtain,

$$z_{int} = (x_{int} - x^{ii}_1)M_{1,int} - (d_1 + d_2) \text{ where,} \tag{16}$$

$$x_{int} = \frac{x^{iii}_0 M_{0,int} - x^{iii}_1 M_{1,int}}{M_{0,int} - M_{1,int}}, \tag{17}$$

$$M_{0,int} = \frac{k_{0,\perp}}{k_{0,||}} = \frac{-\sqrt{k^2_{sub} - \beta^2_0}}{\beta_0}, \tag{18}$$

$$M_{1,int} = \frac{k_{1,-1,\perp}}{k_{1,-1,||}} = \frac{-\sqrt{k^2_{sub} - (\beta_1 - K_D)^2}}{(\beta_1 - K_D)}. \tag{19}$$

Note that the slopes in this case have perpendicular k components traveling in the opposite direction of those in Equations (14) and (15) above. This is because these rays are going down rather than up.

The coordinate, z_{int}, gives the depth *within* the material at which the zero and first-order views separate. This is a higher than necessary upper bound as the two orders still have additional time to separate as they propagate through distance, d_4. However, this provides time for other orders, that may have stronger overlap, to separate and to form a set of view windows that is well-separated at the viewing plane.

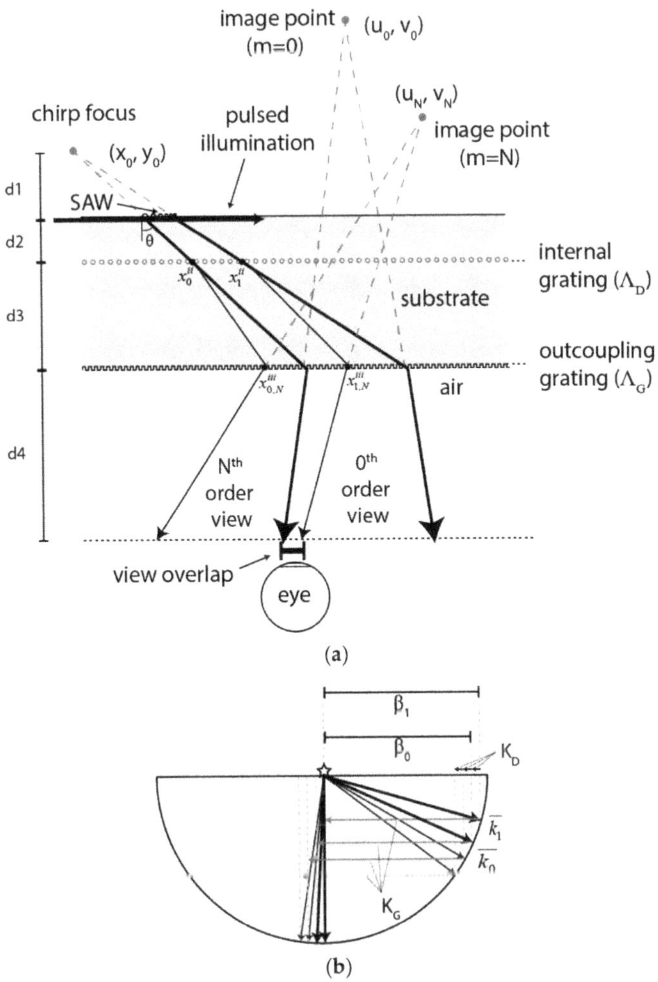

Figure 6. Multiple-order leaky mode concept: (**a**) leaky mode light diffracts into multiple orders within the substrate to form multiple image points at different angles and visible from different angles. (**b**) The K-vector analysis showing momentum change for both the internal grating and the out-coupling grating.

We chose the parameters listed in Table 1 to define our near-eye geometry. We chose a 6 mm (d_2 = 1 mm + d_3 = 5 mm) thick substrate with an internal grating at a 1mm distance from the SAW layer with a period, Λ_D, of approximately 4 μm and an output coupling grating with a period, Λ_G of approximately 300 nm. These values correspond to $K_D \approx 2 * \Delta\beta$ and $K_G \approx \beta_0$, respectively. An approximately 10° bias was assumed to give maximum and minimum leaky mode angles of $\theta_0 = 70°$ and $\theta_1 = 77°$. All of these fit within reasonable fabrication parameters for a two-substrate system.

Table 1. Near-eye parameters.

Parameter	Value	Notes
λ_{air}	633 nm	–
θ_0	70°	10° bias
θ_1	77°	10° bias
K_D	1.58×10^6	$2\Delta\beta$
K_G	2.1×10^7	$\approx \beta_0$
Λ_D	3.99 μm	Internal grating
Λ_G	295 nm	Output grating
N	5	modes (neg.)
n_{ord}	2.2864	LiNbO$_3$
d_1	1 mm	chirp focus
d_2	1 mm	to grating 1
d_3	5 mm	to grating 2
d_4	20 mm	to viewer

3. Results

The results of our analysis, run in Matlab for five negative orders, are shown in Figure 7 and Table 2. The no-overlap condition was met such that there was little or no overlap of view-zones (Figure 7a). The minimum no-overlap window was calculated to be 5.9 mm and the pulse timings to align these windows are given in Table 2. Figure 7b shows that the higher-order diffraction added 52.5° of view zone/field of view to the zero-order view zone/field of view of 4.5° to make a total of 57°—a more than ten-fold increase in the near-eye case. For $K_D \approx 2\Delta\beta$, the angular duty cycle is 50% for one light traveling in only one direction (see discussion).

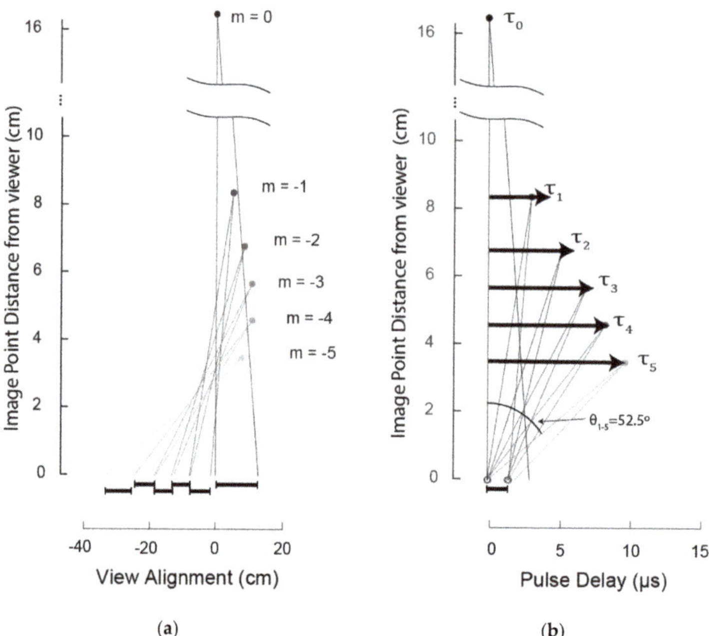

Figure 7. Results of view angle and pulse delay calculations for the parameters in Table 1: (**a**) The five negative orders used show little or no overlap at the viewer. (**b**) The non-zero view angles aggregate to a total of 52.5°. The relative pulse delay times are shown to all be under 10 μs.

Table 2. View window and pulse delay for image points corresponding to diffracted orders.

Order	View Window (mm)	Pulse Delay (µs)
0	13.0	0
−1	6.6	2.1
−2	5.9 [1]	3.6
−3	5.9 [1]	5.1
−4	6.7	6.8
−5	9.2	9.6

[1] These smallest windows define the total view window for the eye.

4. Discussion

The above results show that the multi-order leaky mode devices have the potential to create image points with a ten-fold increase in the view zone and/or field of view. They show that the non-overlap condition can be achieved for the near-eye case with parameters that are within the current fabrication limits for leaky mode devices, femtosecond direct-writing and interference lithography. The minimum view window of 5.9 mm is larger than the average pupil diameter of the eye and can be actively positioned in front of the pupil, if desired. All pulse delay times are under 10 µs and correspond to roughly 10 mm of SAW travel on the device surface. This means that a new pattern can be written at a rate of approximately 100 kHz. Therefore, to address all of these orders, the SAW would have to travel the SAW chirp length (typically between 1 mm and 10 mm) plus 10 mm. This distance is well-matched to the SAW aperture of current leaky mode devices.

The dramatic increase of view zone from 4.5° to 57° has some important context. Because we chose $K_D \approx 2\Delta\beta$ instead of $K_D \approx \Delta\beta$, the resulting ray bundles are not continuous in angle. Instead, the angular views have gaps that result in a 50% angular duty cycle. We could theoretically choose a very high K_D and achieve a nominal view angle approaching 90 or even 180 degrees, but as long as we only use five modes, the active portions of that view angle would only come to a total of between 25° and 30°, regardless of the total view angle. However, there are ways of filling these gaps, aside from using more modes. We could, for example, run a second channel with light traveling in the opposite direction. The result would be the *interlacing* of the output views to create a continuous sweep of views. In this scenario, the additional channel could use the same waveguide, the same grating structures, even the same *transducer*, if desired, (because ITDs are inherently bi-directional). In this scenario, the internal grating could remain the same and the output grating period would be increased (and made simpler to fabricate) so that the fan of angles was symmetric around the surface normal. The result would be a smooth set of views across the 57° sweep without greatly increasing fabrication complexity.

It is worth restating a few of these points. In our methodology, we chose K_G to select which order will be directed normal to the output face. For example, by setting $K_G = \beta_1 - 3K_D$ (the third-order diffraction of the $\overline{k_1}$), we obtain an aggregate sweep that is roughly symmetric and ready for the interlacing described above. If, instead, we choose $K_G = \beta_1$, we obtain a unilateral spread that would be ideal for combining with another unilateral spread. This would be the ideal configuration if the sweep were continuous in angle. However, continuity would require that $K_D \approx \Delta\beta$, which would make it impossible to eliminate view zone overlap. It should be noted, however, that this overlap would be less important for viewers far away from the device (e.g., 500 mm). Therefore, looking forward, this approach might also be of special interest not only to near-eye displays but to large flatscreen displays as well.

5. Conclusions

We have described a new approach to increasing the view zone in leaky mode devices. We have shown the results of an analysis that suggests the existence of at least one set of parameters for a multi-order, near-eye, leaky mode device with viewzones that have little or no overlap. It may now be possible to increase the effective view zone or field of view for of a multi-order leaky mode device more than ten-fold.

Author Contributions: Conceptualization, D.S.; Software, D.P., D.K., D.S.; methodology, D.P., D.S.

Funding: This research was funded by Air Force Research Laboratory contract FA8650-14-C-6571.

Conflicts of Interest: The authors declare no conflict of interest.

References

1. Smalley, D.E.; Smithwick, Q.Y.J.; Bove, V.M.; Barabas, J.; Jolly, S. Anisotropic leaky-mode modulator for holographic video displays. *Nature* **2013**, *498*, 313–317. [CrossRef] [PubMed]
2. Qaderi, K.; Leach, C.; Smalley, D.E. Paired leaky mode spatial light modulators with a 28° total deflection angle. *Opt. Lett.* **2017**, *42*, 1345–1348. [CrossRef] [PubMed]
3. Qaderi, K.; Smalley, D.E. Leaky-mode waveguide modulators with high deflection angle for use in holographic video displays. *Opt. Express* **2016**, *24*, 20831–20841. [CrossRef] [PubMed]
4. McLaughlin, S.; Henrie, A.; Gneiting, S.; Smalley, D.E. Backside emission leaky-mode modulators. *Opt. Express* **2017**, *25*, 20622–20627. [CrossRef] [PubMed]
5. Lee, B.; Li, G.; Hong, J.-Y.; Lee, D.; Yeom, J. Viewing zone enlargement of holographic display using high order terms guided by holographic optical element. In *Propagation through and Characterization of Distributed Volume Turbulence and Atmospheric Phenomena*; Optical Society of America: Washington, DC, USA, 2015; Volume 24, p. JT5A.
6. Li, G.; Jeong, J.; Lee, D.; Yeom, J.; Jang, C.; Lee, S.; Lee, B. Space bandwidth product enhancement of holographic display using high-order diffraction guided by holographic optical element. *Opt. Express* **2015**, *23*, 33170–33183. [CrossRef] [PubMed]
7. Burghoff, J.; Grebing, C.; Nolte, S.; Tünnermann, A. Efficient frequency doubling in femtosecond laser-written waveguides in lithium niobate. *Appl. Phys. Lett.* **2006**, *89*, 081108. [CrossRef]
8. Thomson, R.; Campbell, S.; Blewett, I.; Kar, A.; Reid, D. Optical waveguide fabrication in z-cut lithium niobate (LiNbO$_3$) using femtosecond pulses in the low repetition rate regime. *Appl. Phys. Lett.* **2006**, *88*, 111109. [CrossRef]
9. Wang, C.; Zhang, M.; Chen, X.; Bertrand, M.; Shams-Ansari, A.; Chandrasekhar, S.; Winzer, P.; Loncar, M. Integrated lithium niobate electro-optic modulators operating at CMOS-compatible voltages. *Nature* **2018**, *562*, 101. [CrossRef] [PubMed]
10. Gattass, R.R.; Mazur, E. Femtosecond laser micromachining in transparent materials. *Nat. Photonics* **2008**, *2*, 219. [CrossRef]
11. Savidis, N.; Jolly, S.; Datta, B.; Karydis, T.; Bove, V.M. Fabrication of waveguide spatial light modulators via femtosecond laser micromachining. In *Advanced Fabrication Technologies for Micro/Nano Optics and Photonics IX*; International Society for Optics and Photonics: San Diego, CA, USA, 2016; Volume 9759, p. 97590R.
12. Burghoff, J.; Nolte, S.; Tünnermann, A.J.A.P.A. Origins of waveguiding in femtosecond laser-structured LiNbO$_3$. *Appl. Phys. A* **2007**, *89*, 127–132. [CrossRef]
13. Zelmon, D.E.; Small, D.L.; Jundt, D. Infrared corrected Sellmeier coefficients for congruently grown lithium niobate and 5 mol.% magnesium oxide–doped lithium niobate. *JOSA B* **1997**, *14*, 3319–3322. [CrossRef]

Article

Electronic Tabletop Holographic Display: Design, Implementation, and Evaluation

Jinwoong Kim [1,*], Yongjun Lim [1], Keehoon Hong [1], Hayan Kim [1], Hyun-Eui Kim [1], Jeho Nam [1], Joongki Park [1], Joonku Hahn [2] and Young-ju Kim [3]

[1] Broadcasting and Media Research Laboratory, Electronics and Telecommunications Research Institute, 218 Gajeong-ro, Yuseong-gu, Daejeon 34129, Korea; yongjun@etri.re.kr (Y.L.); khong@etri.re.kr (K.H.); hayankim@etri.re.kr (H.K.); turnn@etri.re.kr (H.-E.K.); namjeho@etri.re.kr (J.N.); jkp@etri.re.kr (J.P.)

[2] School of Electronics Engineering, Kyungpook National University, 80 Daehakro, Bukgu, Daegu 41566, Korea; jhahn@knu.ac.kr

[3] Yunam Optics Inc., 33-7 Eongmalli-ro, Majang-myeon, Icheon-si, Gyeonggi-do 17389, Korea; smallisle@yunamoptics.com

* Correspondence: jwkim@etri.re.kr

Received: 19 January 2019; Accepted: 15 February 2019; Published: 18 February 2019

Abstract: Most of the previously-tried prototype systems of digital holographic display are of front viewing flat panel-type systems having narrow viewing angle, which do not meet expectations towards holographic displays having more volumetric and realistic 3-dimensional image rendering capability. We have developed a tabletop holographic display system which is capable of 360° rendering of volumetric color hologram moving image, looking much like a real object. Multiple viewers around the display can see the image and perceive very natural binocular as well as motion parallax. We have previously published implementation details of a mono color version of the system, which was the first prototype. In this work, we present requirements, design methods, and the implementation result of a full parallax color tabletop holographic display system, with some recapitulation of motivation and a high-level design concept. We also address the important issue of performance measure and evaluation of a holographic display system and image, with initial results of experiments on our system.

Keywords: digital hologram; holographic display; tabletop display; hologram measurement

---

1. Introduction

Digital electroholographic display technology has been believed to be an ultimate solution for 3-dimensional displays due to its theoretically perfect 3-dimensional visual cue support without any conflict among them. There has been active research and development of many prototype systems, though most suffer from critical performance limitations such as small image size or narrow viewing angle. We can categorize the realization of electronic dynamic holographic displays into two groups: one is a flat panel-based direct front viewing type and the other is tabletop type which renders volumetric 3D images floating on or above the display. The former type can render extruding 3-dimensional images in front of or behind the display, which is not much different from conventional stereoscopic or multiview 3D displays. Almost all displays, including TVs, PC monitors, information kiosk, and smartphones, are of the former type and it is natural trying to make a 3-dimensonal display of the same type. However, holograms and holographic displays are expected to have more powerful 3D rendering capability with volumetric 3D images looking much like a real object. This expectation is universally depicted in many SF movies including old classic Star Wars or the very recent Avatar. Moreover, current prototype systems of the former type reported so far have a very narrow viewing angle, so that even the basic requirements, such as supporting motion parallax or binocular parallax, are hardly satisfied.

In the authors' opinion, a tabletop display is the more desirable type of electronic holographic display realization. Tabletop display can show full perspective views of an object or scene from 360 degree around, so that it gives viewers much better perception of the whole shape, volume, and relation among parts of the scene or objects. Miniaturized scene or moving objects on a tabletop display, rendered by holographic principle, will surely give excellent realism that cannot be achieved by front-viewing displays. In this regard, we believe that the tabletop holographic display has much wider application areas.

We have successfully implemented several versions of tabletop holographic display system with commercially available DMD (Digital Micromirror Device) for SLM (Spatial Light Modulator) device. We have previously reported on the first version of the system, which is monocolor one [1]. We have upgraded the system to a full color version, and the system can, for the first time, render color holographic video, which can be watched by multiple viewers at the same time, freely from 360 degrees around the display. In this article, we report on the enhanced version of the system, which is full-color, full parallax with extended vertical viewing range, and address initial experimental result of performance measurement and evaluation of the system. In Section 2, the prior art of the holographic display implementation is briefly reviewed. In Section 3, the overall design concept and methodologies of the target system are capitulated, followed by implementation details of the enhanced parts and experimental results including the system configuration and rendered images are given in Section 4. In Section 5, we address on the performance evaluation issues including definition of metrics, measurements, and evaluation regarding each metric, followed by conclusions and discussions on further issues.

2. Prior Art of Holographic Display Implementation

Since Prof. Benton of MIT Spatial Media Group introduced the first prototype system of electronic holographic display [2], called Mark-1, there have been several different approaches with different SLM device [3–11]. The basic challenge is overcoming the Space Bandwidth Product (SBP) limitation of the SLM device used, by using spatiotemporal multiplexing techniques for wider viewing angle and bigger image size. Table 1 shows the most prominent systems reported so far with their characteristics and specifications.

Table 1. Electronic holographic display systems.

Systems	SLM	Architecture	Characteristics	References
MIT Holo-video display (Mark-I, II, III, IV)	AOM	Multichannel mechanical scanning	P: HPO, C: MC (1992), FC (2012)	[2] [3]
SeeReal VISIO 20	LCD (pixel pitch: ~100 um)	Viewing window tracking	P: HPO, ISD: 20 inch VA: NA (eye tracking)	[4]
SNU Curved holographic display	LCD	Curved spatial tiling of multiple LCDs	P:FP ISD: 0.5 inch VA: 22.8°	[5]
NICT Full color display	4K LCoS	4 × 4 spatial tiling	ISD: 85 mm VA: 5.6°, FR: 20 fps	[6] [7]

Table 1. *Cont.*

Systems	SLM	Architecture	Characteristics	References
TUAT Horizontal scanning holography	DMD	Horizontal scanning with anamorphic system	P: HPO ISD: 6.2 inch VA: 10.2°	[8] [9]
Univ. of Arizona Rewritable printing display	Rewritable photorefractive polymer material	6-ns pulsed laser at a rate of 50 Hz write, LEDs for readout	P: FP ISD: 12 inch VA: NAFR: 0.5 fps	[10]
ETRI	LCD (pixel pitch: ~120 um)	Viewing window tracking	P: FP ISD: 16 inch VA: NA (eye tracking)	[11]

P: Parallax (HPO or Full Parallax (FP)), C: Color (Monocolor(MC) or Full color (FC)), ISD: Image Size Diagonal, VA: Viewing Angle, FR: Frame Rate, DT: Data Throughput, NA: Not Available, MIT: Massachusetts Institute of Technology, SNU: Seoul National University, NICT: National Institute of Information and Communications Technology, TUAT: Tokyo University of Agriculture and Technology.

They used almost all different types of SLMs, including acousto-optic modulator (AOM), liquid crystal display (LCD), digital micromirror device (DMD), liquid crystal on silicon (LCoS), and rewritable photorefractive photopolymer, which reflects that none of them has satisfactory capability to build a high quality holographic display system; for an extensive review and analysis of these systems and similar approaches, please refer to [12].

All of these holographic display systems are of the 'front-viewing' type implementations. Holographic 3-dimensional images are seen in front or behind the display depending on the content to be rendered. Since the viewing angle is small, it can hardly give any volumetric feeling or perception different from conventional multiview displays, while image size and quality is much inferior to them. So, even though the theory tells strength of the holographic display, practical implementations do not yet realize desired superiority of the technology. This situation is depicted in Figure 1. Recently, there have been several other approaches for different type of holographic displays trying to render more volumetric holographic images, either floating in the air [13] or seen from 360 degrees around [14,15]. In reference [13], only one viewer can see the image, and in reference [15], one can see only side views, while the display in reference [14] can show top side views, thus neither one is fulfilling the capability of rendering true volumetric image. On the other hand, there have also been several tabletop display systems demonstrated their quite impressive capability of volumetric 3-dimensional image rendering in the free air, which easily enables some manipulations and interactions with the content [16–18]. These displays shows the usefulness of the tabletop displays, but crucial weakness remains regarding human factor issue, since it cannot support visual accommodation cue of the human 3D vision. This can only be achieved by holographic technology, though super-multiview or integral imaging has also been claimed to have a very similar property [19–21]. In summary, the tabletop holographic display system with rendering capability of more realistic volumetric 3D images is yet to be developed, which is far better than typical 'front-viewing' type ones.

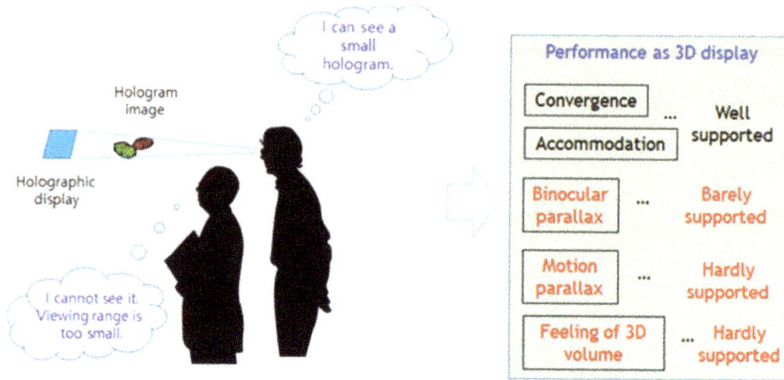

Figure 1. Current status of state-of-the-art holographic displays.

3. Design Concept and Methodologies

In the following are our design goals of tabletop holographic display implementation. What we had in mind from the start is that the system should be designed to evolve eventually to a commercially viable system.

(1) The size of the hologram image should be reasonably big enough so that viewers can perceive the 3-dimentional volume and some details of the image.

(2) The hologram image should be seen by multiple viewers at the same time, and freely from different positions and perspectives so that each viewer can easily perceive the whole shape of rendered objects and all viewers can share the whole scene information in order to discuss about, or collaboratively manipulate, on it.

(3) The image should be rendered in such a way that binocular disparity and motion parallax is perceived to a satisfactory level. In other words, the image should deliver the perception of real objects being there, showing all around perspective views of the object with full parallax.

(4) No physical apparatus of the display should occupy the image space, so that the hologram image is shown in free air, enabling manipulation or interaction with the image as if we do with real objects.

(5) High quality color hologram image with dynamic content, i.e., the hologram video should be presented.

Also note that vergence-accommodation conflict-free property should basically be supported since the system is real holographic display. Performance specification of the display system to be realized is drawn from these requirements, which is summarized in Table 2.

Table 2. Performance specification of the system.

Requirement	→	Specification
1. Size of the hologram image	→	7.63 cm or bigger
2. 360 degree viewable freely by multiple users	→	Horizontal viewing angle 360 degrees Vertical viewing angle 20 degrees
3. Volumetric image with full parallax	→	Floating image in the center of the display with 45 degrees of downward viewing direction
4. Image should not occupy physical space	→	
5. Color hologram video	→	R/G/B full color video rendering

A drawing of the system design goal and schematic diagram of the main system specifications are shown in Figure 2.

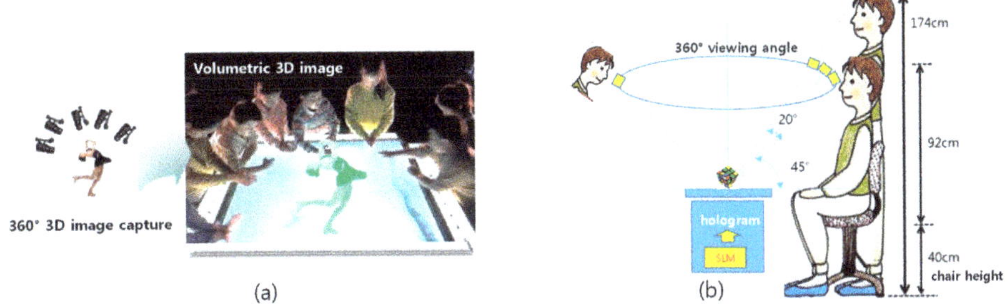

Figure 2. (a) Conceptual drawing of the design goal. (b) Schematics of system specifications.

It is quite challenging to realize these requirements into working display system with currently available commercial SLM devices, since most of them have a physical size of modulation area less than one inch and diffraction angle less than 2 degrees. Figure 3 shows this issue in a schematic diagram. We need to enlarge virtual modulation device to bigger than three inches and also enlarge the field of view to cover 360 degrees.

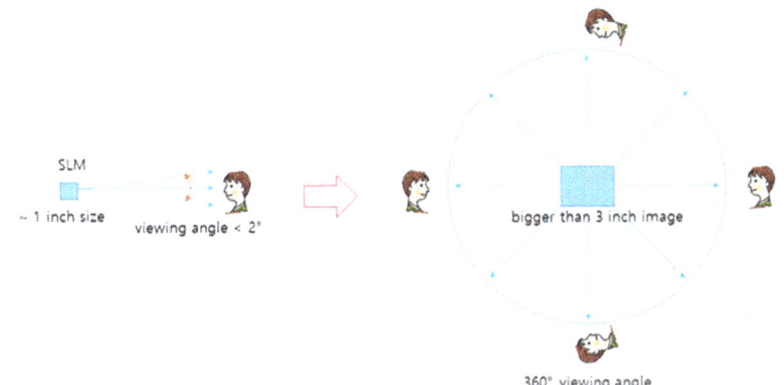

Figure 3. Basic challenge of the system design.

Viewing angle of 360 degrees can only be achieved by either temporally scanning or spatially tiling a small viewing angle of native SLM device. The former method needs temporal multiplexing with fast SLM and the latter one needs spatial tiling with multiple SLMs. The very first thing to do in the system design is to decide on the basic architecture with the selection of appropriate SLM device. We decided to adopt the temporal multiplexing scheme and fast SLM, based on the comparison of expected system complexity and maximum data throughput of SLMs. Usually the performance of an SLM device is denoted by SBP, which is cited most often to compare different SLM devices. SBP simply represents the total pixel numbers of an SLM device, while total data throughput of a device comes from the combination of pixel count and frame rate. Total data throughput actually determines whole system performance in terms of hologram image rendering, thus is more appropriate to judge the performance of an SLM. This quantity is named extended SBP(eSBP), which is defined in Equation (1):

$$eSBP = SBP * frame\ rate\ (or\ refresh\ rate) = (pixel\ resolution) * (frame\ rate) \tag{1}$$

Some of typical SLM devices are compared in terms of SBP and eSBP in Table 3. From the table, we see that although the SBP of DMD is smaller than LCD or LCoS, the eSBP value is much larger.

Table 3. Comparison of SBP and eSBP for different SLM devices.

SLM Type	Pixel Pitch (μm)	Pixel Resolution	SBP (Total Pixels)	Refresh Rate/bit Depth	eSBP (Total Data Rate)	eSBP Ratio	System
LCD	156 × 52	2560 × 2048	5,242,880	60/8	2,516,582,400	0.63	SeeReal
LCoS	4.8 × 4.8	3840 × 2160	8,294,400	60/8	3,981,312,000	1	NICT
DMD	13.68 × 13.68	1024 × 768	786,432	32,000/1	25,165,824,000	6.32	ETRI

We utilize the large eSBP value of a DMD device to spatially distribute fast refreshing hologram images by temporal multiplexing, thus achieving very wide viewing angle. The following details the basic methods that we applied to the display system design for the required specifications.

(1) Image size is first enlarged by simple optical magnification using *4-f* optics, which results in enlarged virtual SLM (from 1.27 cm to 7.62 cm). The price of this image enlargement is reduced diffraction angle of the virtual enlarged SLM (from 2° to 0.3°).

(2) We make a viewing window at viewing distance of 900 mm by adding a field lens in the position of virtual SLM plane, forming the size of 4.7 mm × 4.7 mm rectangular viewing window.

(3) Seamless horizontal viewing zone is made by circularly tiling the viewing windows of the appropriate number of viewpoint-dependent hologram images. In our system, the viewing zone has a circumferential length of ~4000 mm, which can be covered by tiling more than 851 viewing windows. In actual implementation, 1024 viewing windows are tiled with some overlap between adjacent ones. This is achieved by temporal multiplexing with mechanical scanning optics.

Steps (1)–(3) are shown in Figure 4.

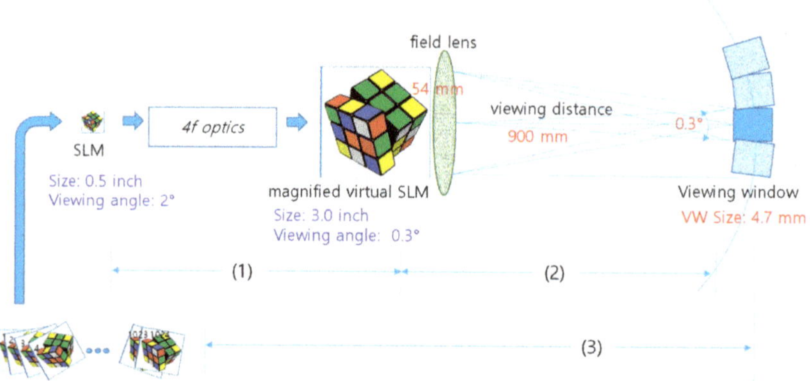

Figure 4. Schematic diagram of the design (1), (2), and (3) denotes the corresponding step in the text.

(4) The viewpoint-dependent hologram images are designed to be seen at the center of the tabletop display, so that coalition of them makes a volumetric 3-dimentional image floating in the air. This is achieved by two stages of optics. In the first stage, the light path of the hologram image is

guided by a combination of flat mirrors into 45-degree slanted direction and in a way to overlap at the axial area of the system. In the second stage, these virtual SLM planes are transported to the top center area of the tabletop display (final hologram image plane in the air above the tabletop of the display) by image transportation function of double parabolic mirrors from one focal area to another. This is shown in Figure 5.

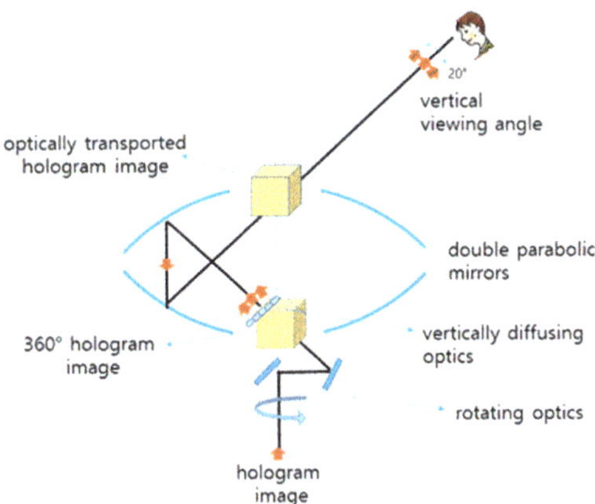

Figure 5. 360-degree volumetric image floating on tabletop display.

Based on this principal design method, we have successfully implemented the first monocolor version of tabletop holographic display systems, which has been reported in the previous publication [22]. In the first prototype system, four DMDs are spatially tiled to make a 2 × 2 multivision style SLM module for enhanced resolution (from 768 × 768 to 1,536 × 1536) and size (1.27 to 2.54 cm). This still small size of SLM is enlarged by a cascade of two *4-f* optics, with the magnification factor of 1.67 and 1.65 each, overall resulting in 2.76 times of magnification effect. For rejection of DC and high order noise, a single-sideband (SSB) filter is inserted in the Fourier plane of the first *4-f* optics. It is monocolor system with a green (wavelength of 632 nm) laser as a light source. For the detail specifications of optical components used in the display system, please refer to [1]. In the second and later implementations of our system, the following are added as well as the enhanced functions.

(1) The SLM module for color hologram generation is designed using three DMDs for each monocolor (R, G, and B) hologram, combining optics consisting of two prisms (one is trichroic and the other is TIR (total internal reflection) type), and three laser light sources with collimated beam output. In this way we preserve the image resolution of the monocolor display in color holographic display. Refer to [22] for some implementation details.

(2) We have thus far a vertically narrow (4.7 mm) viewing stripe around and over the display. A vertical viewing angle of 20° is also required in order to permit the viewers' free position or posture (sitting on a chair, standing, or somewhere in between, as shown in Figure 2b) to view the hologram image, and this is achieved by adding vertical diffuser optics (a lenticular sheet, which is schematically shown in Figure 5) at the virtual SLM plane. Since diffusing optics degrades the spatial focusing capability of hologram somewhat, there is a tradeoff in this regard.

(3) Vertical parallax is added inside this 20° of vertical viewing range (between 35° and 55° from the tabletop). For this, precise and fast pupil tracking of viewers with a dynamic update of the hologram images is implemented for a limited number of viewers. Pupil tracking module consists of multiple IR illuminator and cameras. Two cameras working as a stereo camera covers a viewer in 90° range, and four camera sets are used for 360° coverage, while supporting four viewers simultaneously. Figure 6 shows the working concept of the module and experimental

setup for 180° coverage with reconstructed image at four different vertical positions. Currently, update of pupil position is limited to 10 times per second, which is not sufficient for smooth update of hologram images following viewer's motion.

(4) Capture of 3D information from moving real objects and the 360-degree perspective hologram CGH is implemented, and thus the signal chain from capture-CGH to display is demonstrated.

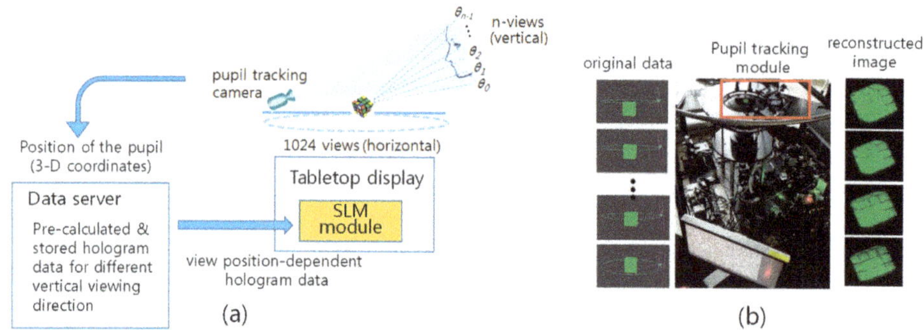

Figure 6. Vertical parallax support by using pupil tracking and dynamic hologram update. (**a**) Conceptual model and (**b**) 180° pupil tracking module and hologram images seen at four different vertical viewing position.

4. Implementation and Experimental Result

The basic system we implemented is a vertically diffused horizontal parallax only (VD-HPO) hologram system. This system is the one without pupil tracking module, having a 20° vertical viewing angle and a nonsupported vertical parallax, which is similar to rainbow hologram. With pupil tracking module added, the system becomes a full parallax hologram system.

Schematic diagram of the basic display system and photo of the final system shape are shown in Figure 7. The system consists of an SLM module consisting of laser light source and three DMDs, *4-f* relay optics for image magnification and noise filtering, temporal multiplexing module consisting of a scanning motor and light path redirecting optics, and volumetric image floating optics consisting of double parabolic mirrors or equivalent optics.

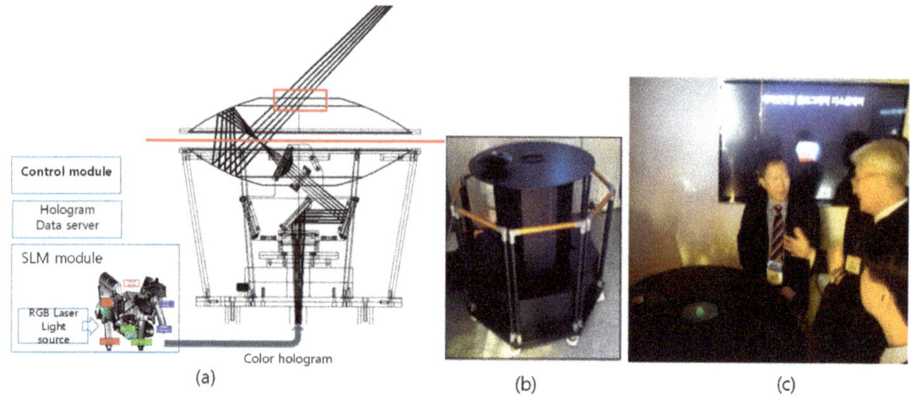

Figure 7. (**a**) Schematic configuration. (**b**) Photo of the implemented system. (**c**) Demonstration.

Three-hundred-and-sixty-degree viewable volumetric color hologram image is rendered in the display by coalition of 1024 view-dependent holograms, which are generated from 3D data consisting of color and depth image for each viewpoint from a CG model or scanning real objects. Figure 8 shows the whole signal chain from 360° 3D data capture from real moving or static objects, possibly including humans, CGH computation, and real-time hologram data transmission via data server with high speed interface up to display system. There are several highly complicated signal processing tasks involved in the chain. The 360° data capture part has two alternative setups: one consists of a single depth sensing camera with a turntable for a static object and the other, for moving objects, consists of multiple synchronized depth sensing cameras arranged 120 degrees apart. Image registration and stitching, smoothing, and hole-filling algorithms are applied to get point cloud data which is optionally converted to 3D mesh model. Finally RGB and depth image data are extracted via virtual camera setting for required number of viewpoints of the tabletop display. This data is input to the CGH computation module, algorithm of which is calculating wave propagation model of the display optics, outputting hologram fringe patterns. Details of this CGH are presented in reference [23]. The hologram data is uploaded to data server and finally transmitted to the SLM module of the display in real time via high speed interface from data server to the display system.

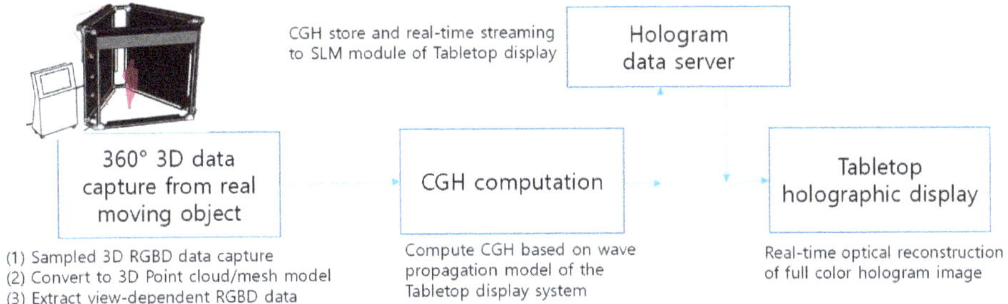

Figure 8. Signal processing chain of the tabletop holographic display system.

Images of optically reconstructed hologram images from the implemented display system are shown in Figure 9.

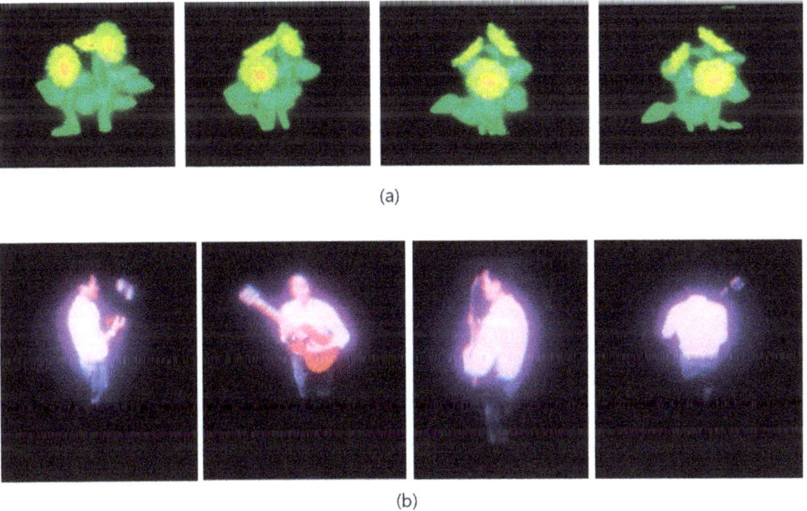

(a)

(b)

Figure 9. Optically-reconstructed hologram image captured from different horizontal viewpoints: (**a**) CG model and (**b**) real human object.

5. Measurements and Evaluation

It is not less important to measure and evaluate the performance and image qualities of the final holographic displays than system implementation itself. A study of how to define display performance measure or image quality has been carried out by only a few researchers so far [24,25].

In order to evaluate performance of a holographic display and compare with different systems, we need to define appropriate performance metrics. Since holographic displays have different characteristics and behavior, existing performance metrics for conventional 2D displays are not sufficient. In Table 4, some essential performance metrics we defined for holographic 3D displays are listed. For the implemented display system, we measured and evaluated several important properties based on this performance metric.

Table 4. Metrics and their definitions for holographic performance measure.

Category	Metrics	Definition/Description
Image volume size	3-dimensional volume (width × height × depth)	Maximally representable image size measured in lateral and depth direction, or equivalent
Viewing angle (VA)	Horizontal/vertical VA	Lateral or circular viewing range measured in angle
Image quality	3D-MTF (image resolution)	MTF measured through depth range of image volume, which represents image resolution
	Color fidelity	Color distortion measured for standard color chart
	Speckle noise	Speckle contrast measured
	Brightness	Perceptive brightness level of the image
Support of accommodation	Degree of Depth of Focus (DOF)	The smaller the DOF, the better the capability of supporting accommodation

5.1. Size of the Image Volume Space

In our tabletop holographic display, a 3-dimensional image is rendered in free space and virtually occupies a volume space, as represented in Figure 10. This image space is made from intersections of adjacent viewing cones, each of which corresponds to a viewpoint and hologram pair. It can be divided into real and virtual image areas, which are seamlessly composing the whole volume space for the hologram image. If we render hologram image inside this image volume, then it is seen much like a real object supporting binocular as well as motion parallax.

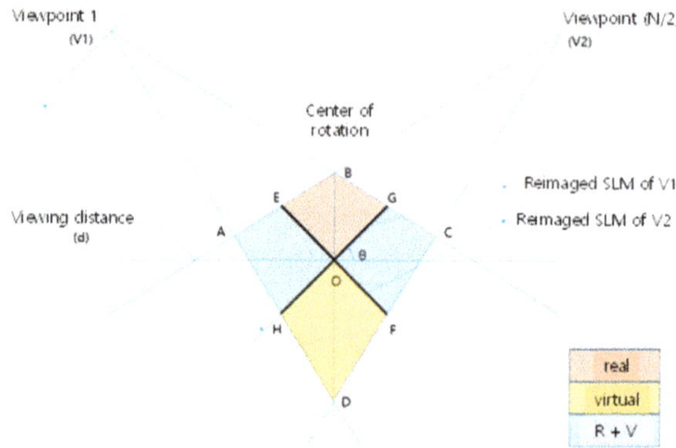

Figure 10. Image volume space.

5.2. Hologram Image Resolution

Since the hologram image is rendered in free space as a volumetric image, simple pixel resolution does not properly represent the resolution or quality of optically reconstructed hologram image.

We defined a new metric for representing the image resolution, which we named 3D-MTF. The modulation transfer function (MTF) is usually used to measure the performance of an optical device like lens in terms of how well it conveys the contrast of an object to the image. MTF is modulation depth denoted as a function of frequency of the test pattern as shown in Equation (2). A 3D-MTF is defined as an extension of conventional MTF by adding depth as another parameter, which is defined as in Equation (3):

$$M = \left[\frac{I_{max} - I_{min}}{I_{max} + I_{min}}\right], \quad MTF = \left[\frac{M_{image}(\xi)}{M_{object}(\xi)}\right] \tag{2}$$

$$3D\ MTF = \{MTF_k\}_{k=1}^{k=N\Delta D} \tag{3}$$

where, M is modulation depth, I_{max} and I_{min} are the maximum and minimum signal intensity of signal, respectively, $M_a(b)$ is the modulation depth of a as a function of b, ΔD is depth resolution, and N is the number of depth layers used in the measurement. MTF_k is the MTF value measured at the k-th depth plane. Defined as such, 3D-MTF is a collection of conventional MTF measured for images at different depth. We measured 3D-MTF value of the implemented display system as shown in Figure 11. For the purpose of accurate measurement, hologram image is captured directly by CCD (Sony ICX834) placed at different depth plane for a fixed viewpoint (Figure 11a). A linear stage and a green laser with a 532 nm wavelength are used in the experiment. An example black and white stripe pattern used in the measurement is shown in Figure 11c, in the order of original shape, its CGH fringe pattern, and reconstructed hologram image. For each stripe pair of the test pattern, modulation depth value is measured and average value of whole stripe pairs is plotted for each depth in Figure 11d. The ranges of measured values for whole stripe pairs at each depth are denoted by vertical bars. MTF changes with image depth, and the average value of MTF is 35.9%@1.2cycles/mm test pattern. More details of measurement and evaluation are given in Appendix A. In reference [26], a holographic stereogram is analyzed for MTF modeling, showing the effect of phase error arising from approximating the wavefront of a 3D image with a wavefront of a projected image at stereogram plane. While a holographic stereogram can be analyzed as a diffraction-limited imaging system with some aberrations, our display system relies on the diffraction to make a real or virtual image, which requires different analysis. Furthermore, binary encoding of computed fringe pattern results in distortions of frequency spectrum, which should be taken care of. This is remained for further study

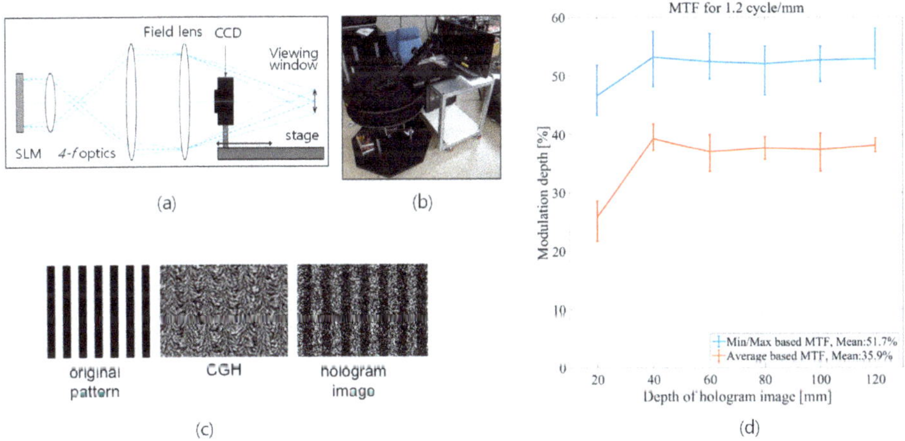

Figure 11. 3D-MTF measurement: (**a**) design of test method, (**b**) environment setup for measurement, (**c**) test pattern, and (**d**) result of the measurement.

5.3. Color Gamut and Color Reproduction Fidelity

In a full color holographic display, color reproduction fidelity is not actually guaranteed even though color gamut of the display system is wider than conventional ones. Figure 11a shows the measured color gamut of our display, which is ~160% wider than CIE 1976 NTSC standard (1953). On the contrary, Figure 12d shows the measured color fidelity for the 24 colors in a standard Macbeth color chart. Blue dots represent desired values and red dots are measured ones. As shown in Figure 12c, optically reconstructed hologram image is corrupted with speckle and other noise, so that colors corresponding to two modulation levels (125 and 255) are not clearly distinguished. As a result, color separation is very bad. Quantification of this result is under study.

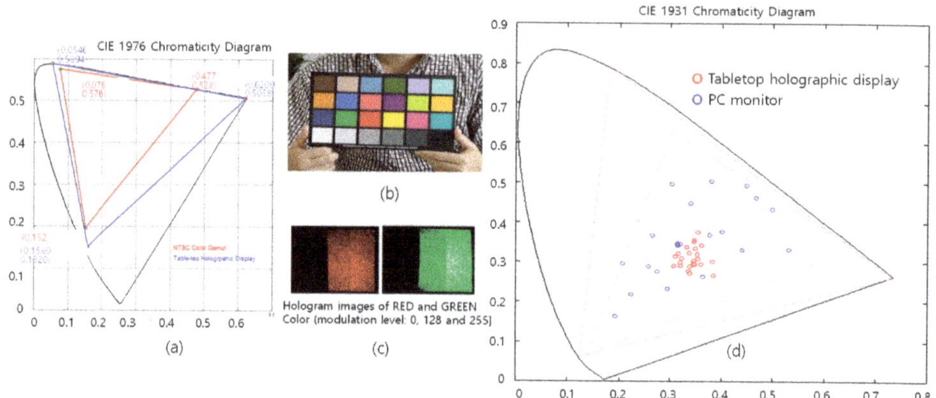

Figure 12. Color gamut and color fidelity of implemented system, (**a**) color gamut, (**b**) 24 color Gretag-Macbeth ColorChecker, (**c**) hologram image, and (**d**) measured color fidelity.

5.4. Speckle Noise

Speckle noise in holographic displays mainly comes from the use of coherent light sources like laser, which is one of major sources of quality degradation of hologram image. Measurement and analysis of speckle noise is well-studied, and speckle contrast is the most used metric to judge the severity of the speckle [27]. There has been quite a lot of researches for speckle noise reduction methods [28,29], and most of them is based on the concept of temporal multiplexing and averaging down overall speckle contrast. We developed an angular spectrum interleaving method, which is verified being able to reduce—by 60%—the level of speckle noise of the system down to 10% [30]. We are currently working on the physical realization of this method for our display system.

5.5. Degree of Depth of Focus (DoF)

As is always mentioned, the holographic display is most distinguished from other conventional 3D displays in the sense that it has the capability of supporting accommodation of human visual system. This means that holographic display can render an image at a specified depth plane, which can be focused as real objects. In previous studies, this has been simply demonstrated by showing two or more objects rendered at different distance, one of which is focused the others become defocused and blurred when captured by a shallow DoF camera. We can evaluate this capability of a display by using a very similar method to measuring 3D-MTF. For an MTF test pattern imaged at a certain depth plane, measuring MTF around the plane in viewing direction can tell how well the image is focused to the plane and defocused elsewhere. Figure 13a–c shows modulation depth values measured for a test pattern imaged at distances of 57 mm, 158 mm, and 213 mm from the virtual SLM plane, respectively. There are discrepancies between the measured image depth and the intended depth rendered in the CGH algorithm, which are 40 mm, 120 mm, and 160 mm. This is believed to be caused from the slight difference between the dimensions of designed and actual optical signal path of the system. It represents varying depth representation capability of the display through the image rendering volume. These modulation depth curves are obtained using the exact same measurement method as is used for obtaining MTF curves. Since for a real hologram image the spherically converging wavefront is supposed to focus at the image plane, defocusing or blur in front or behind this focal plane should be symmetric. So, deviations more than measurement error from this expected symmetry may come from aberrations of the optics of the system.

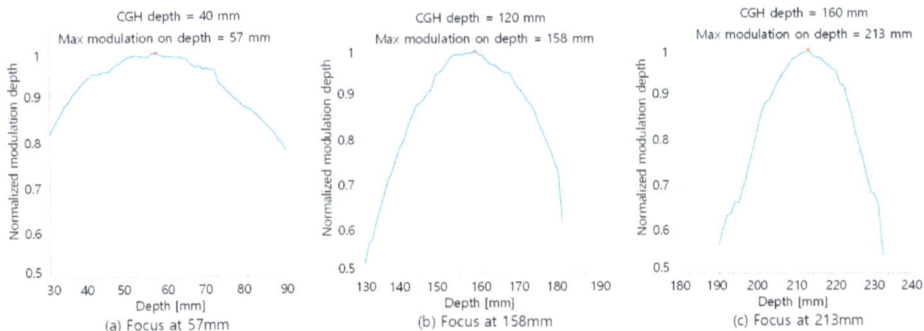

Figure 13. Degree of DoF at different depths, measured with MTF test patterns.

6. Conclusions

As visual media become pervasive in our daily life and more visual information is used for effective communication among people, new tools for more realistic visual content representation and exchange are required. Digital holographic technology is the most wanted one in this regard, since it can perfectly reproduce the visual experience we do meet every day. Due to the rapid progress in every field of technology, commercialization of digital holographic displays and imaging systems becomes visible. Since the working principle and requirements of holographic display systems are quite different from those of conventional displays, we need to explore diverse ways to realize commercially viable systems. In this paper, we presented design concepts, implementation methods of a new type of tabletop holographic display system, and addressed the important issue of measurement and quantitative evaluation for objective comparison of different systems. Especially important is that the quality and resolution of the reconstructed hologram image is measured and analyzed in terms of 3D-MTF, color reproduction fidelity, and degree of DoF. Though we designed each optical component for minimum aberration [20], the system we developed has several elements which degrade the quality of final image, including binary amplitude encoding of the fringe and mechanical movement during image rendering time. Rigorous analysis of the system regarding this issue remains as a future work. We hope this work gives some hint for future development of commercially acceptable holographic display systems.

Author Contributions: Writing—Original Draft Preparation, J.K.; Writing—Review and Editing, Y.L., K.H., H.K., H.-E.K., I.N., I.P., I.H., and Y.K.

Funding: This work was supported by the cross ministry GigaKorea project (GK18D0100, Development of Telecommunications Terminal with Digital Holographic Table-top Display) grant funded by the Korea government (MSIT).

Acknowledgments: This work is a part of the result of collaboration of ETRI and participating universities, companies and research institutes. Participated organizations: LG display, ASTEL, Aoptics, Yunam Optics, Funzin, SiliconWorks, MVTech, KETI, KIST, Konyang University, Korea University, Inha University, Kyungpook University, Sejong University, Kwangwoon University, and Warsaw University of Technology.

Conflicts of Interest: The authors declare no conflict of interest.

Appendix A

In this appendix, more details of 3D-MTF measurements and result of evaluation are presented. Figure A1 shows an optically reconstructed hologram image for a test pattern. Each stripe pair consists of a black and a white vertical bar of width 416.67 µm (1.2 cycles/mm). The CCD sensor (Sony ICX834) we used in the experiment has pixel size of 3.1 µm, so a black (or white) stripe is captured by about 134 pixels in horizontal direction. For each of horizontal pixel positions of the CCD sensor, pixel values are accumulated in vertical direction, which are the raw data we used for MTF calculation. Minmax-based MTF is calculated using maximum value (among 134 values) of white stripe and minimum value of black stripe. Average-based MTF is calculated using the average value of each black and white stripe. Depending on the depth we measure, twelve to fourteen stripe pairs were used, and the average value (yellow box in Table A1) is represented as the MTF value of corresponding depth in Figure 11d, with the range of values denoted by vertical bars in the graph. Standard deviation is also calculated and shown in the last column with heading of 'stdev' in Table A1. The measurement is carried out for a system without the lenticular sheet, which is used for extending vertical viewing angle. Measurements for Figure 13 are done with the same method.

Table A1. Measured values for MTF calculations.

a. Minmax_based MTF.

depth	20	40	60	80	100	120
max	0.519	0.576	0.572	0.551	0.551	0.581
mean	0.467	0.533	0.525	0.521	0.527	0.529
min	0.433	0.482	0.494	0.468	0.490	0.511

b.

SP#	1	2	3	4	5	6	7	8	9	10	11	12	13	14	average	Stdev
20	0.459	0.458	0.447	0.461	0.454	0.469	0.475	0.498	0.484	0.519	0.433	0.453	0.449	0.474	0.467	0.022
40	0.527	0.545	0.550	0.543	0.576	0.523	0.542	0.521	0.482	0.519	0.529	0.537	0.530		0.533	0.022
60	0.518	0.507	0.532	0.526	0.529	0.504	0.572	0.528	0.494	0.528	0.531	0.505	0.561	0.509	0.525	0.022
80	0.497	0.551	0.523	0.533	0.516	0.536	0.541	0.546	0.484	0.518	0.468	0.542			0.521	0.026
100	0.545	0.536	0.517	0.547	0.526	0.551	0.515	0.499	0.519	0.529	0.536	0.544	0.490		0.527	0.019
120	0.520	0.521	0.524	0.541	0.517	0.524	0.527	0.523	0.581	0.511	0.532	0.526			0.529	0.018

c. Average_based MTF.

depth	20	40	60	80	100	120
max	0.286	0.418	0.400	0.396	0.402	0.393
mean	0.259	0.393	0.370	0.377	0.374	0.381
min	0.217	0.372	0.336	0.357	0.336	0.369

d.

SP#	1	2	3	4	5	6	7	8	9	10	11	12	13	14	average	Stdev
20	0.286	0.260	0.276	0.247	0.284	0.271	0.243	0.262	0.279	0.252	0.251	0.260	0.217	0.231	0.259	0.020
40	0.405	0.418	0.400	0.413	0.400	0.393	0.394	0.380	0.378	0.382	0.372	0.383	0.391		0.393	0.014
60	0.375	0.369	0.400	0.363	0.383	0.371	0.388	0.370	0.359	0.373	0.366	0.336	0.365	0.360	0.370	0.015
80	0.359	0.396	0.376	0.387	0.381	0.382	0.386	0.382	0.365	0.363	0.357	0.385			0.377	0.013
100	0.388	0.370	0.378	0.402	0.380	0.368	0.392	0.361	0.383	0.368	0.363	0.368	0.336		0.374	0.017
120	0.369	0.370	0.382	0.384	0.387	0.393	0.383	0.382	0.387	0.370	0.383	0.382			0.381	0.008

SP#: stripe pair number.

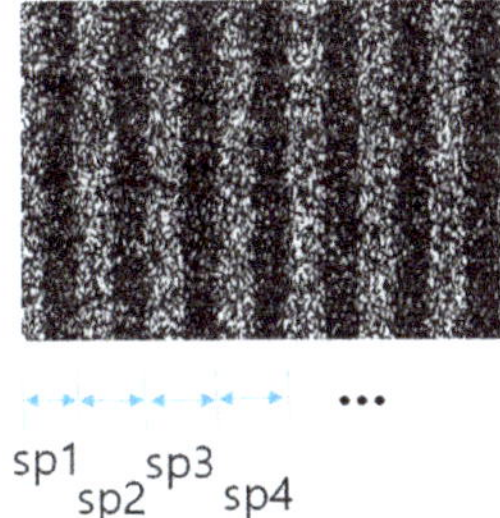

Figure A1. Reconstructed hologram image at 80 mm depth (brightness level is enhanced for clear presentation).

References

1. Lim, Y.; Hong, K.; Kim, H.; Kim, H.-E.; Chang, E.-Y.; Lee, S.; Kim, T.; Nam, J.; Choo, H.-G.; Kim, J.; et al. 360-degree tabletop electronic holographic display. *Opt. Express* **2016**, *24*, 24999–25009. [CrossRef] [PubMed]
2. St-Hilaire, P.; Benton, S.A.; Lucente, M.E.; Jepsen, M.L.; Kollin, J.; Yoshikawa, H.; Underkoffler, J.S. Electronic display system for computational holography. In Proceedings of the SPIE, Los Angeles, CA, USA, 18–19 January 1990; pp. 174–182.
3. Smalley, D.E.; Smithwick, Q.Y.J.; Bove, V.M.; Barabas, J.; Jolly, S. Anisotropic leaky-mode modulator for holographic video displays. *Nature* **2013**, *498*, 313–317. [CrossRef] [PubMed]
4. Häussler, R.; Reichelt, S.; Leister, N.; Zschau, E.; Missbach, R.; Schwerdtner, A. Large real-time holographic displays: From prototypes to a consumer product. *IS&T/SPIE Electron. Imaging* **2009**, *7237*, 72370.
5. Hahn, J.; Kim, H.; Lim, Y.; Park, G.; Lee, B. Wide viewing angle dynamic holographic stereogram with a curved array of spatial light modulators. *Opt. Express* **2008**, *16*, 12372–12386. [CrossRef] [PubMed]
6. Senoh, T.; Mishina, T.; Yamamoto, K.; Oi, R.; Kurita, T. Viewing-Zone-Angle-Expanded Color Electronic Holography System Using Ultra-High-Definition Liquid Crystal Displays with Undesirable Light Elimination. *J. Display Technol.* **2011**, *7*, 382–390. [CrossRef]
7. Oi, R.; Sasaki, H.; Wakunami, K.; Ichihashi, Y.; Senoh, T.; Yamamoto, K. Large size three-dimensional video by electronic holography using multiple spatial light modulators. *Sci. Rep.* **2014**, *4*, 6177.
8. Takaki, Y.; Okada, N. Hologram generation by horizontal scanning of a high speed spatial light modulator. *Appl. Opt.* **2009**, *48*, 3255–3260. [CrossRef] [PubMed]
9. Takaki, Y.; Matsumoto, Y.; Nakajima, T. Color image generation for screen-scanning holographic display. *Opt. Express* **2015**, *23*, 26986–26998. [CrossRef] [PubMed]
10. Blanche, P.-A.; Bablumian, A.; Voorakaranam, R.; Christenson, C.; Lin, W.; Gu, T.; Flores, D.; Wang, P.; Hsieh, W.-Y.; Kathaperumal, M.; et al. Holographic three-dimensional telepresence using large-area photorefractive polymer. *Nature* **2010**, *468*, 80–83. [CrossRef] [PubMed]
11. Park, M.; Chae, B.G.; Kim, H.; Hahn, J.; Moon, K.; Moon, K.; Kim, J. Digital Holographic Display System with Large Screen Based on Viewing Window Movement for 3D Video Service. *ETRI J.* **2014**, *36*, 232–241. [CrossRef]
12. Blinder, D.; Ahar, A.; Bettens, S.; Birnbaum, T.; Symeonidou, A.; Ottevaere, H.; Schretter, C.; Schelkens, P. Signal processing challenges for digital holographic video display systems. *Signal Process. Image Commun.* **2019**, *70*, 114–130 [CrossRef]
13. Kawashima, T.; Kakue, T.; Nishitsuji, T.; Suzuki, K.; Shimobaba, T.; Ito, T. Aerial projection of three-dimensional motion pictures by electro-holography and parabolic mirrors. *Sci. Rep.* **2015**, *5*, 11750.
14. Inoue, T.; Takaki, Y. Table screen 360 degree holographic display using circular viewing-zone scanning. *Opt. Express* **2015**, *23*, 6533–6542. [CrossRef] [PubMed]
15. Sando, Y.; Barada, D.; Yatagai, T. Holographic 3D display observable for multiple simultaneous viewers from all horizontal directions by using a time division method. *Opt. Lett.* **2014**, *39*, 5555. [CrossRef] [PubMed]
16. Jones, A.; McDowall, I.; Yamada, H.; Bolas, M.; Debevec, P. Rendering for an Interactive 360° Light Field Display. *ACM Trans. Graph.* **2007**, *26*, 40.

17. Butler, A.; Hilliges, O.; Izadi, S.; Hodges, S.; Molyneaux, D.; Kim, D.; Kong, D. Vermeer: Direct Interaction with a 360° Viewable 3D Display. In Proceedings of the ACM Symposium on User Interface Software and Technology, Santa Barbara, CA, USA, 16–19 October 2011.
18. Yoshida, S. fVisiOn: 360-degree viewable glasses-free tabletop 3D display composed of conical screen and modular project arrays. *Opt. Express* **2016**, *24*, 13194–13203. [CrossRef] [PubMed]
19. Takaki, Y. High-Density Directional Display for Generating Natural Three-Dimensional Images. *Proc. IEEE* **2006**, *94*, 654–663. [CrossRef]
20. Takaki, Y.; Nago, N. Multi-projection of lenticular displays to construct a 256-view super multi-view display. *Opt. Express* **2010**, *18*, 8824–8835. [CrossRef]
21. Hua, H.; Javidi, B. A 3D integral imaging optical see-through head-mounted display. *Opt. Express* **2014**, *22*, 13484. [CrossRef]
22. Kim, J.; Hong, K.; Lim, Y.; Kim, J.-H.; Park, M. Design options for 360 degree viewable table-top digital color holographic displays. In Proceedings of the SPIE Commercial + Scientific Sensing and Imaging, Orlando, FL, USA, 15–19 April 2018.
23. Chang, E.-Y.; Choi, J.; Lee, S.; Kwon, S.; Yoo, J.; Choo, H.-G.; Kim, J. 360-degree Color Hologram Generation for Real 3-D Object. *Dig. Holography Three-Dimens. Imaging* **2018**, *57*, A91–A100. [CrossRef]
24. Seo, W.; Song, H.; An, J.; Seo, J.; Sung, G.; Kim, Y.-T.; Choi, C.-S.; Kim, S.; Kim, H.; Kim, Y.; et al. Image Quality Assessment for Holographic Display. In Proceedings of the IS&T International Symposium on Electronic Imaging 2017, Burlingame, CA, USA, 29–31 January 2017.
25. Yoshikawa, H.; Yamaguchi, T. Image Quality Evaluation of a Computer-Generated Hologram. In Proceedings of the Digital Holography & 3-D Imaging Meeting, Orlando, FL, USA, 25–28 June 2018.
26. Hilaire, P.S. Modulation transfer function and optimum sampling of holographic stereograms. *Appl. Opt.* **1994**, *33*, 768–774. [CrossRef]
27. Ong, D.C.; Solanki, S.; Liang, X.; Xu, X. Analysis of laser speckle severity, granularity, and anisotropy using the power spectral density in polar-coordinate representation. *Opt. Eng.* **2012**, *51*, 054301. [CrossRef]
28. Takaki, Y.; Yokouchi, M. Speckle-free and grayscale hologram reconstruction using time-multiplexing technique. *Opt. Express* **2011**, *19*, 7567–7579. [CrossRef] [PubMed]
29. Makowski, M. Minimized speckle noise in lens-less holographic projection by pixel separation. *Opt. Express* **2013**, *21*, 29205–29216. [CrossRef] [PubMed]
30. Lim, Y.; Park, J.; Hahn, J.; Kim, H.; Hong, K.; Kim, J. Reducing speckle artifacts in digital holography by the use of programmable filtration. *ETRI J.* **2019**. [CrossRef]

Article

Optical Design for Novel Glasses-Type 3D Wearable Ophthalmoscope

Cheng-Mu Tsai [1], Tzu-Chyang King [2], Yi-Chin Fang [3,*], Nai-Wie Hsueh [3] and Che-Wei Lin [3]

[1] Department of Applied Physics, National Pingtung University, Taichung 402, Taiwan; jmutsai@email.nchu.edu.tw
[2] Graduate Institute of Precision Engineering, National Chung Hsing University, Pingtung County 900, Taiwan; tcking@mail.nptu.edu.tw
[3] Department of Mechanical and Automation Engineering, National Kaohsiung First Univ. of Science and Technology, Kaohsiung 824, Taiwan; s3715748@gmail.com (N.-W.H.); U0357811@nkfust.edu.tw (C.-W.L.)
* Correspondence: yfang@nkfust.edu.tw; Tel.: +886-7-6011000 (ext. 32290)

Received: 24 November 2018; Accepted: 31 January 2019; Published: 19 February 2019

Featured Application: This proposed miniature glasses-type 3D wearable ophthalmoscope presents a novel optical design, which is aimed to functionally improve the current glasses-type ophthalmoscope in the market by cooperating with 3D image technology, infrared spectrum technology, future medical diagnostics, the cloud and big data analysis.

Abstract: This paper proposes a new optical design that will cooperate with 3D image technology, infrared spectrum technology, future medical diagnostics, the cloud, and big data analysis. We first conducted image recognition experiments to compare the pros and cons of 2D and 3D frameworks in order to make sure that the optical and mechanical framework of a glasses-type 3D ophthalmoscope would be a better choice. The experimental results showed that a 3D image recognition rate (90%) was higher than a 2D image recognition rate (84%), and hence the 3D mechanism design was selected. The glasses-type 3D ophthalmoscope design is primarily based on the specification of indirect ophthalmoscope requirements and two working spectrums: a near infrared and a visible spectrum. The design is a 2.5x magnification fixed focal telecentric relay system with a right-angle prism, which uses a large aperture to increase the amount of incident light ($F/\# = 2.0$). As the infrared spectrums that have better transmittance towards human eye tissue are 965 nm and 985 nm, so that we took account of the visible spectrum and the near-infrared spectrum simultaneously to increase the basis of the physician's diagnosis. In this research, we conclude that a wearable ophthalmoscope can be designed optically and mechanically with 3D technology, an infrared and a visible working spectrum and further, possibly in cooperation with the cloud and big data analysis.

Keywords: ophthalmoscope; lens design; medical optics and biotechnology

---

1. Introduction

Owing to the rapid increase in the ageing population, more ophthalmic diseases than ever require regular diagnosis and early prevention; thus, the demand for ophthalmic medicals has increased. As technology has progressed, ophthalmic diagnostic equipment technology has been enhanced [1,2]; from the early stages, when doctors diagnosed using eye observation, it has evolved into the current indirect tomography scans. An ophthalmoscope is one of the most important pieces of equipment in ophthalmology clinics. To enhance the convenience of a diagnosis, the ophthalmoscope has evolved from a handheld instrument into an indirect ophthalmoscope and a headset ophthalmoscope. Since the requirements for optical system specifications have increased, system module expansions requiring complicated operation are necessary to obtain good imaging quality [3,4]. However, so far there has been no breakthrough that obviates the need for the traditional ophthalmoscope. Therefore, the new generation of optical design requires more innovative breakthrough ideas for ophthalmic medical applications. At the same time, current electronic equipment is gradually raising the system requirements and moving toward creating business opportunities for wearable electronics, which allow users to take advantage of science and technology. This goal is a shared vision for academics and manufacturers to study and hopefully make breakthroughs. Ophthalmoscope equipment is mostly stored in hospitals for use by doctors; thus, patients need to go to a hospital for diagnosis and treatment. However, the development of the cloud and big data technology has made it technically feasible for today's patients to receive medical care at home. If it were possible to use an ophthalmoscope at home, patients could operate the equipment themselves and transfer the data to the hospital using the cloud. This would improve the convenience of being diagnosed by an ophthalmologist. Starting with the demand for a home-use ophthalmoscope, this paper proposes a novel glasses-type 3D ophthalmoscope design. The glasses-type 3D ophthalmoscope design is primarily based on the specifications of indirect ophthalmoscope requirements and two working spectrums: one near infrared and the other the visible spectrum. The ophthalmoscope is based on right angle prism to design a 2.5x magnification telecentric relay system with fixed focal length and F/# = 2.0. As a result, the infrared spectrums at 965 nm and 985 nm have better transmittance towards human eye tissue. The visible and the near-infrared spectra are applied to assist increasing the basis of a physician's diagnosis. In this research, we conclude that a wearable ophthalmoscope can be designed optically and mechanically with 3D technology, infrared and visible working spectrum and further, possibly in cooperation with the cloud and big data analysis.

2. Image Recognition Experiment

We first conducted a human eye image recognition experiment by comparing the pros and cons of 2D and 3D image recognition. Twenty people, randomly chosen, aged between 20 and 25 years, participated in this experiment. According to professional ophthalmologists, students aged around 20 are at the peak of human vision. In addition, their visual acuity is statistically consistent. Compared with many subjects, the elderly are more vulnerable to psychological factors, working environment, or imminent death. For example, the U.S. military conducted large-scale visual testing experiments in the 1950s; all subjects were young soldiers. Their visual acuities after correction were greater than 0.8 and none of the participants had any form of eye disease. The display monitor was a 55-inch 3D liquid-crystal display (LCD) television, adopting shutter 3D display technology with a resolution of 1920×1080 pixels and a fixed brightness (display brightness was 34 cd/m^2). The distance from the human eye to the screen was 250 cm. The experimental images were circular figures with a diameter of 50 cm on the screen. Experimental conditions are listed in Table 1.

Table 1. Experimental conditions.

Display Monitor	3D Television	
Image type	2D	3D
Participant number	20 people	
Display duration	5 s per image	
Background illuminance	10 cd	
Distance between eyes and screen	250 cm	
Circular figure diameter	50 cm	

2.1. Experimental Imaging Setting

Fairhurst and Lettington proposed using geometric figures to replace complex images in image recognition experiments in 1998 and 2000 [5,6]. The images used in this experiment, as shown in Tables 2–4, were circles with diameters of 50 cm, formed by geometric figures. In this experiment, the control variables were (1) two types of 2D and 3D images; (2) shapes divided into variations of seven types, namely, round, vertical rectangle, horizontal rectangle, triangle, square, pentagon, and hexagon, as shown in Table 2; (3) low contrasts divided into variations of six types, namely 20%, 10%, 8%, 6%, 4%, and 2%, as shown in Table 3; and (4) spatial frequency, which is the amount of frequency variation contained within a unit space [7,8]. In this study, spatial frequency is defined as the figure changes in periodic numbers on the circumference of the rings on the screen, being seen from a position of 30 cm in front of the human eye within a width of 2 cm. In this study, spatial frequencies of eight types were used: 0.9, 1.1, 1.2, 1.5, 1.8, 2.5, 3.7, and 5.4, as shown in Table 4. There were $7 \times 6 \times 8 = 336$ types of image in total for every participant in 2D or 3D test.

Table 2. Geometric figures.

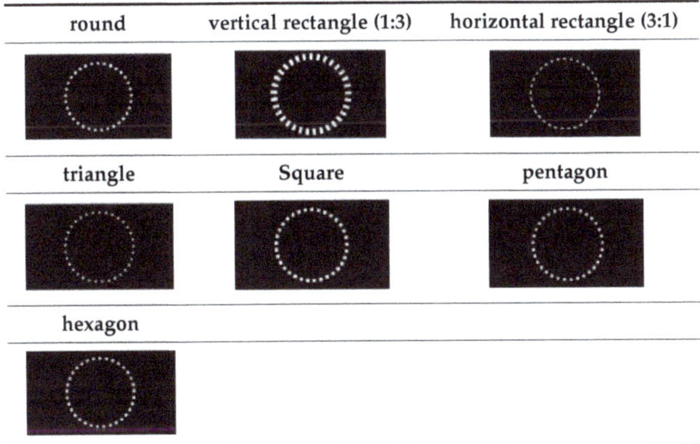

round	vertical rectangle (1:3)	horizontal rectangle (3:1)
triangle	Square	pentagon
hexagon		

Table 3. Low contrast variations.

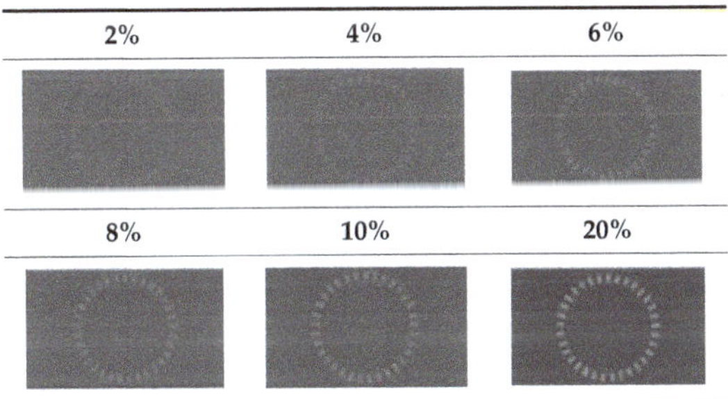

2%	4%	6%
8%	10%	20%

Table 4. Spatial frequency variations.

0.9	1.1	1.2
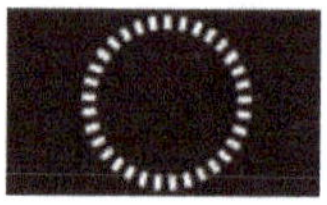	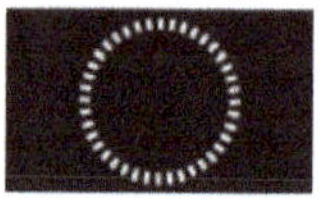	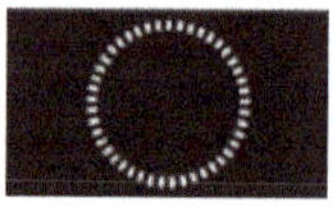
1.5	**1.8**	**2.5**
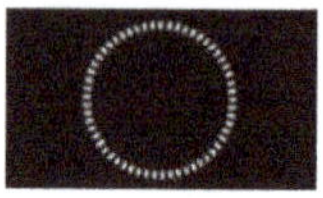		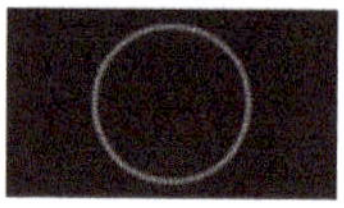
3.7	**5.4**	

2.2. Image Recognition Results

The results of image recognition are shown in Table 5, where N is the number of participants, M is the average of correctly recognized images. "T" is the parameter of t-test. The "p" is the significance, obviously distinguishable when $p < 0.05$. The η is semi-partial correlation. The SD is standard deviation, which is a measure to quantify the amount of variation or dispersion of a set of data values. Compared to 3D experiment and 2D ones, the SD of 3D experiment is obviously smaller, which indicates that the 3D vision is more accurate than 2D. The results show that the 3D image recognition rate ($302.05/336 \approx 90\%$) is higher than the 2D image ($281.85/336 \approx 84\%$). Using an independent sample test, the difference between the two groups achieved a statistical significance ($p < 0.01$), and the strength of association measures ($2 > 0.14$) in large effect size [9].

Table 5. Test of effects.

Image Type	N	M	SD	$T\ (s)$	p	η^2
3D	20	302.050	11.587	2.751	0.009	0.166
2D	20	281.850	30.720			

3. Glasses-Type 3D Ophthalmoscope Design

In this experiment, an optical system was simulated and optimized by two different spectra, visible and near-infrared (close to 950 nm). It is because near infrared spectrum around 950 nm have less absorption for water so that we might derive some near infrared pictures from human eyes. Part of the near infrared light can penetrate the eye so the optical system may be able to see through the internal structure of human eye. Therefore, this optical system can replicate the 3D surface of the eye, including part of the internal structure. Therefore, the data can be uploaded anytime by cloud technology to provide long-term analysis and tracking of personal eye health via medical center big data analysis system.

The glasses-type 3D ophthalmoscope design allows patients to use the equipment at home, obtain timely images of the fundus of the eyes, and reduce the complexity of medical diagnosis by transferring big data to the hospital through the cloud technology. In the design, the position of the ophthalmoscope's lens inside the glasses frame is aligned with the human eye. Figure 1 is a product schematic, plotted using Solid Work. This design utilizes four lenses in total; every two-lens structure comes with a single 3D image-taking module.

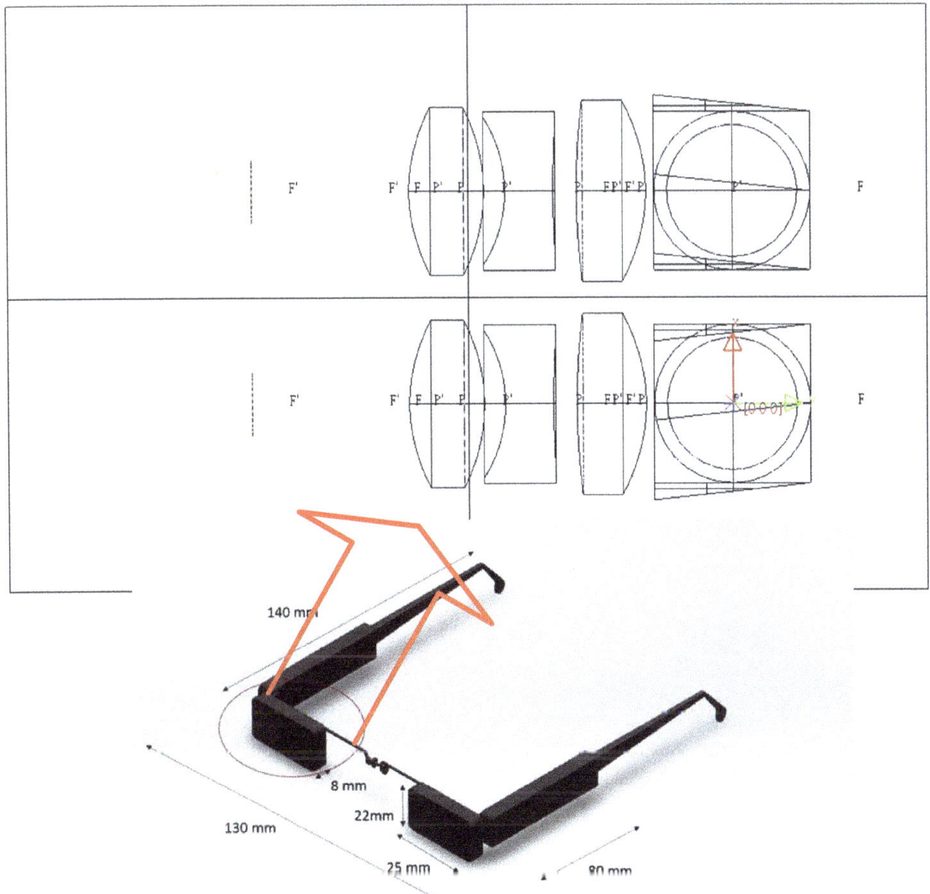

Figure 1. A conceptual design: the left diagram illustrates the detailed structure inside a single-glass frame of the right diagram; that is, a single 3D image-taking module.

The glasses-type ophthalmoscope lens mechanism is designed to be rotatable. This means that, when required to take images of the fundus of the eyes, the 3D image-taking modules are rotated to the front of the eyes for shooting, to ensure that the lens can accurately capture clear images of the fundus of the eyes. When the 3D image-taking modules are rotated, so that they are enclosed within the frames, they can be used as ordinary glasses, as shown in Figure 2.

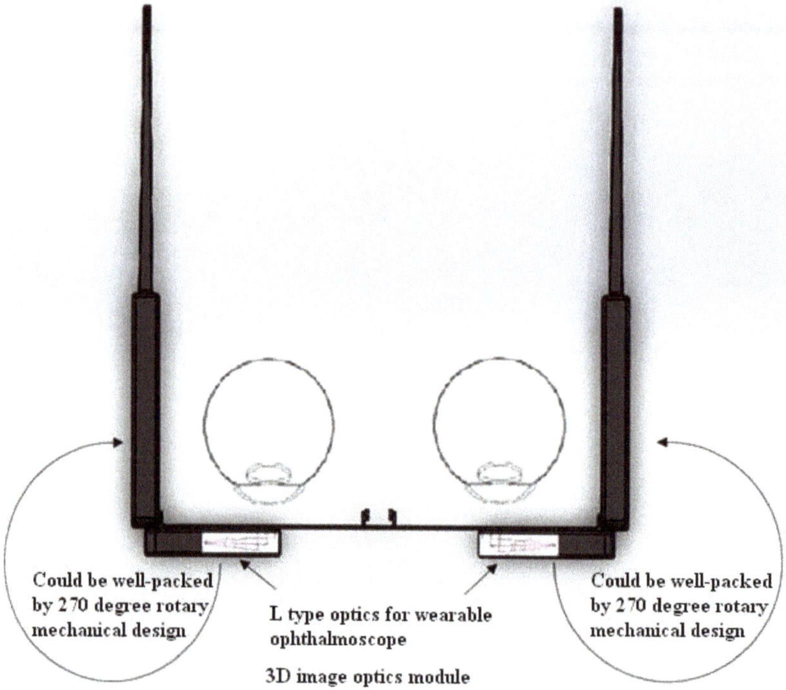

Figure 2. Enclosed concept for the ophthalmoscope lens groups; 3D image-taking modules are rotated 270 degrees and enclosed in the sides of the glasses for use as ordinary glasses.

3.1. Optical Design for Glasses-Type 3D Ophthalmoscope Lens

In accordance with the glasses-type 3D ophthalmoscope framework, this study analyzed the specifications of the lens design, in which visible lights and near-infrared lights (985 nm and 965 nm) were used for optimization with the simulation design. These two segments of near-infrared light wavelength spectrum have good transmission through the crystal body and the organ in the human eye of the retina, effectively increasing the macular image of the retina and enabling the symptoms to be observed via the infrared wavelength penetration [10]. This design uses a large aperture to increase the amount of incident light. Figure 3 shows a right-angle prism inverted fixed focal design for visible light and the near-infrared wavelengths. Its specifications are listed in Table 6.

Table 6. Right angle prism inverted fixed focal system specifications.

Initial Conditions of Design	
Image Height	2.4 mm
Source wavelength	Photonic 5 with 481, 546, 656, 965, 985 nm
Focal Length	10 mm
Magnification	2.5x
F/#	2.0
Overall Length	20 mm
Optical Distortion	<3%
MTF	>20–40% (100 lp/mm)

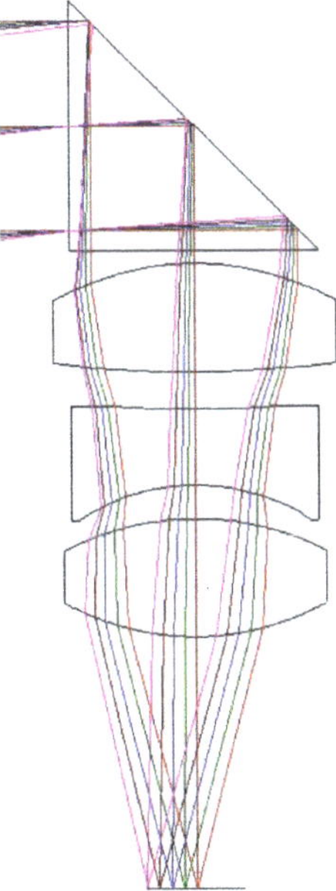

Figure 3. Right angle prism inverted fixed focal lens simulation design.

3.2. Imaging Performance Assessment (Spots, Visual Field Diagrams, Aberrations, Modulation Transfer Function)

We employ the optical software CODE V, a professional optical design software from the 1980s, to design the lens and implement the simulation. The CODE V is well-known for its accuracy of pupil grid function so that CODE V is able to predict pupil curve so precisely compared to other software. The important is that pupil curve directly reflect the aberrations of whole system except chromatic aberrations. Other software without precise description of CODE V might provide higher MTF because its pupil is close to perfect circle, which is simply a placebo. Besides, CODE V is very powerful with regard to optimization and extended optimization. LightTool is a non-image optimization software based on CODE V's optimization engine, which started from 2000. LightTool is different from traditional non-image optical software ASAP. LightTool is great for optimization but ASAP takes advantage of light tracking analysis.

The imaging qualities of a right angle prism fixed focal optical design system are as shown in Figures 4–7. Images with Modulation Transfer Function (MTF) less than 20% at specific spatial frequencies will be no longer visible to the human eye if there is no electric noise available such as film camera. Image system with Complementary Metal-Oxide-Semiconductor (CMOS) or other electric sensor might require up to MTF 40% in order to guarantee that target at that specific spatial frequency is visible due to inherent electric noise. That's the reason why the 20% to 40% MTF is suggested, which depends on noise level of electric detector. Human eyes might detect higher frequencies in some cases. However, this kind of miniature lens employed in this case has its limitation up to 20% to 40% at 100 lp/mm according to some experiments. Their performance is limited due to chromatic aberrations, total length restriction, and actual front and rear size. The reason given is that very few plastic optical materials could be selected for this kind of optical design so that its MTF performance might be limited.

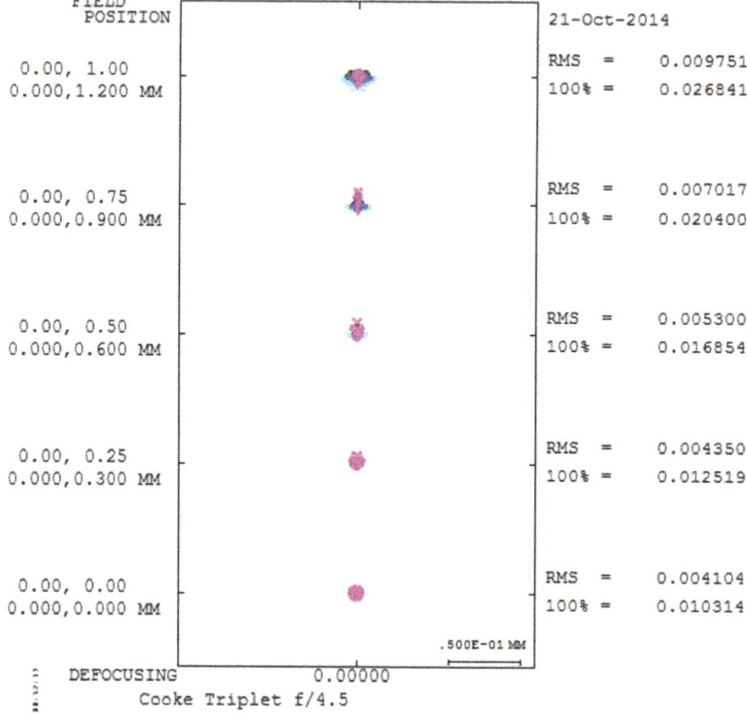

Figure 4. Spots diagram.

In Figure 4, we see that lateral color aberration, coma aberration, and astigmatism aberration play a role at the full field, although the diameter of spot diagram has been reduced to the minimum. The reason why the performance of the edge field image is not as good as the central image is given: first, the aperture stop of this lens is at the front of prism in order to get the telecentric effect, which is critical for digitalized image system. This kind of optical design might complicate the aberrations so that it is very difficult to minimize the three aberrations mentioned above. More elements have to be added if aberrations are to be further eliminated. However, the overall length of this lens is limited so that there is no room for more elements.

Distortion has been well controlled, as shown in Figure 5. Generally speaking, 2% of distortion is maximum for human vision. In this case, the distortion is under 0.20%. With regard to field curvature, it will be balanced with astigmatism so that both have to be evaluated by modulation transfer function (MTF) in Figure 6. The minimum MTF requirement of this lens will be 40% at 40 lp/mm and the ideal performance will be 20% at 100 lp/mm. According to simulated diffraction MTF, it is concluded that our MTF surpasses the minimum requirements of this kind optical design so all MTF pass the ideal MTF requirement except full field. At full field, astigmatism combined with field curvature appears, which reflects that this optical element for this design is not sufficient to reach ideal performance. Three reasons are given: firstly, the restricted overall length of the optics. Secondly, there is a very wide range of spectrum, from 481 nm to 985 nm. The wider the optical spectrum is, the more difficult optical design will be. Thirdly, according to the wavelength weight in Figure 6, most credit of optical element has been assigned to eliminate chromatic aberration. Without a sufficient optical element, it is not possible to obtain perfect optics. In conclusion, this optical design is close to ideal performance (roughly equivalent to diffraction limitation of optics).

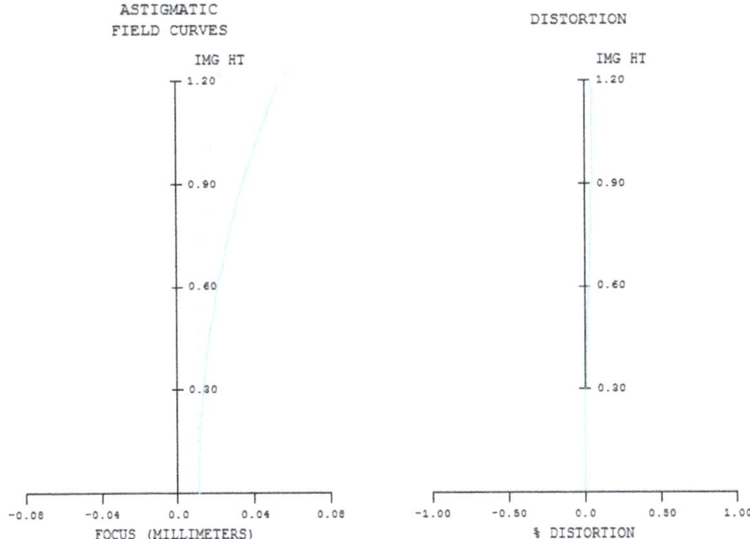

Figure 5. Visual field diagrams.

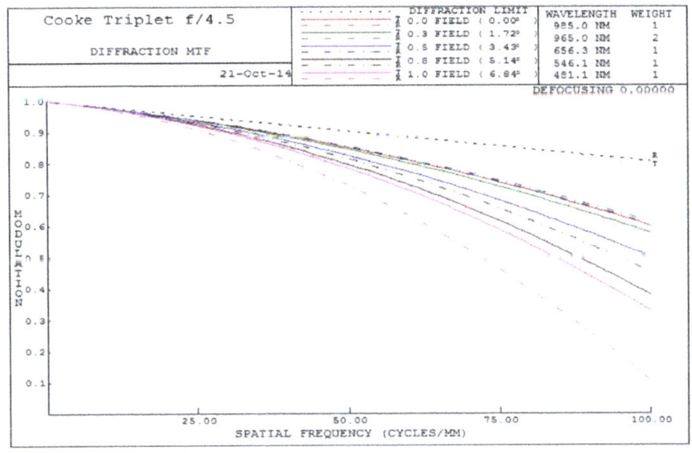

Figure 6. MTF curves of proposed optics.

From Figure 7, we can conclude that this lens suffers severe lateral color aberration and astigmatism, which might be inherent in such optical designs with prism and location of aperture stop at first place and whose working spectrum is from 483 nm to 985 nm. We believe that the optical design in this paper is close to ideal performance. Further improvement will be dependent on the discovery of new optical materials and power optimization software.

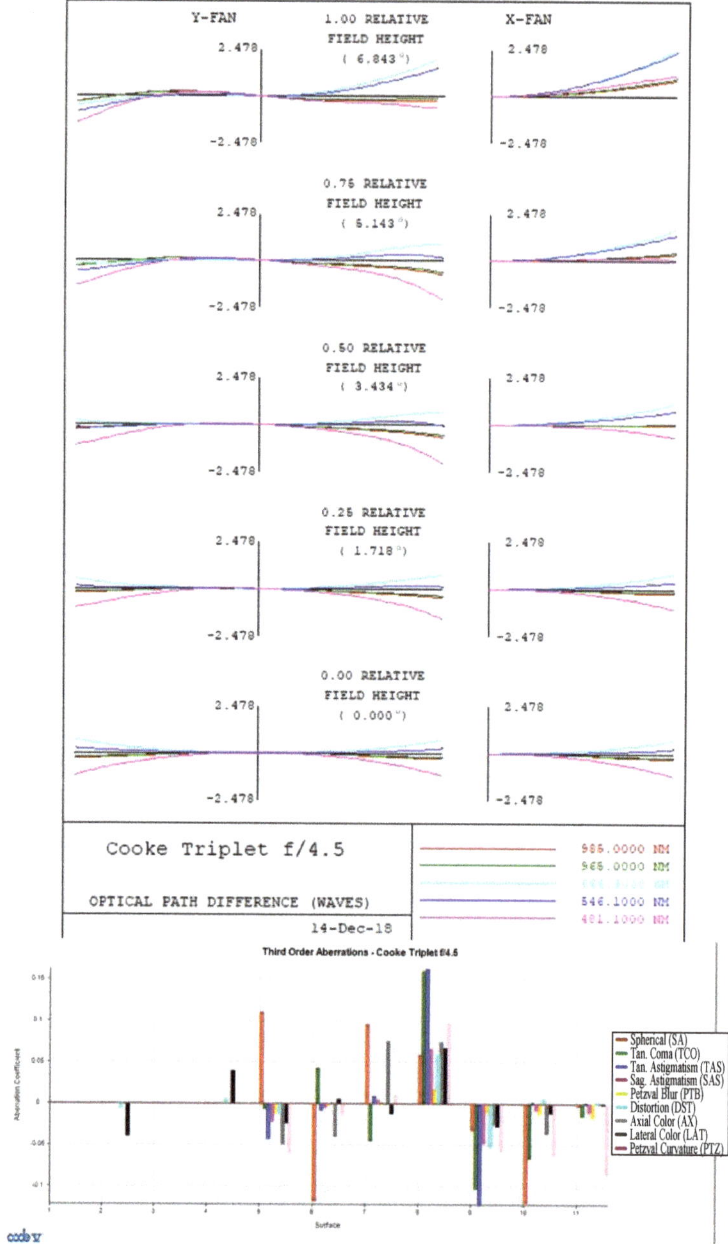

Figure 7. Aberration diagrams of proposed optics.

3.3. 3D Image-Taking Structure Design

In order to create 3D stereoscopic effects when the doctor views the images of the eye fundus through an 8K computer screen, the positional relationship of the fundus image screen presented to the doctor's eyes has to be consistent with the positional relationship of the patient's eye fundus when using the 3D image-taking lenses, as shown in Figure 8. The normal distance between the doctor's eyes and the computer is 300 mm, resulting in an image depth of 0.5 mm. According to the "Stereo-image design for flat-screen" proposed by Liu [11] in 2001, the angles of 3D image-taking are presented. After shrinking the scale of the human eye computer-viewing distance, usage and other things, the simulation of the lenses shooting the image of the eye fundus would compute the structural relationship of "the human eye toward the computer" and the "lenses toward the eye fundus."

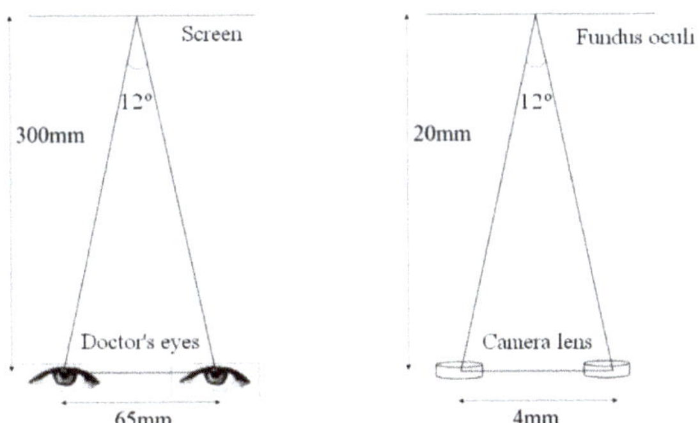

Figure 8. Structural relationship between human eyes and 3D ophthalmoscope glasses.

The imaging optics software Code V was used to complete the lens group design, and the non-imaging simulation optics software of Light Tools was inserted to simulate the two groups of lens modes, the angle, and the distance of the 3D structure. Figure 9 shows a schematic of the established 3D optical mechanism.

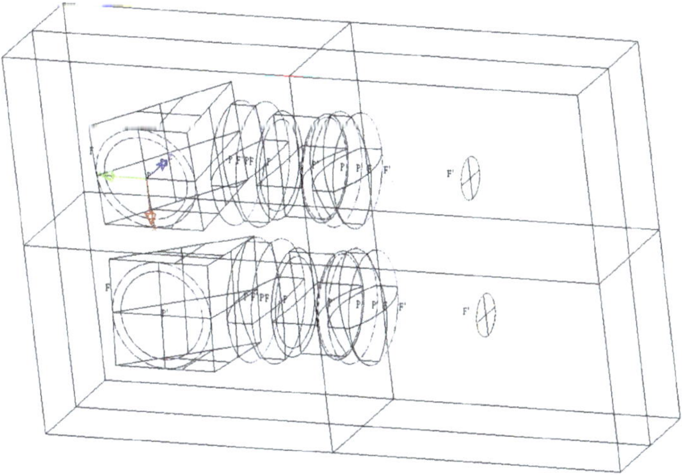

Figure 9. Schematic of the 3D optical mechanism.

4. Conclusions and Recommendations

This study establishes an ophthalmoscope system that integrates the cloud and big data with a wearable-type ophthalmoscope, double working spectrum, and 3D image-taking technology in the same hardware design. This design allows patients to transfer the captured images to their ophthalmic hospital via the cloud method using big data, thus shortening the diagnosis time and increasing doctors' basis for diagnosis. Currently, the main research outcomes are as follows:

(1) Image recognition experiment: The 3D image recognition rate (90%) was higher than the 2D image recognition rate (84%); this reached obvious significance and, thus, the 3D mechanism design was selected.

(2) A glasses-type 3D ophthalmoscope lens mechanism design: A wearable ophthalmoscope with 3D image-taking features and visible-infrared working spectrum was developed. These features are suitable for the diagnostic tracking of eye fundus' imaging using a rotatable mechanism via a 3D optical structure design.

(3) Ophthalmoscope lens design: Using imaging simulation technology, a set of ophthalmoscope lenses was designed to fit a glasses-type 3D ophthalmoscope; then, with the mechanism design, the specifications of wearable ophthalmoscope glasses were significantly reduced.

(4) This proposed design and its mechanism can be connected with wireless networks, Bluetooth, or a cloud processing system in order to transfer the diagnostic images to the hospital and assist in the doctors' diagnosis via cloud technology and big data statistics.

Optical zoom employed in wearable glasses will be in even higher demand in the future. The relay lens can be adapted to invert the images in the position of the glasses frame and will provide sufficient space to accommodate zoom optics, which may be three times larger than fixed focal optics.

Author Contributions: C.-M.T. and Y.C.F. conceived and designed the experiments; N.-W.H. and C.-W.L. performed the experiments. T.-C.K. analyzed the data. Y.C.F. contributed reagents/materials/analysis tools. Y.C.F., C.-M.T., N.-W.H. and C.-W.L. wrote the paper.

Acknowledgments: The authors wish to thank the anonymous reviewers for their valuable suggestions. This study was supported by the Ministry of Science and Technology (MOST) of Taiwan under Contract No.: MOST 107-2221-E-005-050.

Conflicts of Interest: The authors declare no conflict of interest.

References

1. Meadway, A.; Girkin, C.A.; Zhang, Y.H. A dual-modal retinal imaging system with adaptive optics. *Opt. Express* **2013**, *24*, 29792–29807. [CrossRef] [PubMed]

2. Felberer, F.; Kroisamer, J.S.; Hitzenberger, C.K.; Pircher, M. Lens based adaptive optics scanning laser ophthalmoscope. *Opt. Express* **2012**, *16*, 17297–17310. [CrossRef] [PubMed]

3. Wu, B.W. Optical computing for application to reducing the thickness of high-power-composite lenses. *Appl. Opt.* **2014**, *53*, H7–H13. [CrossRef] [PubMed]

4. Sun, J.H. Tolerance reallocation of an optical zoom lens to meet multiperformance criteria. *Appl. Opt.* **2014**, *53*, 233–238. [CrossRef] [PubMed]

5. Fairhurst, A.M.; Lettington, A.H. The effect of visual perception on the required performance of imaging systems. *J. Mod. Opt.* **2000**, *8*, 1435–1446. [CrossRef]

6. Fairhurst, A.M.; Lettington, A.H. Method of predicting the probability of human observers recognizing targets in simulated thermal images. *Opt. Eng.* **1998**, *3*, 744–751. [CrossRef]

7. Wu, B.W.; Fang, Y.C.; Chang, L.S. Studies of Human Vision Recognition: Some Improvements. *J. Mod. Opt.* **2009**, *57*, 107–114. [CrossRef]

8. Fang, Y.C.; Wu, B.W. Prediction of the Thermal Imaging Minimum Resolvable (Circle) Temperature Difference with Neural Network Application. *IEEE Trans. Pattern Anal. Mach. Intell.* **2008**, *30*, 2218–2228. [CrossRef] [PubMed]

9. Yust, B.G.; Mimun, L.C.; Sardar, D.K. Optical absorption and scattering of bovine cornea, lens, and retina in the near-infrared region. *Lasers Med. Sci.* **2011**, *2*, 413–422. [CrossRef] [PubMed]

10. Cohen, J. *Statistical Power Analysis for the Behavioral Science*; Academic Press: New York, NY, USA, 1977.

11. Liu, R.Z. Stereo-Imaging Design for Flat-Screen. Master's Thesis, Institute of Optical Sciences, National Central University, Taoyuan, Taiwan, 2001.

Article

Holographic Three-Dimensional Virtual Reality and Augmented Reality Display Based on 4K-Spatial Light Modulators

Hongyue Gao †, Fan Xu †, Jicheng Liu *,†, Zehang Dai , Wen Zhou , Suna Li and Yingjie Yu and Huadong Zheng

Ultra-precision Optoelectronic Metrology and Information Display Technologies Research Center, Department of Precision Mechanical Engineering, School of Mechatronic Engineering and Automation, Shanghai University, Shanghai 200072, China; gaohylet@shu.edu.cn (H.G.); fanxu1023@163.com (F.X.); dai457975764@163.com (Z.D.); 18800375810@163.com (W.Z.); sophe_cbd1234@163.com (S.L.); yingjieyu@staff.shu.edu.cn (Y.Y.); bluenote2008@shu.edu.cn (H.Z.)

* Correspondence: liujicheng@shu.edu.cn

† These authors contributed equally to this work.

Received: 11 January 2019; Accepted: 11 March 2019; Published: 20 March 2019

Abstract: In this paper, we propose a holographic three-dimensional (3D) head-mounted display based on 4K-spatial light modulators (SLMs). This work is to overcome the limitation of stereoscopic 3D virtual reality and augmented reality head-mounted display. We build and compare two systems using 2K and 4K SLMs with pixel pitches 8.1 μm and 3.74 μm, respectively. One is a monocular system for each eye, and the other is a binocular system using two tiled SLMs for two eyes. The viewing angle of the holographic head-mounted 3D display is enlarged from 3.8° to 16.4° by SLM tiling, which demonstrates potential applications of true 3D displays in virtual reality and augmented reality.

Keywords: holography; holographic display; digital holography; virtual reality

1. Introduction

Virtual reality (VR) and augmented reality (AR) have been hot topics recently, and 3D technology is an important part of VR technology [1]. VR is widely used in various fields, such as architecture, gaming, and education, where it has a promising future in the development of VR. Compared to holographic displays, conventional optics present several disadvantages for near-to-eye 3D products for VR and AR [2]. Although 3D techniques are popular in current VR and AR applications, most 3D head-mounted displays (HMD) are based on stereoscopic 3D display technology, i.e., left and right eyes get two images with binocular parallax. The human brain can combine these two images into a 3D image. However, these images in the viewing angle of the stereoscopic 3D display are separate two-dimensional (2D) images, which are different from the continuous wavefront of 3D objects or 3D digital models. When images were displayed outside the range of accommodation, visual fatigue was induced [3–6]. The stereoscopic 3D display is not suitable for everyone to watch for a long time, because not everyone's pupil distance is within the applicable range. Therefore, 3D techniques should be improved to overcome the drawback in VR and AR. It is a conventional 3D technology in VR head-mounted displays, as shown in Figure 1.

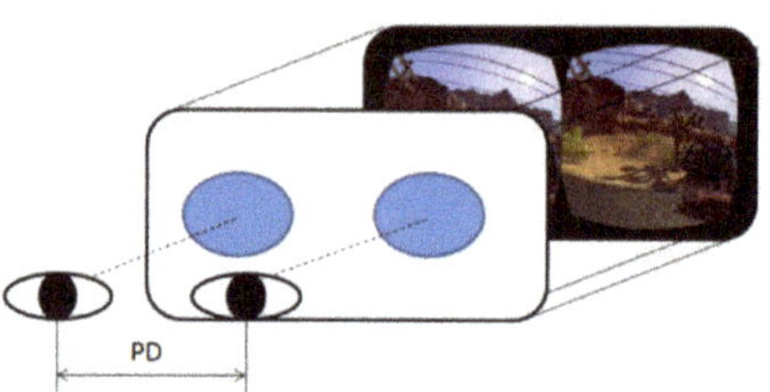

Figure 1. Stereoscopic display. PD: pupil distance.

To overcome the limitation of the stereoscopic 3D technique, some researchers have studied some other 3D techniques in HMD. Eunkyong Moon et al. used an LED light source instead of lasers as the reading light source in the holographic HMD and produced a nice effect [7]. Han-Ju Yeom et al. did some work on astigmatism aberration compensation of the holographic HMD, and their system can present holographic 3D images in a see-through fashion clearly in a wide depth range [8]. J. Han et al. presented a compact waveguide display system integrating free-form elements and volume holograms; this design makes the system more compact because of the waveguide [9]. Mei-Lan Piao et al. used a transparent volume hologram for a holographic projection HMD and filmed the holograms to show that the see-through holographic projection HMD system can present three-dimensional images clearly [10]. Galeotti et al. designed real-time tomographic holography for AR, which used a large-aperture holographic optical element (HOE) and obtained good results, using a large-aperture HOE to project an off-axis, viewpoint-independent virtual image 1 m away [11]. H.-E. Kim et al. used an active shutter for HMD application to get a more precise system [12]. J. Piao et al. used photopolymer for a full-color waveguide-type HMD and obtained a compact system, and it provided wide angular selectivity and can fabricate high-quality full-color HOEs [13]. Gang Li et al. used a mirror-lens to propose a see-through AR display and achieve an optimized optical recording condition of the mirror-lens HOE [14]. Jong-Young Hong et al. used an index-matched anisotropic crystal lens to propose a holographic see-through near-eye display, and the system showed the possibility of a holographic display with a large field of view [15]. Peng Sun et al. proposed a holographic near-eye display system based on a double-convergence light algorithm and provided a promising solution in future 3D AR realization [16]. Jisoo Hong et al. proposed a near-eye foveated holographic display [17]. Among these 3D techniques, holography is thought to be a good candidate for true 3D VR, because digital 3D holography based on spatial light modulators (SLMs) can provide the holographic 3D property of wavefront reconstruction and realize real-time video rate dynamic display [18–21]. In this paper, we present a holographic 3D HMD and study its 3D properties, such as viewing angle and depth of field dependent on the pixel size of SLMs. We believe that with the development of SLMs and digital information technology, holographic 3D VR and AR can be applied in the future.

2. Monocular System

Holographic VR is a true 3D display, which is different from that based on stereoscopic 3D technology. Stereoscopic displays can cause visual fatigue, which is not suitable for most people to watch for a long time. In holographic 3D VR, the image has a certain depth of field, which is the distance between the nearest and the furthest objects that are in acceptably-sharp focus in an image [22], and the maximum diffraction angle depends on the pixel pitch of SLM, while the viewing angle of the reconstructed images, θ, is twice the maximum diffraction angle, which is the maximum angle at which an image can be viewed with acceptable visual performance. Equation (1) shows the relationship between the pixel pitch pand the viewing angle θ.

$$\theta = 2\arcsin\frac{\lambda}{2p} \tag{1}$$

where λ is the wavelength of reading light. For instance, here, the wavelength of reading light is 532.8 nm. When the pixel pitch of SLMs is 8.1 µm and 3.74 µm, respectively, obtained by calculation, the viewing angle should be 3.81° (θ_1) and 8.2° (θ_2), respectively, as shown in Figure 2a. It is obvious that the smaller pixel pitch of SLM leads to a larger viewing angle, as shown in Figure 2b.

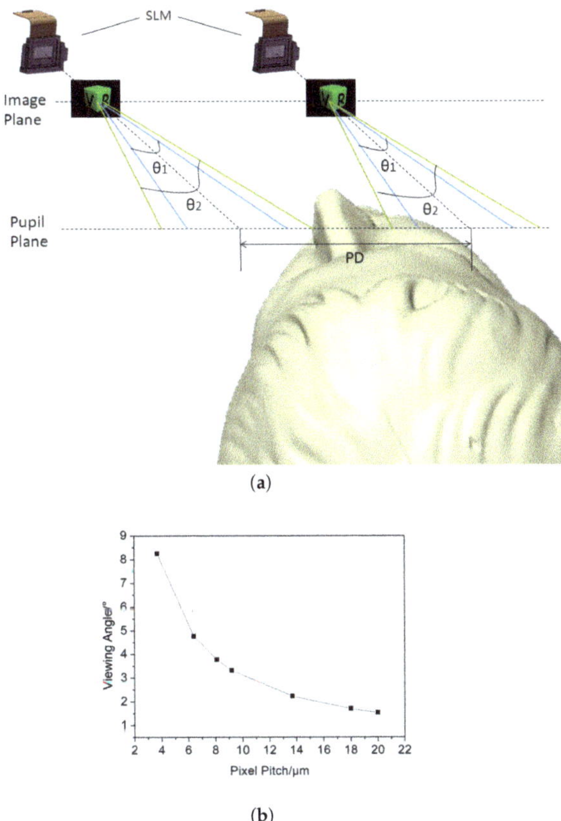

Figure 2. (**a**) Viewing angles of spatial light modulators (SLMs). (**b**) Dependence of the viewing angle on the pixel pitch of SLM.

In Figure 3a, a is the distance between image and eyes and b is the width in the pupil plane. The image plane is the hologram plane, and it is a Fresnel hologram. Equation (2) shows the relationship between a, b, and θ.

$$a = b/(2\tan(\theta/2)) \tag{2}$$

The pupillary distance is the distance between a human's two pupils, which is 58–64 mm for adults, and 52–64 mm for children. It is different for different humans. The difference in the range of pupillary distance between children and adults is 12 mm. We compare distances between displayed images and eyes based on SLM with a pixel pitch of 3.74 µm and 8.1 µm, respectively. Setting the pupillary distance of d and the difference in the range of pupillary distance of Δd, Δb = (Δd)/2 = 6 mm, because the two eyes are symmetrical, obtained by calculation; a_1 is 42.1 mm, and a_2 is 90.2 mm. Obviously, the distance between the displayed image and eyes in the system with a 3.74-µm pixel pitch SLM can be much smaller than that with the 8.- µm pixel pitch SLM. Figure 3b shows the dependence of the distance between the image and eyes on the pixel pitch of SLM. Figure 4 shows the schematic layout of the monocular system, which consists of two SLMs (SLM1 and SLM2) for two eyes, two beam splitters (BS1 and BS2), a half wave plate, etc. Some mirrors were used to make the whole system more compact.

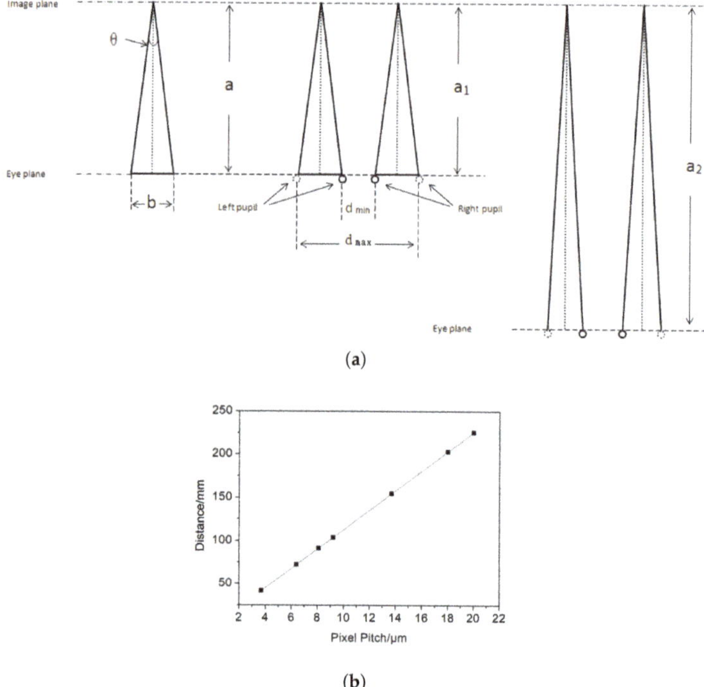

Figure 3. (**a**) Schematic diagram of the viewing angle and pupillary distance range difference. (**b**) Dependence of the distance between the image and eyes on the pixel pitch of SLM.

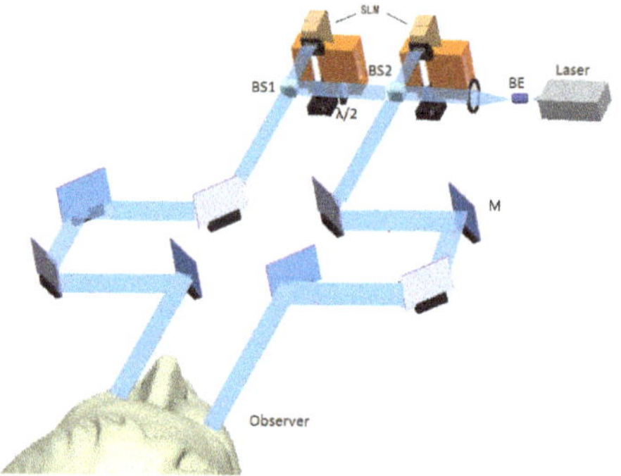

Figure 4. Schematic layout of the monocular system, BS: beam splitter, BE: beam expander, $\lambda/2$: half wave plate, M: mirror.

In Figure 5a, the image plane and convergence plane of the stereoscopic virtual reality HMD are shown. If they are in different planes, once all the elements of the HMD are fixed, there will be difficulty in making them be in the same plane. The image plane and convergence plane of the holographic virtual reality HMD can easily be in the same plane, as shown in Figure 5b, because the depth of field of the holographic 3D display can be set in the process of digital hologram computation (we changed some parameters in our MATLAB program of the computer-generated hologram to change the depth of the holographic reconstruction image). The computer-generated holograms are displayed in SLMs. This can improve the interaction and can eliminate the limitation of stereoscopic VR HMD in Figure 5a.

(a)   (b)

Figure 5. (a) Image plane and convergence plane of the stereoscopic display. (b) Image plane and convergence plane of the holographic display (the convergence plane is the same as the interaction plane). HMD, head-mounted display.

3. Binocular System

Although holography is a true 3D display, the limitation of the holographic device, such as a big pixel pitch of SLM, leads to a small viewing angle. To increase the viewing angle, it is necessary to tile multiple SLMs. Therefore, we propose a binocular holographic HMD system using two tiled SLMs for holographic HMD. As shown in Figure 6, there are two SLMs tiled in a curved configuration with tiled angle α in order to ensure accurate tiling. When α is smaller than θ, the viewing angle is increased to $\theta + \alpha$ in this case [23], otherwise, the reconstructed image is not seen continuously. Under ideal conditions, i.e., $\alpha = \theta$, the dependence of the viewing angle on the pixel pitch of SLMs is as shown in Figure 7a. The dependence of the distance between the image and eyes on the pixel pitch of SLMs is shown in Figure 7b.

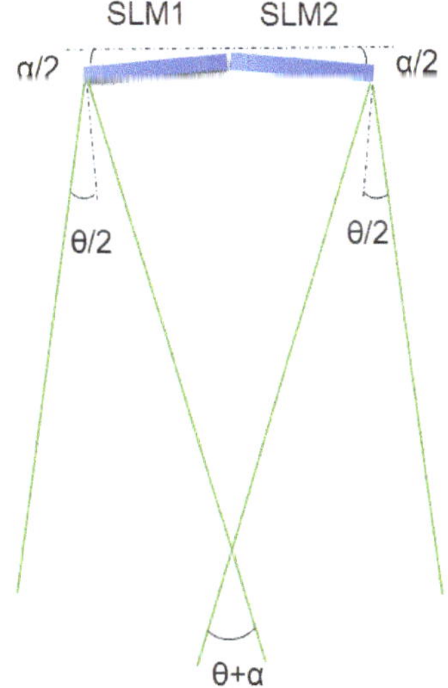

Figure 6. Viewing angle of two tiled SLMs.

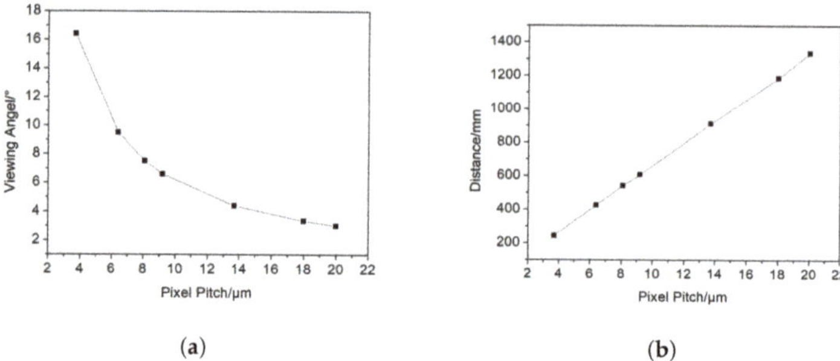

Figure 7. (**a**) Dependence of the viewing angle on the pixel pitch of SLMs. (**b**) Dependence of the distance between the image and eyes on the pixel pitch of SLMs.

Figure 8 shows a schematic diagram of the proposed binocular holographic HMD system. The tiled SLM unit consists of two SLMs (SLM1 and SLM2 with the same pixel pitch) and a beam splitter (BS). Some of the laser beam is reflected by the BS to SLM1 and reads out one computer-generated hologram display in SLM1, and then, the reconstructed image transmits through the BS and combines with the reconstructed image from SLM2, which is reflected by the BS after it is read out from the other computer-generated hologram display in SLM2 by the laser beam, which transmits through the BS. These two holograms displayed in SLM1 and SLM2 have different viewing angles of an object or a scene.

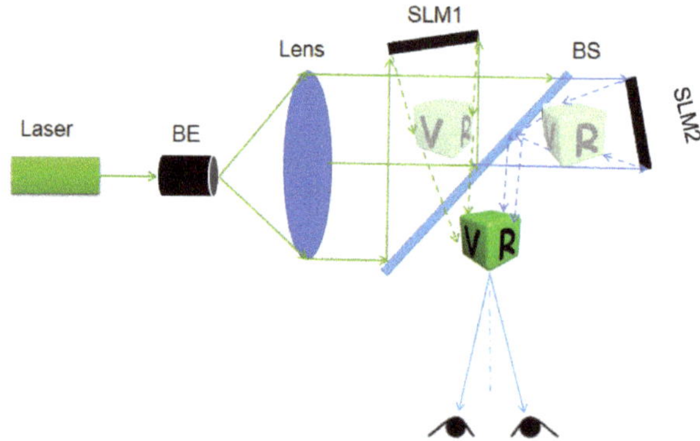

Figure 8. Schematic diagram of the proposed binocular holographic head-mounted display system.

4. Experiments and Results

In this experiment, we took pictures of the reconstruction from the hologram at different viewpoints, which were left view, middle view, and right view, using a single SLM with pixel pitch of 8.1 μm and 3.74 μm, respectively. The SLMs were liquid-crystal-on-silicon (LCoS) SLMs, Holoeye Pluto (resolution: 1920 × 1080, pixel pitch: 8.1 μm), and Jasper (resolution: 4094 × 2400, pixel pitch: 3.74 μm). In Figure 9, the image is captured from the left, middle, and right viewpoints in the optical system using an SLM with a pixel pitch of 8.1 μm, and the reconstructed image with the viewing angle, 3.8°, is a cube with the letters V and R; the measured distance from the image to the camera was about 300 mm, and the lateral motion of the camera was about 20 mm. Figure 10 shows the captured image from three viewpoints from the left to the right in the optical system, which used an SLM with a 3.74-μm pixel pitch. The viewing angle of this system was up to 8.2°, and the measured distance from the image to the camera was about 300 mm, while the lateral motion of the camera was about 43 mm.

By comparing Figures 9 and 10, the reconstruction quality of Figure 10 is better than that in Figure 9, because the SLM in Figure 10 has a higher resolution and smaller pixel pitch.

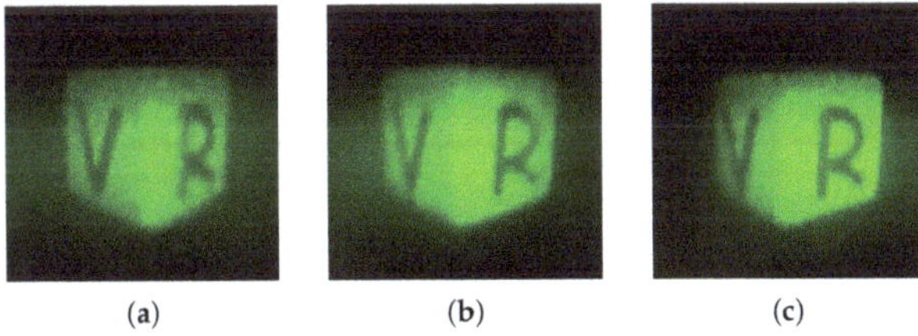

Figure 9. Pictures of the reconstructed cube based on SLM with a pixel pitch of 8.1 μm from different viewpoints. (**a**) Left viewpoint (−1.9°); (**b**) middle viewpoint (0°); (**c**) right viewpoint (1.9°).

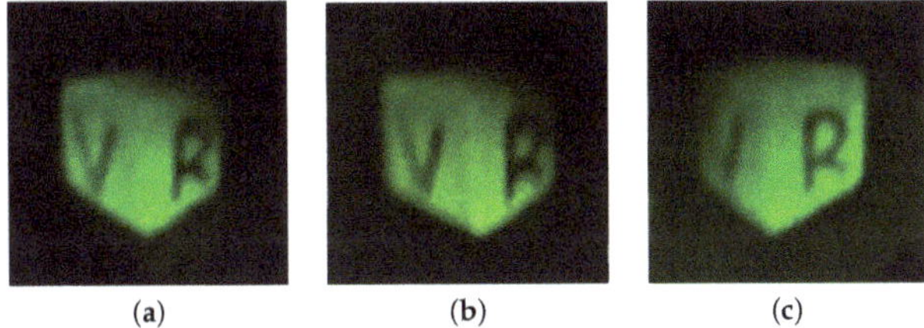

Figure 10. Pictures of the reconstructed cube based on SLM with a pixel pitch of 3.74 μm from different viewpoints. (**a**) Left viewpoint (−4.1°); (**b**) middle viewpoint (0°); (**c**) right viewpoint (4.1°).

We tiled two SLMs with a pixel pitch of 3.74 μm and used a full-color light source, red, green, and blue lasers at wavelengths of 632.8 nm, 532.8 nm, and 473 nm, respectively. The reconstructed color image is shown in Figure 11, and the measured distance from the image to the camera was about 300 mm, while the lateral motion of the camera was about 86 mm. Compared to the system based on single SLM, its viewing angle increased significantly. It was demonstrated that large viewing angles can be obtained in this binocular holographic HMD.

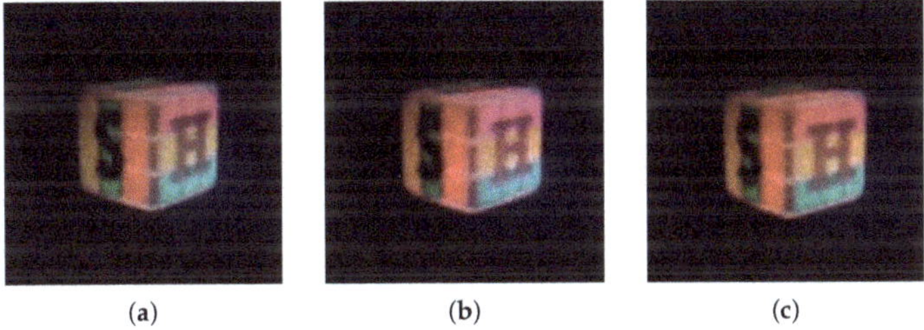

Figure 11. Pictures of the reconstructed full-color cube from different viewpoints in the binocular system. (**a**) Left viewpoint (−8.2°); (**b**) middle viewpoint (0°); (**c**) right viewpoint (8.2°).

According to Equation (1), the viewing angle can reach 16.4°, in the two tiled SLMs with a pixel pitch of 3.74 μm, and it can be further increased, if the 4f system can be used in the display system to reduce the pixel pitch. The 4f system consists of two convex lenses. When the focal length of the lens close to the SLM is larger than the focal length of the lens close to the human eye, the 3D image viewing angle can be enlarged.

Through the above work, we proved the advantages of holographic display applications in HMD, such as more realistic images, meeting human perception habits. Compared to current 3D VR and AR, holography has more feasibility for future near-eye 3D display.

5. Feasibility Analysis

Currently, VR and AR glasses on the market usually use liquid crystal displays (LCD) or organic light-emitting diodes (OLED), which are very good for 2D display, but cannot be used as SLM. LCoS is a kind of display module in SLM, and it has a compact structure, which is suitable for 3D HMDs; because its size is about on inch, its resolution is currently up to 4K, and its minimum pixel pitch reaches 3.74 μm. The viewing angle of holographic HMD can be 16.4°, which is wide enough for near-eye display. Furthermore, with the development of lasers and other light sources, it is easy to get enough high intensity of readout light with a small size and low energy consumption. Therefore, true 3D HMD with a high brightness, high resolution, large viewing angle, and depth cue will be easy to obtain by holography with the development of SLMs. We believe that holographic 3D display for VR and AR can meet the requirements of near-eye display.

6. Conclusions

In this paper, holographic 3D HMD suitable for observers with different pupillary distance is proposed. We built a monocular system and a binocular system using SLMs with pixel pitches of 3.74 μm and 8.1 μm for 3D HMD. Based on the experimental results, we demonstrated that the viewing angle of the reconstructed 3D image of the holographic HMD can be improved by tiled SLMs and the 4f system. In our binocular system, it reached 16.4°, which was two-times wider than the original viewing angle formed by the monocular system. This is meaningful for holographic application in VR and AR. Holographic 3D VR and AR displays can be good candidates for true 3D near-eye display, and compared with stereoscopic display, they can give viewers a more realistic visual experience with continuous wavefront reconstruction. Thus, we demonstrate that holographic technology can be combined with VR and AR to achieve true 3D HMD, providing technical reference for 3D VR and AR applications in various fields.

Author Contributions: F.X. structured and wrote the paper; H.G. guided other authors to complete the paper; J.L. selected the topic; W.Z. supervised the progress of the whole work; Z.D. and S.L. supplied the materials to the paper; Y.Y. and H.Z. helped to proofread the manuscript.

Funding: This work was financially supported by the National Natural Science Foundation of China (Grant Nos. 11004037, 61005073, and 11474194), the Creative Research Fund of Shanghai Municipal Education Commission (14YZ009), and the Shanghai Natural Science Foundation (Grant Nos. 14ZR1415700, and 14ZR1415500), and it was also supported in part by the Open Research Fund of the Chinese Academy of Sciences (Grant No. SKLST201104).

Conflicts of Interest: The authors declare no conflict of interest.

References

1. Zhao, X.P. Summarize of virtual reality. *Chin. Sci.* **2009**, *39*, 2–26,
2. Russo, J.M. Dimov, F.; Padiyar, J.; Coe-Sullivan, S. Mass production of holographic transparent components for augmented and virtual reality applications. *Light Energy Environ.* **2017**. [CrossRef]
3. Yano, S.; Emoto, M.; Mitsuhashi, T. Two factors in visual fatigue caused by stereoscopic HDTV images. *Displays* **2004**, *25*, 141–150. [CrossRef]
4. Maimone, A.; Wetzstein, G.; Hirsch, M.; Lanman, D.; Raskar, R.; Fuchs, H. Focus 3D: Compressive accommodation display. *ACM Trans. Graph.* **2013**, *32*, 1–13. [CrossRef]
5. Deng, H.; Wang, Q.H.; Luo, C.G.; Liu, C.L.; Li, C. Accommodation and convergence in integral imaging 3D display. *J. SID* **2014**, *22*, 158–162. [CrossRef]
6. Hiura, H.; Mishina, T.; Arai, J.; Iwadate, Y. Accommodation response measurements for integral 3D image. *Proc. SPIE* **2014**, *9011*. [CrossRef]
7. Moon, E.; Kim, M.; Roh, J.; Kim, H.; Hahn, J. Holographic head-mounted display with RGB light emitting diode light source. *Opt. Express* **2014**, *22*, 6526–6534. [CrossRef]

8. Yeom, H.; Kim, H.; Kim, S.; Zhang, H.; Li, B.; Ji, Y.; Kim, S.; Park, J. 3D holographic head mounted display using holographic optical elements with astigmatism aberration compensation. *Opt. Express* **2015**, *23*, 32025–32034. [CrossRef]

9. Han, J.; Liu, J.; Yao, X.; Wang, Y. Portable waveguide display system with a large field of view by integrating free-form elements and volume holograms. *Opt. Express* **2015**, *23*, 3534–3549. [CrossRef]

10. Piao, M.L.; Wu, H.Y.; Kim, N. Holographic projection head mounted display with transparent volume hologram. *Imaging Appl. Opt. Congress* **2016**. [CrossRef]

11. Galeotti, J.M.; Siegel, M.; Stetten, G. Real-time tomographic holography for augmented reality. *Opt. Lett.* **2010**, *35*, 2352–2354. [CrossRef] [PubMed]

12. Kim, H.E.; Kim, N.; Song, H.; Lee, H.S.; Park, J.H. Three-dimensional holographic display using active shutter for head mounted display application. *Proc. SPIE* **2011**, *7863*. [CrossRef]

13. Piao, J.; Li, G.; Piao, M.; Kim, N. Full Color Holographic Optical Element Fabrication for Waveguide-type Head Mounted Display Using Photopolymer. *J. Opt. Soc. Korea* **2013**, *17*, 242–248. [CrossRef]

14. Li, G.; Lee, D.; Jeong, Y.; Cho, J.; Lee, B. Holographic display for see-through augmented reality using mirror-lens holographic optical element. *Opt. Lett.* **2016**, *41*, 2486–2489. [CrossRef] [PubMed]

15. Hong, J.; Li, G.; Lee, B. See-through optical combiner for augmented reality head-mounted display: index-matched anisotropic crystal lens. *Sci. Rep.* **2017**, *7*, 2753. [CrossRef] [PubMed]

16. Sun, P.; Chang, S.; Liu, S.; Tao, X.; Wang, C.; Zheng, Z. Holographic near-eye display system based on double-convergence light Gerchberg-Saxton algorithm. *Opt. Express* **2018**, *26*, 10140–10151. [CrossRef]

17. Hong, J.; Kim, Y.; Hong, S.; Shin, C.; Kang, H. Near-eye foveated holographic display. *Imaging Appl. Opt.* **2018**. [CrossRef]

18. Gang, J. Three-dimensional display technologies. *Adv. Opt. Photonics* **2013**, *5*, 456–535. [CrossRef]

19. Hiura, H.; Mishina, T.; Arai, J.; Iwadate, Y. Anisotropic leaky-mode modulator for holographic video displays. *Nature* **2013**, *498*, 313.

20. Onural, B.; Yaras, F.; Kang, H. Digital holographic threedimensional video displays. *Proc. IEEE* **2011**, *99*, 576–589. [CrossRef]

21. Hiura, H.; Mishina, T.; Arai, J.; Iwadate, Y. Wide viewing angle dynamic holographic stereogram with a curved array of spatial light modulators. *Opt. Express* **2008**, *16*, 12372–12386.

22. Nanette, S.; Leslie, S. *Basic Photographic Materials and Processes*; Taylor and Francis: Abingdon, UK, 2009; Volume 110; ISBN 978-0-240-80984-7.

23. Zeng, Z.; Zheng, H.; Yu, Y.; Asundi, A.K.; Valyukh, S. Full-color holographic display with increased-viewing-angle. *Appl. Opt.* **2017**, *56*, 112–120. [CrossRef] [PubMed]

Article

Dual-View Three-Dimensional Display Based on Direct-Projection Integral Imaging with Convex Mirror Arrays

Hee-Min Choi [1], Jae-Gwan Choi [2,3] and Eun-Soo Kim [4,*]

[1]   HoloDigilog Human Media Research Center, Nano Device Application Center, Kwangwoon University, 21, Gwangun-ro, Nowon-Gu, Seoul 01890, Korea; hmchoi@kw.ac.kr
[2]   HoloDigilog Human Media Research Center, Kwangwoon University, 21, Gwangun-ro, Nowon-Gu, Seoul 01890, Korea; uest37@naver.com
[3]   Intelligent Medical Platform Research Center, Kyunghee University, 1732, Deogyeong-daero, Giheung-gu, Yongin-si, Gyeonggi-do 17104, Korea
[4]   HoloDigilog Human Media Research Center, 3D Display Research Center, Kwangwoon University, 21, Gwangun-ro, Nowon-Gu, Seoul 01890, Korea
*   Correspondence: eskim@kw.ac.kr

Received: 27 March 2019; Accepted: 11 April 2019; Published: 16 April 2019

Abstract: Dual three-dimensional (3-D) view displays have been attracting much attention in many practical application fields since they can provide two kinds of realistic 3-D images with different perspectives to the viewer. Thus, in this paper, a new type of the dual-view 3-D display system based on direct-projection integral imaging using a convex-mirror-array (CMA) is proposed. Two elemental image arrays (EIAs) captured from each of the two 3-D objects are synthesized into a single dual-view EIA (DV-EIA) with a selective sub-image mapping scheme. The divergent beam of the projector containing the information of the DV-EIA is projected onto the CMA. On each convex mirror of the CMA, left and right-view components of the DV-EIA are separated and reflected back into their viewing directions. Two different 3-D scene images are then integrated and displayed on their respective viewing zones. Ray-optical analysis with the parallel-ray-approximation method and experiments with the test 3-D objects on the implemented 22″ DV 3-D display prototype confirm the feasibility of the proposed system in the practical application.

Keywords: 3-D optical imaging processing; 3-D optical display; 3-D projection integral imaging; holographic display; three-dimensional dual-view display

---

1. Introduction

Thus far, many kinds of dual-view (DV) displays have been developed due to the high interest in various application fields including car navigation, multi-vision, medicine and digital signage [1–9]. For example, a DV car navigation system presents live traffic information to the driver while showing movies to the passenger [4–6]. Moreover, the DV digital signage system can provide two kinds of advertising videos to each of the two groups of customers standing on the street at different viewing directions [3]. In this way, the need for supplying respective displays for each of the driver and passenger, and each group of the customers can be removed, which then results in saving associated cost and reduction space occupied by the other displays [4–6]. Of course, each individual in the car and each customer group on the street can enjoy the same video contents simultaneously in their positions.

In fact, there exists a very close similarity between the DV and stereoscopic displays in terms of generating two different views. That is, the DV display provides two different views to the two observers located at different viewing directions, while the stereoscopic display delivers two perspectives of an input 3-D scene with a binocular disparity to each of the left and right eyes of an observer [9]. This functional similarity implies that the operational principle of the DV display might be closely related to that of the stereoscopic display. Thus, just like the conventional stereoscopic displays employing the polarizing glasses or shutter glasses, as well as the lenticular sheets or parallax barriers for their generation of two or multi-views, DV displays also use the related optical elements for generation of their two different views [10–12].

Thus, DV displays can be largely classified into spatial and time-multiplexed systems [13–19]. In the spatial-multiplexed system, the main pixel of the DV LCD (liquid crystal device) panel consists of a right sub-pixel (RSP) and a left sub-pixel (LSP), and those two different sub-pixels project two different images to the viewers at different directions. Spatial-multiplexed DV LCD displays have been successfully demonstrated with parallax barriers and lenticular sheets [16]. However, in these systems, additional optical elements must be attached to their LCD panels, the absolute spatial-resolution of the displayed images must be reduced in half, and crosstalk may occur between two displayed images [16]. On the other hand, in the time-multiplexed system, two different images are time-sequentially provided to two viewers with the absolute full spatial-resolution of the LCD panel [18]. For this, this system requires a directional backlight module with complex reflectors, and an LCD panel with faster response time. In addition, it may suffer from crosstalk and flicker-noise problems [18].

Meanwhile, since the 3-D display can provide a realistic 3-D image with different perspectives to the viewer, some research on the DV 3-D display was recently carried out [19]. In most conventional DV 3-D display systems, time or spatial-multiplexed stereoscopic images are used for providing two different stereoscopic images to the two different viewers. However, those systems require four times of spatial and time-multiplexing processes, compared to the 2-D display system. It may result in a one-quarter reduction of the spatial resolution in the spatial-multiplexed system and four-fold acceleration of the LCD response time in the time-multiplexed system, respectively [19]. These practical problems prevent the stereoscopic technique from being widely applied for the DV 3-D display.

As an alternative, an integral imaging-based DV 3-D display was proposed [20–23]. Wu and et al. suggested a DV 3-D LCD system based on the lens array-based off-axis integral imaging method [19]. However, since the narrow viewing-zone of the ordinary lens array-based integral imaging system was shared by two viewing-zones, sufficient viewing-areas for the DV 3-D display could not be offered. Thus, they tried to employ a polarization parallax barrier to improve the viewing angle [21], but it suffered from the degraded resolution and brightness of the displayed 3-D images due to the polarization barriers blocking light coming from the elemental image array (EIA).

In fact, for the practical application of the 3-D DV display, two viewing-zones must be completely separated and have relatively large viewing-angles to experience the perspective changes of 3-D images. For this, J. Jeong et al. proposed a projection-type DV 3-D display based on integral imaging [23]. They used a series of optical elements, such as a collimator, a lenticular lens with a vertical diffuser, and convex lens array for generation of a collimated beam, separation of DV images with enhanced viewing-angles, and integration of 3-D images to be displayed in their viewing zones, respectively. These additional optical elements, however, may degrade the quality of the reconstructed 3-D images. Moreover, there exists an overlapped viewing-zone between the left and right 3-D images with some distortion due to the crossing of the left and right rays. The scale of this system may also depend on the practical sizes of the lenticular lens and collimator. Furthermore, this convex-lens-array-based projection integral imaging (CLA-PII) system shows a narrower viewing angle than the of the convex-mirror-array-based system [23].

To alleviate those problems, in this paper, we propose a practical type of a scalable DV 3-D display system based on convex-mirror-array-based direct projection integral imaging (CMA-DPII). The proposed system is simply composed of a projector and a CMA, where two kinds of EIAs for each of the two different input 3-D scenes are generated with the on-axis pickup integral imaging system and synthesized into a single EIA, which is called DV-EIA, based on the selective sub-image mapping (SSIM) method. Here, the DV-EIA contains all information of the intensities, perspectives, viewing zones of those two different input 3-D scenes. The divergent beam of the projector containing the information of the DV-EIA is directly projected onto the surface of the CMA without any additional optical elements. On every convex mirror, each of the left and right-scene components of the DV-EIs are instantaneously separated and reflected back into their viewing directions. Two different 3-D scene images are then integrated and displayed in their viewing zones with their viewing angles.

In fact, the viewing angle of the proposed CMA-DPII system would be much more enhanced than that of the conventional CLA-PII system since each convex mirror can be made to have a much smaller f-number than the corresponding convex lens [24]. Moreover, contrary to the conventional system, each of the two viewing-zones and angles of the proposed system can be made changeable with the SSIM process, as well as a scalable DV 3-D display without flip-image and Moiré disturbance problems can be implemented [25]. To confirm the feasibility of the proposed system, ray-optical analysis with the parallel-ray-approximation (PRA) method, as well as optical experiments with test 3-D objects on the implemented 22″ DV 3-D display prototype is performed, and the results are compared with those of the conventional system.

2. Methods

Figure 1 shows an overall block diagram of the proposed system, which is composed of three-step processes. At the 1st step, two kinds of EIAs for each of the two different 3-D objects were generated based on the on-axis pickup integral imaging system, which are called left-EIA (L-EIA) and right-EIA (R-EIA) corresponding to each of the left and right views, respectively. In the 2nd step, these two EIAs of the L-EIA and R-EIA were multiplexed into a single EIA, which is called DV-EIA, based on the selective sub-image mapping (SSIM) method. At the 3rd step, the divergent beam of the projector containing the information of the DV-EIA was projected onto the CMA. Then, on each convex mirror of the CMA, the left and right-view components of the DV-EIA were instantaneously separated and reflected back into their viewing directions with their viewing angles. Finally, two different 3-D scenes were integrated and displayed in their viewing zones.

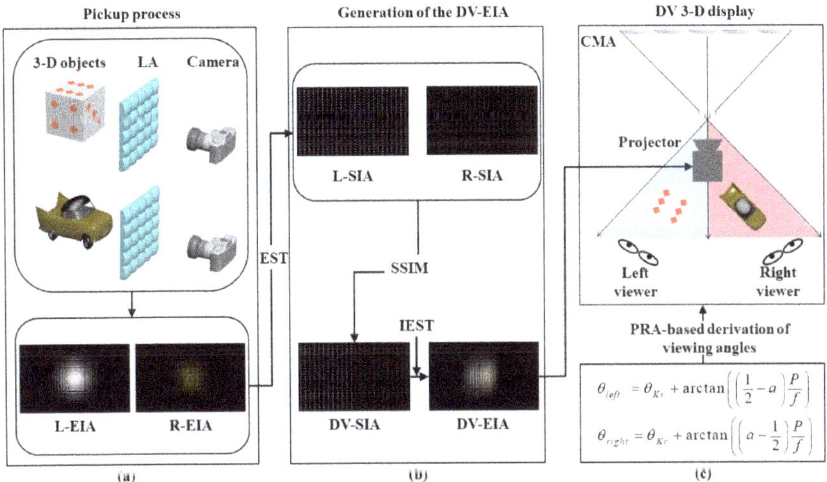

$$\theta_{left} = \theta_{K_l} + \arctan\left[\left(\frac{1}{2} - a\right)\frac{P}{f}\right]$$

$$\theta_{right} = \theta_{K_r} + \arctan\left[\left(a - \frac{1}{2}\right)\frac{P}{f}\right]$$

Figure 1. Overall block-diagram of the proposed system: (**a**) Pickup of two elemental image arrays (EIAs), (**b**) generation of the dual-view elemental image array (DV-EIA) based on the selective sub-image mapping (SSIM) method, (**c**) reconstruction of the dual-view (DV) 3-D images in their viewing zones.

2.1. Capturing of Two Kinds of EIAs Based on on-Axis Integral Imaging

Figure 2 shows two types of optical configurations for pickup and display of the EIA of a 3-D object in integral imaging, such as the on-axis and off-axis integral imaging systems. In the off-axis integral imaging system of Figure 2a, the normal axis of the EI surface and optical axis of the elemental lens were made to be offset. In other words, the optical axis of the pickup lens array was somewhat shifted from the normal axis of the EIA plane [23]. As seen in Figure 2, rays coming from the object points of O_1 and O_2, could be picked up on the different locations of the EIA depending on the relative position of the elemental lens. These two picked-up EIAs were then synthesized into a single DV-EIA. For the reconstruction of the DV-EIA, the same type of the off-axis integral imaging system was used. This off-axis integral imaging system has been used in the conventional LCD panel-type and projection-type integral imaging DV 3-D systems [20–23].

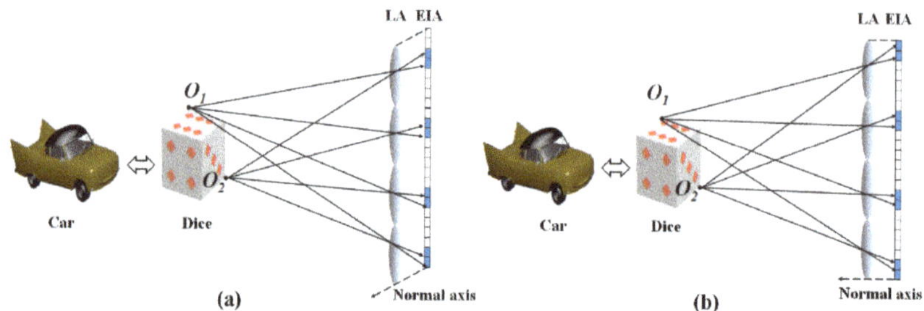

Figure 2. Two types of the EIA pickup systems: (**a**) Off-axis pickup, (**b**) on-axis pickup.

On the other hand, Figure 2b shows the on-axis integral imaging system, which looks like the ordinary integral imaging system. As seen in Figure 2b, the normal axis of the EI surface and optical axis of the elemental lens were on the same axis. The same on-axis integral imaging system was also used in the display process just like the ordinary integral imaging system. Unlike the conventional CLA-PII system, the proposed CMA-DPII system was implemented on the ordinary on-axis integral imaging system. For instance, Figure 3 shows two kinds of EIAs captured for each of the two different 3-D objects of 'Dice' and 'Car' based on the on-axis integral imaging system of Figure 2b.

Figure 3. Two captured EIAs from each of the two different 3-D objects of 'Dice' and 'Car' on the on-axis integral imaging system (**a**) captured EIA from the 'Dice' (**b**) captured EIA from the 'Car'.

2.2. Synthesis of Two EIAs into a Single DV-EIA

To generate the DV zones corresponding to each of the left and right views, two EIAs captured from each of the two different 3-D objects were multiplexed into a single EIA, which is called DV-EIA, based on the SSIM method, as mentioned above. Figure 4 shows a conceptual diagram of the SSIM method, which is composed of a four-step process.

At the 1st step, two EIA were captured from each of the two different 3-D objects, which were assigned to the L-EIA and R-EIA, respectively. At the second step, L-EIA and R-EIA were transformed into the corresponding sub-image arrays (SIAs), which are called L-SIA and R-SIA, respectively, based on the EIA-to-SIA transformation (EST) method [26]. At the 3rd step, odd and even number of components of the EIs were selectively chosen from each of the L-SIA and R-EIA and mapped into their corresponding left-half and right-half parts of the dual-view sub-image array (DV-SIA), respectively.

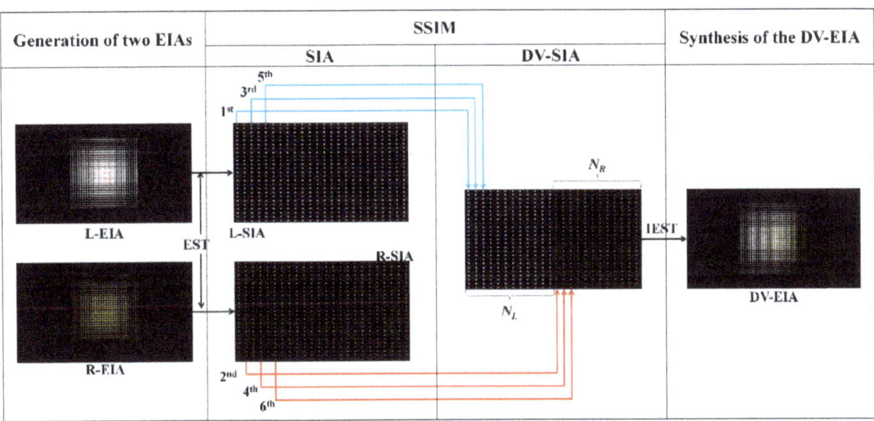

Figure 4. Generation of the DV-EIA with the left elemental image array (L-EIA) and left elemental image array (R-EIA) based on the SSIM method.

Here, N_L and N_R, respectively, represent the total number of SIs selected from the L-SIA and R-SIA, and allocated to their corresponding left and right-half parts of the DV-EIA. N_{DV} ($=N_L + N_R$) also denotes the total number of SIs mapped into the DV-SIA along the horizontal direction, where the number of pixels in each EI was equal to the number of SIs. This DV-SIA was then transformed into its corresponding DV-EIA using the inverse-EST (IEST) method. At the 4th step, the DV-EIAs for the two L-EIA and R-EIA were finally generated and used for separate displaying of two different 3-D objects into their viewing zones. Meanwhile, in the conventional system, the left and right viewing zones are set to be equal and physically fixed by the employed optical elements, such as the lens array and lenticular lens array, whereas in the proposed system, a parameter of α can be employed for controlling the relative sizes of the left and right viewing zones, where α is defined as a relative portion of the N_L in N_{DV} and expressed by Equation (1).

$$\alpha = \frac{N_L}{N_{DV}} \tag{1}$$

According to the relationship between the EIA and SIA, N_{DV} and N_L become equal to the total number of pixels of the DV-EI and number of pixels of the L-EI allocated to the left-hand side of the DV-EI, respectively. Thus, N_L turn out to be $\alpha \times N_{DV}$ and the other portion of the N_{DV}, N_R becomes $(1 - \alpha) \times N_{DV}$, which represents the number of pixels of the R-EI allocated to the right-hand side of the DV-EI. This scalability of the left and right viewing zones is another property of the proposed system in the practical application.

2.3. Direct Projection of the DV-EIA onto the CMA for the DV 3-D Display

Figure 5 shows an optical configuration of the direct-projection integral imaging system for the DV 3-D display of the synthesized DV-EIA with the CMA, where straight lines denote the real rays in the real space, whereas dashed lines represent the virtual rays generated by the reflected beams on the surface of the CMA. In Figure 5, the beam projector, where the synthesized DV-EIA was loaded, was located far from the CMA, enabling the performance analysis of the proposed system using parallel-ray-approximation (PRA) under the far-field condition, which is to be discussed in the next section. In the conventional projection DV 3-D display system, however, this far field condition may not be considered since the diverging beam coming from the beam projector is set to be parallelized by using a collimator. Here, in the proposed system, the divergent beam of the projector containing the information of the DV-EIA was projected onto the CMA. This diverging beam carrying the DV-EIA was reflected on the surface of the CMA and divided into the left and right EIAs (L-EIA and R-EIA) on each component of the CMA in the real space. From each of the L-EIA and R-EIA, the left and right viewing zones were generated in the space. Further, each of the L-EIA and R-EIA was integrated into each viewing zone, and two different 3-D objects were reconstructed in their viewing zones.

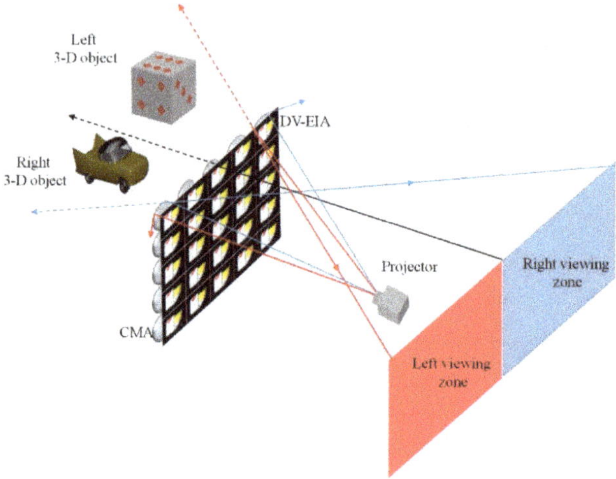

Figure 5. Optical configuration of the direct-projection integral imaging system for the DV 3-D display.

2.3.1. Parallel-Ray Approximation Method

In the field of the array antenna, the parallel-ray approximation (PRA) method has been used in computing the frequency response of the array antenna rather than the ray-tracing method [27–30]. Here, the array antenna is built with a 2-D periodic array structure. Since this array antenna looks very similar to the convex mirror array (CMA), the PRA method can be applied for the performance analysis of the proposed CMA-DPII system under the far-field condition.

Thus, in this paper, the PRA method was employed for the ray-optical analysis of the proposed system. Figure 6 shows an optical geometry for analyzing the proposed system based on the PRA method. In Figure 6, P and K represent the pitch of an elemental convex mirror and the maximum number of convex mirrors along the x-direction. The center of the 0th convex mirror was set to be the origin of the vertical x-coordinate and a point source whose vertical coordinate was the same as that of the 0th convex mirror. L denotes the distance between the point source and CMA. Here, two red and blue-color rays coming from the point source were assumed to arrive at the upper and bottom edges of the Kth convex mirror, respectively, where R and θ_{Red} denote the optical path length of the red-color ray and its incidence angle to the Kth convex mirror, while r and θ_{Blue} denote the optical path length of the blue-color ray and its incidence angle to the Kth convex mirror, respectively. In addition, θ_{Center} represents the angle of the ray incoming to the center of the Kth convex mirror.

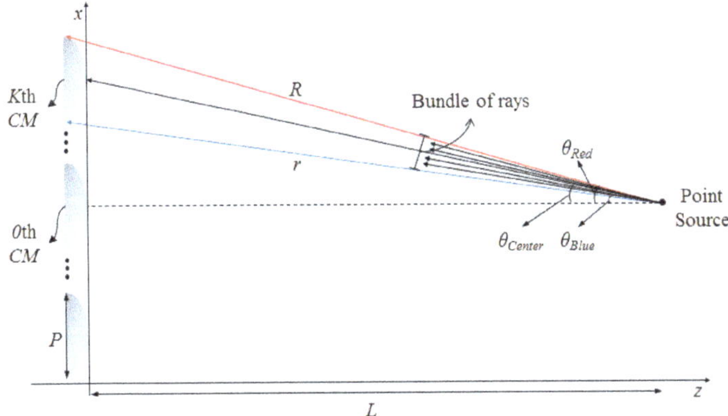

Figure 6. Optical geometry for the ray-optical analysis of the proposed system based on the parallel-ray-approximation (PRA) method.

The proposed CMA-DPII-based DV 3-D display system can be made under the far-field condition, which enables the use of the PRA method for the analysis of its operational performance. Under the far-field condition of $P/L < 0.01$, the optical path lengths of the red and blue-color rays from the point source to the Kth convex mirror, R and r, as well as the incidence angles of the red and blue-color ray to the Kth convex mirror, θ_{Red} and θ_{Blue}, can be approximated to be equal. With this relationship of $\theta_{Red} \approx \theta_{Blue}$, red and blue-color rays can be assumed to be parallel rays with the same propagation angles to the Kth convex mirror. Thus, the propagation angles of all those rays in a bundle can be given by $\theta_{Center} \approx \theta_{Red} \approx \theta_{Blue}$, where θ_{Center} represents the incidence angle of the center ray. In other words, the incidence angles of rays to the Kth convex mirror can be modeled as a single representative angle of θ_{Center}, which is given by Equation (2).

$$\tan(\theta_{Center}) = \frac{KP}{L} \tag{2}$$

Equation (2) means that the incidence angle of the center ray to each convex mirror depends on the vertical location of the convex mirror (KP) and the distance between the point source and the CMA plane (L). Since the CMA consists of a number of same-sized convex mirrors with regular intervals along the x and y-directions, each bundle of rays incoming to each convex mirror has different incidence angles. According to Equation (2), as the number of convex mirrors of K increases, its corresponding incidence angle of θ_{Center} can be increased up to 90°. Now, when a point source is replaced with a beam projector, all rays coming from the projector carrying the DV-EIA can be approximated to be a set of bundles of parallel rays with corresponding DV-EIs, and the operational performance of the proposed system, which is simply composed of a beam projector and a CMA, can be analyzed based on the PRA method.

2.3.2. Analysis of the Dual-Viewing Zones and Angles

Thus, under this circumstance, dual-viewing zones and angles of the proposed system can be analyzed. If these parallel rays containing the information data of the DV-EIA, are projected into the surfaces of each convex mirror, left and right-view components of the DV-EIs are separated and reflected back into their viewing directions with their viewing angles. Then, two different 3-D object images are integrated from all those surfaces of the CMA and displayed in their viewing zones.

Figure 7 shows an optical configuration of the bundle of parallel rays coming from the projector and incident to the Kth convex mirror based on the ray-tracing model. As seen in Figure 7, a bundle of red-color parallel rays with information of one component of the DV-EIA, is assumed to be incident onto the Kth convex mirror. Green-color rays represent the reflected beams on the surface of the Kth convex-mirror for their corresponding incident rays, forming rays through the backward extension of the reflected ones. Here, f_m denotes a focal length of the convex mirror, and θ_p represents the half beam angle of the projector being located far away from the CMA plane.

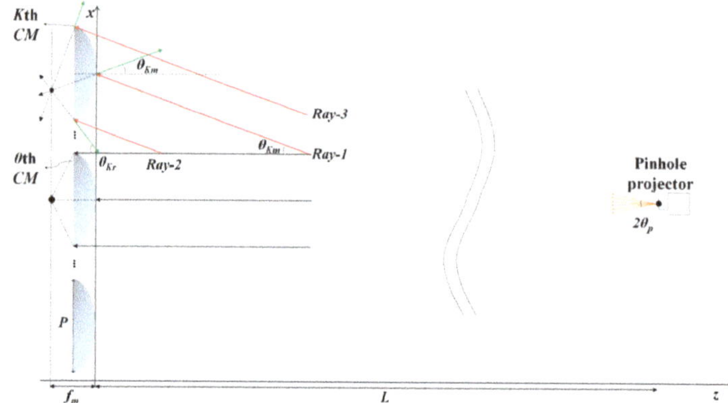

Figure 7. Optical configuration of the bundle of parallel rays incident to the Kth convex mirror from the pinhole projector.

The *Ray-1*, representing the center ray of the bundle of rays incident to the Kth convex mirror with the incidence angle of θ_{Km}, is reflected back from the surface of the Kth convex mirror to make the reflected ray with the reflection angle of θ_{Km}. This reflection of the incident ray is equivalent to its virtual propagation into the convex mirror as seen in Figure 7. The *Ray-2* and *Ray-3* also represent the parallel rays incident to both edges of the Kth convex mirror. Just like the *Ray-1* case, *Ray-2* and *Ray-3* are also reflected back from the convex mirror to make their reflected rays, which are equivalent to their virtually propagated rays into the convex lens. All those virtually propagated rays into the convex lens come across on the same points called a focal point of the convex lens as seen In Figure 7.

Based on the ray-tracing model, all virtually propagated rays for each convex mirror are focused on its focal point at the focal plane of the CMA as seen in Figure 7. Since the angle of a bundle of parallel rays incident to the Kth convex mirror is related to the relative position of the convex mirror (KP) and the distance from the projector to the CMA (L) under the far-field condition, the incidence angle of the center ray to the Kth convex mirror, θ_{Km} can be calculated based on ray-optics, and given by Equation (3).

$$\theta_{Km} = \arctan\left(\frac{KP}{L}\right) \tag{3}$$

Even though the incidence angle to the corresponding convex mirror can be increased as the number of K increases, its maximum angle value may be practically bounded by the diverging angle of the projector, $2\theta_p$. The whole bunch of rays between the *Ray-2* and *Ray-3* incident to the Kth convex mirror corresponds to one DV-EI of the DV-EIA. This DV-EI is composed of a pair of L-EI and R-EI. Each of the left and right viewing zones can be generated by the reflected propagations of the L-EI at the Kth convex mirror and R-EI at the Kth convex mirror, respectively. Thus, the virtual propagation angle of the *Ray-2* needs to be calculated for calculating the viewing zones of the proposed system because the reflection angle of the *Ray-2* affects the viewing angle for each viewing zone. Here, the reflection angle θ_{Kr} of the *Ray-2* on the Kth convex mirror can be derived as Equation (4).

$$\theta_{Kr} = \arctan\left(\tfrac{1}{2f\#} - \tan\theta_{Km}\right), \quad \theta_{Km} \leq \arctan\left(\tfrac{1}{2f\#}\right)$$
$$\theta_{Kr} = \text{Nonavailable}, \quad\quad\quad \theta_{Km} > \arctan\left(\tfrac{1}{2f\#}\right) \tag{4}$$

Equation (4) shows that the reflection angle, θ_{Kr} depends on the incidence angle, θ_{Km}. In particular, in case the incidence angle of the ray becomes less than or equal to the value of arctan(1/(2*f*#)), the focal point exists within the mirror pitch of P. Otherwise, the focal point may be placed beyond the mirror pitch of P, which means that dual-viewing zones cannot be properly generated in this case. Thus, the maximum value of one-half of the diverging beam angle of the projector must be less than or equal to arctan(1/(2*f*#)). Here, it is noted that the viewing zone is defined as the movable range in the viewing space where the viewer can see a full-resolution image [31].

Figure 8 shows an optical configuration for analyzing the dual-viewing zones of the proposed system generated from the DV-EIA and CMA under the far-field condition. As mentioned above, the DV-EIA is generated by a combined use of the L-EIA and R-EIA with the parameter of α. As seen in Figure 8, the red and blue areas represent the ray zones reflected from the L-EI and R-EI, respectively. In addition, black rays represent the boundary rays of the left and right viewing zones. When the total number of convex mirrors is set to be $2K + 1$, each of the left and right viewing angles can be decided by the Kth and $- K$th CM, respectively.

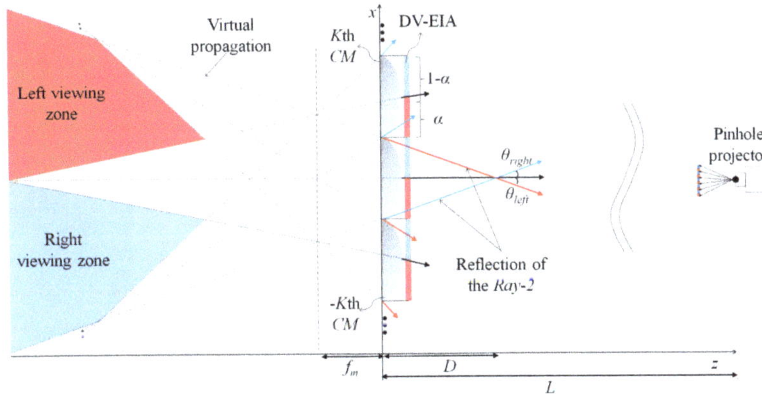

Figure 8. Optical configuration for analyzing the dual-viewing zones of the proposed system.

The left viewing zone can be generated by overlapping of all L-EIs of the DV-EIA and its boundary can be decided by the virtual propagation of the *Ray*-2 on the Kth convex mirror and boundary ray on the 0th convex mirror, whereas the right viewing zone is generated by overlapping of all R-EIs of the DV-EIA and its boundary is decided by the virtual propagation of the *Ray*-2 on the $- K$th convex mirror and boundary ray on the 0th convex mirror. Thus, the left and right viewing angles of θ_{left} and θ_{right} can be given by Equation (5).

$$\theta_{left} = \theta_{Kr} + \arctan\left(\left(\alpha - \tfrac{1}{2}\right)\tfrac{P}{f}\right)$$
$$\theta_{right} = \theta_{Kr} + \arctan\left(\left(\tfrac{1}{2} - \alpha\right)\tfrac{P}{f}\right)$$

(5)

As seen in Equation (5), viewing angles depend on the value of α. That is, as the value of α increases, the relative portion of the L-EI in the DV-EI increases, which results in an increase of the corresponding left viewing angle. Whereas when the value of α decreases, the right viewing angle gets extended. For the case of $\alpha = 0.5$, the left and right viewing angles become exactly the same, which corresponds to the conventional display system. In Figure 8, D represents the distance from the CMA for the left and right viewing zones to be separated along the z-direction in the real space, which can be derived by Equation (6).

$$D = \frac{(K - 0.5)P}{\tan(\theta_{Kr})}$$

(6)

As seen in Equation (6), the separation distance of D mostly depends on the pitch of the convex mirror (K) and the reflection angle of the *Ray-2* on the Kth convex mirror (θ_{Kr}). Thus, viewers standing beyond the separation distance in the real space can see each of those reconstructed 3-D objects on the two different directions.

2.4. Reconstruction of Dual 3-D Views with their Viewing Angles

Figure 9 shows the optical configuration of the reconstructed dual 3-D views in their viewing zones. The aperture of a pinhole projector is positioned at the distance of L from the CMA along the z-direction. Aperture image point (AIP) is generated by the pinhole of the projector acting as a point light source [32]. Here, z_{aip} represents the position of the AIP along the z-direction and the total number of convex mirrors is set to be ($2K + 1$), which are located along the x-direction.

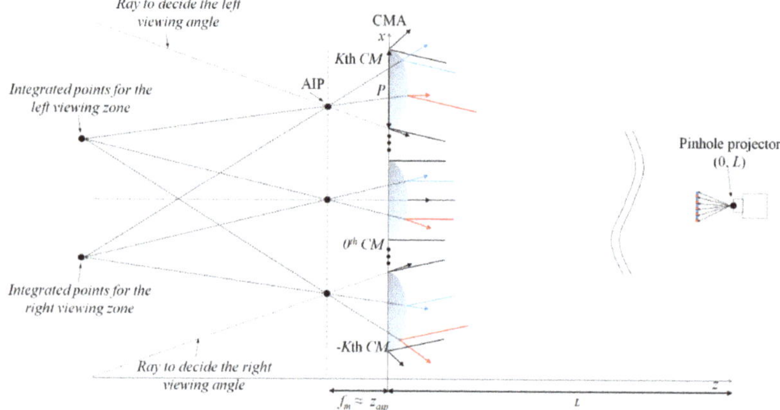

Figure 9. Optical configuration of the integrated points of 3-D object images in the left/right viewing zones.

The divergent beam of the projector containing the information of the DV-EIA is projected onto the CMA. Since the distance of L is much larger than the pitch of the convex mirror, each ray incoming to the corresponding convex mirror is assumed to be parallel rays with the same incidence angles under the far-field condition. Incoming rays with the L-EI and R-EI are reflected on the corresponding convex mirror and generate a single AIP depending on the convex mirror because a single pinhole projector is used in the display system. Here, the position of the AIP z_{aip} becomes f_m because incoming rays are approximated to be the parallel rays based on the PRA method. The number of AIPs is the same as the total number of convex mirrors. All those AIPs generated in each CM are integrated in the left and right viewing zones, which are decided by the rays incoming from the bottom of the Kth convex mirror and from the top of the Kth convex mirror. Thus, different 3-D senses in the left and right viewing zones can be reconstructed from these integrated AIPs.

3. Results

3.1. Experimental Setup

Figure 10 shows an experimental setup of the proposed system, which is simply composed of a pair of the CMA and beam projector. As a projector, the LG beam projector (Model: LG PF85K) whose resolution, brightness and size were 1920×1080 pixels, 1000 ANSI-lumens and $275 \times 219 \times 45$ mm, respectively, was used. The total diverging angle ($2\theta_p$) and depth-of-field (DoF) of this projector were estimated to be 39°, and ranged from 62 cm to 372 cm, respectively. In addition, for the experiments, a 22″CMA whose pitch and curvature radius were 7.47 mm and 15.50 mm, respectively, was fabricated and located at the distance of 800 mm from the beam projector. Here, the focal length of an elemental CM of the CMA was calculated to be 7.32 mm.

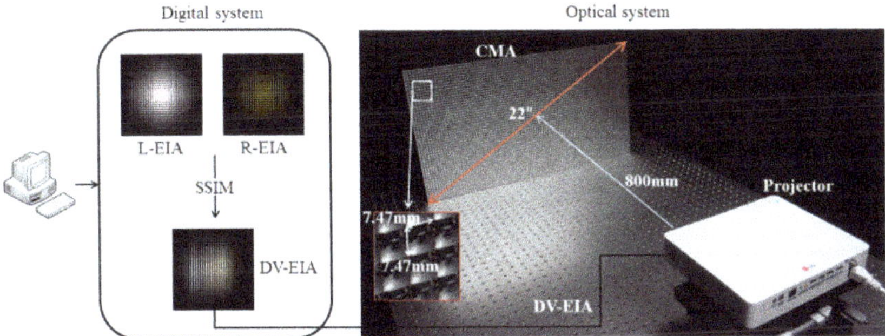

Figure 10. Experimental setup of the proposed system.

As mentioned above, to test the feasibility of the proposed system in the practical application, a 22″CMA was specially designed and fabricated. That is, the CMA was manufactured by coating the aluminum particles being spread from the thermal evaporator on the surface of a base convex lens array to make a high-reflector aluminum layer. Here, aluminum was used as a coating material because the reflectance for the visible light becomes much more than 90%. Thus, the convex lens array with a thin aluminum layer acted as a convex mirror array. Each convex mirror had the same curvature and pitch as the base convex lens, but its focal length became much shorter than that of the base convex lens. In the experiment, pitches and focal lengths of the base convex lens and convex mirror were 7.47 mm, 29.88 mm, and 7.47 mm, 7.32 mm, respectively. That is, the focal length of the convex mirror became 4.08-fold shorter than that of the corresponding convex lens. In fact, as the focal length of the convex mirror got shorter, its f-number ($f\#$) also decreased, which then resulted in an increase of the viewing angle of the corresponding CMA by a factor of $2 \times \arctan(1/2(f\#))$ [24]. Table 1 shows the detailed specification of the pickup and display setups.

Table 1. Detailed specifications of the pickup and display setups.

	- Resolution	1920 × 1080 pixels
Elemental Image Array (EIA)	- Number of EIs	77 × 44
	- Resolution of each EI	25 × 25 pixels
Projector	- Diverging angle ($2\theta_p$)	39°
(LG PF85K)	- Distance to the CMA	800 mm
	- Pitch & focal length	7.47 mm and 7.32 mm
Convex mirror array (CMA)	- $\arctan(1/2(f\#))$	27.0°
	- Number of CMAs	77 × 44
	- Pitch & focal length	1.63 mm and 3.13 mm
Pickup lens array	- Number of lenses	77 × 44
	- Distance from the camera	160 mm
Viewing angle	- Left viewing-angle	20.0°
	- Right viewing-angle	20.0°
Test object ('VR')	- Front depth	50 mm
	- Rear depth	100 mm
Test object ('3D')	- Front depth	50 mm
	- Real depth	100 mm

3.2. Capturing of the L-EIA and R-EIA from Each of the Two Test 3-D Objects

As the test objects, two kinds of 3-D objects were computationally generated with the 3D Max, which were composed of two pairs of 2-D images with different depths. As seen in Figure 11a, one of them was the 3-D object composed of two English alphabetical letters of 'V' and 'R', which were located at the depth planes of 50mm and 100mm, respectively, and the other one was the 3-D object composed of a numeric letter of '3' and an English alphabetical letter of 'D', which were also located at the same depth planes of 50 mm and 100 mm, respectively, from the pickup lens array.

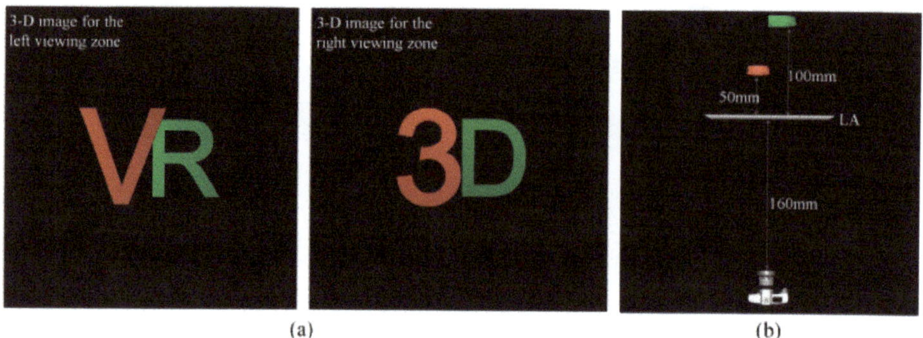

Figure 11. (a) Two test 3-D objects composed of two pairs of 2-D images with different depths ('VR' and '3D') and (b) on-axis integral imaging pickup system.

Here, 'V' and '3' images were colored with red, whereas 'R' and 'D' images were colored with green for the visual separation. In addition, two 3-D objects of 'VR' and '3D' were assigned to each of the left and right views for the DV 3-D display, respectively.

These two test 3-D objects were computationally picked up with the on-axis integral imaging system of Figure 11b. As mentioned above, two English alphabetical and numeric images of 'V' and '3' were located at the distances of 50 mm, while two other English alphabetical images of 'R' and 'D' were located at the distances of 100 mm from the pick-up lens array, where depth differences between them are given by 50 mm. Here, the pickup camera was set to be located at the distance of 160 mm from the lens array. In the pickup process, the number of lenses of the lens array and the pitch of the lens array and focal length of an elemental lens are in the Table 1, respectively.

Figure 12 shows two kinds of EIAs captured from two 3-D objects of 'VR' and '3D', which were designated to the L-EIA and R-EIA, respectively. That is, the 'VR' image was used for the left viewing zone, while the '3D' image was used for the right viewing zone.

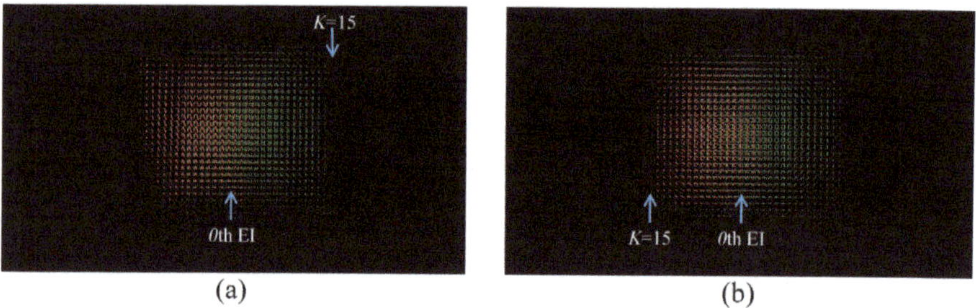

Figure 12. Two EIAs captured from each 3-D object of the 'VR' and '3D': (a) L-EIA captured from the 'VR', (b) R-EIA captured from the '3D'.

3.3. Generation of the DV-EIA from the L-EIA and R-EIA Based on the SSIM Method

Then, two captured EIAs of the L-EIA and R-EIA were synthesized into a single DV-EIA, based on the SSIM method on the sub-image plane. Figure 13 shows the synthesizing process of the DV-EIA with the picked-up L-EIA and R-EIA.

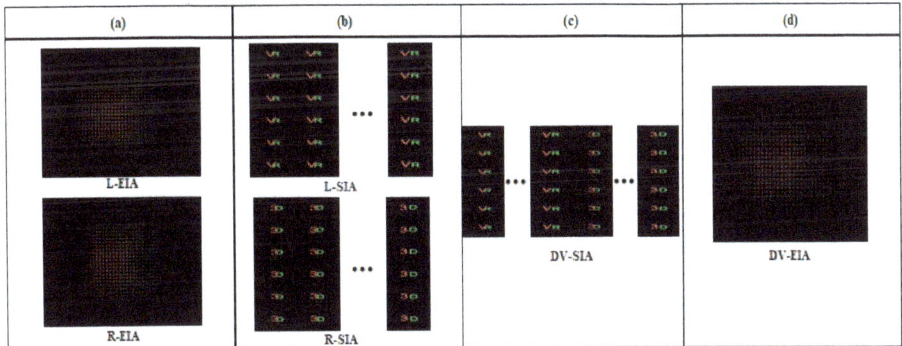

Figure 13. Generation of the DV-EIA with L-EIA and R-EIA based on SSIM: (**a**) L-EIA and R-EIA, (**b**) left sub-image array (L-SIA) and right sub-image array (R-SIA), (**c**) dual-view right sub-image array (DV-SIA), and (**d**) DV-EIA.

As seen in Figure 13a,b, these two captured EIAs of the L-EIA and R-EIA were transformed into their corresponding sub-image arrays (SIAs), such as the L-SIA and R-SIA, by using the EIA-to-SIA transformation (EST) method. Now, the DV-SIA could be synthesized from the L-SIA and R-SIA based on the SSIM method of Figure 4. That is, odd and even-number components of the L-SIs and R-SIs were selectively chosen from each of the L-SIA and R-SIA and mapped into the left and right half parts of the dual-view SIA (DV-SIA), respectively.

Figure 13c shows the synthesized DV-SIA from the L-SIA and R-SIA, where the numbers of L-SIs and R-SIs consisting of the DV-SIA are 12 and 13, respectively along the x-direction, which means the value of α was set to be 0.5 from Equation (1). This DV-SIA was then transformed into the corresponding DV-EIA by using the inverse EST (IEST) method, which is shown in Figure 13d. Since the DV-EIA was a multiplexed EIA with two kinds of EIAs, such as the L-EIA and R-EIA, the resolution and number of EIs of the DV-EIA became the same as those of the L-EIA and R-EIA.

3.4. Dual-View 3-D Display of the DV-EIA

The DV-EIA can be reconstructed into two different 3-D object images of 'VR' and '3D' in the left and right viewing directions, respectively, on the proposed CMA-DPII system, which is shown in Figure 10.

As seen in Figure 10, the fabricated 22''CMA was set to be located at a distance of 800 mm from the beam projector (Model: LG PF85K). Since the pitch of the elemental convex mirror and the distance between the projector and CMA were 7.47 mm and 800 mm, respectively, the far-field condition of $P/L = 0.009 < 0.01$ for the PRA method could be satisfied. In addition, in the experiments, the parameters of α and K were set to be 0.5 and 15, respectively, and arctan(1/2(f#)) of the CM was calculated to be 27.0°. Thus, with these experimental parameters, each of the left and right viewing angles was calculated to be 20° from Equation (5).

Thus, the DV-EIA generated from the pickup system was loaded on the projector. Then, the divergent beam of the projector containing the information of the DV-EIA was projected onto the CMA. On every convex mirror of the CMA, the left and right-view components of the DV-EIs were separated and reflected back into their viewing directions, and two different 3-D scenes were integrated and displayed on their viewing zones. Figure 14 shows two kinds of optically reconstructed 3-D images of 'VR' and '3D' at two different viewing zones on the proposed display system of Figure 14 from the DV-EIA. That is, Figures 14a–c and 14d–f show three kinds of optically reconstructed 3-D images of 'VR' and '3D' at the left and right viewing zones, which are viewed from each of the different viewing angles of −19.5°, −10.0°, −3.0 and 3.0, 10.0°, 19.5°, respectively.

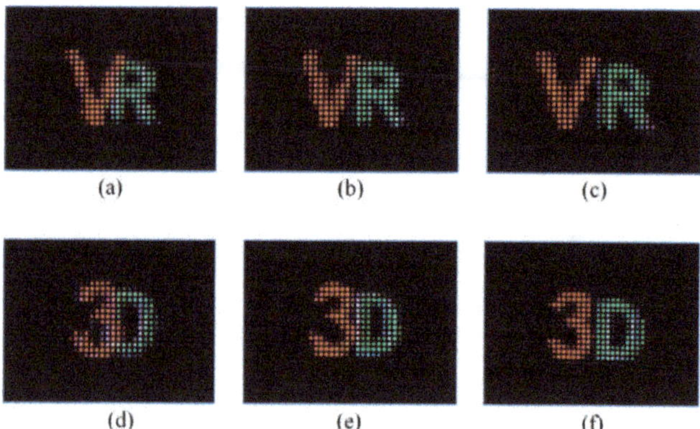

Figure 14. 3-D images of 'VR' viewed from three different viewing angles of (**a**) −21.0°, (**b**) −10.0° and (**c**) −3.0°, and those of '3D' viewed from those of (**d**) 3.0°, (**e**) 10.0° and (**f**) 21.0°, respectively.

As we can see, in the left-hand 3-D image of 'VR' viewed from the angle of −19.5°, 'V' and 'R' images are partially overlapped, where a part of the 'R' image is blocked by the image of 'V' because the 'R' object was originally located back of the 'R' object by 50 mm in the pickup process. It means that the reconstructed image of 'VR' is a form of 3-D data with depth. In the case of the center image of 'VR' of Figure 14b, which is viewed from the angle of −10.0° at the left viewing zone, 'V' and 'R' images look aligned in parallel, whereas in case of the right-hand image of 'VR', 'V' and 'R' images get a little bit separated from each other since they have different depths. That is, as we move from the left to the center and right directions, the distance difference between the two object images of 'V' and 'R' increases, which confirms that these object images have depth information. In addition, Figure 14d–f also show the optically reconstructed 3-D images of '3D' at the right viewing zone from the DV-EIA. Those reconstructed '3' and 'D' images at three different directions look almost the same as those images in the case of the 'VR'. These good experimental results of Figure 14 confirm that the proposed system can be applied to the practical dual-view 3-D display.

Moreover, viewing angles for the left and right viewing zones have been measured to range from −19.5° to −3.0° in the left viewing zone, and from 3.0° to 19.5° in the right viewing zone, respectively. These results may also confirm that the proposed system can provide relatively large viewing-angles with a simple optical configuration composed of only a pair of projectors and CMA.

3.5. Experiments with Two Volumetric 3-D Objects

Now, to confirm the feasibility of the proposed system in the practical application, two volumetric 3-D objects of 'Dice' and 'Car' were used as the test objects in the experiments. Figure 15 shows two kinds of test volumetric 3-D objects of 'Dice' and 'Car' whose sizes are 25 mm × 25 mm × 25 mm and 25 mm × 21 mm × 50 mm, respectively.

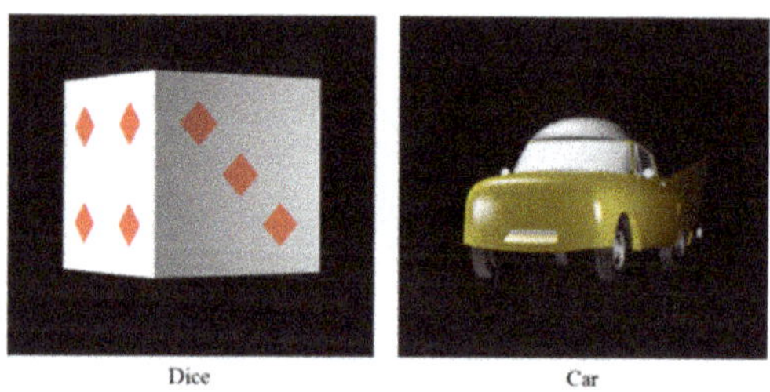

Figure 15. Two test volumetric 3-D objects of '**Dice**' and '**Car**'.

In the experiments, the same pickup lens array was employed. Thus, the pitch and focal length of the pickup lens array were also 1.63 mm and 3.13 mm, respectively, and they were located at the distance of 50 mm from each of the 3-D objects of 'Dice' and 'Car'. Here, the 'Dice' and 'Car' objects were used for the left and right views, respectively, in the dual 3-D display.

As seen in Equation (5), the viewing angle may change depending on the θ_{kr}. Thus, under the condition that the pitch and focal length of the convex mirror are given by 7.32 mm and 7.47 mm, and the CMA is located at the distance of 800 mm from the projector, the maximum K value is 38 since it is limited by half the number of lenses of 77 along the x-direction. As the K value increases, the corresponding viewing angle decreases, whereas it increases as the K value decreases, which may confirm that the K value is inversely related to the viewing angle.

Figure 16 shows the synthesized DV-EIA based on the SSIM method from the captured L-EIA and R-EIA from each of the 'Dice' and 'Car', respectively, for three cases of K = 15, 11 and 8.

Here, the DV-EIA with 1920 × 1080 pixels is composed of 77 × 44 EIs, where each EI has the resolution of 25 × 25 pixels. Depending on the K values, three kinds of DV-EIAs are synthesized based on the SSIM method, which are shown in Figure 16.

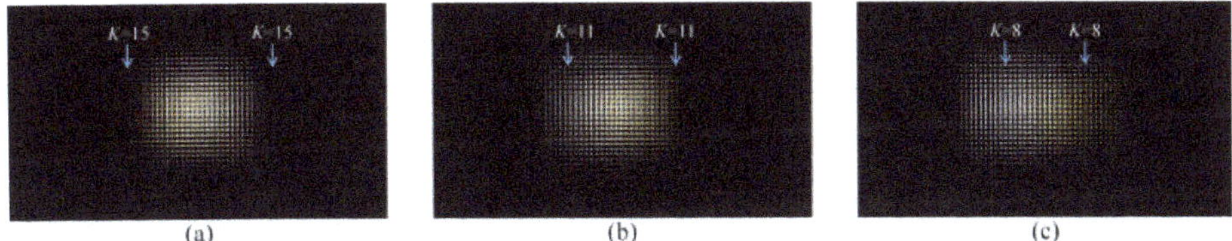

Figure 16. Synthesized DV-EIA for the 'Dice' and 'Car' objects with different K values: (**a**) K = 15, (**b**) K = 11 and (**c**) K = 8.

Now, these DV-EIAs can be reconstructed on the proposed display system of Figure 17.

Figure 17 also shows the experimental results of the volumetric 3-D object images of 'Dice' and 'Car' reconstructed at the left and right viewing zones, respectively, for three cases of the K values.

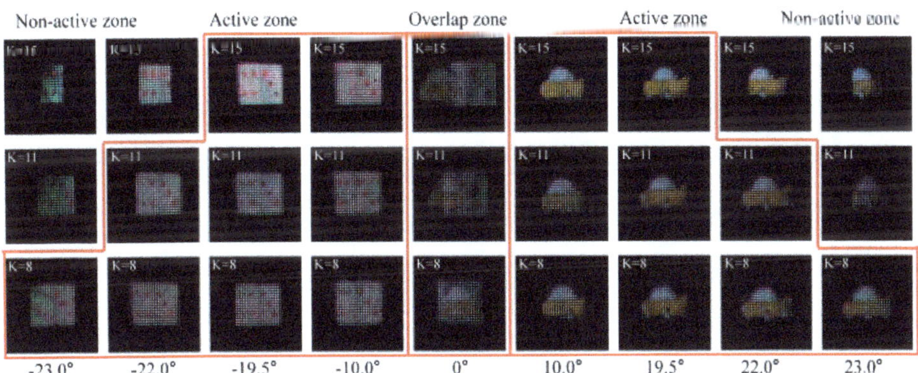

Figure 17. Reconstructed dual-view 3-D object images of 'Dice' and 'Car' at the left and right viewing zones, respectively, for three cases of the K values.

As seen in Figure 17, for the case of K = 15, the reconstructed object images of 'Dice' and 'Car' at each of the left and right viewing zones are getting truncated and disappeared just beyond the viewing angle of −19.5° and 19.5°, respectively. These results show that the viewing angles of this dual-view 3-D display system have been measured to be ranged from −19.5° to −3.0° and from 3.0° to 19.5° for the left and right viewing zones, respectively. These measured left and right viewing angles of −19.5°

and 19.5° have been found to be almost the same as those calculated values of −20° and 20°, where the error ratio between the calculated and measured viewing angles has been calculated to around 3.0%. Moreover, as we move from the left to the center and center to right directions, different perspectives of the object images of 'Dice' and 'Car' can be viewed, as seen in Figure 17, which confirms the volumetric 3-D reconstruction of the 'Dice' and 'Car' objects in the proposed system.

For the case of $K = 11$, left and right viewing angles are measured to range from −22.0° to −3.0° and from 3.0° to 22.0°, respectively, as seen in Figure 17, where those values are calculated to be −22.2° and 22.2°, respectively, using Equation (5). Thus, the error ratio between the calculated and measured viewing angles has been calculated to be around 3.6%. Here, it must be noted that the left and right viewing angles increased by 3° as the K value decreased to 11 from 15.

For the case of $K = 8$, left and viewing angles are calculated to be −23.5° and 23.5°, respectively. In addition, those values have been measured to range from -23.0° to -3.0° and from 3.0° to 23.0°, respectively. Thus, these measured left and right viewing angles of −23.0° and 23.0° look almost the same as those of the calculated values of −23.5° and 23.5°. Thus, for the case of $K = 8$, the error ratio between the calculated and measured viewing angles has been calculated to around 2.1%.

As seen in Figure 17, there are three kinds of viewing zones, such as active, inactive and overlapped regions. Here, the active zones marked with red color represent the DV zones, where we are able to watch the displayed 3-D objects. On the other hand, in other zones including the inactive and overlapped regions, a 3-D object cannot be properly viewed. These experimental results confirm that two different volumetric 3-D object images with their changing perspectives can be successfully reconstructed in both of the left and right viewing zones just like in the case of the previous experiments, and each of the left and right viewing angles can be changed depending on the K value.

Figure 18 visually shows that two observers sitting at each of the left and right viewing directions are separately watching the two different 3-D object images of 'Dice' and 'Car' reconstructed from the proposed system, where the projector is located at the distance of 800mm from the CMA, and two observers are seated at the distance of 1200 mm from the CMA.

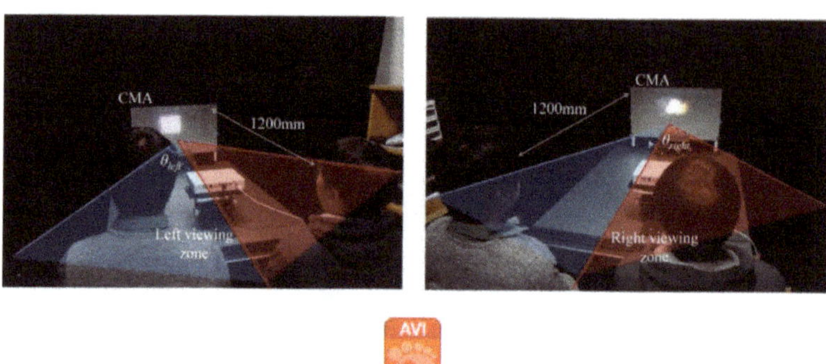

Video_dice and car.avi

Figure 18. Photos of two observers watching two different 3-D object images of 'Dice' and 'Car' at the left and right viewing zones with the optical setup of the proposed system.

As seen in Figure 18, each of the left and right viewers can see only the 'Dice' or 'Car' object images reconstructed at their viewing zones, which means that two different 3-D object images can be separately viewed by two observers who are distinctly sitting in their left and right viewing zones. Experimental results on two test 3-D object images of 'Dice' and 'Car' with the proposed system have been taken as video files and attached in Figure 18 as a video clip. In the video clip, volumetric 3-D object images of the 'Dice' and 'Car' with their changing perspectives, which are viewed from −23.0° to −3.0° and from 3.0° to 23.0°, respectively, have been recorded for the case of $K = 11$. Here, the video frame rate and total recording time are set to be 24 fps (frames per second) and 34 s, respectively.

In the video files, each of the left and right 3-D images of the 'Dice' and 'Car' has been recorded for 17 s, respectively.

As seen in the experimental results of Figures 17 and 18, the resolution and visibility of the reconstructed 3-D images still look somewhat degraded since the fill factor of the elemental convex mirror is low [33]. Thus, using elemental mirrors with much-enhanced fill factors can enhance the resolution and visibility of those 3-D images.

4. Conclusions

In this paper, a new type of the viewing angle-enhanced 3-D dual-view display based on the CMA-DPII system is proposed. Two kinds of elemental image arrays (EIAs) are captured from each of the two 3-D objects and synthesized into a single dual-view EIA (DV-EIA) with the selective sub-image mapping (SSIM) scheme. Then, a divergent beam of the projector containing this DV-EIA image is directly projected onto the CMA, where left and right-view components of the DV-EIA are separately reflected back into their viewing directions and finally form the two different 3-D images displayed on their viewing zones. From the ray-optical analysis with the parallel-ray-approximation method and successful experimental results with the test 3-D objects on an implemented 22''DV 3-D display prototype, the feasibility of the proposed system is confirmed in the practical application.

Author Contributions: Conceptualization, H.-M.C., J.-G.C., E.-S.K.; methodology, H.-M.C., J.-G.C.; validation H.-M.C., J.-G.C.; formal analysis, H.-M.C.; writing-original draft preparation, H.-M.C., J.-G.C.; writing-review and editing, E.-S.K.; funding acquisition, E.-S.K.

Funding: This research was partially funded by Basic Science Research Program through the National Research Foundation of Korea (NRF) supported by the Ministry of Education (No. 2018R1A6A1A03025242) and supported by the MSIT (Ministry of Science and ICT), Korea, under the ITRC support program (IITP-2017-01629) supervised by the IITP.

Conflicts of Interest: The authors declare no conflict of interest.

References

1. Gulick, S., Jr. Dual-View Imaging Product. U.S. Patent 5757550 A, 26 May 1998.
2. Kuhlman, F.F.; Harbach, A.P.; Parker, R.D.; Sarma, D.H.R. Dual View Display System Using a Transparent Display. U.S. Patent 8362992 B2, 29 January 2013.
3. Yamamoto, H.; Kimura, T.; Matsumoto, S.; Suyama, S. Viewing-Zone Control of Light-Emitting Diode Panel for Stereoscopic Display and Multiple Viewing Distances. *J. Disp. Technol.* **2010**, *6*, 359–366. [CrossRef]
4. Kuhlman, F.F.; Harbach, A.P.; Sarma, D.H.R.; Sultan, M.F. Dual View Display System. U.S. Patent 8363325 B2, 29 January 2013.
5. Yamane, T.; Konishi, T.; Nagamoto, S.; Tanaka, S.; Uehara, S.; Kamoto, M.; Ohshima, T. Fall 2005 DUAL AVN development. *FUJITSU Ten Tech. J.* **2006**, *26*, 17–22.
6. Mather, J.; Montgomery, D.J.; Winlow, R.; Bourhill, G.; Barrett, N.W. Multiple-View Directional Display. U.S. Patent US7580186 B2, 25 August 2009.
7. Sharp. Available online: http://www.sharp-world.com/corporate/news/0507.html (accessed on 19 March 2019).
8. Mather, J.; Barratt, N.; Kean, D.U.; Walton, E.J.; Bourhill, G.; Powell, T.W. Directional Backlight, a Multiple View Display and a Multi-Direction Display. U.S. Patent 8154686 B2, 10 April 2012.
9. Mather, J.; Smith, N.J. Multiple View Display. U.S. Patent 9274345 B2, 1 March 2016.
10. Chen, C.Y.; Hsieh, T.Y.; Deng, Q.L.; Su, W.C.; Cheng, Z.S. Design of a novel symmetric microprism array for dual-view display. *Displays* **2010**, *31*, 99–103. [CrossRef]
11. Lanman, D.; Wetzstein, G.; Hirsch, M.; Heidrich, W.; Raskar, R. Polarization Fields: Dynamic Light Field Display using Multi-Layer LCDs. In Proceedings of the 2011 SIGGRAPH Asia Conference, Hong Kong, China, 12–15 December 2011; pp. 1–9.
12. Mather, J.; Jones, L.P.; Gass, P.; Imai, A.; Takatani, T.; Yabuta, K. Potential improvements for dual directional view displays. *Appl. Opt.* **2014**, *53*, 769–776. [CrossRef] [PubMed]
13. Hsieh, C.T.; Shu, J.N.; Chen, H.T.; Huang, C.Y.; Tian, C.J.; Lin, C.H. Dual-view liquid crystal display fabricated by patterned electrodes. *Opt. Express* **2012**, *20*, 8641–8648. [CrossRef]

14. Hsieh, C.T.; Li, G.Y.; Wu, T.T.; Huang, C.Y.; Tien, C.J.; Lo, K.Y.; Lin, C.H. Twisted Nematic Dual-View Liquid Crystal Display Based on Patterned Electrodes. *IEEE J. Disp. Technol.* **2014**, *10*, 464–469. [CrossRef]

15. Tang, P.; Cui, J.-P.; Liang, D.; Wang, Q.-H. Spatial-multiplexed dual-view display using blue phase liquid crystal. *SID Symp. Dig. Tech. Pap.* **2014**, *45*, 1389–1391. [CrossRef]

16. Kean, D.U.; Montgomery, D.J.; Mather, J.; Bourhill, G.; Jones, G.R. Parallax Barrier and Multiple View Display. U.S. Patent 7154653 B2, 26 December 2006.

17. Krijn, M.; de Zwart, S.T.; de Boer, D.K.G.; Willemsen, O.H.; Sluijter, M. 2D/3D displays based on switchable lenticulars. *J. SID* **2008**, *16*, 847–855.

18. Cui, J.P.; Li, Y.; Yan, J.; Cheng, H.C.; Wang, Q.H. Time-multiplexed dual-view display using a blue phase liquid crystal. *J. Disp. Technol.* **2013**, *9*, 87–90. [CrossRef]

19. Silva, V.N.H.; Stoenescu, D.; Nassour, T.C.; de la Rivière, J.-B.; Tocnaye, J.-L. Ghosting Impingements in 3D Dual-View Projection Systems. *J. Disp. Technol.* **2014**, *10*, 540–547. [CrossRef]

20. Wu, F.; Deng, H.; Luo, C.G.; Li, D.H.; Wang, Q.H. Dual-view integral imaging three-dimensional display. *Appl. Opt.* **2013**, *52*, 4911–4914. [CrossRef]

21. Wu, F.; Wang, Q.H.; Luo, C.G.; Li, D.H.; Deng, H. Dual-view integral imaging 3D display using polarizer parallax barriers. *Appl. Opt.* **2014**, *53*, 2037–2039. [CrossRef]

22. Wang, Q.-H.; Ji, C.-C.; Li, L.; Deng, H. Dual-view integral imaging 3D display by using orthogonal polarizer array and polarization switcher. *Opt. Express* **2016**, *24*, 9–16. [CrossRef]

23. Jeong, J.; Lee, C.K.; Hong, K.; Yeom, J.; Lee, B. Projection-type dual-view three-dimensional display system based on integral imaging. *Appl. Opt.* **2014**, *53*, 12–18. [CrossRef]

24. Jang, J.-S.; Javidi, B. Three-dimensional projection integral imaging using micro-convex-mirror arrays. *Opt. Express* **2004**, *12*, 1077–1086. [CrossRef] [PubMed]

25. Okui, M.; Arai, J.; Nojiri, Y.; Okano, F. Optical screen for direct projection of integral imaging. *Appl. Opt.* **2006**, *45*, 9132–9139. [CrossRef] [PubMed]

26. Kang, H.-H.; Lee, J.-H.; Kim, E.-S. Enhanced compression rate of integral images by using motion-compensated residual images in three-dimensional integral-imaging. *Opt. Express* **2012**, *20*, 5440–5459. [CrossRef]

27. Xiaoya, Z.; Dong, X.; Yongsheng, W.; He, S. A simple parallel ray approximation based stochastic channel model for MIMO UWB systems with measurement verification. In Proceedings of the 2010 International Conference on Communications and Mobile Computing, Shenzhen, China, 12–14 April 2010; pp. 102–106.

28. Tiberi, G.; Bertini, S.; Malik, W.Q.; Monorchio, A.; Edwards, D.J.; Manara, G. Analysis of realistic ultrawideband indoor communication channels by using an efficient ray-tracing based method. *IEEE Trans. Antennas Propag.* **2009**, *57*, 777–785. [CrossRef]

29. Miron, D.B. *Small Antenna Design*; NEWNES: Boston, MA, USA, 2006; pp. 255–256.

30. Sard, A. *Linear Approximation*; American Mathematical Society: Providence, RI, USA, 1963; p. 83.

31. Stern, A.; Javidi, B. Three-dimensional image sensing, visualization, and processing using integral imaging. *Proc. IEEE* **2006**, *94*, 591–607. [CrossRef]

32. Jang, J.-Y.; Shin, D.; Lee, B.-G.; Kim, E.-S. Multi-projection integral imaging by use of a convex mirror array. *Opt. Lett.* **2014**, *39*, 2853–2856. [CrossRef] [PubMed]

33. Park, S.; Song, B.-S.; Min, S.-W. Analysis of Image Visibility in Projection-type Integral Imaging System without Diffuser. *J. Opt. Soc. Korea* **2010**, *14*, 121–126. [CrossRef]

Review

High-Resolution Episcopic Microscopy (HREM): Looking Back on 13 Years of Successful Generation of Digital Volume Data of Organic Material for 3D Visualisation and 3D Display

Stefan H. Geyer and Wolfgang J. Weninger *

Division of Anatomy & MIC, Medical University of Vienna, 1090 Vienna, Austria;
stefan.geyer@meduniwien.ac.at
* Correspondence: wolfgang.weninger@meduniwien.ac.at; Tel.: +43-1-40160-37560

Received: 18 August 2019; Accepted: 9 September 2019; Published: 12 September 2019

Abstract: High-resolution episcopic microscopy (HREM) is an imaging technique that permits the simple and rapid generation of three-dimensional (3D) digital volume data of histologically embedded and physically sectioned specimens. The data can be immediately used for high-detail 3D analysis of a broad variety of organic materials with all modern methods of 3D visualisation and display. Since its first description in 2006, HREM has been adopted as a method for exploring organic specimens in many fields of science, and it has recruited a slowly but steadily growing user community. This review aims to briefly introduce the basic principles of HREM data generation and to provide an overview of scientific publications that have been published in the last 13 years involving HREM imaging. The studies to which we refer describe technical details and specimen-specific protocols, and provide examples of the successful use of HREM in biological, biomedical and medical research. Finally, the limitations, potentials and anticipated further improvements are briefly outlined.

Keywords: imaging; 3D; high-resolution episcopic microscopy; episcopic; phenotyping; HREM

---

1. Introduction

Imaging plays a central role in all areas of modern science and is one of the most expanding fields of biomedical research. Therefore, a plethora of highly sophisticated imaging methods spanning from ground-penetrating radar to Brillouin microscopy were developed in the last century [1–3].

In particular, the life sciences had a massive benefit from the availability of novel imaging modalities designed for 3D visualisation of living humans and animals, harvested embryos and tissue samples. Techniques such as single-photon emission computed tomography (SPECT) and positron emission tomography (PET), micro-computed topography (µCT) and optical coherence tomography (OCT) are already used as routine tools for the diagnosis of pathologies in the clinical routine [4–9]. Others, such as micromagnetic resonance tomography (µMRI), atomic force microscopy (AFM) and optical projection tomography (OPT), are still restricted to the scientific analysis and characterisation of biological samples [10–16]. Their focus rests on the interpretations of the three-dimensional (3D) arrangement of organs, tissues, cells and molecules in healthy and diseased organisms, which is the basis for understanding the genetic, epigenetic and functional aspects of developmental processes, the genesis of pathologies and the effect of strategies to treat diseases.

Each imaging method has unique advantages for a small field of applications. But due to technical constraints, most are restricted to highly specific research fields and a narrow selection of specimen types. A method that is highly potent, especially in structural 3D visualisation of embryos of biomedical models and biopsy material, is high resolution episcopic microscopy (HREM).

HREM [17] is an episcopic imaging method which generates a series of inherently aligned digital images of histologically processed and embedded specimens. The images are virtually stacked to be analysed by scrolling through the original or virtual resections and to lend themselves to volume rendering or segmentation and surface rendering. Thus, HREM is in line with other episcopic imaging techniques, such as serial block-face scanning electron microscopy (SBFSEM) [18], serial reconstruction technique [19–21], Epi-3D [22], surface imaging microscopy [23,24], imaging cryomicrotome [25], serial block-face imaging [26] and episcopic fluorescent image capturing (EFIC) [27], although it is optimised for a low microscopic level of resolution.

In a nutshell, materials are harvested and processed as for traditional histology. During dehydration, they are stained with eosin red mixtures, and resin (JB4) dyed with eosin or eosin/acridine orange is used for infiltration and as an embedding material. The resulting resin blocks are sectioned on a microtome, while digital images of subsequently exposed block surfaces are captured in fluorescence-mode using yellow fluorescent protein (YFP) (excitation 500/20 nm, emission 535/30 nm) or green fluorescent protein (GFP) (excitation 470/40, emission 525/50) filter sets for visualising eosin-contrasted structures. Cells and tissues specifically stained with LacZ or NBT/BCIP (nitro-blue tetrazolium and 5-bromo-4-chloro-3′-indolyphosphate) appear heavily contrasted when also using a Texas Red filter system as shown in Figure 1. Several protocol variations optimised for various materials are already published and extensively discussed [28–32].

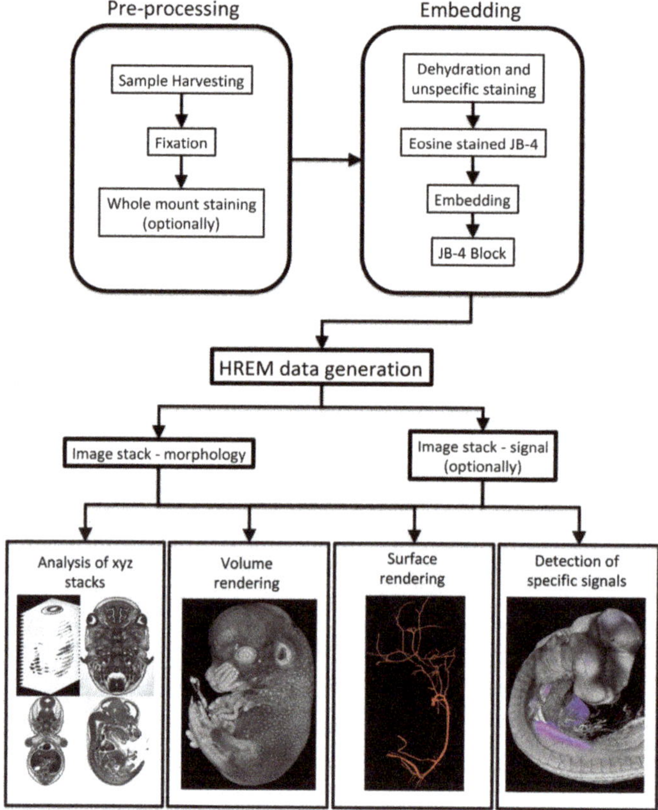

Figure 1. Workflow. The flow chart briefly summarizes the steps of the HREM-workflow from sample harvesting to 3D visualisation and data analysis.

HREM data typically have a voxel size of $2 \times 2 \times 2 \ \mu m^3$ and are created from specimens with a volume of $6 \times 6 \times 6 \ mm^3$, although larger and smaller samples and higher or lower resolutions are possible. Data contrasts fit for identifying cells and nuclei, nerve fibres, capillaries and collagen bundles in the context of overall morphology and tissue architecture. Data generation varies according to specimen size, but it takes approximately 4–5 h for data comprised of 2000 single section images. Since numerous publications comprehensively describe the method [33–36] and ready-to-go HREM apparatuses are already commercially available (OHREM, Indigo Scientific Ltd., Baldock, England), this paper will not provide protocols or technical descriptions, but will focus on discussing published scientific results and potential applications.

2. Applications of HREM

HREM has been employed in various areas of the life sciences and for visualising a number of different organic materials. Its chief domains of application are phenotyping of embryos of biomedical model organisms and structural analysis of human biopsy material. Besides those, there are a number of unconventional applications.

2.1. HREM in Embryo Research

HREM was used for analysing whole embryos or embryonic organs and tissues of various species, including humans. However, the highest impact in this field was gained from phenotyping mouse embryos and from the selection of HREM as the tool for phenotyping mouse mutants harvested at embryonic day (E) 14.5 in the Deciphering the Mechanisms of Developmental Disorders (DMDD) program.

2.1.1. Mouse Embryos

Moderate breeding costs, short generation times and the existence of sophisticated molecular tools to manipulate the mouse genome facilitate efficient engineering of mutant mouse lines by random or targeted disruption of specific genes. Due to the conservation of basic developmental mechanisms and gene function, such lines are then used for studying normal mammalian gene function and the genetic components of diseases.

Identifying morphological defects of mouse embryos of genetically engineered mouse lines is an important step to characterise the function of the disrupted gene during embryo development and to define its role in the genesis of hereditary diseases and malformations. HREM was developed to permit the production of 3D volume data of whole E14.5 mouse embryos in a resolution and quality to fit for a holistic qualitative and quantitative analysis of embryo morphology, organ, blood vessel and nerve topology and tissue architecture as shown in Figure 2. In the scope of DMDD [37], which was linked with the International Mouse Phenotyping Consortium (IMPC), HREM was employed to build a virtual resource of fully annotated phenotype data of over 200 normal and over 500 mutant mouse embryos harvested at E14.5 stemming from 87 individual single knock-out or knock-down lines producing lethal or subvital homozygous offspring.

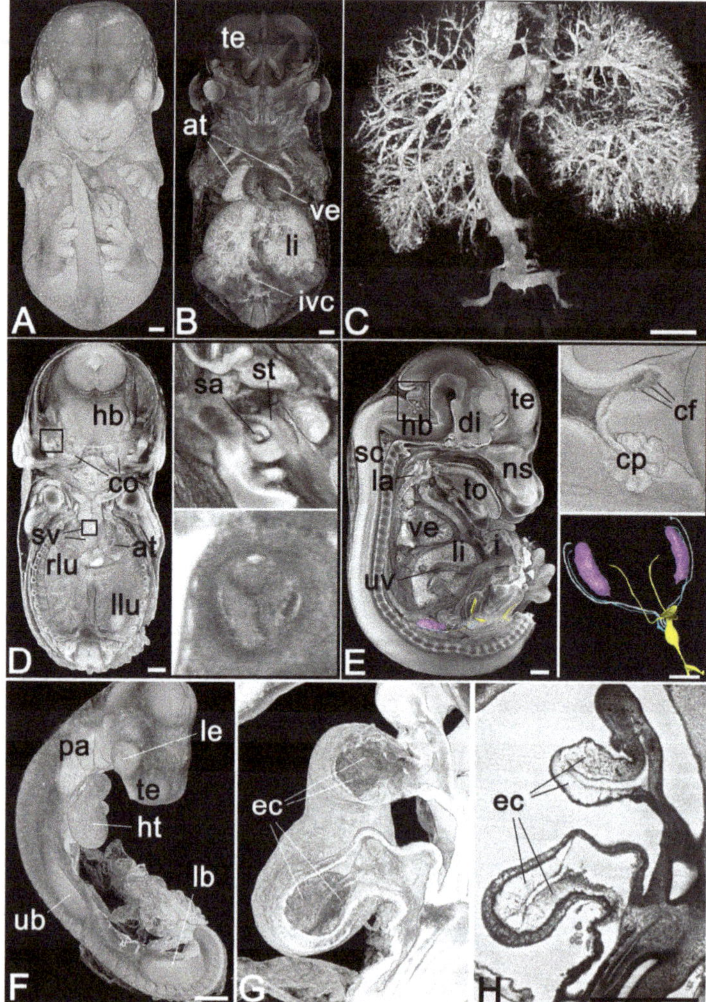

Figure 2. 3D visualisation of embryo material. (**A–E**). Embryos harvested at embryonic day 14.5 in the scope of the Deciphering the Mechanisms of Developmental Disorders (DMDD) project. Opaque and semitransparent volume models from ventral view (**A–D**). Note the intrinsic contrast of the blood-filled liver vasculature inferior vena cava (ivc) and the atria (at) in (**B**) and (**C**). In (**D**), the opaque volume model is coronally sectioned and the stapedial artery (sa) penetrating the developing stapes (st) is zoomed in on in the inlay to the top. The bottom inlay provides the aortic valve as detailed as it appears in the coronal 2D resection used as a sectioning plane. (**E**) shows a sagittally sectioned semitransparent volume model from lateral, integrating surface models of the urogenital tract. Choroid plexus (cp) and commissural fibres (cf) are detailed in the zoom-in displayed as an inlay to the top right. The bottom inlay shows a cranial and oblique view of the surface models of the ovaries (violet), ureter, urinary bladder and urethra (yellow), Müller duct (light blue) and Wolff duct (blue). (**F–H**). Chick embryo at developmental stage 18 according to Hamburger and Hamilton. Opaque volume model from the ventrolateral in (**F**) and (**G**) with a zoom-in on the virtually sectioned heart tube (ht) in (**G**). (**H**) shows the virtual 2D resection of the cutting plane in (**G**). ve, ventricles; li, liver; hb, hindbrain; co, cochlea; sv, semilunar valves; rlu, right lung; llu, left lung; te, telencephalon; di, diencephalon; sc, spinal chord; ns, nasal septum; to, tongue; la, larynx; uv, umbilical vein; I, intestine; pa, pharyngeal arches; le, lens; ub, bud of upper limb; lb, bud of lower limb; ec, endocardial cushion. Scale bars: 500 μm.

In the scope of this program, a standardised and ergonomic protocol for the comprehensive scoring of the morphologic phenotype of E14.5 embryos [38], a novel system for defining developmental substages of E14.5 mouse embryos [39], and reference data for the correct interpretation of phenotype abnormalities and for distinguishing variations from abnormalities [40] were created. Approximately 200 novel mouse phenotype (MP) terms were added to the MP ontology list as a result of HREM allowing for the detection of details that could not be detected with alternative imaging techniques. Numerous publications in highly ranked journals were produced based on HREM data generated in DMDD [38–48].

DMDD independent stand-alone studies researching genetic regulation of mouse development include the analysis of: normal and abnormal cardiovascular development [49–61]; the development of the limbs, cloaca and pancreas [42,62–64]; ciliopathies [65]; and the characterisation of cardiac defects in Down syndrome mouse models [66,67]. Despite its value for analysing such processes in embryos between embryonic day E8.5 and E14.5, when organogenesis is already finished, HREM also proved its value for analysing developmental processes in very early embryos immediately after implantation [68].

HREM has proved an excellent fit for pure descriptive but also metric analysis, which will have a great impact on researching the influence of biomechanical forces on prenatal tissue and organ formation and remodelling. Simple descriptive studies allowed, for example, the visualisation of pharyngeal arch artery development, proved the transitory existence of a 5th pharyngeal arch artery in the mouse and defined the dimensions of large arteries in mouse embryos bred on various genetic backgrounds [69–74]. Precise visualisation of septation and cardiac cushion, the outflow tract and valve development were used as a basis for learning the concepts underlying associated malformations [75–86]. Besides its importance for research, digital 3D models are of course eagerly anticipated teaching aids, as they can exemplify complex developmental remodelling processes.

Simple metric data were produced for defining the dimensions of the great intrathoracic arteries of embryos bred on various genetic backgrounds. However, based on HREM data, more sophisticated models could be calculated, which could mathematically define the description of the heart looping process [52] or use 3D fractal analysis for defining the complexity of trabeculation during ventricular development of the heart [87–89].

2.1.2. Chick Embryos

Chick embryos of very early to late developmental stages were already examined, as shown in Figure 2. Embryos of the blastoderm stage [90,91] were used for researching the effect that the storage of eggs has on embryonic development and viability. Embryos of late developmental stages have been successfully used for analysing the effect of hemodynamic alterations on the topology and remodelling of the cardiac outflow [72], and for producing metric definition of the dimensions of the great intrathoracic arteries to serve as reference data for such studies [73]. In addition, HREM analysis of an embryo with cephalothoracopagus malformation was successfully conducted, which permitted the formulation of a novel concept explaining the genesis of this abnormality [92].

2.1.3. Embryos/Fetuses of Other Species

HREM is not restricted to research mouse and chick embryos. It also has been shown to work on a broad variety of embryos of biomedical models. This includes quail and *Xenopus* embryos, where the proof of principle was published [17,28], but also includes the zebrafish, for which HREM was shown to fit for visualisation and mathematical characterisation of complex developmental processes, such as dentation and cardiac jogging [17,93].

Yet, HREM is not restricted to small biomedical models. 3D visualisation of voluminous fetal material is also possible. In particular, joint development was visualised in horse fetuses [94]. Furthermore, the potential of HREM to analyse the heart morphology of human fetal hearts in the first trimester has been shown [95,96]. Also, other (data not yet published) fetal materials fit as well for HREM imaging, but due to restricted availability of human embryos for morphological studies, collecting such materials is rather problematic and limited to highly specific scientific projects.

2.2. HREM for Visualising Adult Material

Though originally developed for imaging embryonic samples, protocols are already available for preparing biopsy material from adult individuals of various species for HREM imaging, as shown in Figure 3. Examples are materials stemming from rodents, zebrafish, turquoise killfish, ferrets and the fruit fly, for which there are preliminary data, but no publications yet exist.

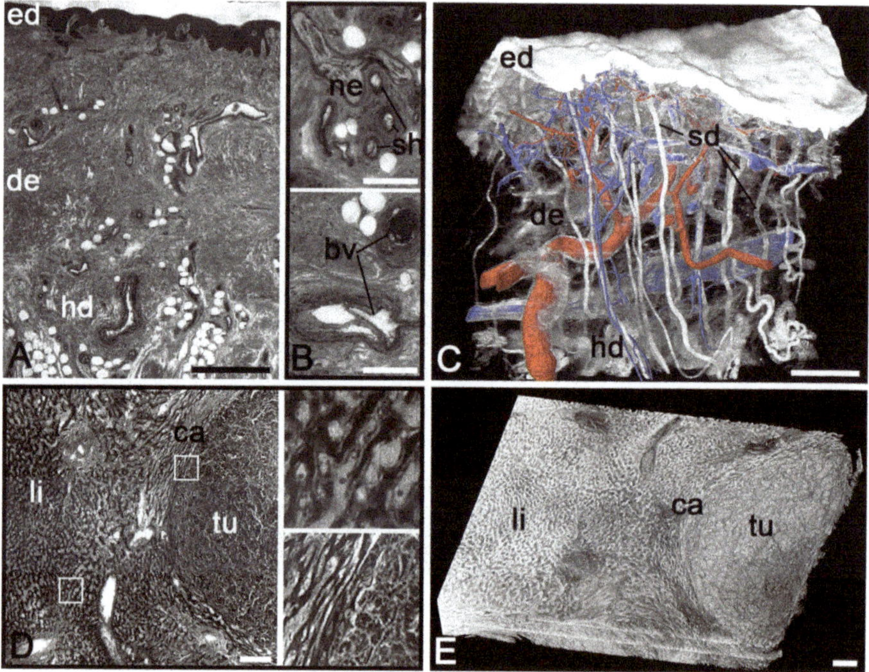

Figure 3. HREM data created from material harvested from adult human (**A**–**C**) skin biopsies. Virtual resection perpendicular to the original HREM section plane (**A**). (**B**) shows details from the original HREM sections to demonstrate HREM's ability to visualize small dermal nerves (ne) and a Suquet-Hoyer canal (sh) in the top image and small dermal blood vessels (bv) in the bottom image. (**C**) shows a semitransparent 3D volume model of the sample combined with surface models of the lumina of the dermal arteries (red) and veins (blue). (**D**,**E**) show metastatic material (tu) in a liver (li) biopsy. Original 2D HREM section in (**D**). Boxed areas are zoomed-in views provided as inlays to the right. (**E**) shows a 3D volume model. ed, epidermis; de, dermis; hd, hypodermis; ca, fibrous tumour capsule; sd, sweat duct. Scale bars 500 μm in (**A**), (**C**), (**D**), (**E**); 150 μm in (**B**).

Published data does exist for the pig and the mouse, in which vascularisation in wound healing was researched [97,98], and for humans. In humans, HREM was successfully employed to develop a novel concept for dermal vascularisation in thick [99,100] and thin skin [101] and to analyse the topology of arteries and nerves in the auricle [102]. Besides this, it recently permitted the characterisation of plaques in the coronary arteries [103].

HREM could also show the structure of dermal matrix skin substitutes with and without seeded keratinocytes, and the effect of dermal matrix skin substitutes in combination with skin graft transplants on vascularisation in a porcine wound model. Recently, HREM was an integral part of a multimodal imaging pipeline for comprehensively characterising the vascularisation of murine tumour models [104].

2.3. HREM for Other Organic Materials

The materials primarily subjected to HREM imaging are sourced from animals. However, very recently, HREM expanded its application into the realm of plant sciences. More concretely, the morphology of wild-type and genetically altered tomato plants were visualised in order to study abnormal leaf axil patterning [105]. In pilot studies, the capacity of HREM for testing paper quality [28,106] and the fibre arrangement and the formation of keratinocytes seeded on skin replacement material [107] were evaluated. These experiments demonstrated that HREM is not restricted to biomedical research, but has yet-unknown capacities to aid research in many other fields of modern science.

3. Conclusions and Further Perspectives

HREM was developed in 2006, and for about a decade, operated on self-assembled apparatuses. In 2015, a fully operable, all-inclusive HREM machine became commercially available under the trade name of 3D Optical HREM imaging (OHREM). This increased the number of projects using the HREM technique, which was illustrated by a quadruplication of publications based on HREM data over the last ten years, and it boosted the development of new protocols and technical advancements. However, HREM is still to be considered as a technique in its beginnings with great potential for refinement and improvement. As pilot data has shown, it has a yet-unexplored high potential for visualising a broad variety of organic materials in unmatched resolution and data quality.

3.1. Stitching

One of the strengths of HREM is its ability to image volumes of up to $6 \times 6 \times 12$ mm^3 in a numeric resolution of $3 \times 3 \times 3$ μm^3. Increasing the numeric resolution is possible, but requires focusing on a smaller volume to be scanned [34,108]. First prototypes of HREM machines, which scan several images of the same block-surface and incorporate stitching algorithms for combining them, overcome these limitations and are already in use. It is to be expected that this promising approach will soon advance to become the HREM routine.

3.2. Pipelines

Even by using block scanning and stitching, the use of HREM is limited to exploring the microanatomy of specimens of a relatively small size. For large specimens, HREM imaging has to be combined with other imaging modalities to gain HREM detail in the context of overall specimen information. Since this is quite simple with techniques such μMRI, the first imaging pipeline that includes HREM dates back to the very early phase of HREM imaging [51]. Large batches of embryos were scanned with high-throughput μMRI in moderate resolution. Interesting specimens or specimen parts were then identified, selected and subjected to HREM imaging for providing tissue detail. This approach was modified and expanded in the last years [109], and the first publications providing protocols and demonstrating the benefits of high complex multimodal, multiscale imaging pipelines integrating HREM with imaging modalities such as μMRI, US, μCT, OCT, PAT and histopathology are already in preparation [104].

3.3. Specific Stainings

HREM data volumes comprise thousands of digital images, each resembling a greyscale image of a hematoxylin–eosine-stained histological section. Thus, HREM offers 3D information of morphological details of organic materials in a straightforward and simple way. Attempts to expand HREM to permit 3D visualisation of specifically labelled structures and molecular signals as well are as old as the HREM technique itself. Even the first publication included an example for visualising LacZ-stained tissues in mouse embryos and NBT/BCIP signals after whole-mount in situ hybridisation in zebrafish embryos [17]. However, despite these efforts, visualising specific stained tissues is still experimental and restricted to small specimens and specimens composed of loose and easy-to-penetrate tissues [103,110,111]. Protocols for late embryos and dense tissues are eagerly anticipated and will open new fields of applications for HREM. We are confident that with the growing community of HREM developers and users, solutions for this problem will be presented within the next few years.

Author Contributions: Conceptualization, S.H.G. and W.J.W.; writing—original draft preparation, S.H.G. and W.J.W.; writing—review and editing, S.H.G. and W.J.W.; visualization, S.H.G. and W.J.W.; supervision, W.J.W.

Funding: This research received no external funding.

Acknowledgments: The authors thank Barbara Maurer-Gesek and all the participants in the DMDD program.

Conflicts of Interest: The authors declare no conflict of interest.

References

1. Annan, A. GPR—History, trends, and future developments. *Subsurf. Sens. Technol. Appl.* **2002**, *3*, 253–270. [CrossRef]
2. Nikolic, M.; Conrad, C.; Zhang, J.; Scarcelli, G. Noninvasive Imaging: Brillouin Confocal Microscopy. *Adv. Exp. Med. Biol.* **2018**, *1092*, 351–364. [CrossRef] [PubMed]
3. Yun, S.H.; Chernyak, D. Brillouin microscopy: Assessing ocular tissue biomechanics. *Curr. Opin. Ophthalmol.* **2018**, *29*, 299–305. [CrossRef] [PubMed]
4. Peix, A.; Mesquita, C.T.; Paez, D.; Pereira, C.C.; Felix, R.; Gutierrez, C.; Jaimovich, R.; Ianni, B.M.; Soares, J., Jr.; Olaya, P.; et al. Nuclear medicine in the management of patients with heart failure: Guidance from an expert panel of the International Atomic Energy Agency (IAEA). *Nucl. Med. Commun.* **2014**, *35*, 818–823. [CrossRef] [PubMed]
5. Wei, W.; Ehlerding, E.B.; Lan, X.; Luo, Q.; Cai, W. PET and SPECT imaging of melanoma: The state of the art. *Eur. J. Nucl. Med. and Mol. Imaging* **2018**, *45*, 132–150. [CrossRef] [PubMed]
6. Hutchinson, J.C.; Shelmerdine, S.C.; Simcock, I.C.; Sebire, N.J.; Arthurs, O.J. Early clinical applications for imaging at microscopic detail: Microfocus computed tomography (micro-CT). *Br. J. Radiol.* **2017**, *90*, 20170113. [CrossRef] [PubMed]
7. Spaide, R.F.; Fujimoto, J.G.; Waheed, N.K.; Sadda, S.R.; Staurenghi, G. Optical coherence tomography angiography. *Prog. Retin Eye Res.* **2018**, *64*, 1–55. [CrossRef]
8. Levine, A.; Wang, K.; Markowitz, O. Optical Coherence Tomography in the Diagnosis of Skin Cancer. *Dermatol. Clin.* **2017**, *35*, 465–488. [CrossRef]
9. Liu, M.; Maurer, B.; Hermann, B.; Zabihian, B.; Sandrian, M.G.; Unterhuber, A.; Baumann, B.; Zhang, E.Z.; Beard, P.C.; Weninger, W.J.; et al. Dual modality optical coherence and whole-body photoacoustic tomography imaging of chick embryos in multiple development stages. *Biomed. Opt. Express* **2014**, *5*, 3150–3159. [CrossRef]
10. Cleary, J.O.; Modat, M.; Norris, F.C.; Price, A.N.; Jayakody, S.A.; Martinez-Barbera, J.P.; Greene, N.D.E.; Hawkes, D.J.; Ordidge, R.J.; Scambler, P.J.; et al. Magnetic resonance virtual histology for embryos: 3D atlases for automated high-throughput phenotyping. *Neuroimage* **2011**, *54*, 769–778. [CrossRef]
11. Zamyadi, M.; Baghdadi, L.; Lerch, J.P.; Bhattacharya, S.; Schneider, J.E.; Henkelman, R.M.; Sled, J.G. Mouse embryonic phenotyping by morphometric analysis of MR images. *Physiol. Genom.* **2010**, *42*, 89–95. [CrossRef] [PubMed]
12. Stylianou, A.; Kontomaris, S.V.; Grant, C.; Alexandratou, E. Atomic Force Microscopy on Biological Materials Related to Pathological Conditions. *Scanning* **2019**, *2019*, 8452851. [CrossRef] [PubMed]
13. Dufrene, Y.F.; Ando, T.; Garcia, R.; Alsteens, D.; Martinez-Martin, D.; Engel, A.; Gerber, C.; Muller, D.J. Imaging modes of atomic force microscopy for application in molecular and cell biology. *Nat. Nanotechnol.* **2017**, *12*, 295–307. [CrossRef] [PubMed]
14. Ban, S.; Cho, N.H.; Min, E.; Bae, J.K.; Ahn, Y.; Shin, S.; Park, S.A.; Lee, Y.; Jung, W. Label-free optical projection tomography for quantitative 3D anatomy of mouse embryo. *J. Biophotonics* **2019**, *12*, e201800481. [CrossRef] [PubMed]

15. Sharpe, J. Optical projection tomography as a new tool for studying embryo anatomy. *J. Anat.* **2003**, *202*, 175–181. [CrossRef] [PubMed]
16. Sharpe, J.; Ahlgren, U.; Perry, P.; Hill, B.; Ross, A.; Hecksher-Sorensen, J.; Baldock, R.; Davidson, D. Optical projection tomography as a tool for 3D microscopy and gene expression studies. *Science* **2002**, *296*, 541–545. [CrossRef] [PubMed]
17. Weninger, W.J.; Geyer, S.H.; Mohun, T.J.; Rasskin-Gutman, D.; Matsui, T.; Ribeiro, I.; Costa Lda, F.; Izpisua-Belmonte, J.C.; Muller, G.B. High-resolution episcopic microscopy: A rapid technique for high detailed 3D analysis of gene activity in the context of tissue architecture and morphology. *Anat. Embryol.* **2006**, *211*, 213–221. [CrossRef] [PubMed]
18. Denk, W.; Horstmann, H. Serial block-face scanning electron microscopy to reconstruct three-dimensional tissue nanostructure. *PLoS Biol.* **2004**, *2*, e329. [CrossRef]
19. Odgaard, A. Quantification of Connectivity in Cancellous Bone, with Special Emphasis on 3-D Reconstructions. *Bone* **1993**, *14*, 173–182. [CrossRef]
20. Odgaard, A.; Andersen, K.; Melsen, F.; Gundersen, H.J. A direct method for fast three-dimensional serial reconstruction. *J. Microsc.* **1990**, *159*, 335–342. [CrossRef]
21. Odgaard, A.; Andersen, K.; Ullerup, R.; Frich, L.H.; Melsen, F. Three-dimensional reconstruction of entire vertebral bodies. *Bone* **1994**, *15*, 335–342. [CrossRef]
22. Weninger, W.J.; Meng, S.; Streicher, J.; Müller, G.B. A new episcopic method for rapid 3-D reconstruction: Applications in anatomy and embryology. *Anat. Embryol.* **1998**, *197*, 341–348. [CrossRef]
23. Gerneke, D.A.; Sands, G.B.; Ganesalingam, R.; Joshi, P.; Caldwell, B.J.; Smaill, B.H.; Legrice, I.J. Surface imaging microscopy using an ultramiller for large volume 3D reconstruction of wax- and resin-embedded tissues. *Microsc. Res. Tech.* **2007**, *70*, 886–894. [CrossRef] [PubMed]
24. Ewald, A.J.; McBride, H.; Reddington, M.; Fraser, S.E.; Kerschmann, R. Surface imaging microscopy, an automated method for visualizing whole embryo samples in three dimensions at high resolution. *Dev. Dyn.* **2002**, *225*, 369–375. [CrossRef] [PubMed]
25. Spaan, J.A.; ter Wee, R.; van Teeffelen, J.W.; Streekstra, G.; Siebes, M.; Kolyva, C.; Vink, H.; Fokkema, D.S.; VanBavel, E. Visualisation of intramural coronary vasculature by an imaging cryomicrotome suggests compartmentalisation of myocardial perfusion areas. *Med. Biol. Eng. Comput.* **2005**, *43*, 431–435. [CrossRef] [PubMed]
26. Sivaguru, M.; Fried, G.A.; Miller, C.A.; Fouke, B.W. Multimodal optical microscopy methods reveal polyp tissue morphology and structure in Caribbean reef building corals. *J. Vis. Exp.* **2014**, *91*, e51824. [CrossRef] [PubMed]
27. Weninger, W.J.; Mohun, T. Phenotyping transgenic embryos: A rapid 3-D screening method based on episcopic fluorescence image capturing. *Nat. Genet.* **2002**, *30*, 59–65. [CrossRef]
28. Geyer, S.H.; Maurer-Gesek, B.; Reissig, L.F.; Weninger, W.J. High-resolution Episcopic Microscopy (HREM)—Simple and Robust Protocols for Processing and Visualizing Organic Materials. *J. Vis. Exp.* **2017**, *125*, e56071. [CrossRef]
29. Mohun, T.J.; Weninger, W.J. Generation of volume data by episcopic three-dimensional imaging of embryos. *Cold Spring Harb. Protoc.* **2012**, *2012*, 681–682. [CrossRef]
30. Mohun, T.J.; Weninger, W.J. Embedding embryos for high-resolution episcopic microscopy (HREM). *Cold Spring Harb. Protoc.* **2012**, *2012*, 678–680. [CrossRef]
31. Mohun, T.J.; Weninger, W.J. Episcopic three-dimensional imaging of embryos. *Cold Spring Harb. Protoc.* **2012**, *2012*, 641–646. [CrossRef] [PubMed]
32. Zhang, H.; Huang, J.; Liu, X.; Zhu, P.; Li, Z.; Li, X. Rapid Acquisition of 3D Images Using High-resolution Episcopic Microscopy. *J. Vis. Exp.* **2016**, *117*, e54625. [CrossRef]
33. Weninger, W.J.; Maurer-Gesek, B.; Reissig, L.F.; Prin, F.; Wilson, R.; Galli, A.; Adams, D.J.; White, J.K.; Mohun, T.J.; Geyer, S.H. Visualising the Cardiovascular System of Embryos of Biomedical Model Organisms with High Resolution Episcopic Microscopy (HREM). *J. Cardiovasc. Dev. Dis.* **2018**, *5*, 58. [CrossRef] [PubMed]

34. Geyer, S.H.; Mohun, T.J.; Weninger, W.J. Visualizing vertebrate embryos with episcopic 3D imaging techniques. *Sci. World J.* **2009**, *9*, 1423–1437. [CrossRef] [PubMed]

35. Mohun, T.J.; Weninger, W.J. Imaging heart development using high-resolution episcopic microscopy. *Curr. Opin Genet. Dev.* **2011**, *21*, 573–578. [CrossRef] [PubMed]

36. Mohun, T.; Weninger, W.J.; Bhattacharya, S. Imaging Cardiac Developmental Malformations in the Mouse Embryo. In *Heart Development and Regeneration*; Rosenthal, N., Harvey, R.P., Eds.; Academic Press: London, UK, 2010; pp. 779–791.

37. Mohun, T.; Adams, D.J.; Baldock, R.; Bhattacharya, S.; Copp, A.J.; Hemberger, M.; Houart, C.; Hurles, M.E.; Robertson, E.; Smith, J.C.; et al. Deciphering the Mechanisms of Developmental Disorders (DMDD): A new programme for phenotyping embryonic lethal mice. *Dis. Model. Mech.* **2013**, *6*, 562–566. [CrossRef] [PubMed]

38. Weninger, W.J.; Geyer, S.H.; Martineau, A.; Galli, A.; Adams, D.J.; Wilson, R.; Mohun, T.J. Phenotyping structural abnormalities in mouse embryos using high-resolution episcopic microscopy. *Dis. Model. Mech.* **2014**, *7*, 1143–1152. [CrossRef] [PubMed]

39. Geyer, S.H.; Reissig, L.; Rose, J.; Wilson, R.; Prin, F.; Szumska, D.; Ramirez-Solis, R.; Tudor, C.; White, J.; Mohun, T.J.; et al. A staging system for correct phenotype interpretation of mouse embryos harvested on embryonic day 14 (E14.5). *J. Anat.* **2017**, *230*, 710–719. [CrossRef]

40. Geyer, S.H.; Reissig, L.F.; Husemann, M.; Hofle, C.; Wilson, R.; Prin, F.; Szumska, D.; Galli, A.; Adams, D.J.; White, J.; et al. Morphology, topology and dimensions of the heart and arteries of genetically normal and mutant mouse embryos at stages S21-S23. *J. Anat.* **2017**, *231*, 600–614. [CrossRef]

41. Collins, J.E.; White, R.J.; Staudt, N.; Sealy, I.M.; Packham, I.; Wali, N.; Tudor, C.; Mazzeo, C.; Green, A.; Siragher, E.; et al. Common and distinct transcriptional signatures of mammalian embryonic lethality. *Nat. Commun.* **2019**, *10*, 2792. [CrossRef]

42. De Franco, E.; Watson, R.A.; Weninger, W.J.; Wong, C.C.; Flanagan, S.E.; Caswell, R.; Green, A.; Tudor, C.; Lelliott, C.J.; Geyer, S.H.; et al. A Specific CNOT1 Mutation Results in a Novel Syndrome of Pancreatic Agenesis and Holoprosencephaly through Impaired Pancreatic and Neurological Development. *Am. J. Hum. Genet.* **2019**, *104*, 985–989. [CrossRef] [PubMed]

43. Henkelman, R.M.; Friedel, M.; Lerch, J.P.; Wilson, R.; Mohun, T. Comparing homologous microscopic sections from multiple embryos using HREM. *Dev. Biol.* **2016**, *415*, 1. [CrossRef] [PubMed]

44. Perez-Garcia, V.; Fineberg, E.; Wilson, R.; Murray, A.; Mazzeo, C.I.; Tudor, C.; Sienerth, A.; White, J.K.; Tuck, E.; Ryder, E.J.; et al. Placentation defects are highly prevalent in embryonic lethal mouse mutants. *Nature* **2018**, *555*, 463–468. [CrossRef]

45. Reissig, L.F.; Herdina, A.N.; Rose, J.; Maurer-Gesek, B.; Lane, J.L.; Prin, F.; Wilson, R.; Hardman, E.; Galli, A.; Tudor, C.; et al. The Col4a2(em1(IMPC)Wtsi) mouse line: Lessons from the Deciphering the Mechanisms of Developmental Disorders program. *Biol. Open* **2019**, *8*, bio042895. [CrossRef] [PubMed]

46. Wilson, R.; Geyer, S.H.; Reissig, L.; Rose, J.; Szumska, D.; Hardman, E.; Prin, F.; McGuire, C.; Ramirez-Solis, R.; White, J.; et al. Highly variable penetrance of abnormal phenotypes in embryonic lethal knockout mice. *Wellcome Open Res.* **2016**, *1*, 1. [CrossRef] [PubMed]

47. Wilson, R.; McGuire, C.; Mohun, T.; Project, D. Deciphering the mechanisms of developmental disorders: Phenotype analysis of embryos from mutant mouse lines. *Nucleic Acids Res.* **2016**, *44*, D855–D861. [CrossRef] [PubMed]

48. Dickinson, M.E.; Flenniken, A.M.; Ji, X.; Teboul, L.; Wong, M.D.; White, J.K.; Meehan, T.F.; Weninger, W.J.; Westerberg, H.; Adissu, H.; et al. High-throughput discovery of novel developmental phenotypes. *Nature* **2016**, *537*, 508–514. [CrossRef]

49. Notari, M.; Hu, Y.; Sutendra, G.; Dedeic, Z.; Lu, M.; Dupays, L.; Yavari, A.; Carr, C.A.; Zhong, S.; Opel, A.; et al. iASPP, a previously unidentified regulator of desmosomes, prevents arrhythmogenic right ventricular cardiomyopathy (ARVC)-induced sudden death. *Proc. Natl. Acad. Sci. USA* **2015**, *112*, E973–E981. [CrossRef]

50. Zhou, Z.; Wang, J.; Guo, C.; Chang, W.; Zhuang, J.; Zhu, P.; Li, X. Temporally Distinct Six2-Positive Second Heart Field Progenitors Regulate Mammalian Heart Development and Disease. *Cell Rep.* **2017**, *18*, 1019–1032. [CrossRef]

51. Pieles, G.; Geyer, S.H.; Szumska, D.; Schneider, J.; Neubauer, S.; Clarke, K.; Dorfmeister, K.; Franklyn, A.; Brown, S.D.; Bhattacharya, S.; et al. microMRI-HREM pipeline for high-throughput, high-resolution phenotyping of murine embryos. *J. Anat.* **2007**, *211*, 132–137. [CrossRef]

52. Le Garrec, J.F.; Dominguez, J.N.; Desgrange, A.; Ivanovitch, K.D.; Raphael, E.; Bangham, J.A.; Torres, M.; Coen, E.; Mohun, T.J.; Meilhac, S.M. A predictive model of asymmetric morphogenesis from 3D reconstructions of mouse heart looping dynamics. *Elife* **2017**, *6*, e28951. [CrossRef] [PubMed]

53. Bailey, K.E.; MacGowan, G.A.; Tual-Chalot, S.; Phillips, L.; Mohun, T.J.; Henderson, D.J.; Arthur, H.M.; Bamforth, S.D.; Phillips, H.M. Disruption of embryonic ROCK signaling reproduces the sarcomeric phenotype of hypertrophic cardiomyopathy. *JCI Insight* **2019**, *4*, e125172. [CrossRef] [PubMed]

54. Dupays, L.; Shang, C.; Wilson, R.; Kotecha, S.; Wood, S.; Towers, N.; Mohun, T. Sequential Binding of MEIS1 and NKX2-5 on the Popdc2 Gene: A Mechanism for Spatiotemporal Regulation of Enhancers during Cardiogenesis. *Cell Rep.* **2015**, *13*, 183–195. [CrossRef] [PubMed]

55. Dupays, L.; Towers, N.; Wood, S.; David, A.; Stuckey, D.J.; Mohun, T. Furin, a transcriptional target of NKX2-5, has an essential role in heart development and function. *PLoS ONE* **2019**, *14*, e0212992. [CrossRef] [PubMed]

56. Ivins, S.; Chappell, J.; Vernay, B.; Suntharalingham, J.; Martineau, A.; Mohun, T.J.; Scambler, P.J. The CXCL12/CXCR4 Axis Plays a Critical Role in Coronary Artery Development. *Dev. Cell* **2015**, *33*, 455–468. [CrossRef] [PubMed]

57. Zak, J.; Vives, V.; Szumska, D.; Vernet, A.; Schneider, J.E.; Miller, P.; Slee, E.A.; Joss, S.; Lacassie, Y.; Chen, E.; et al. ASPP2 deficiency causes features of 1q41q42 microdeletion syndrome. *Cell Death Differ.* **2016**, *23*, 1973–1984. [CrossRef]

58. Garcia-Canadilla, P.; Cook, A.C.; Mohun, T.J.; Oji, O.; Schlossarek, S.; Carrier, L.; McKenna, W.J.; Moon, J.C.; Captur, G. Myoarchitectural disarray of hypertrophic cardiomyopathy begins pre-birth. *J. Anat.* **2019**, in press. [CrossRef]

59. Lescroart, F.; Mohun, T.; Meilhac, S.M.; Bennett, M.; Buckingham, M. Lineage tree for the venous pole of the heart: Clonal analysis clarifies controversial genealogy based on genetic tracing. *Circ. Res.* **2012**, *111*, 1313–1322. [CrossRef]

60. Breckenridge, R.A.; Piotrowska, I.; Ng, K.E.; Ragan, T.J.; West, J.A.; Kotecha, S.; Towers, N.; Bennett, M.; Kienesberger, P.C.; Smolenski, R.T.; et al. Hypoxic regulation of hand1 controls the fetal-neonatal switch in cardiac metabolism. *PLoS Biol.* **2013**, *11*, e1001666. [CrossRef]

61. Rana, M.S.; Theveniau-Ruissy, M.; De Bono, C.; Mesbah, K.; Francou, A.; Rammah, M.; Dominguez, J.N.; Roux, M.; Laforest, B.; Anderson, R.H.; et al. Tbx1 coordinates addition of posterior second heart field progenitor cells to the arterial and venous poles of the heart. *Circ. Res.* **2014**, *115*, 790–799. [CrossRef]

62. Huang, Y.C.; Chen, F.; Li, X. Clarification of mammalian cloacal morphogenesis using high-resolution episcopic microscopy. *Dev. Biol.* **2016**, *409*, 106–113. [CrossRef] [PubMed]

63. Delaurier, A.; Burton, N.; Bennett, M.; Baldock, R.; Davidson, D.; Mohun, T.J.; Logan, M.P. The Mouse Limb Anatomy Atlas: An interactive 3D tool for studying embryonic limb patterning. *BMC Dev. Biol.* **2008**, *8*, 83. [CrossRef] [PubMed]

64. Hasson, P.; DeLaurier, A.; Bennett, M.; Grigorieva, E.; Naiche, L.A.; Papaioannou, V.E.; Mohun, T.J.; Logan, M.P. Tbx4 and tbx5 acting in connective tissue are required for limb muscle and tendon patterning. *Dev. Cell* **2010**, *18*, 148–156. [CrossRef] [PubMed]

65. Kim, Y.J.; Osborn, D.P.; Lee, J.Y.; Araki, M.; Araki, K.; Mohun, T.; Kansakoski, J.; Brandstack, N.; Kim, H.T.; Miralles, F.; et al. WDR11-mediated Hedgehog signalling defects underlie a new ciliopathy related to Kallmann syndrome. *EMBO Rep.* **2018**, *19*, 269–289. [CrossRef]

66. Dunlevy, L.; Bennett, M.; Slender, A.; Lana-Elola, E.; Tybulewicz, V.L.; Fisher, E.M.; Mohun, T. Down's syndrome-like cardiac developmental defects in embryos of the transchromosomic Tc1 mouse. *Cardiovasc. Res.* **2010**, *88*, 287–295. [CrossRef] [PubMed]

67. Lana-Elola, E.; Watson-Scales, S.; Slender, A.; Gibbins, D.; Martineau, A.; Douglas, C.; Mohun, T.; Fisher, E.M.; Tybulewicz, V. Genetic dissection of Down syndrome-associated congenital heart defects using a new mouse mapping panel. *Elife* **2016**, *5*, e11614. [CrossRef] [PubMed]

68. Gershon, E.; Hadas, R.; Elbaz, M.; Booker, E.; Muchnik, M.; Kleinjan-Elazary, A.; Karasenti, S.; Genin, O.; Cinnamon, Y.; Gray, P.C. Identification of Trophectoderm-Derived Cripto as an Essential Mediator of Embryo Implantation. *Endocrinology* **2018**, *159*, 1793–1807. [CrossRef] [PubMed]

69. Bamforth, S.D.; Chaudhry, B.; Bennett, M.; Wilson, R.; Mohun, T.J.; Van Mierop, L.H.; Henderson, D.J.; Anderson, R.H. Clarification of the identity of the mammalian fifth pharyngeal arch artery. *Clin. Anat.* **2013**, *26*, 173–182. [CrossRef] [PubMed]

70. Geyer, S.H.; Maurer, B.; Potz, L.; Singh, J.; Weninger, W.J. High-resolution episcopic microscopy data-based measurements of the arteries of mouse embryos: Evaluation of significance and reproducibility under routine conditions. *Cells Tissues Organs* **2012**, *195*, 524–534. [CrossRef] [PubMed]

71. Geyer, S.H.; Weninger, W.J. Metric characterization of the aortic arch of early mouse fetuses and of a fetus featuring a double lumen aortic arch malformation. *Ann. Anat.* **2013**, *195*, 175–182. [CrossRef] [PubMed]

72. Maurer-Gesek, B. Malformations of the Great Intrathoracic Arteries caused by Hemodynamic Alterations in Chick Embryos. Ph.D. Thesis, Medical University of Vienna, Vienna, Austria, 2016.

73. Weninger, W.J.; Maurer, B.; Zendron, B.; Dorfmeister, K.; Geyer, S.H. Measurements of the diameters of the great arteries and semi-lunar valves of chick and mouse embryos. *J. Microsc.* **2009**, *234*, 173–190. [CrossRef] [PubMed]

74. Geyer, S.H.; Weninger, W.J. Some mice feature 5th pharyngeal arch arteries and double-lumen aortic arch malformations. *Cells Tissues Organs* **2012**, *196*, 90–98. [CrossRef] [PubMed]

75. Aiello, V.D.; Spicer, D.E.; Anderson, R.H.; Brown, N.A.; Mohun, T.J. The independence of the infundibular building blocks in the setting of double-outlet right ventricle. *Cardiol. Young* **2017**, *27*, 825–836. [CrossRef] [PubMed]

76. Anderson, R.H.; Brown, N.A.; Mohun, T.J. Insights regarding the normal and abnormal formation of the atrial and ventricular septal structures. *Clin. Anat.* **2016**, *29*, 290–304. [CrossRef] [PubMed]

77. Anderson, R.H.; Chaudhry, B.; Mohun, T.J.; Bamforth, S.D.; Hoyland, D.; Phillips, H.M.; Webb, S.; Moorman, A.F.; Brown, N.A.; Henderson, D.J. Normal and abnormal development of the intrapericardial arterial trunks in humans and mice. *Cardiovasc. Res.* **2012**, *95*, 108–115. [CrossRef] [PubMed]

78. Anderson, R.H.; Jensen, B.; Mohun, T.J.; Petersen, S.E.; Aung, N.; Zemrak, F.; Planken, R.N.; MacIver, D.H. Key Questions Relating to Left Ventricular Noncompaction Cardiomyopathy: Is the Emperor Still Wearing Any Clothes? *Can. J. Cardiol.* **2017**, *33*, 747–757. [CrossRef]

79. Anderson, R.H.; Mohun, T.J.; Brown, N.A. Clarifying the morphology of the ostium primum defect. *J. Anat.* **2015**, *226*, 244–257. [CrossRef]

80. Anderson, R.H.; Mohun, T.J.; Sanchez-Quintana, D.; Mori, S.; Spicer, D.E.; Cheung, J.W.; Lerman, B.B. The anatomic substrates for outflow tract arrhythmias. *Heart Rhythm* **2019**, *16*, 290–297. [CrossRef]

81. Anderson, R.H.; Mori, S.; Spicer, D.E.; Brown, N.A.; Mohun, T.J. Development and Morphology of the Ventricular Outflow Tracts. *World J. Pediatric Congenit. Heart Surg.* **2016**, *7*, 561–577. [CrossRef]

82. Anderson, R.H.; Spicer, D.E.; Brown, N.A.; Mohun, T.J. The development of septation in the four-chambered heart. *Anat. Rec.* **2014**, *297*, 1414–1429. [CrossRef]

83. Anderson, R.H.; Spicer, D.E.; Mohun, T.J.; Hikspoors, J.; Lamers, W.H. Remodeling of the Embryonic Interventricular Communication in Regard to the Description and Classification of Ventricular Septal Defects. *Anat. Rec.* **2019**, *302*, 19–31. [CrossRef]

84. Sizarov, A.; Lamers, W.H.; Mohun, T.J.; Brown, N.A.; Anderson, R.H.; Moorman, A.F. Three-dimensional and molecular analysis of the arterial pole of the developing human heart. *J. Anat.* **2012**, *220*, 336–349. [CrossRef] [PubMed]

85. Spicer, D.E.; Bridgeman, J.M.; Brown, N.A.; Mohun, T.J.; Anderson, R.H. The anatomy and development of the cardiac valves. *Cardiol. Young* **2014**, *24*, 1008–1022. [CrossRef] [PubMed]

86. Tretter, J.T.; Steffensen, T.; Westover, T.; Anderson, R.H.; Spicer, D.E. Developmental considerations with regard to so-called absence of the leaflets of the arterial valves. *Cardiol. Young* **2017**, *27*, 302–311. [CrossRef] [PubMed]

87. Captur, G.; Ho, C.Y.; Schlossarek, S.; Kerwin, J.; Mirabel, M.; Wilson, R.; Rosmini, S.; Obianyo, C.; Reant, P.; Bassett, P.; et al. The embryological basis of subclinical hypertrophic cardiomyopathy. *Sci. Rep.* **2016**, *6*, 27714. [CrossRef] [PubMed]

88. Captur, G.; Wilson, R.; Bennett, M.F.; Luxan, G.; Nasis, A.; de la Pompa, J.L.; Moon, J.C.; Mohun, T.J. Morphogenesis of myocardial trabeculae in the mouse embryo. *J. Anat.* **2016**, *229*, 314–325. [CrossRef] [PubMed]

89. Paun, B.; Bijnens, B.; Cook, A.C.; Mohun, T.J.; Butakoff, C. Quantification of the detailed cardiac left ventricular trabecular morphogenesis in the mouse embryo. *Med. Image Anal.* **2018**, *49*, 89–104. [CrossRef] [PubMed]

90. Pokhrel, N.; Ben-Tal Cohen, E.; Genin, O.; Sela-Donenfeld, D.; Cinnamon, Y. Cellular and morphological characterization of blastoderms from freshly laid broiler eggs. *Poult. Sci.* **2017**, *96*, 4399–4408. [CrossRef]

91. Pokhrel, N.; Cohen, E.B.; Genin, O.; Ruzal, M.; Sela-Donenfeld, D.; Cinnamon, Y. Effects of storage conditions on hatchability, embryonic survival and cytoarchitectural properties in broiler from young and old flocks. *Poult. Sci.* **2018**, *97*, 1429–1440. [CrossRef] [PubMed]

92. Maurer, B.; Geyer, S.H.; Weninger, W.J. A chick embryo with a yet unclassified type of cephalothoracopagus malformation and a hypothesis for explaining its genesis. *Anat. Histol. Embryol.* **2013**, *42*, 191–200. [CrossRef]

93. Bruneel, B.; Matha, M.; Paesen, R.; Ameloot, M.; Weninger, W.J.; Huysseune, A. Imaging the zebrafish dentition: From traditional approaches to emerging technologies. *Zebrafish* **2015**, *12*, 1–10. [CrossRef] [PubMed]

94. Jenner, F.; van Osch, G.J.; Weninger, W.; Geyer, S.; Stout, T.; van Weeren, R.; Brama, P. The embryogenesis of the equine femorotibial joint: The equine interzone. *Equine Vet. J.* **2015**, *47*, 620–622. [CrossRef] [PubMed]

95. Matsui, H.; Mohun, T.; Gardiner, H.M. Three-dimensional reconstruction imaging of the human foetal heart in the first trimester. *Eur. Heart J.* **2010**, *31*, 415. [CrossRef] [PubMed]

96. Gindes, L.; Matsui, H.; Achiron, R.; Mohun, T.; Ho, S.Y.; Gardiner, H. Comparison of ex-vivo high-resolution episcopic microscopy with in-vivo four-dimensional high-resolution transvaginal sonography of the first-trimester fetal heart. *Ultrasound Obstet. Gynecol.* **2012**, *39*, 196–202. [CrossRef] [PubMed]

97. Ertl, J.; Pichlsberger, M.; Tuca, A.C.; Wurzer, P.; Fuchs, J.; Geyer, S.H.; Maurer-Gesek, B.; Weninger, W.J.; Pfeiffer, D.; Bubalo, V.; et al. Comparative study of regenerative effects of mesenchymal stem cells derived from placental amnion, chorion and umbilical cord on dermal wounds. *Placenta* **2018**, *65*, 37–46. [CrossRef] [PubMed]

98. Wiedner, M.; Tinhofer, I.E.; Kamolz, L.P.; Seyedian Moghaddam, A.; Justich, I.; Liegl-Atzwanger, B.; Bubalo, V.; Weninger, W.J.; Lumenta, D.B. Simultaneous dermal matrix and autologous split-thickness skin graft transplantation in a porcine wound model: A three-dimensional histological analysis of revascularization. *Wound Repair Regen.* **2014**, *22*, 749–754. [CrossRef] [PubMed]

99. Geyer, S.H.; Nohammer, M.M.; Matha, M.; Reissig, L.; Tinhofer, I.E.; Weninger, W.J. High-resolution episcopic microscopy (HREM): A tool for visualizing skin biopsies. *Microsc. Microanal.* **2014**, *20*, 1356–1364. [CrossRef]

100. Geyer, S.H.; Nohammer, M.M.; Tinhofer, I.E.; Weninger, W.J. The dermal arteries of the human thumb pad. *J. Anat.* **2013**, *223*, 603–609. [CrossRef]

101. Tinhofer, I.E.; Zaussinger, M.; Geyer, S.H.; Meng, S.; Kamolz, L.P.; Tzou, C.H.; Weninger, W.J. The dermal arteries in the cutaneous angiosome of the descending genicular artery. *J. Anat.* **2018**, *232*, 979–986. [CrossRef]

102. Razlighi, B.D.; Kampusch, S.; Geyer, S.H.; Hoang Le, V.; Thurk, F.; Brenner, S.; Szeles, J.C.; Weninger, W.J.; Kaniusas, E. In-Silico Ear Model Based on Episcopic Images for Percutaneous Auricular Vagus Nerve Stimulation. In Proceedings of the 2018 EMF-Med 1st World Conference on Biomedical Applications of Electromagnetic Fields (EMF-Med), Split, Croatia, 10–13 September 2018.

103. Franck, G.; Even, G.; Gautier, A.; Salinas, M.; Loste, A.; Procopio, E.; Gaston, A.T.; Morvan, M.; Dupont, S.; Deschildre, C.; et al. Haemodynamic stress-induced breaches of the arterial intima trigger inflammation and drive atherogenesis. *Eur. Heart J.* **2019**, *40*, 928–937. [CrossRef]

104. Walter, A. Correlated Multimodal Imaging of Tumour Vasculature. (manuscript in preparation).

105. Izhaki, A.; Alvarez, J.P.; Cinnamon, Y.; Genin, O.; Liberman-Aloni, R.; Eyal, Y. The Tomato BLADE ON PETIOLE and TERMINATING FLOWER Regulate Leaf Axil Patterning Along the Proximal-Distal Axes. *Front. Plant Sci.* **2018**, *9*, 1126. [CrossRef]

106. Wiltsche, M.; Donoser, M.; Kritzinger, J.; Bauer, W. Automated serial sectioning applied to 3D paper structure analysis. *J. Microsc.* **2011**, *242*, 197–205. [CrossRef]

107. Geyer, S.H.; Tinhofer, I.E.; Lumenta, D.B.; Kamolz, L.P.; Branski, L.; Finnerty, C.C.; Herndon, D.N.; Weninger, W.J. High-resolution episcopic microscopy (HREM): A useful technique for research in wound care. *Ann. Anat.* **2015**, *197*, 3–10. [CrossRef]

108. Weninger, W.J.; Geyer, S.H. Episcopic 3D Imaging Methods: Tools for Researching Gene Function. *Curr. Genom.* **2008**, *9*, 282–289. [CrossRef]

109. Desgrange, A.; Lokmer, J.; Marchiol, C.; Houyel, L.; Meilhac, S.M. Standardised imaging pipeline for phenotyping mouse laterality defects and associated heart malformations, at multiple scales and multiple stages. *Dis. Model. Mech.* **2019**, *12*, dmm038356. [CrossRef]
110. Weninger, W.J.; Kamolz, L.P.; Geyer, S.H. 3D Visualisation of Skin Substitutes. In *Dermal Replacements in General, Burn, and Plastic Surgery*; Kamolz, L.P., Lumenta, D.B., Eds.; Springer: Vienna, Austria, 2013.
111. Weninger, W.J.; Mohun, T.J. Three-dimensional analysis of molecular signals with episcopic imaging techniques. *Methods Mol. Biol.* **2007**, *411*, 35–46. [CrossRef]

applied sciences

Article

3D Display System Based on Spherical Wave Field Synthesis

Claas Falldorf [1,*], Ping-Yen Chou [2], Daniel Prigge [1,3] and Ralf B. Bergmann [1,3]

[1]   Bremer Institut für angewandte Strahltechnik GmbH (BIAS), Klagenfurter Strasse 5, 28359 Bremen, Germany; danielprigge96@gmail.com (D.P.); bergmann@bias.de (R.B.B.)

[2]   Department of Photonics, College of Electrical and Computer Engineering, National Chiao Tung University, Hsinchu 30010, Taiwan; pac@cc.nctu.edu.tw

[3]   Faculty of Physics and Electrical Engineering, University of Bremen, 28359 Bremen, Germany

*   Correspondence: falldorf@bias.de; Tel.: +49-421-218-58013

Received: 15 August 2019; Accepted: 12 September 2019; Published: 14 September 2019

Abstract: We present a novel concept and first experimental results of a new type of 3D display, which is based on the synthesis of spherical waves. The setup comprises a lens array (LA) with apertures in the millimeter range and a liquid crystal display (LCD) panel. Each pixel of the LCD creates a spherical wave cutout that propagates towards the observer. During the displaying process, the curvature of the spherical waves is dynamically changed by either changing the distance between LA and LCD or by adapting the focal lengths of the lenses. Since the system, similar to holography, seeks to approximate the wavefront of a natural scene, it provides true depth information to the observer and therefore avoids any vergence–accommodation conflict (VAC).

Keywords: 3D imaging system; wave field synthesis; vergence–accommodation conflict; lens array; focus cue; light field display

---

1. Introduction

Recently, display technologies [1,2] have grown dramatically, and flat-panel displays based on LCDs or organic LEDs dominate the market. Especially in the last decade, three-dimensional (3D) display systems [3–5] have greatly advanced. Compared with a flat panel display, 3D technologies can not only provide images, but also deliver depth information to create a more immersive vision experience. According to whether the observer needs to wear 3D glasses or not, 3D displays can be classified as stereoscopic and autostereoscopic systems, which have different applications in human life. Traditional 3D displays are based on the principle of binocular parallax [6,7], in order to give the observer two different images for both eyes to merge as one 3D image. According to the human visual experiment [8,9], the accommodation of the eye lens is focused on the display plane, which is different from the vergence, which is directed towards the reconstructed object. This mismatch is called vergence–accommodation conflict (VAC). It induces dizziness and makes the observer feel uncomfortable. Therefore, floating image display systems, which can reconstruct images in space with real depth cues in order to solve the VAC issue, are one of the prime goals of 3D display technology. One way to implement a floating image display is to synthesize a wave field, so that it appears to be scattered by a natural scene or object. Currently, two different approaches are reported, which can be classified under this category: holographic displays and light field (LF) displays.

Holographic displays are often considered as the gold standard of 3D display systems. They can, in principle, create any arbitrary wave field and therefore provide the most comfortable experience to the observer. Because the shape of the wavefield can be fully controlled, a well-made hologram can appear indistinguishable from a real object. However, while holography works great for static scenes using holographic film material, up to now there exists no practical solution for dynamic displays.

The reason for this is that holography relies on forming wavefields based on the physical principle of diffraction. This calls for a huge space bandwidth product (i.e., number of pixels on the order of 10^{11}) which is required to generate the fine diffractive structures across the area of a macroscopic display screen [10–12].

On the other hand, the structure of LF display technologies [13–20] is simple. It just consists of a micro-lens array (MLA) and a display panel. The aim of an LF system is to approximate the light field of a natural scene. From a wave field perspective, the light field can be described by a set of plane waves. However, the number of plane waves in an LF system is limited by the number of display pixels, and the diameter of the plane waves typically equals the diameter of the involved micro-lenses, i.e., is comparably small. This leads to a strongly fragmented representation of the ideal wave field. As a consequence, light field systems usually have to find a trade-off between spatial and angular resolution and can only display very limited scenes. In addition, the depth range around the central depth plane (CDP) [21], which is the plane of the highest spatial resolution, is usually narrow.

In this publication, we present a new approach to a floating image display system, which is based on synthesizing spherical waves instead of plane waves in order to overcome these drawbacks. Similar to an LF system, the corresponding setup consists of a display panel and a lens array (LA). Yet, compared with the LF system, the aperture size of the individual lenses is much larger. Additionally, in order to control the curvature of the spherical waves, the distance between the display and the LA is dynamically changed. The benefit of this approach can be understood in the wave field picture, since using spherical waves instead of plane waves adds more degrees of freedom to the base functions of the synthesis. It, therefore, represents one step further towards true holography, which offers the highest degree of wavefront complexity. Yet, in contrast to holography, the method can be implemented with already existing technology. As an additional benefit, the LA is a refractive element with very little dispersion, so that the system operates almost wavelength-independently and can, therefore, display colored scenes.

2. Principle and Design of Wave Field Synthesis 3D Imaging System

2.1. Optical System Principle

In a real environment, the surface of an object can be assumed to be composed of a huge number of dipoles. Each of them is excited when illuminated and becomes the source of dipole radiation. Because the dipoles are unordered, the scattered light can be approximated as a set of mutually incoherent spherical waves, as shown in Figure 1. This simple gedankenexperiment shows that spherical waves can be very useful as a polynomial basis to imitate and synthesize wave fields generated by natural scenes.

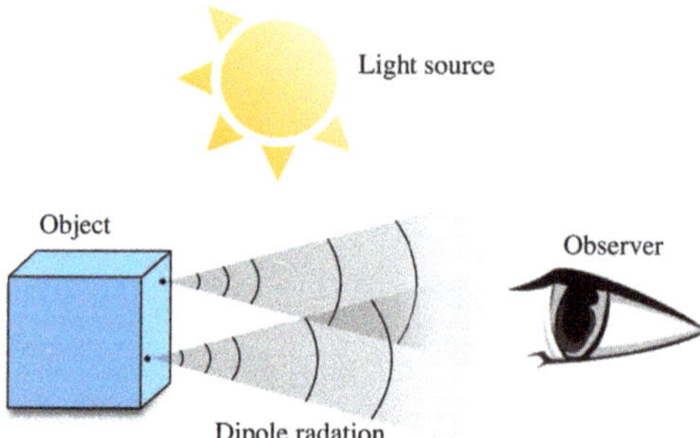

Figure 1. The light source irradiates an object. The surface dipoles start to oscillate and emit mutually incoherent spherical waves as dipole radiation that can be detected by the observer.

We will employ this principle by using the display scheme depicted by Figure 2. It consists of a liquid crystal display (LCD) and an lens array (LA). The lateral pixel position is given by x_n while the lateral position of the virtual object point is defined as x_0. The wavefront behind the lens propagates in the direction determined by α.

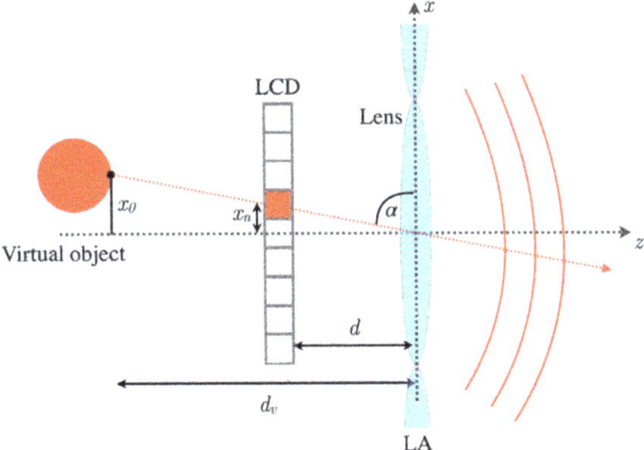

Figure 2. Schematic 3D display system and geometrical quantities. The lens manipulates the wavefront of the light emitted from the red pixel in a way, that the generated spherical wave cutout seems to originate from a point in the virtual distance d_v and propagates in the direction determined by α. Multiple points can be used to create the sensation of a virtual object or even a scene.

In the system, light originating from each LCD pixel will form a spherical wave. The curvature of the spherical waves, which also defines the distance to its origin, can be controlled by either changing the optical path between the LCD and the LA (the distance d) or, more conveniently, by changing the focal length of the lenses using a dynamic LA. The setup resembles an LF display. However, the basic principle is completely different. While an LF display seeks to generate rays (i.e., plane waves), we use a set of spherical waves to synthesize a wave field. Compared with rays, spherical waves exhibit the additional parameter of curvature. Therefore, the curvature has to be dynamically varied during the display process using one of the above-mentioned techniques. In our work, we change the curvature by mechanically varying the optical path between LCD and LA. Additionally, since the aim is to generate wide-field spherical waves rather than rays (narrow plane waves), the individual lenses of the LA exhibit a much larger aperture.

If one of the spherical waves hits the eye of the observer, the eye lens adjusts properly to image the origin of the spherical wave, thus creating the sensation of a point floating in space. Multiple points can then be used to create surfaces of objects or even complex scenes. Different depth layers can be addressed by changing the curvature, while lateral coordinates can be selected by the pixel position on the LCD panel. For the rendering, we use a ray trace approach and follow the line that intersects the center of a lens and the center of a corresponding LCD pixel. Whether the pixel is switched on or not depends on the virtual distance d_v to the center of the spherical wave. If it is in close vicinity to an object's surface and nothing is blocking the path (object occlusion), the pixel is switched on.

For the renderer, it is, therefore, necessary to calculate d_v from the distance d between the LCD and the LA and the focal length f of the lenses. The separability of the problem allows us to treat it in 2 dimensions, x, and z. To further simplify the calculation, the parabolic approximation is used to describe the complex amplitude U_S of light originating from a specific LCD pixel in front of its associated lens [22], which is given by

$$U_S(x,y) = \frac{A_n}{d}\exp\left[ikd + i\frac{k}{2d}(x - x_n)^2\right].\tag{1}$$

The origin is located at the corresponding lens center, x_n indicates the pixel coordinate relative to the lens center, A_n represents the amplitude provided by the pixel and $k = 2\pi/\lambda$ is the wave number. In addition, the lens modulation based on the thin lens approximation can be described as

$$M_L(x,y) = \exp\left[-i\frac{k}{2f}x^2\right].\tag{2}$$

By multiplying Equation (1) and (2), the perceived wave function of the pixel after passing through the lens can be written as

$$U(x,z) = U_S \cdot M_L = \frac{A_n}{d}\exp\left[i\frac{k}{2}\left(\frac{1}{d}-\frac{1}{f}\right)x^2 + ikd\right] \cdot \exp\left[-i\frac{kx_n}{d}x\right],\tag{3}$$

where the first exponential term with square in x indicates the curvature, and the second term linear in x represents the average propagation direction of the corresponding spherical wave cutout, which forms behind the lens. By comparing Equation (1) and (3), the spherical wave can be assumed to originate from a point source in a distance d_v behind the lens array, which depends on d and f according to

$$d_v = \frac{fd}{f-d}.\tag{4}$$

Please note that due to their limited aperture, the lenses only produce a cutout of a spherical wave rather than the entire wave field. This effect is an important requirement for the spherical wave field synthesis. It enables creating the effect of object occlusions, i.e., when one object in the front blocks parts of the spherical waves originating from an object behind it. At this point, the main direction in which the spherical wave cutout propagates shall be derived. According to Figure 2, one can find the geometrical relation

$$\cot \alpha = \frac{x_n}{d} = \frac{x_0}{d_v}.\tag{5}$$

Within the paraxial approximation, this is equivalent to

$$\cos \alpha = \frac{x_n}{d}.\tag{6}$$

Inserting Equation (6) turns the second exponential term in Equation (3) into a plane wave travelling along the direction α with respect to the lens plane. The angle α therefore defines the direction in which the spherical wave cutout propagates. The Equations (4) and (6) are crucial for the design of the desired wave field to depict an object in a virtual distance d_v behind the lens' plane. Finally, in our experiments, each LCD pixel was associated to the lens in front of it (in the direction of the optical axis). We avoided crosstalk between neighboring lenses by a 3D-printed parallax barrier.

2.2. Optical Components

A lens array (LA) (model Stock no.63-231 from Edmund Optics GmbH) is adapted as the refractive component of our system, whose specifications are shown in Table 1. The arrangement as well as the aperture of the lenses is rectangular and the substrate is made from B270 material with high transmittance in the range of visual wavelengths. The working area is 58 mm by 60 mm. Since the fundamental principle of our system is spherical wave field synthesis, a larger size, 5.4 mm by 7.0 mm, of the lenses compared with a MLA is selected to avoid strong fragmentation of the wavefronts due to small lens apertures. In addition, the focal length of the LA is 41.90 mm.

Table 1. Specifications of the lens array.

Dimensions (mm)	58.0×60.0
Size of lenslet (mm)	5.4×7.0
Effective focal length EFL (mm)	41.90
Radius of lenslet (mm)	22.0
Substrate	B270
Wavelength range (nm)	400 ... 700
Thickness (mm)	3.0

A commercial iPhone 7 plus with 401 ppi is chosen as the display device and its specifications are shown in Table 2. Hence the system is capable of working with partially coherent light, which is more convenient compared to the requirements of holographic displays. The two main factors of selecting the specific display are the high resolution and the wide color gamut, which affect imaging quality of the reconstructed images directly.

Table 2. Specifications of the LCD panel.

Dimensions (mm)	158.2×77.9
Type	IPS LCD
Resolution (pixel)	1920×1080
Pixel size (μm)	63.34

3. Experiments and Results

To verify the concept of a 3D display by means of spherical wave synthesis, a prototype was set up. Figure 3 illustrates the optical components and configuration of the proposed system, which just consists of an LA and an LCD panel. For the prototype setup, all-optical components are fixed on a stage with all six degrees of freedom to adjust the relative position between the LA and the LCD precisely. Moreover, a stepper motor is employed to adjust the distance d along the z axis fast and accurately. The depicted images can be recorded with a camera, which is mounted vertically above the LA

In preparation for the experiment, a color image and a depth image of a 3D model (dog) were rendered by using the stereo modeling software Blender, as seen from Figure 4. The depth range of the dog is set from 80 mm to 180 mm underneath the LA and the model was sliced into 11 depth planes. By using the ray-tracing based rendering approach described in Section 2.1, the patterns to be displayed by the LCD were generated for all depth planes, as shown in Figure 5. By following Equation (4), the distance d between LCD and LA was controlled during the display process using the stepper motor.

We used a CCD camera to imitate the eye of the observer. The system works in full color, yet we used a monochrome camera to better inspect fine details of the object representation. Interested readers can find the full-color Video S1 of the display in the Supplementary Material. To demonstrate focusing effects, the camera is equipped with a high numerical aperture objective with F# 0.95. The 3D display scheme is based on the effect of visual persistence [23], i.e., all recorded CCD images are added/integrated up while the distance d is scanned through all 11 depth planes.

To show focus blur, we have recorded multiple images with the objective being focused on different parts of the displayed dog. The results are seen in Figure 6. In the results, it can be easily recognized that the tail of the dog is behind the head, as expected from the model. In addition, when the focusing plane of the objective is set at 80 mm, the dog's head is much clearer than the tail, and the captured image at the unfocused areas exhibits a gradual blur from neck to tail, which indicates that the reconstructed image provides continuous depth information. When the focusing plane is subsequently moved from 80 mm to 180 mm, the focused area would be changed from head to neck, forefoot, body, tail, and hindfoot, where the blurred areas are also shifted correspondingly. This phenomenon strongly confirms that the reconstructed images in our system contain real depth information with focus cue, similar to what is expected from a hologram.

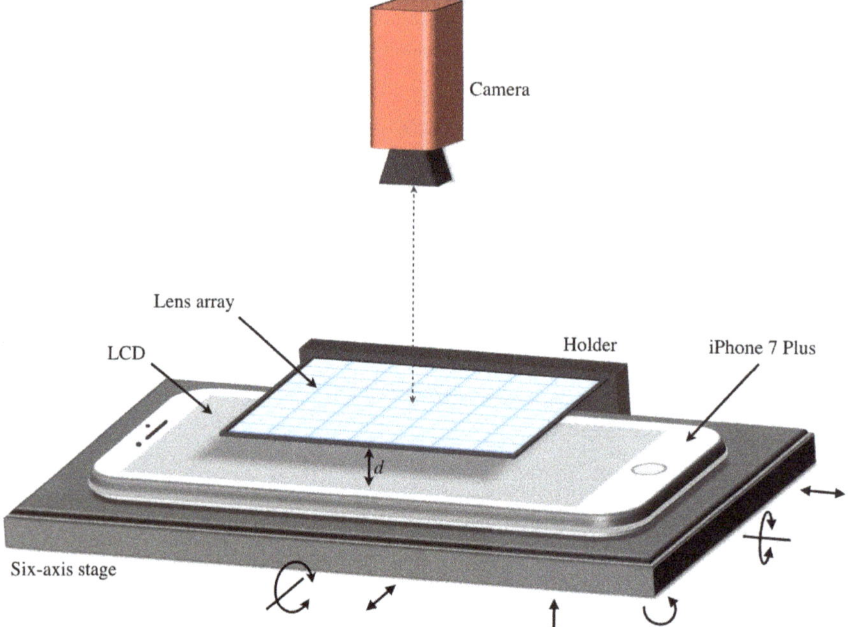

Figure 3. Optical components and configuration of the wave field synthesis 3D imaging system. The lens array is placed inside the holder in a vertical distance d to the LCD. The arrows indicate the movability of the six-axis stage. The depicted images can be captured by a camera placed vertically above the lens array.

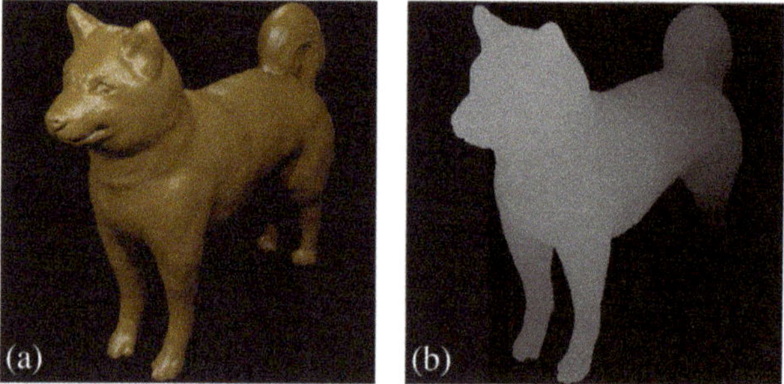

Figure 4. (**a**) Color image and (**b**) depth image of the 3D model.

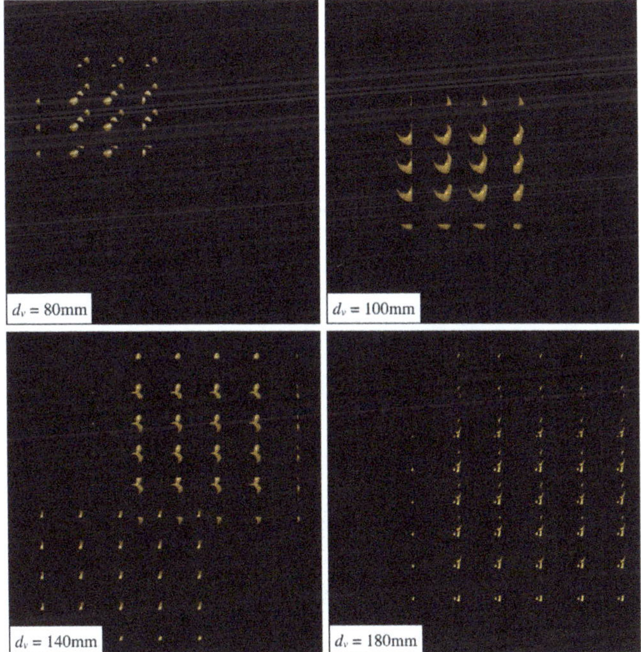

Figure 5. Displayed patterns of the dog for different exemplary depth planes.

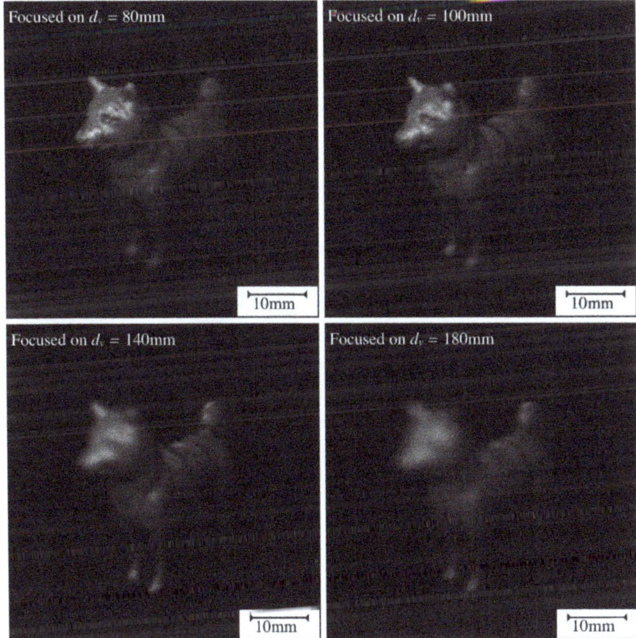

Figure 6. 3D model of the dog displayed with the proposed system. For any of the displayed images, we added up 11 individually captured depth images. The camera's objective focuses on the corresponding depth plane.

To demonstrate the effect of parallax, two letters in different depth planes are depicted, which is shown in Figure 7. A letter "A" at the virtual distance $d_v = 50\,\text{mm}$ and a letter "B" at $d_v = 150\,\text{mm}$. For the used LA with a focal length of $f = 41.90\,\text{mm}$, this results in a vertical distance between the LCD and the LA of $d = 22.80\,\text{mm}$ and $d = 32.75\,\text{mm}$, respectively. The alignment of the LCD and the LA is kept the same. The camera, placed in a vertical distance of $750\,\text{mm}$ above the LA, was horizontally moved between capturing three images along the x-axis to change the viewing angle while the camera's objective kept being focused on the depth plane of the "A".

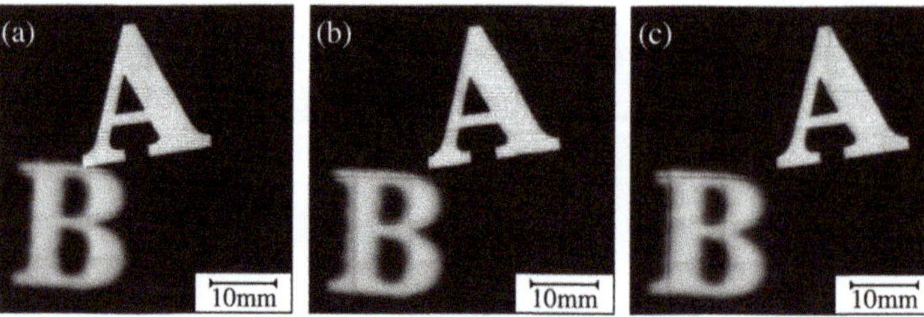

Figure 7. Capturing two letters in different depth planes to show the parallax effect. (**a**) Viewing angle of 3.1°, (**b**) camera placed on the optical axis, and (**c**) viewing angle of −3.1°.

Figure 7a was captured while the camera was horizontally shifted by $40\,\text{mm}$ from the optical axis in the positive x-direction (to the right), resulting in a change of the viewing angle by 3.1°. In Figure 7b, the camera is placed on the optical axis. Figure 7c shows the case where the camera was moved by $−40\,\text{mm}$ in the other direction (to the left), which changed the viewing angle by −3.1°. As expected, the letter closer to the viewer ("A") seems to move horizontally relatively to the letter further in the background ("B"). If the camera is placed on the right side of the optical axis, the "A" appears to be laterally closer to the "B" and if the camera is placed on the left side, the "A" is laterally farther away from the "B". This experiment clearly shows the inherent effect of parallax that the system is able to provide as a crucial depth cue. The Video S2 in the Supplementary Materials also shows parallax and focus blur of the display.

4. Conclusions

We have presented a new approach for the 3D display of natural scenes, which is based on the synthesis of spherical waves. From a wave field perspective, the method can be categorized between LF displays, which seek to recreate a scene by synthesizing plane waves (rays), and holography, which can generate arbitrary wave fields but is technically very challenging. In this sense, our method represents a significant step towards true, full color, dynamic holography, yet with the great benefit that it can be implemented with already existing technology. As a proof of concept, we have shown that a system based on the method recreates crucial depth cues, such as full parallax and depth blurring and by design avoids any vergence-accomodation conflict.

The main drawback of the current implementations of the method is the slow frame rate. Due to technical limitations of the moving stages, it requires several seconds to capture a single 3D frame. In the future, we will therefore concentrate on improving the response time of the stages or on rapidly varying the focal length of the lens array in order to achieve multiple frames per second, and thus creating an immersive experience for the observer.

Supplementary Materials: The following is available online at http://www.mdpi.com/2076-3417/9/18/3862/s1, Video S1: Experimental results showing the color performance of the display, Video S2: Experimental results showing the parallax and the focus blur of the display.

Author Contributions: conceptualization, C.F.; formal analysis, C.F. and D.P.; investigation, P.-Y.C. and D.P.; methodology, C.F.; resources, R.B.B.; supervision, C.F. and R.B.B.; validation, P.-Y.C. and D.P.; visualization, P.-Y.C. and D.P.; writing—original draft, C.F., P.-Y.C. and D.P.; writing—review and editing, R.B.B.

Funding: This research was funded by the Deutsche Forschungsgemeinschaft (DFG) grant number 250959575, and Ministry of Science and Technology (MOST) in Taiwan grant number MOST 107-2221-E-009-115-MY3.

Acknowledgments: The authors would like to thank Reiner Klattenhoff for valuable help with the experiments. We are also grateful to the Deutsche Forschungsgemeinschaft (DFG) for funding this work under the grant 250959575 (RELPH-II).

Conflicts of Interest: The authors declare no conflict of interest.

Abbreviations

The following abbreviations are used in this manuscript:

3D	Three-dimensional
LA	Lens array
LCD	Liquid crystal display
VAC	Vergence–accommodation conflict
LF	Light field
MLA	Micro-lens array
CDP	Central depth plane
CCD	Charge coupled device

References

1. Hainich, R.R.; Bimber, O. *Displays–Fundamentals & Applications*, 2nd ed.; CRC Press: Boca Raton, FL, USA, 2017.
2. Souk, J. (Ed). *Flat Panel Display Manufacturing*; Wiley: Hoboken, NJ, USA, 2018.
3. Geng, J. Three-dimensional display technologies. *Adv. Opt. Photonics* **2013**, *5*, 456–535. [CrossRef] [PubMed]
4. Holliman, N. *3D Display Systems*; Department of Computer Science: Durham, UK 2005; pp. 456–535.
5. Hill, L.; Jacobs, A. 3-D Liquid Crystal Displays and Their Applications; *Proc. IEEE* **2006**, *94*, 575–590. [CrossRef]
6. Wheatstone, C. Contributions to the physiology of vision.–Part the first. On some remarkable, and hitherto unobserved, phenomena of binocular vision. *Phil. Trans. R. Soc.* **1838**, *128*, 371–394.
7. Lappe, M.; Bremmer, F.; den Berg, A.V. Perception of self-motion from visual flow. *Trends Cogn. Sci.* **1999**, *3*, 329–336. [CrossRef]
8. Kim, J.; Kane, D.; Banks, M.S. The rate of change of vergence-accommodation conflict affects visual discomfort. *Vis. Res.* **2014**, *105*, 159–165. [CrossRef] [PubMed]
9. Burke, R.; Brickson, L. Focus cue enabled head-mounted display via microlens array. *TOG* **2013**, *32*, 220.
10. Agour, M.; Falldorf, C.; Bergmann, R.B. Holographic display system for dynamic synthesis of 3D light fields with increased space bandwidth product. *Opt. Express* **2016**, *24*, 14393–14405. [CrossRef] [PubMed]
11. Blanche, P.A.; Bablumian, A.; Voorakaranam, R.; Christenson, C.; Lin, W.; Gu, T.; Flores, D.; Wang, P.; Hsieh, W.Y.; Kathaperumal, M.; et al. Holographic three-dimensional telepresence using large-area photorefractive polymer. *Nature* **2010**, *458*, 80–83. [CrossRef] [PubMed]
12. Häussler, R.; Gritsai, Y.; Zschau, E.; Missbach, R.; Sahm, H.; Stock, M.; Stolle, H. Large real-time holographic 3d displays: enabling components and results. *Appl. Opt.* **2017**, *56*, F45–F52. [CrossRef] [PubMed]
13. Huang, F.C.; Luebke, D.P.; Wetzstein, G. The light field stereoscope. *Acm Trans. Graph.* **2015**, *34*, 60:1–60:12. [CrossRef]
14. Chou, P.Y.; Wu, J.Y.; Huang, S.H.; Wang, C.P.; Qin, Z.; Huang, C.T.; Hsieh, P.Y.; Lee, H.H.; Lin, T.H.; Huang, Y.P. Hybrid light field head-mounted display using time-multiplexed liquid crystal lens array for resolution enhancement. *Opt. Express* **2019**, *27*, 1164–1178. [CrossRef] [PubMed]
15. Jang, J.S.; Javidi, B. Three-dimensional integral imaging of micro-objects. *Opt. Lett.* **2004**, *29*, 1230–1232. [CrossRef] [PubMed]
16. Levoy, M.; Zhang, Z.; McDowall, I. Recording and controlling the 4D light field in a microscope using microlens arrays. *J. Microsc.* **2009**, *235*, 144–162. [CrossRef] [PubMed]
17. Huang, H.; Hua, H. Systematic characterization and optimization of 3D light field displays. *Opt. Express* **2017**, *25*, 18508–18525. [CrossRef] [PubMed]
18. Lanman, D.; Luebke, D. Near-eye light field displays. *ACM Trans. Graph.* **2013**, *32*, 220.1–220.10. [CrossRef]

19. Xiao, X.; Javidi, B.; Martinez-Corral, M.; Stern, A. Advances in three-dimensional integral imaging: Sensing, display, and applications. *Appl. Opt.* **2013**, *52*, 546–560. [CrossRef] [PubMed]
20. Balram, N.; Tošić, I. Light-field imaging and display systems. *Inf. Display* **2019**, *32*, 6–13. [CrossRef]
21. Kim, C.J.; Chang, M.H.; Lee, M.Y.; Kim, J.O.; Won, Y.H. Depth plane adaptive integral imaging using a varifocal liquid lens array. *Appl. Opt.* **2015**, *54*, 2565–2571. [CrossRef] [PubMed]
22. Goodmann, J. *Introduction to Fourier Optics*, 2nd ed.; McGraw-Hill: New York, NY, USA, 1996.
23. Johansson, G. Visual perception of biological motion and a model for its analysis. *Percept. Psychophys.* **1973**, *14*, 201–211. [CrossRef]

MDPI

St. Alban-Anlage 66

4052 Basel

Switzerland

Tel. +41 61 683 77 34

Fax +41 61 302 89 18

www.mdpi.com

Applied Sciences Editorial Office

E-mail: applsci@mdpi.com

www.mdpi.com/journal/applsci